Table of Contents

Contents, *continued*

Contents, *continued*

Contents, *continued*

Preface and Acknowledgements

Concrete's versatility, durability, and economy have made it the world's most used construction material. It is used in highways, streets, parking lots, parking garages, bridges, high-rise buildings, dams, homes, floors, sidewalks, driveways, and numerous other applications.

Design and Control of Concrete Mixtures has been the cement and concrete industry's primary reference on concrete technology, since the first U.S. edition was published in the mid 1920s.

This fully revised 7th Canadian edition was written to provide a concise, current reference on concrete, including the many advances that occurred since the last edition was published in 1995. The text is backed by over 85 years of research. It reflects the latest information on standards, specifications, and test methods of the Canadian Standards Association.

Acknowledgements. The authors wish to acknowledge contributions made by many individuals and organizations who provided valuable assistance in the writing and publishing of the 7th edition. A special thanks to Ken Hover, Cornell University, for extensive technical recommendations; Howard "Buck" Barker, RVT Engineering Services, for photography and text edits; Cheryl Taylor, Consultant, for months of desktop layout; and John Bickley, John Bickley and Associates, Ltd., for contributions to Chapter 17 on High Performance Concrete.

The authors have tried to make this edition of *Design and Control of Concrete Mixtures* a concise and current reference on concrete technology. Readers are encouraged to submit comments to improve future printings and editions of this book.

The authors also wish to acknowledge contributions made by the staff at the Portland Cement Association, the Cement Association of Canada, the Construction Technology Laboratories, and finally the Canadian Standards Association, the American Concrete Institute, and the American Society for Testing and Materials, all of whose publications and articles are frequently cited.

With the permission of the Canadian Standards Association, material is reproduced from CSA Standards and Test Methods, which are copyrighted by CSA, 178 Rexdale Boulevard, Rexdale, Ontario, Canada, M9W 1R3. While use of this material has been authorized, CSA shall not be responsible for the manner in which the information is presented, nor for any interpretations thereof. This CSA material may not be updated to reflect amendments made to the original content. For up-to-date information, contact CSA. Copies of the material can be purchased from CSA, tel. 416-747-4044, fax 416-747-2475.

The authors have tried to make this edition of *Design and Control of Concrete Mixtures* a concise and current reference on concrete technology. Readers are encouraged to submit comments to improve future prints and editions of this book.

CHAPTER 1
Fundamentals of Concrete

Concrete is basically a mixture of two components: aggregates and paste. The aggregate component is normally comprised of sand and gravel or crushed stone. The paste component is normally comprised of cementing materials, (portland cement with or without supplementary cementing materials), water, and air. The paste, acting like a glue, hardens due to a chemical reaction between the cement and water and binds the aggregates together into a rock-like mass—which is known as concrete (Fig. 1-1)*.

Aggregates are generally divided into two groups: fine and coarse. Fine aggregates consist of natural or manufactured sand with particle sizes ranging up to 10 mm; coarse aggregates are particles retained on the 1.25 mm sieve and ranging up to 150 mm in size. The most commonly used maximum size of coarse aggregate is 20 mm.

The paste is usually composed of cementing materials, water, and entrapped air or purposely entrained air.

Fig. 1-1. Concrete components: cement, water, fine aggregate and coarse aggregate, are combined to form concrete. (55361)

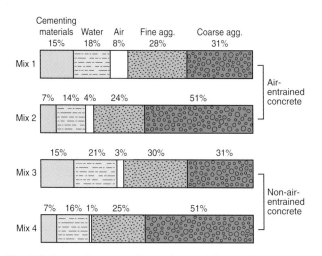

Fig. 1-2. Range in proportions of materials used in concrete, by absolute volume. Bars 1 and 3 represent rich mixes with small size aggregates. Bars 2 and 4 represent lean mixes with large size aggregates.

The paste ordinarily constitutes about 25% to 40% of the total volume of concrete. Fig. 1-2 shows that the absolute volume of cementing materials is usually between 7% and 15% and the water between 14% and 21%. Air content in air-entrained concrete ranges from about 4% to 8% of the volume of the concrete.

Since aggregates make up about 60% to 75% of the total volume of concrete, their selection is important. Aggregates should consist of particles with adequate strength and resistance to exposure conditions and should not contain materials that will cause deterioration of the concrete. A continuous gradation of aggregate particle sizes is desirable for efficient use of the paste. Throughout this text, it will be assumed that suitable aggregates are being used, except where otherwise noted.

The quality of the concrete depends upon the quality of the paste and aggregate, and the bond between the two. In properly made concrete, each and every particle of aggregate is completely coated with paste and all of the spaces between aggregate particles are completely filled with paste, as illustrated in Fig. 1-3.

---

* This text addresses the utilization of portland cement in the production of concrete. The term "portland cement" pertains to a calcium silicate hydraulic cement produced by heating materials containing calcium, silicon, aluminum, and iron. The term "cement" used throughout the text pertains to portland cement or blended hydraulic cement unless otherwise stated. The term "cementing materials" means portland cement, with or without supplementary cementing materials.

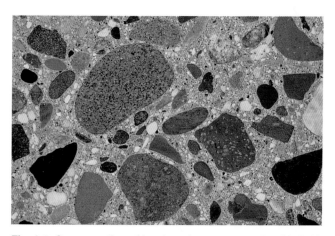
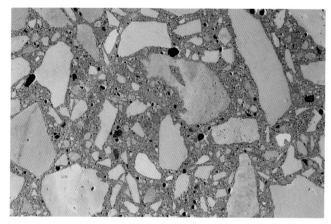

Fig. 1-3. Cross section of hardened concrete made with (left) rounded siliceous gravel and (right) crushed limestone. Cement-and-water paste completely coats each aggregate particle and fills all spaces between particles. (1051, 1052)

For any particular set of materials and conditions of curing, the quality of hardened concrete is strongly influenced by the amount of water used in relation to the amount of cement (Fig. 1-4). Unnecessarily high water contents dilute the cement paste (the glue of concrete). Following are some advantages of reducing water content:

- Increased compressive and flexural strength
- Lower permeability, thus lower absorption and increased watertightness
- Increased resistance to weathering
- Better bond between successive layers and between the concrete and reinforcement
- Reduced drying shrinkage and cracking
- Less volume change from wetting and drying

The less water used, the better the quality of the concrete—provided the mixture can be consolidated properly. Smaller amounts of mixing water result in stiffer mixtures; but with vibration, stiffer mixtures can be easily placed. Thus, consolidation by vibration permits improvement in the quality of concrete.

The freshly mixed (plastic) and hardened properties of concrete may be changed by adding chemical admixtures to the concrete, usually in liquid form, during batching. Chemical admixtures are commonly used to (1) adjust setting or hardening time, (2) reduce water demand, (3) increase workability, (4) intentionally entrain air, and (5) adjust other fresh or hardened concrete properties.

After completion of proper proportioning, batching, mixing, placing, consolidating, finishing, and curing, concrete hardens into a strong, noncombustible, durable, abrasion-resistant, and watertight building material that requires little or no maintenance. Furthermore, concrete is an excellent building material because it can be formed into a wide variety of shapes, colours, and textures for use in an unlimited number of applications.

FRESHLY MIXED CONCRETE

Freshly mixed concrete should be plastic or semifluid and generally capable of being moulded by hand. A very wet concrete mixture can be moulded in the sense that it can be cast in a mould, but this is not within the definition of "plastic"—that which is pliable and capable of being moulded or shaped like a lump of modeling clay.

In a plastic concrete mixture all grains of sand and pieces of gravel or stone are encased and held in suspension. The ingredients are not apt to segregate during transport; and when the concrete hardens, it becomes a homogeneous mixture of all the components. Concrete of plastic consistency does not crumble but flows sluggishly without segregation.

In construction practice, thin concrete members and heavily reinforced concrete members require workable, but never soupy, mixes for ease of placement. A plastic mixture is required for strength and for maintaining homogeneity during handling and placement. While a plastic mixture is suitable for most concrete work, plasticizing admixtures may be used to make concrete more flowable in thin or heavily reinforced concrete members.

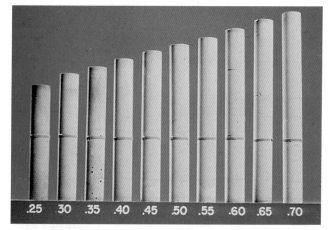

Fig. 1-4. Ten cement-paste cylinders with water-cement ratios from 0.25 to 0.70. The band indicates that each cylinder contains the same amount of cement. Increased water dilutes the effect of the cement paste, increasing volume, reducing density, and lowering strength. (1071)

Mixing

In Fig. 1-1, the basic components of concrete are shown separately. To ensure that they are combined into a homogeneous mixture requires effort and care. The sequence of charging ingredients into the mixer can play an important part in the uniformity of the finished product. The sequence, however, can be varied and still produce a quality concrete. Different sequences require adjustments in the time of water addition, the total number of revolutions of the mixer drum, and the speed of revolution. Other important factors in mixing are the size of the batch in relation to the size of the mixer drum, the elapsed time between batching and mixing, and the design, configuration, and condition of the mixer drum and blades. Approved mixers, correctly operated and maintained, ensure an end-to-end exchange of materials by a rolling, folding, and kneading action of the batch over itself as the concrete is mixed.

Workability

The ease of placing, consolidating, and finishing freshly mixed concrete and the degree to which it resists segregation is called workability. Concrete should be workable but the ingredients should not separate during transport and handling (Fig. 1-5).

The degree of workability required for proper placement of concrete is controlled by the placement method, type of consolidation, and type of concrete. Different types of placements require different levels of workability.

Factors that influence the workability of concrete are: (1) the method and duration of transportation; (2) quantity and characteristics of cementing materials; (3) concrete consistency (slump); (4) grading, shape, and surface texture of fine and coarse aggregates; (5) percentage of entrained air; (6) water content; (7) concrete and ambient air temperatures; and (8) admixtures. A uniform distribution of aggregate particles and the presence of air entrainment significantly help control segregation and improve workability. Fig. 1-6 illustrates the effect of casting temperature on the consistency or slump, and potential workability of concrete mixtures.

Properties related with workability include consistency, segregation, mobility, pumpability, bleeding, and finishability. Slump, a measure of consistency, is considered a close indication of workability. A low-slump concrete has a stiff consistency. If the consistency is too dry and

Fig. 1-5. Workable concrete should flow sluggishly into place without segregation. (59292)

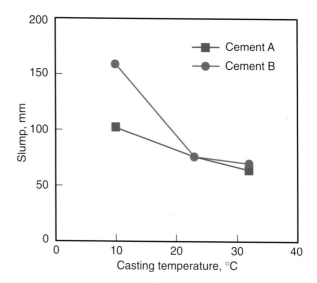

Fig. 1-6. Effect of casting temperature on the slump (and relative workability) of two concretes made with different cements (Burg 1996).

harsh, the concrete will be difficult to place and compact and larger aggregate particles may separate from the mix. However, it should not be assumed that a more fluid mix is necessarily more workable. If the mix is too fluid, segregation and honeycombing can occur. The slump should be the stiffest consistency practicable for placement using the available consolidation equipment. See Powers (1932) and Scanlon (1994).

Bleeding and Settlement

Bleeding is the development of a layer of water at the top or surface of freshly placed concrete. It is caused by the settlement of solid particles (cementing materials and aggregate) and the simultaneous upward migration of water (Fig. 1-7). Bleeding is normal and it should not diminish the quality of properly placed, finished, and cured concrete. Some bleeding is helpful to control plastic shrinkage cracking.

Fig. 1-7. Bleed water on the surface of a freshly placed concrete slab. (P29992)

Excessive bleeding increases the water-cement ratio near the top surface; a weak top layer with poor durability may result, particularly if finishing operations take place while bleed water is present. A water pocket or void can develop under a prematurely finished surface.

After evaporation of all bleed water, the hardened surface will be slightly lower than the freshly placed surface. This reduction in volume or vertical dimension from time of placement to initial set is called settlement shrinkage.

The bleeding rate and bleeding capacity (total settlement per unit of original concrete height) increases with initial water content, concrete height, and pressure. Use of properly graded aggregate, certain chemical admixtures, air entrainment, supplementary cementing materials, and finer ground cements, reduce bleeding. Concrete used to fill voids, provide support, or provide watertightness with a good bond should have low bleeding properties to avoid formation of water pockets. See Powers (1939), Steinour (1945), and Kosmatka (1994).

Fig. 1-8. (left) Good consolidation is needed to achieve a dense and durable concrete. (right) Poor consolidation can result in early corrosion of reinforcing steel and low compressive strength. (70016, 68806)

Consolidation

Vibration sets into motion the particles in freshly mixed concrete, reducing friction between them and giving the mixture the mobile qualities of a thick fluid. The vibratory action permits use of a stiffer mixture containing a larger proportion of coarse and a smaller proportion of fine aggregate. The larger the maximum size aggregate in concrete with a well-graded aggregate, the less volume there is

to fill with paste and the less aggregate surface area there is to coat with paste; thus less water and cement are needed. Concrete with an optimally graded aggregate will be easier to consolidate and place (Fig. 1-8 left). Consolidation of coarser as well as stiffer mixtures results in improved quality and economy. On the other hand, poor consolidation can result in porous, weak concrete (Fig. 1-9) with poor durability (Fig. 1-8 right).

Mechanical vibration has many advantages. Vibrators make it possible to economically place mixtures that are impractical to consolidate by hand under many conditions. As an example, Fig. 1-10 shows concrete of a stiff consistency (low slump). This concrete was mechanically vibrated in forms containing closely spaced reinforcement. With hand rodding, a much more fluid consistency would have been necessary.

Fig. 1-10. Concrete of a stiff consistency (low slump). (44485)

Hydration, Setting Time, and Hardening

The binding quality of portland cement paste is due to the chemical reaction between the cement and water, called hydration.

Portland cement is not a simple chemical compound, it is a mixture of many compounds. Four of these make up 90% or more of the mass of portland cement: tricalcium silicate, dicalcium silicate, tricalcium aluminate, and tetracalcium aluminoferrite. In addition to these major compounds, several others play important roles in the hydration process. Different types of portland cement contain the same four major compounds, but in different proportions.

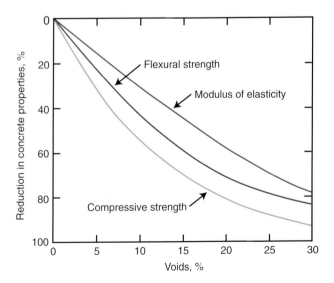

Fig. 1-9. Effect of voids in concrete due to a lack of consolidation on modulus of elasticity, compressive strength, and flexural strength.

When clinker (the kiln product that is ground to make portland cement) is examined under a microscope, most of the individual cement compounds can be identified and their amounts determined. However, the smallest grains elude visual detection. The average diameter of a typical cement particle is approximately 15 µm. If all cement particles were average, portland cement would contain about 300 billion particles per kilogram, but in fact there are some 16,000 billion particles per kilogram because of the broad range of particle sizes. The particles in a kilogram of portland cement have a surface area of approximately 400 square metres.

The two calcium silicates, which constitute about 75% of the mass of portland cement, react with water to form two new compounds: calcium hydroxide and *calcium silicate hydrate*. The latter is by far the most important cementing component in concrete. The engineering properties of concrete—setting and hardening, strength, and dimensional stability—depend primarily on calcium silicate hydrate. It is the heart of concrete.

The chemical composition of calcium silicate hydrate is somewhat variable, but it contains lime (CaO) and silicate (SiO_2) in a ratio on the order of 3 to 2. The surface area of calcium silicate hydrate is some 300 square metres per gram. In hardened cement paste, the calcium silicate hydrate forms dense, bonded aggregations between the other crystalline phases and the remaining unhydrated cement grains; they also adhere to grains of sand and to pieces of coarse aggregate, cementing everything together (Copeland and Schulz 1962). The formation of this structure is the paste's cementing action and is responsible for setting, hardening, and strength development.

As concrete hardens, its gross volume remains almost unchanged, but hardened concrete contains pores filled with water and air that have no strength. The strength is in the solid part of the paste, mostly in the calcium silicate hydrate and crystalline compounds.

The less porous the cement paste, the stronger the concrete. Therefore, when mixing concrete, use no more water than is absolutely necessary to make the concrete plastic and workable. Even then, the water used is usually more than is required for complete hydration of the cement. Excluding evaporable water, the water to portland cement ratio (by mass) of completely hydrated cement is approximately 0.22 to 0.25. To ensure enough water is available to achieve complete hydration, about 0.4 grams of water per gram of cement are required (Powers 1948 and 1949). However, complete hydration is rare in field concrete due to a lack of moisture and the long period of time (decades) required for cement to reach this state.

Knowledge of the amount of heat released as cement hydrates can be useful in planning construction. In winter, the heat of hydration will help protect the concrete against damage from freezing temperatures. However, the heat may be harmful in massive structures such as dams because it may produce undesirable temperature differentials resulting in thermal cracking.

Knowledge of the rate of reaction between cement and water is important because the rate determines the rate of hardening. The initial reaction must be slow enough to allow time for the concrete to be transported and placed. Once the concrete has been placed and finished, however, rapid hardening is desirable. Gypsum, added at the cement mill when clinker is ground, acts as a regulator of the initial rate of setting of portland cement. Other factors that influence the rate of hydration include cement fineness, admixtures, amount of water added, and temperature of the materials at the time of mixing. Fig. 1-11 illustrates the setting properties of a concrete mixture at different temperatures.

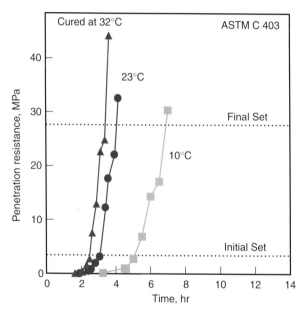

Fig. 1-11. Initial and final set times for a concrete mixture at different temperatures (Burg 1996).

HARDENED CONCRETE

Curing

Increase in strength with age continues provided (1) unhydrated cement is still present, (2) the concrete remains moist or has a relative humidity above approximately 80% (Powers 1948), (3) the concrete temperature remains favorable, and (4) sufficient space is available for hydration products to form. When the relative humidity within the concrete drops to about 80%, or the temperature of the concrete drops below freezing, hydration and strength gain virtually stop. Fig. 1-12 illustrates the relationship between strength gain and moist curing, while Fig. 1-13 illustrates the relationship between curing temperature and strength gain.

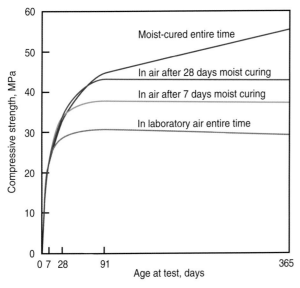

Fig. 1-12. Concrete strength increases with age as long as moisture and a favorable temperature are present for hydration of cement (Gonnerman and Shuman 1928).

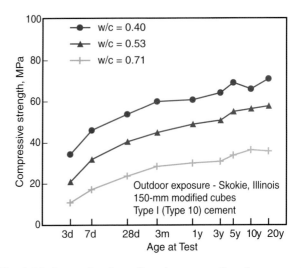

Fig. 1-14. Concrete strength gain versus time for concrete exposed to outdoor conditions. Concrete continues to gain strength for many years when moisture is provided by rainfall and other environmental sources (Wood 1992).

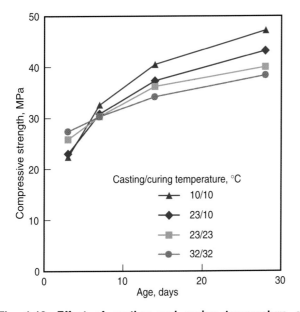

Fig. 1-13. Effect of casting and curing temperature on strength development. Note that cooler temperatures result in lower early strength and higher later strength (Burg, 1996).

If concrete is resaturated after a drying period, hydration is resumed and strength will again increase. However, it is best to moist-cure concrete continuously from the time it is placed until it has attained the desired quality; once concrete has dried out it is difficult to resaturate. Fig. 1-14 illustrates the long-term strength gain of concrete in an outdoor exposure. Outdoor exposures often continue to provide moisture through ground moisture and rainfall. Indoor concretes often dry out after curing and do not continue to gain strength (Fig. 1-12).

Drying Rate of Concrete

Concrete does not harden or cure by drying. Concrete (or more precisely, the cementing materials in it) needs moisture to hydrate and harden. When concrete dries out, it ceases to gain strength; the fact that it is dry is no indication that it has undergone sufficient hydration to achieve the desired physical properties.

Knowledge of the rate of drying is helpful in understanding the properties or physical condition of concrete. For example, as mentioned, concrete must continue to hold enough moisture throughout the curing period for the cement to hydrate to the extent that desired properties are achieved. Freshly cast concrete usually has an abundance of water, but as drying progresses from the surface inward, strength gain will continue at each depth only as long as the relative humidity at that point remains above 80%.

A common illustration of this is the surface of a concrete floor that has not had sufficient moist curing. Because it has dried quickly, concrete at the surface is weak and traffic on it creates dusting. Also, when concrete dries, it shrinks as it loses water (Fig. 1-15), just as wood and clay do (though not as much). Drying shrinkage is a primary cause of cracking, and the width of cracks is a function of the degree of drying, spacing or frequency of cracks, and the age at which the cracks occur.

While the surface of a concrete element will dry quite rapidly, it takes a much longer time for concrete in the interior to dry. Fig. 1-15 (top) illustrates the rate of drying at various depths within concrete cylinders exposed to laboratory air. Field concrete elements would have different drying profiles due to environmental conditions, size effects, and concrete properties.

6

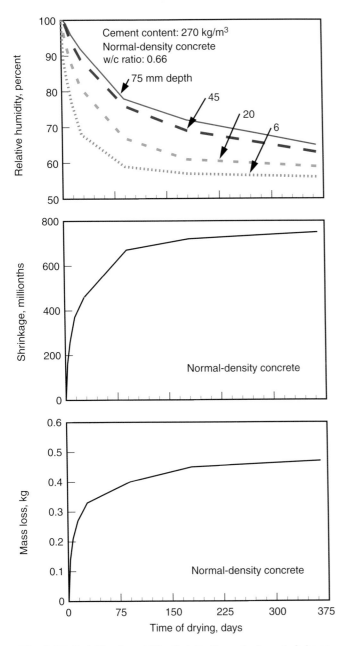

Fig. 1-15. Relative humidity distribution, drying shrinkage, and mass loss of 150 x 300-mm cylinders moist cured for 7 days followed by drying in laboratory air at 23°C (Hanson 1968).

The moisture content of concrete depends on the concrete's constituents, original water content, drying conditions, and the size of the concrete element (Hedenblad 1997 and 1998). After several months of drying in air with a relative humidity of 50% to 90%, moisture content is about 1% to 2% by mass of the concrete. Fig. 1-15 illustrates moisture loss and resulting shrinkage.

Size and shape of a concrete member have an important bearing on the rate of drying. Concrete elements with large surface area in relation to volume (such as floor slabs) dry faster than voluminous concrete members with relatively small surface areas (such as bridge piers).

Many other properties of hardened concrete also are affected by its moisture content; these include elasticity, creep, insulating value, fire resistance, abrasion resistance, electrical conductivity, frost resistance, scaling resistance, and resistance to alkali-aggregate reactivity.

Strength

Compressive strength may be defined as the measured maximum resistance of a concrete or mortar specimen to axial loading. It is generally expressed in megapascals (MPa) at an age of 28 days. One megapascal equals the force of one newton per square millimetre (N/mm²) or 1,000,000 N/m². Other test ages are also used; however, it is important to realize the relationship between the 28-day strength and other test ages. Seven-day strengths are often estimated to be about 75% of the 28-day strength and 56-day and 90-day strengths are about 10% to 15% greater than 28-day strengths as shown in Fig. 1-16. The specified compressive strength is designated by the symbol f_c', and ideally is exceeded by the actual compressive strength f_c.

The compressive strength that a concrete achieves, f_c, results from the water-cement ratio (or water-cementing materials ratio), the extent to which hydration has progressed, the curing and environmental conditions, and the age of the concrete. The relationship between strength and water-cement ratio has been studied since the late 1800s and early 1900s (Feret 1897 and Abrams 1918). Fig. 1-17 shows 28-day compressive strengths for a range of concrete mixtures and water-cement ratios. Tests were made on 150 x 300-mm cylinders. Note that the strength increases as the water-cement ratio decreases. This factor also affects the flexural and tensile strengths and bond of concrete to steel.

The water-cement ratio-compressive strength relationships in Fig. 1-17 are for typical non-air-entrained concretes. When more precise values for concrete are required, graphs

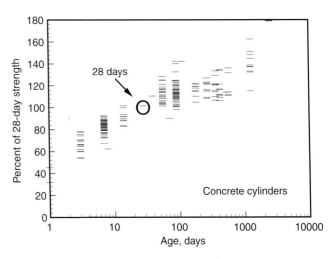

Fig. 1-16. Compressive strength development of various concretes illustrated as a percentage of the 28-day strength (Lange 1994).

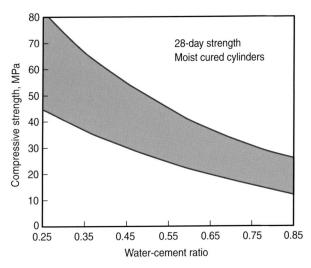

Fig. 1-17. Range of typical strength to water-cement ratio relationships of portland cement concrete based on over 100 different concrete mixtures cast between 1985 and 1999.

Fig. 1-18. Testing a 150 x 300-mm concrete cylinder in compression. The load on the test cylinder is registered on the display. (68959)

should be developed for the specific materials and mix proportions to be used on the job.

For a given workability and a given amount of cementing materials, air-entrained concrete requires less mixing water than non-air-entrained concrete. The lower water-cement ratio possible for air-entrained concrete tends to offset the somewhat lower strengths of air-entrained concrete, particularly in lean to medium cement content mixes.

To determine compressive strength, tests are made on specimens of mortar or concrete; in Canada, unless otherwise specified, compression tests of mortar are made on 50-mm cubes, while compression tests of concrete are made on cylinders 100 mm in diameter and 200 mm high. In the United States, however, 150 x 300-mm cylinders are also used (see Fig. 1-18).

Compressive strength of concrete is a primary physical property and frequently used in design calculations for bridges, buildings, and other structures. Most general-use concrete has a compressive strength between 20 and 35 MPa. High-strength concrete by definition, has a compressive strength of at least 70 MPa. Compressive strengths up to 140 MPa have been used in special bridge and high-rise building applications.

The flexural strength or modulus of rupture of concrete is used to design pavements and other slabs on ground. Compressive strength, which is easier to measure than flexural strength, can be used as an index of flexural strength, once the empirical relationship between them has been established for the materials to be used in the mix and the size of the element involved. The flexural strength or modulus of rupture of normal-density concrete is often approximated as 0.6 to 0.8 times the square root of the compressive strength in megapascals. Wood (1992) illustrates the relationship between flexural strength and compressive strength for concretes exposed to moist curing, air curing, and outdoor exposure.

The direct tensile strength of concrete is about 8% to 12% of the compressive strength and is often estimated as 0.4 to 0.7 times the square root of the compressive strength in megapascals. Splitting tensile strength is 8% to 14% of the compressive strength (Hanson 1968). Splitting tensile strength versus time is presented by Lange (1994).

The torsional strength for concrete is related to the modulus of rupture and the dimensions of the concrete element. Hsu (1968) presents torsional strength correlations.

The shear strength of concrete is about 5% of the compressive strength. The correlation between compressive strength and flexural, tensile, torsional, and shear strength varies with concrete ingredients and environment.

Modulus of elasticity, denoted by the symbol E, may be defined as the ratio of normal stress to corresponding strain for tensile or compressive stresses below the proportional limit of a material. For normal-density concrete, E ranges from 14,000 to 41,000 MPa. For concrete strengths between 20 MPa and 40 MPa, E can be approximated as 5,000 times the square root of the compressive strength. Like other strength relationships, the modulus of elasticity to compressive strength relationship is mix-ingredient specific and relationships should be verified in a laboratory (Wood 1992).

Density

Conventional concrete, normally used in pavements, buildings, and other structures, has a density (unit weight) in the range of 2200 to 2400 kg/m³. The density of concrete varies, depending on the amount and relative density of the aggregate, the amount of air that is entrapped or purposely entrained, and the water and cementing materials content, which in turn are influenced by the maximum size of the aggregate. Reducing the cement paste content (increasing aggregate volume) increases density. Values of the density of fresh concrete are given in Table 1-1. In the design of reinforced concrete structures, the combination of conventional concrete and reinforcing bars is commonly assumed to have a density of 2400 kg/m³.

The mass of dry concrete equals the mass of the freshly mixed concrete ingredients less the mass of mix water that evaporates into the air. Some of the mix water combines chemically with the cement during the hydration process, converting the cement into cement gel. Also, some of the water remains tightly held in pores and capillaries and does not evaporate under normal conditions. The amount of mix water that will evaporate from concrete exposed to ambient air at 50% relative humidity is about ½% to 3% of the concrete mass; the actual amount depends on initial water content of the concrete, absorption characteristics of the aggregates, and size and shape of the concrete element.

Aside from conventional concrete, there is a wide spectrum of concretes to meet various needs; they range from low-density insulating concretes with a density of 240 kg/m³ to high-density concrete with a density of up to 6000 kg/m³ used for counterweights or radiation shielding.

Permeability and Watertightness

Concrete used in water-retaining structures or exposed to weather or other severe exposure conditions must be virtually impermeable or watertight. Watertightness is often referred to as the ability of concrete to hold back or retain water without visible leakage. Permeability refers to the amount of water migration through concrete when the water is under pressure or to the ability of concrete to resist penetration by water or other substances (liquid, gas, ions, etc.). Generally, the same properties of concrete that make concrete less permeable also make it more watertight.

The overall permeability of concrete to water is a function of: (1) the permeability of the paste; (2) the permeability and gradation of the aggregate; (3) the quality of the paste and aggregate transition zone; and (4) the relative proportion of paste to aggregate. Decreased permeability improves concrete's resistance to freezing and thawing, resaturation, sulphate and chloride-ion penetration, and other chemical attack.

The permeability of the paste is particularly important because the paste envelops all constituents in the concrete. Paste permeability is related to water-cementing materials

ratio and the degree of cement hydration or length of moist curing. A low-permeability concrete requires a low water-cementing materials ratio and an adequate moist-curing period. Air entrainment aids watertightness but has little effect on permeability. Permeability increases with drying.

The permeability of mature hardened cement paste kept continuously moist ranges from 0.1 x 10⁻¹² to 120 x 10⁻¹² cm per sec. for water-cement ratios ranging from 0.3 to 0.7 (Powers and other 1954).

The permeability of rock commonly used as concrete aggregate varies from approximately 1.7 x 10⁻⁹ to 3.5 x 10⁻¹³ cm per sec. The permeability of mature, good-quality concrete is approximately 1 x 10⁻¹⁰ cm per sec.

The relationship between permeability, water-cement ratio, and initial curing for 100 x 200-mm cylindrical concrete specimens tested after 90 days of air drying and subjected to 20 MPa of water pressure is illustrated in Fig. 1-19. Although permeability values would be different for other liquids

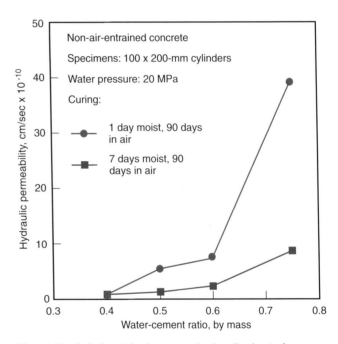

Fig. 1-19. Relationship between hydraulic (water) permeability, water-cement ratio, and initial curing on concrete specimens (Whiting 1989).

Table 1-1. Observed Average Density of Fresh Concrete*

Maximum size of aggregate, mm	Air content, percent	Water, kg/m³	Cement, kg/m³	Density, kg/m³**				
				Relative density of aggregate†				
				2.55	2.60	2.65	2.70	2.75
19	6.0	168	336	2194	2227	2259	2291	2323
38	4.5	145	291	2259	2291	2339	2371	2403
76	3.5	121	242	2307	2355	2387	2435	2467

* Source: U.S. Bureau of Reclamation 1981, Table 4.
** Air-entrained concrete with indicated air content.
† On saturated surface-dry basis. Multiply relative density by 1000 to obtain density of aggregate in kg/m³.

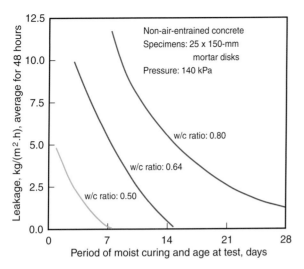

Fig. 1-20. Effect of water-cement ratio (w/c) and curing duration on permeability of mortar. Note that leakage is reduced as the water-cement ratio is decreased and the curing period increased (McMillan and Lyse 1929 and PCA Major Series 227).

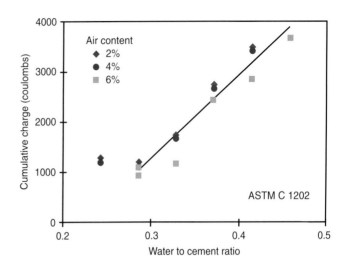

Fig. 1-21. Total charge at the end of the rapid chloride permeability test as a function of water to cement ratio (Pinto and Hover 2001).

and gases, the relationship between water-cement ratio, curing period, and permeability would be similar.

Test results obtained by subjecting 25 mm-thick non-air-entrained mortar disks to 140 kPa water pressure are given in Fig. 1-20. In these tests, there was no water leakage through mortar disks that had a water-cement ratio of 0.50 by mass or less and were moist-cured for seven days. Where leakage occurred, it was greater in mortar disks made with high water-cement ratios. Also, for each water-cement ratio, leakage was less as the length of the moist-curing period increased. In disks with a water-cement ratio of 0.80, the mortar still permitted leakage after being moist-cured for one month. These results clearly show that a low water-cement ratio and a period of moist curing significantly reduce permeability.

Fig. 1-21 illustrates the effect of different water to cement ratios on concrete's resistance to chloride ion penetration as indicated by electrical conductance. The total charge in coulombs was significantly reduced with a low water to cement ratio. Also, the results showed that a lower charge passed when the concrete contained a higher air content.

A low water-cementing materials ratio also reduces segregation and bleeding, further contributing to watertightness. Of course, concrete must also be free from cracks, honeycomb, or other large visible voids.

Occasionally, pervious concrete, like a no-fines concrete that readily allows passage of water, is designed for special applications. In these concretes, the fine aggregate is greatly reduced or completely removed, producing a high volume of air voids. Pervious concrete has been used in tennis courts, pavements, parking lots, greenhouses, and drainage structures. Pervious concrete has also been used in buildings because of its thermal insulation properties.

Abrasion Resistance

Floors, pavements, and hydraulic structures are subjected to abrasion; therefore, in these applications concrete must have a high abrasion resistance. Test results indicate that abrasion resistance is closely related to the compressive strength of concrete. Strong concrete has more resistance to abrasion than does weak concrete. Since compressive strength depends on water-cementing materials ratio and curing, a low water-cementing materials ratio and adequate curing are necessary for abrasion resistance. The

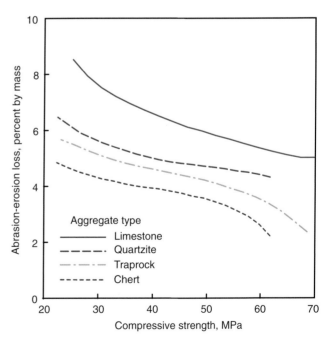

Fig. 1-22. Effect of compressive strength and aggregate type on the abrasion resistance of concrete (ASTM C 1138). High-strength concrete made with a hard aggregate is highly resistant to abrasion (Liu 1981).

type of aggregate and surface finish or treatment used also have a strong influence on abrasion resistance. Hard aggregate is more wear resistant than soft aggregate and a steel-troweled surface resists abrasion better than a surface that had not been troweled.

Fig. 1-22 shows results of abrasion tests on concretes of different compressive strengths and aggregate types. Fig. 1-23 illustrates the effect hard steel troweling and surface

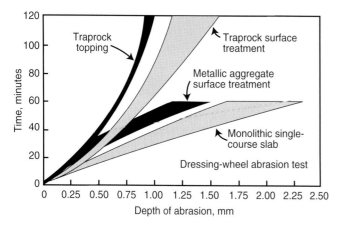

Fig. 1-23. Effect of hard steel troweling and surface treatments on the abrasion resistance of concrete (ASTM C 779). Base slab compressive strength was 40 MPa at 28 days. All slabs were steel troweled (Brinkerhoff 1970).

Fig. 1-24. Test apparatus for measuring abrasion resistance of concrete. The machine can be adjusted to use either revolving disks or dressing wheels. With a different machine, steel balls under pressure are rolled over the surface of the specimen. The tests are described in ASTM C 779. (44015)

treatments, such as metallic or mineral aggregate surface hardeners have on abrasion resistance. Abrasion tests can be conducted by rotating steel balls, dressing wheels, or disks under pressure over the surface (ASTM C 779). One type of test apparatus is pictured in Fig. 1-24. Other types of abrasion tests are also available (ASTM C 418, C 944, and C 1138). CSA A23.1 requires that concrete exposed to severe abrasion or scouring have a minimum 28-day specified compressive strength of 35 MPa and be made with abrasion-resistant aggregates.

Volume Stability and Crack Control

Hardened concrete changes volume slightly due to changes in temperature, moisture, and stress. These volume or length changes may range from about 0.01% to 0.08%. Thermal volume changes of hardened concrete are about the same as those for steel.

Concrete under stress will deform elastically. Sustained stress will result in additional deformation called creep. The rate of creep (deformation per unit of time) decreases with time.

Concrete kept continually moist will expand slightly. When permitted to dry, concrete will shrink. The primary factor influencing the amount of drying shrinkage is the water content of the freshly mixed concrete. Drying shrinkage increases directly with increases in this water content. The amount of shrinkage also depends upon several other factors, such as: (1) amounts of aggregate used; (2) properties of the aggregate; (3) size and shape of the concrete mass element; (4) relative humidity and temperature of the environment; (5) method of curing; (6) degree of hydration; and (7) time.

Two basic causes of cracks in concrete are: (1) stress due to applied loads, and (2) stress due to temperature or moisture changes when concrete is restrained. Thermal stress due to fluctuations in temperature also can cause cracking, particularly at an early age. Concrete increases in volume slightly with an increase in moisture and contracts or shrinks with a loss of moisture.

Concrete shrinkage cracks can occur because of restraint. If there was no restraint, the concrete would not crack. Restraint however comes from several sources. Drying shrinkage is always greater near the surface of concrete; the moist inner portions restrain the concrete near the surface, which can cause cracking. Other sources of restraint are reinforcing steel embedded in concrete, the interconnected parts of a concrete structure, and the friction of the subgrade on which concrete is placed.

Drying shrinkage is an inherent, unavoidable property of concrete; therefore, attention should be given to methods of controlling potential random cracking. The use of joints to predetermine and control the location of cracks, or the use of properly sized and positioned reinforcing steel to reduce crack widths or increase joint spacing are some common methods.

Joints. Joints are the most effective method of controlling unsightly cracking. If a sizable expanse of concrete (a wall, slab, or pavement) is not provided with properly spaced joints to accommodate drying shrinkage and temperature contraction, the concrete will crack in a random manner.

Contraction (shrinkage control) joints are grooved, formed, or sawed into sidewalks, driveways, pavements, floors, and walls so that cracking will occur in these joints rather than in a random manner. Contraction joints permit movement in the plane of a slab or wall. They extend to a depth of approximately one-quarter the concrete thickness.

Isolation joints separate a slab from other parts of a structure and permit horizontal and vertical movements of the slab. They should be used at the junction of floors with walls, columns, footings, and other points where restraint can occur. They extend the full depth of the slab and include a premoulded joint filler.

Construction joints occur where concrete work is concluded for the day; they separate areas of concrete placed at different times. In slabs-on-ground, construction joints usually align with, and function as, control or isolation joints. They may require dowels for load transfer.

Fig. 1-25. Air-entrained concrete (bottom bar) is highly resistant to repeated freeze-thaw cycles. (P25542)

DURABILITY

The durability of concrete may be defined as the ability of concrete to resist weathering action, chemical attack, and abrasion while maintaining its desired engineering properties with minimal loss of mass in an aggressive environment. Different concretes require different degrees of durability depending on the exposure environment and the properties desired. The concrete ingredients, proportioning of those ingredients, interactions between the ingredients, and placing and curing practices determine the ultimate durability and life expectancy of the concrete. CSA A23.1 - Section 15 and its relative clauses, specifically address the durability requirements of concrete.

Resistance to Freezing and Thawing

Concrete used in structures and pavements is expected to have long life and low maintenance. It must have good durability to resist anticipated exposure conditions. The most destructive weathering factor is freezing and thawing while the concrete is wet, particularly in the presence of deicing chemicals. Deterioration is caused by the freezing of the water and subsequent expansion in the paste, the aggregate particles, or both.

With air entrainment, concrete is highly resistant to this deterioration as shown in Fig. 1-25. During freezing, the water displaced by ice formation in the paste is accommodated so that it is not disruptive; the microscopic air bubbles in the paste provide chambers for the water to enter and thus relieve the hydraulic pressure generated.

When freezing occurs in concrete containing saturated aggregate, disruptive hydraulic pressures can also be generated within the aggregate. Water displaced from the aggregate particles during the formation of ice cannot escape fast enough to the surrounding paste to relieve pressure. However, under most exposure conditions, a paste of good quality (low water-cementing materials ratio) will prevent most aggregate particles from becoming saturated. Also, if the paste is air-entrained, it will accommodate the small amounts of excess water that may be expelled from aggregates, thus protecting the concrete from freeze-thaw damage.

Fig. 1-26 illustrates, for a range of water-cement ratios, that (1) air-entrained concrete is much more resistant to freeze-thaw cycles than non-air-entrained concrete, (2) concrete with a low water-cement ratio is more durable than concrete with a high water-cement ratio, and (3) a drying period prior to freeze-thaw exposure substantially benefits the freeze-thaw resistance of air-entrained concrete. Air-entrained concrete with a low water-cement ratio and an air content of 4% to 8% will withstand a great number of cycles of freezing and thawing without distress (Woods 1956).

Freeze-thaw durability can be determined by laboratory test procedure ASTM C 666, *Standard Test Method for Resistance of Concrete to Rapid Freezing and Thawing*. From the test, a durability factor is calculated that reflects the number of cycles of freezing and thawing required to produce a certain amount of deterioration. Deicer-scaling resistance can be determined by ASTM C 672, *Standard Test Method for Scaling Resistance of Concrete Surfaces Exposed to Deicing Chemicals*.

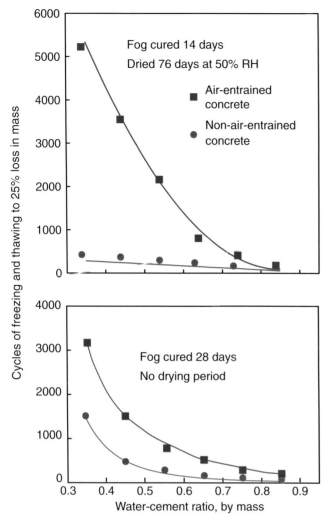

Fig. 1-26. Relationship between freeze-thaw resistance, water-cement ratio, and drying for air-entrained and non-air-entrained concretes made with Type I (Type 10) cement. High resistance to freezing and thawing is associated with entrained air, low water-cement ratio, and a drying period prior to freeze-thaw exposure (Backstrom and others 1955).

Fig. 1-27. Cracking, joint closing, spalling, and lateral offset are caused by severe alkali-silica reactivity in this parapet wall. (56586)

Alkali-Aggregate Reactivity

Alkali-aggregate reactivity (AAR) is a reaction between the active mineral constituents of some aggregates and the sodium and potassium alkali hydroxides in the concrete. The reactivity is potentially harmful only when it produces significant expansion. Alkali-aggregate reactivity occurs in two forms—alkali-silica reaction and alkali-carbonate reaction. Alkali-silica reaction is of more concern than alkali-carbonate reaction because the occurrence of aggregates containing reactive silica minerals is more common.

Indications of the presence of deleterious alkali-aggregate reactivity may be in the form of a network of cracks (map cracking), closed or spalling joints, or displacement of different portions of a structure (Fig. 1-27). Because deterioration due to alkali-aggregate reaction is a slow process,

the risk of catastrophic failure is low. In fact, in Canada much concrete made with reactive aggregates remains in service. However, concrete with alkali-aggregate reaction can cause serviceability problems and exacerbate other deterioration mechanisms, such as those that occur in frost, deicer, or sulphate exposures.

Current practices to mitigate the detrimental effects of alkali-silica reactivity include the use of non-reactive aggregates, reducing the alkali content of the concrete by using low-alkali cement where available, and by using supplementary cementing materials or blended cements proven by testing to control the reaction. Supplementary cementing materials include fly ash, ground granulated blast-furnace slag, silica fume, and natural pozzolans (Fig.

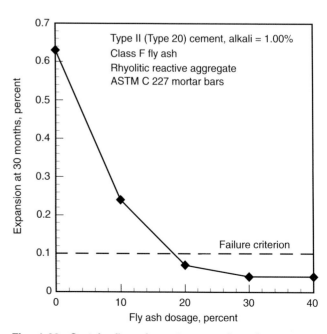

Fig. 1-28. Certain fly ashes at appropriate dosages can control alkali-silica reactivity.

1-28). Blended cements also contain these materials to control alkali-silica reactivity. CSA A864-00 provides in-depth information on the identification and mitigation of alkali silica reactivity in structures.

The use of low-alkali cement, blended cements, or supplementary cementing materials does not control alkali-carbonate reaction. Fortunately this reaction is rare. If aggregate testing indicates that an aggregate is susceptible to alkali-carbonate reactivity, the reaction can be controlled through the use of selective quarrying, aggregate blending, reducing maximum aggregate size, or using special compounds that inhibit the reaction. The most practical preventative measure for this reaction is to avoid the use of these aggregates.

See CSA Standard A23.1, Appendix B, *Alkali-Aggregate Reaction* and Farny and Kosmatka (1997) for more information on alkali-silica reaction and alkali-carbonate reaction.

Carbonation

Carbonation of concrete is a process by which carbon dioxide in the ambient air penetrates the concrete and reacts with the hydroxides, such as calcium hydroxide, to form carbonates. In the reaction with calcium hydroxide, calcium carbonate is formed. Carbonation and rapid drying of fresh concrete may affect surface durability, but this is prevented by proper curing. Carbonation of hardened concrete does not harm the concrete mixture. However, carbonation significantly lowers the alkalinity (pH) of the concrete. High alkalinity is needed to protect embedded steel from corrosion; consequently, concrete should be resistant to carbonation to help prevent steel corrosion.

The amount of carbonation is significantly increased in concrete that has a high water to cementing materials ratio, low cement content, short curing period, low strength, and highly permeable (porous) paste. The depth of carbonation in good-quality, well-cured concrete is generally of little practical significance as long as embedded steel has adequate concrete cover (Fig. 1-29). Depending on the concrete properties, ingredients, age, and environmental exposure, concrete with finished surfaces tend to have less carbonation than with formed surfaces. Carbonation depth with finished surfaces is often observed to a depth of 1 to 10 mm and for formed surfaces, between 2 and 20 mm after several years of exposure. (Campbell, Sturm, and Kosmatka 1991).

Chloride Resistance and Steel Corrosion

Concrete protects embedded steel from corrosion through its highly alkaline nature. The high pH environment in concrete (usually greater than 12.5) causes a passive and noncorroding protective oxide film to form on steel. However, the presence of chloride ions from deicers or seawater can destroy or penetrate the film. Once the chloride corrosion threshold (about 0.15% water-soluble chloride by mass of cement) is reached, an electric cell is formed along the steel or between steel bars and the electrochemical process of corrosion begins. Some steel areas along the bar become the anode discharging current in the electric cell; from there the iron goes into solution. Steel areas that receive current are the cathodes where hydroxide ions are formed. The iron and hydroxide ions form iron hydroxide, FeOH, which further oxidizes to form rust (iron oxide). Rusting is an expansive process—rust expands up to four times its original volume—which induces internal stress and eventual spalling of the concrete over reinforcing steel. The cross-sectional area of the steel can also be significantly reduced.

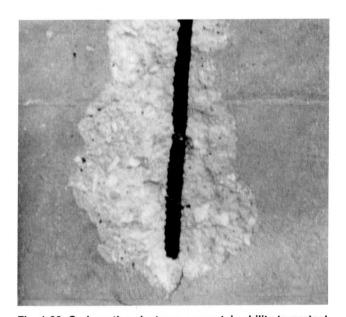

Fig. 1-29. Carbonation destroys concrete's ability to protect embedded steel from corrosion. All concrete carbonates to a small depth, but reinforcing steel must have adequate concrete cover to prevent carbonation from reaching the steel. This reinforcing bar in a wall had less than 10 mm of concrete cover. After years of outdoor exposure, the concrete carbonated to the depth of the bar, allowing the steel to rust and spall the concrete surface. (68340)

Fig. 1-30. Epoxy-coated reinforcing steel used in a bridge deck. (69915)

Once it starts, the rate of steel corrosion is influenced by the concrete's electrical resistivity, moisture content, and the rate at which oxygen migrates through the concrete to the steel. Chloride ions alone can also penetrate the passive film on the reinforcement; they combine with iron ions to form a soluble iron chloride complex that carries the iron into the concrete for later oxidation (rust).

The resistance of concrete to chloride is good; however, it can be increased by using a low water-cementing materials ratio, at least seven days of moist curing, and supplementary cementing materials, such as silica fume, to reduce permeability. Increasing the concrete cover over the steel also helps slow down the migration of chlorides. CSA A23.1 outlines requirements for concrete subjected to various classes of exposure (C Classes pertain to chloride exposure) and should be followed to ensure adequate durability and performance.

Other methods of reducing steel corrosion include the use of epoxy-coated reinforcing steel (ASTM D 3963), surface treatments, corrosion inhibiting admixtures, concrete overlays, and cathodic protection.

Epoxy-coated reinforcing steel works by preventing chloride ions from reaching the steel (Fig 1-30). Surface treatments and concrete overlays attempt to stop or reduce chloride ion penetration at the concrete surface. Silanes, siloxanes, methacrylates, epoxies, and other materials are used as surface treatments.

Impermeable materials, such as most epoxies, should not be used on slabs on ground or other concrete where moisture can freeze under the coating. The freezing water can cause surface delamination under the impermeable coating. Latex-modified portland cement concrete, low-slump concrete, and concrete with silica fume are used in overlays to reduce chloride-ion ingress.

Cathodic protection methods reverse the corrosion current flow through the concrete and reinforcing steel. This is done by inserting a nonstructural anode in the concrete and forcing the steel to be the cathode by electrically charging the system. The anode is connected to the positive pole of a rectifier. Since corrosion occurs where the current leaves the steel, the steel cannot corrode if it is receiving the induced current.

Chloride present in plain concrete (not containing steel) is generally not a durability concern.

Corrosion of nonferrous metals in concrete is discussed in Kerkhoff (2001).

Chemical Resistance

Portland cement concrete is resistant to most natural environments; however, concrete is sometimes exposed to substances that can attack and cause deterioration. Concrete in chemical manufacturing and storage facilities is especially prone to chemical attack. The effect of sulphates and chlorides was discussed previously. Acids attack concrete by dissolving cement paste and calcareous aggregates. In addition to using concrete with a low permeability, surface treatments can be used to keep the aggressive substance from coming in contact with the concrete. *Effects of Substances on Concrete and Guide to Protective Treatment* (Kerkhoff 2001) discusses the effects of hundreds of chemicals on concrete and provides a list of treatments to help control chemical attack.

Sulphate Attack

Excessive amounts of sulphates in soil or water can attack and destroy a concrete that is not properly designed. Sulphates (for example calcium sulphate, sodium sulphate, and magnesium sulphate) can attack concrete by reacting with hydrated compounds in the hardened cement paste. These reactions can induce sufficient pressure to disrupt the cement paste, resulting in disintegration of the concrete (loss of paste cohesion and strength). Calcium sulphate attacks calcium aluminate hydrate and forms ettringite. Sodium sulphate reacts with calcium hydroxide and calcium aluminate hydrate forming ettringite and gypsum. Magnesium sulphate attacks in a manner similar to sodium sulphate and forms ettringite, gypsum, and also brucite (magnesium hydroxide). Brucite forms primarily on the concrete surface; it consumes calcium hydroxide, lowers the pH of the pore solution, and then decomposes the calcium silicate hydrates (Santhanam and others 2001).

Thaumasite may form during sulphate attack in moist conditions at temperatures usually between 0°C and 10°C and it occurs as a result of a reaction between calcium silicate hydrate, sulphate, calcium carbonate, and water (Report of the Thaumasite Expert Group 1999). In concretes where deterioration is associated with excess thaumasite formation, cracks can be filled with thaumasite and haloes of white thaumasite are present around aggregate particles. At the concrete/soil interface the surface concrete layer can be "mushy" with complete replacement of the cement paste by thaumasite (Hobbs 2001).

Like natural rock formations such as limestone, porous concrete is susceptible to weathering caused by salt crystallization. These salts may or may not contain sulphates and they may or may not react with the hydrated compounds in concrete. Examples of salts known to cause weathering of field concrete include sodium carbonate and sodium sulphate (laboratory studies have also related saturated solutions of calcium chloride and other salts to concrete deterioration). The greatest damage occurs with drying of saturated solutions of these salts, often in an environment with specific cyclic changes in relative humidity and temperature that alter mineralogical phases. In permeable concrete exposed to drying conditions, salt solutions can rise to the surface by capillary action and subsequently—as a result of surface evaporation—the solution phase becomes supersaturated and salt crystallization occurs, sometimes generating pressures large enough to cause cracking. If the rate of migration of the salt solution through the pores is lower than the rate of evaporation, a drying zone forms beneath the surface and salt crystallization occurs in the pores causing expansion and scaling (Mehta 2000). Both aggregate particles and cement paste can be attacked by salts.

Fig. 1-31. Sulphate attack is often the most severe at the location of the most wetting and drying, which is usually near the soil line. Here concrete posts have been attacked by sulphates near the soil line. Also see the inset in Fig. 1-32. The concrete is in better condition deep within the soil where it is moist. (43093)

Sulphate attack and salt crystallization are more severe at locations where the concrete is exposed to wetting and drying cycles, than continuously wet exposures. This is often seen in concrete posts where the concrete has deteriorated only a few millimetres above and below the soil line. The portion of concrete deep in the soil (where it is continuously wet) is in good condition (Fig. 1-31 and 1-32). However, if the sulphate exposure is severe enough, with time even the continuously moist sections can be attacked by sulphates if the concrete is not properly designed.

For the best defense against external sulphate attack: (1) design concrete with a low water to cementing materials ratio (around 0.4), and (2) use cements specially formulated for sulphate environments, such as CSA A5 Type 20, Moderate sulphate resistant cement, or Type 50 Sulphate-resistant. Type 20 and 50 cements are shown in Fig. 1-33.

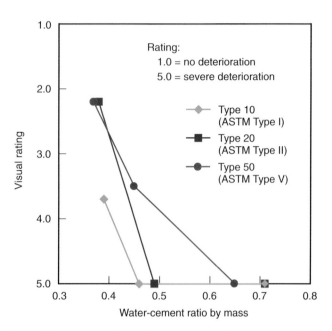

Fig. 1-33. Average 16-yr ratings of concrete beams for three portland cements at various water-cement ratios (Stark 2002).

Fig. 1-32. Concrete beams after seven years of exposure to sulphate-rich wet soil in a Sacramento, California, test plot. The beams in better condition have low water-cementing materials ratios, and most have sulphate resistant cement. The inset shows two of the beams tipped on their side to expose decreasing levels of deterioration with depth and moisture level. (66900, 58499)

Seawater Exposures

Concrete has been used in seawater exposures for decades with excellent performance. However, special care in mix design and material selection is necessary for these severe environments. A structure exposed to seawater or seawater spray is most vulnerable in the tidal or splash zone where there are repeated cycles of wetting and drying and/or freezing and thawing. Sulphates and chlorides in seawater require the use of low permeability concrete to minimize steel corrosion and sulphate attack (Fig. 1-34).

Portland cements with tricalcium aluminate (C_3A) contents that range from 4% to 10% have been found to provide satisfactory protection against seawater sulphate attack, as well as protection against corrosion of reinforcement by chlorides. Water-cementing materials ratios should not exceed 0.40, a satisfactory air-void system meeting the requirements of CSA A23.1, and cover over reinforcement of at least 60 mm (and preferably more) should be provided. High strength concrete should be considered where large ice formations abrade the structure. See Stark (1995 and 2001), Farny (1996), and PCA (2001).

Fig. 1-34. Concrete bridges in seawater exposure must be specially designed for durability. (68667)

Ettringite and Heat Induced Delayed Expansion

Ettringite, one form of calcium sulphoaluminate, is found in all portland cement paste. Calcium sulphate sources, such as gypsum, are added to portland cement during final grinding at the cement mill to prevent rapid setting and improve strength development. Sulphate is also present in supplementary cementing materials and admixtures. Gypsum and other sulphate compounds react with calcium aluminate in cement to form ettringite within the first few hours after mixing with water. Most of the sulphate in cement is normally consumed to form ettringite or

Fig. 1-35. White secondary ettringite deposits in void. Field width 64 μm. (69547)

calcium monosulphate within 24 hours (Klemm and Miller 1997). At this stage ettringite is uniformly and discretely dispersed throughout the cement paste at a submicroscopic level (less than a micrometre in cross-section). This ettringite is often called primary ettringite.

If concrete is exposed to moisture for long periods of time (many years), the ettringite can slowly dissolve and reform in less confined locations. Upon microscopic examination, harmless white needle-like crystals of ettringite can be observed lining air voids. This reformed ettringite is usually called secondary ettringite (Fig. 1-35).

Concrete deterioration accelerates the rate at which ettringite leaves its original location in the paste to go into solution and recrystallize in larger spaces such as air voids or cracks. Both water and sufficient space must be present for the crystals to form. Cracks can form due to damage caused by frost action, alkali-aggregate reactivity, drying shrinkage, thermal effects, strain due to excessive stress, or other mechanisms.

Ettringite crystals in air voids and cracks are typically two to four micrometres in cross section and 20 to 30 micrometres long. Under conditions of extreme deterioration or decades in a moist environment, the white ettringite crystals can appear to completely fill voids or cracks. However, secondary ettringite, as large needle-like crystals, should not be interpreted as being harmful to the concrete (Detwiler and Powers-Couche 1997).

Heat Induced Delayed Expansion. Heat induced delayed expansion (HIDE)—also called delayed ettringite formation (DEF)—refers to a rare condition of internal sulphate attack* in which mature concretes undergo expansion and cracking. Only concretes of particular chemical makeup are affected when they have achieved high temperatures,

---

*Internal sulphate attack refers to deterioration mechanisms occurring in connection with sulphate that is in the concrete at the time of placement.

usually after the first few hours of placement (between 70°C and 100°C depending on the concrete ingredients and the time the temperature is achieved after casting). This can occur because the high temperature decomposes any initial ettringite formed and holds the sulphate and alumina tightly in the calcium silicate hydrate (C-S-H) gel of the cement paste. The normal formation of ettringite is thus impeded.

In the presence of moisture, sulphate desorbs from the confines of the C-S-H and reacts with calcium monosulphoaluminate to form ettringite in cooled and hardened concrete. After months or years of desorption, ettringite forms in confined locations within the paste. Such ettringite can exert crystallization pressures because it forms in a limited space under supersaturation. One theory: since concrete is rigid and if there are insufficient voids to accommodate the ettringite volume increase, expansion and cracks can occur. In addition, some of the initial (primary) ettringite may be converted to monosulphoaluminate at high temperatures and upon cooling revert back to ettringite. Because ettringite takes up more space than monosulphoaluminate from which it forms, the transformation is an expansive reaction. The mechanism causing expansion in the paste is not fully understood at this time; the true influence of ettringite formation on this expansion is still being investigated. Some research indicates that there is little relationship between ettringite formation and expansion.

As a result of an increase in paste volume, separation of the paste from the aggregates is usually observed with heat induced delayed expansion. It is characterized by the development of rims of ettringite around the aggregates (Fig. 1-36). At early stages of heat induced delayed expansion, the voids between paste and aggregate are empty (no

ettringite present). It should be noted that concrete can sustain a small amount of this expansion without harm. Only extreme cases result in cracking, and often heat induced delayed expansion is associated with other deterioration mechanisms, especially alkali-silica reactivity.

Only concretes in massive elements that retain the heat of hydration or elements exposed to very high temperatures at an early age are at risk of HIDE; and of these only a few have the chemical makeup or temperature profile to cause detrimental expansion. Normal sized concrete elements cast and maintained near ambient temperatures cannot experience HIDE when sound materials are used.

Fly ash and slag may help control heat induced delayed expansion, along with control over early-age temperature development. For more information, see Lerch (1945), Day (1992), Klemm and Miller (1997), Thomas (1998), and Famy (1999).

REFERENCES

ACI Committee 201, *Guide to Durable Concrete*, ACI 201.2R-92, American Concrete Institute, Farmington Hills, Michigan, 1992.

ACI Committee 318, *Building Code Requirements for Structural Concrete and Commentary*, ACI 318-99, American Concrete Institute, Farmington Hills, Michigan, 1999.

ACI Manual of Concrete Practice, American Concrete Institute, Farmington Hills, Michigan, 2001.

Abrams, D. A., *Design of Concrete Mixtures*, Lewis Institute, Structural Materials Research Laboratory, Bulletin No. 1, PCA LS001, http://www.portcement.org/pdf_files/LS001.pdf, 1918, 20 pages.

Abrams, M. S., and Orals, D. L., *Concrete Drying Methods and Their Effect on Fire Resistance*, Research Department Bulletin RX181, Portland Cement Association, http://www.portcement.org/pdf_files/RX181.pdf, 1965.

Backstrom, J. E.; Burrow, R. W.; and Witte, L. P., *Investigation into the Effect of Water-Cement Ratio on the Freezing-Thawing Resistance of Non-Air- and Air-Entrained Concrete*, Concrete Laboratory Report No. C-810, Engineering Laboratories Division, U.S. Department of the Interior, Bureau of Reclamation, Denver, November 1955.

Brinkerhoff, C. H., "Report to ASTM C-9 Subcommittee III-M (Testing Concrete for Abrasion) Cooperative Abrasion Test Program," University of California and Portland Cement Association, 1970.

Bureau of Reclamation, *Concrete Manual*, 8th Edition, U.S. Bureau of Reclamation, Denver, 1981, page 33.

Burg, Ronald G., *The Influence of Casting and Curing Temperature on the Properties of Fresh and Hardened Concrete*, Research and Development Bulletin RD113, Portland Cement Association, 1996, 20 pages.

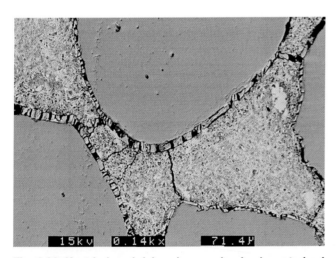

Fig. 1-36. Heat induced delayed expansion is characterized by expanding paste that becomes detached from non-expansive components, such as aggregates, creating gaps at the paste-aggregate interface. The gap can subsequently be filled with larger opportunistic ettringite crystals as shown here. (69154)

Campbell, Donald H.; Sturm, Ronald D.; and Kosmatka, Steven H., "Detecting Carbonation," *Concrete Technology Today*, PL911, Portland Cement Association, http://www.portcement.org/pdf_files/PL911.pdf, March 1991, 5 pages.

Canadian Standards Association, CSA Standard A23.1-00, *Concrete Materials and Methods of Concrete Construction*, and CAN/CSA Standard A23.2-00, *Methods of Test for Concrete*, Canadian Standards Association, Toronto, 2000.

Canadian Standards Association, CSA Standard A864-00, *Guide to the Evaluation and Management of Concrete Structures Affected by Alkali-Aggregate Reaction*, Canadian Standards Association, Toronto, 2000.

Copeland, L. E., and Schulz, Edith G., *Electron Optical Investigation of the Hydration Products of Calcium Silicates and Portland Cement*, Research Department Bulletin RX135, Portland Cement Association, http://www.portcement.org/pdf_files/RX135.pdf, 1962, 12 pages.

Day, Robert L., *The Effect of Secondary Ettringite Formation on the Durability of Concrete: A Literature Analysis*, Research and Development Bulletin RD108, Portland Cement Association, 1992, 126 pages.

Detwiler, Rachel J., and Powers-Couche, Laura, "Effect of Ettringite on Frost Resistance," *Concrete Technology Today*, PL973, Portland Cement Association, http://www.portcement.org/pdf_files/PL973.pdf, 1997, pages 1 to 4.

Famy, Charlotte, *Expansion of Heat-Cured Mortars*, thesis, University of London, 1999, 310 pages.

Farny, Jamie, "Treat Island, Maine—The Army Corps' Outdoor Durability Test Facility," *Concrete Technology Today*, PL963, http://www.portcement.org/pdf_files/PL963.pdf, December 1996, pages 1 to 3.

Farny, James A., and Kosmatka, Steven H., *Diagnosis and Control of Alkali-Aggregate Reactions in Concrete*, IS413, Portland Cement Association, 1997, 24 pages.

Feret, R., "Etudes Sur la Constitution Intime Des Mortiers Hydrauliques" [Studies on the Intimate Constitution of Hydraulic Mortars], *Bulletin de la Societe d'Encouragement Pour Industrie Nationale*, 5th Series, Vol. 2, Paris, 1897, pages 1591 to 1625.

Gonnerman, H. F., and Shuman, E. C., "Flexure and Tension Tests of Plain Concrete," Major Series 171, 209, and 210, *Report of the Director of Research*, Portland Cement Association, November 1928, pages 149 and 163.

Hanson, J. A., *Effects of Curing and Drying Environments on Splitting Tensile Strength of Concrete*, Development Department Bulletin DX141, Portland Cement Association, http://www.portcement.org/pdf_files/DX141.pdf, 1968, page 11.

Hedenblad, Göran, *Drying of Construction Water in Concrete*, T9, Swedish Council for Building Research, Stockholm, 1997.

Hedenblad, Göran, "Concrete Drying Time," *Concrete Technology Today*, PL982, Portland Cement Association, http://www.portcement.org/pdf_files/PL982.pdf, 1998, pages 4 to 5.

Hobbs, D. W., "Concrete deterioration: causes, diagnosis, and minimizing risk," *International Materials Review*, 2001, pages 117 to 144.

Hsu, Thomas T. C., *Torsion of Structural Concrete—Plain Concrete Rectangular Sections*, Development Department Bulletin DX134, Portland Cement Association, http://www.portcement.org/pdf_files/DX134.pdf, 1968.

Kerkhoff, Beatrix, *Effects of Substances on Concrete and Guide to Protective Treatments*, IS001, Portland Cement Association, 2001, 36 pages.

Kirk, Raymond E., and Othmer, Donald F., eds., "Cement," *Encyclopedia of Chemical Technology*, 3rd ed., Vol. 5, John Wiley and Sons, Inc., New York, 1979, pages 163 to 193.

Klemm, Waldemar A., and Miller, F. MacGregor, "Plausibility of Delayed Ettringite Formation as a Distress Mechanism—Considerations at Ambient and Elevated Temperatures," Paper 4iv059, *Proceedings of the 10th International Congress on the Chemistry of Cement*, Gothenburg, Sweden, June 1997, 10 pages.

Kosmatka, Steven H., "Bleeding," *Significance of Tests and Properties of Concrete and Concrete-Making Materials*, STP 169C, American Society for Testing and Materials, West Conshohocken, Pennsylvania, 1994, pages 88 to 111. [Also available as PCA RP328].

Kosmatka, Steven H., *Portland, Blended, and Other Hydraulic Cements*, IS004, Portland Cement Association, 2001, 31 pages.

Lange, David A., *Long-Term Strength Development of Concrete*, RP326, Portland Cement Association, 1994.

Lerch, William, *Effect of SO_3 Content of Cement on Durability of Concrete*, R&D Serial No. 0285, Portland Cement Association, 1945.

Liu, Tony C., "Abrasion Resistance of Concrete," *Journal of the American Concrete Institute*, Farmington Hills, Michigan, September-October 1981, pages 341 to 350.

McMillan, F. R., and Lyse, Inge, "Some Permeability Studies of Concrete," *Journal of the American Concrete Institute, Proceedings*, vol. 26, Farmington Hills, Michigan, December 1929, pages 101 to 142.

McMillan, F. R., and Tuthill, Lewis H., *Concrete Primer*, SP-1, 3rd ed., American Concrete Institute, Farmington Hills, Michigan, 1973.

Mehta, P. Kumar, "Sulfate Attack on Concrete: Separating Myth from Reality," *Concrete International*, Farmington Hills, Michigan, August 2000, pages 57 to 61.

Pinto, Roberto C. A., and Hover, Kenneth C., *Frost and Scaling Resistance of High-Strength Concrete*, Research and Development Bulletin RD122, Portland Cement Association, 2001, 70 pages.

Powers, T. C., "Studies of Workability of Concrete," *Journal of the American Concrete Institute*, Vol. 28, Farmington Hills, Michigan, February 1932, page 419.

Powers, T. C., *The Bleeding of Portland Cement Paste, Mortar, and Concrete*, Research Department Bulletin RX002, Portland Cement Association, http://www.portcement.org/pdf_files/RX002.pdf, 1939.

Powers, T. C., *A Discussion of Cement Hydration in Relation to the Curing of Concrete*, Research Department Bulletin RX025, Portland Cement Association, http://www.portcement.org/pdf_files/RX025.pdf, 1948, 14 pages.

Powers, T. C., *The Nonevaporable Water Content of Hardened Portland Cement Paste—Its Significance for Concrete Research and Its Method of Determination*, Research Department Bulletin RX029, Portland Cement Association, http://www.portcement.org/pdf_files/RX029.pdf, 1949, 20 pages.

Powers, T. C., *Topics in Concrete Technology*, Research Department Bulletin RX174, Portland Cement Association, http://www.portcement.org/pdf_files/RX174.pdf, 1964.

Powers, T. C., *The Nature of Concrete*, Research Department Bulletin RX196, Portland Cement Association, http://www.portcement.org/pdf_files/RX196.pdf, 1966.

Powers, T. C., and Brownyard, T. L., *Studies of the Physical Properties of Hardened Portland Cement Paste*, Research Department Bulletin RX022, Portland Cement Association, http://www.portcement.org/pdf_files/RX022.pdf, 1947.

Powers, T. C.; Copeland, L. E.; Hayes, J. C.; and Mann, H. M., *Permeability of Portland Cement Pastes*, Research Department Bulletin RX053, Portland Cement Association, http://www.portcement.org/pdf_files/RX053.pdf, 1954.

Report of the Thaumasite Expert Group, *The thaumasite form of sulfate attack: Risks, diagnosis, remedial works and guidance on new construction*, Department of the Environment, Transport and the Regions, DETR, London, 1999.

Santhanam, Manu; Cohen, Menahi D.; and Olek, Jan, "Sulfate attack research—whither now?," *Cement and Concrete Research*, 2001, pages 845 to 851.

Scanlon, John M., "Factors Influencing Concrete Workability," *Significance of Tests and Properties of Concrete and Concrete-Making Materials*, STP 169C, American Society for Testing and Materials, West Conshohocken, Pennsylvania, 1994, pages 49 to 64.

Stark, David, *Durability of Concrete in Sulfate-Rich Soils*, Research and Development Bulletin RD097, Portland Cement Association, http://www.portcement.org/pdf_files/RD097.pdf, 1989, 16 pages.

Stark, David, *Long-Time Performance of Concrete in a Seawater Exposure*, RP337, Portland Cement Association, 1995, 58 pages.

Stark, David, *Long-Term Performance of Plain and Reinforced Concrete in Seawater Environments*, Research and Development Bulletin RD119, Portland Cement Association, 2001, 14 pages.

Stark, David, *Performance of Concrete in Sulfate Environments*, PCA Serial No. 2248, Portland Cement Association, 2002.

Steinour, H. H., *Further Studies of the Bleeding of Portland Cement Paste*, Research Department Bulletin RX004, Portland Cement Association, http://www.portcement.org/pdf_files/RX004.pdf, 1945.

Taylor, Peter C.; Whiting, David A.; and Nagi, Mohamad A., *Threshold Chloride Content for Corrosion of Steel in Concrete: A Literature Review*, PCA Serial No. 2169, Portland Cement Association, http://www.portcement.org/pdf_files/SN2169.pdf, 2000.

Thomas, M. D. A., *Delayed Ettringite Formation in Concrete—Recent Developments and Future Directions*, University of Toronto, 1998, 45 pages.

VanGeem, Martha G., and Whiting, David A., *Chloride Limits in Reinforced Concrete*, PCA Serial No. 2438, Portland Cement Association, http://www.portcement.org/pdf_files/SN2438.pdf, 2000.

Verbeck, G. J., *Carbonation of Hydrated Portland Cement*, Research Department Bulletin RX087, Portland Cement Association, http://www.portcement.org/pdf_files/RX087.pdf, 1958, 20 pages.

Whiting, D., *Origins of Chloride Limits for Reinforced Concrete*, PCA Serial No. 2153, Portland Cement Association, http://www.portcement.org/pdf_files/SN2153.pdf, 1997.

Whiting, D., "Permeability of Selected Concretes," *Permeability of Concrete*, SP108, American Concrete Institute, Farmington Hills, Michigan, 1989, pages 195 to 222.

Wood, Sharon L., *Evaluation of the Long-Term Properties of Concrete*, Research and Development Bulletin RD102, Portland Cement Association, 1992, 99 pages.

Woods, Hubert, *Observations on the Resistance of Concrete to Freezing and Thawing*, Research Department Bulletin RX067, Portland Cement Association, http://www.portcement.org/pdf_files/RX067.pdf, 1956.

CHAPTER 2

Portland, Blended, and Other Hydraulic Cements

Fig. 2-1. Portland cement is a fine powder that when mixed with water becomes the glue that holds aggregates together in concrete. (58420)

Portland cements are hydraulic cements composed primarily of hydraulic calcium silicates (Fig. 2-1) that set and harden by reacting chemically with water. This chemical reaction is called hydration. When the paste (cement, air, and water) is added to aggregates (sand and gravel, crushed stone, or other granular material) it acts as an adhesive and binds the aggregates together to form a stonelike mass, concrete, the world's most versatile and most widely used construction material.

Hydration begins as soon as cement comes in contact with water. This chemical reaction produces cement hydrates, a product of hydration, which form on the surface of each cement particle. These cement hydrates gradually grow and spread until they interlock with others attached to adjacent cement particles or adhere to other nearby substances. This ongoing hydration process results in progressive stiffening, hardening and strength development. Hydration continues as long as moisture and temperature conditions are favourable (curing) and space for hydration products is available.

The stiffening of concrete can be recognized by a loss of workability that usually occurs within three hours of mixing, but is dependent upon the composition and fineness of the cement, the admixtures, the mixture proportions, and the temperature conditions. Subsequently, the concrete sets and becomes hard. Most of the hydration and strength development take place within the first month of concrete's life cycle, but they continue, though more slowly, for a long time with adequate moisture and temperature; continuous strength increases exceeding 30 years have been recorded (Washa and Wendt 1975 and Wood 1992).

THE BEGINNING OF AN INDUSTRY

Early builders used clay to bind stones together into a solid structure for shelter and protection. The oldest concrete discovered so far dates from around 7000 BC and was found in 1985 when a concrete floor was uncovered during the construction of a road at Yiftah El in Galilee, Israel. It consisted of a lime concrete, made from burning limestone to produce quicklime, which when mixed with water and stone, hardened to form concrete (Brown 1996 and Auburn 2000).

A cementing material was used between the stone blocks in the construction of the Great Pyramid at Giza in Ancient Egypt around 2500 BC. Some reports say it was a lime mortar while others say the cementing material was made from burnt gypsum. By 500 BC, the art of making lime-based mortar arrived in Ancient Greece. The Greeks used lime-based materials as a binder between stone and brick and as a rendering material over porous limestones commonly used in the construction of their temples and palaces.

Examples of early Roman concrete have been found dating back to 300 BC. The very word concrete is derived from the Latin word "concretus" meaning grown together or compounded. The Romans perfected the use of pozzolan as a cementing material. Sometime during the second century BC the Romans quarried a volcanic ash near Pozzuoli, thinking it was sand, they mixed it with lime and found the mixture to be much stronger than they had produced previously. This discovery was to have a significant effect on construction as the material was not sand, but a fine volcanic ash containing silica and alumina, which combined chemically with the lime to produce what became known as pozzolanic cement. This material was used by builders of the famous Roman walls, aqueducts and other historic structures including the Theatre at Pompeii (seating 20,000 spectators), the Colosseum and the Pantheon in Rome. Pozzolan seems to have been ignored during the Middle Ages when building practices were much less refined than earlier and the quality of cementing materials deteriorated. The practice of burning lime and the use of pozzolan was not introduced again until the 1300s.

Efforts to determine why some limes possess hydraulic properties while those made from essentially pure limestones do not were not made until the 18th century. John Smeaton, often referred to as the "father of civil engineering in England," concentrated his work in this field. He found that an impure, soft limestone, containing clay minerals made the best hydraulic cement. This combined with a pozzolan, imported from Italy, was used in his project to rebuild the Eddystone Lighthouse in the English Channel, southwest of Plymouth England. The project took three years to complete and began operation in 1759; it was recognized as a significant accomplishment in the development of the cement industry. A number of discoveries followed as efforts within a growing natural cement industry were now directed to the production of a consistent quality material.

The difference between a hydraulic lime and natural cement is a function of the temperature attained during calcination. Furthermore, a hydraulic lime can hydrate in a "lump" form, whereas natural cements must be crushed and finely ground before hydration can take place. Natural cement is stronger than hydraulic lime but weaker than portland cement. Records indicate that lime and hydraulic cement were first produced in Canada as early as 1830 to 1840 at a plant in Hull, Quebec.

The development of portland cement was the result of persistent investigation by science and industry to produce a superior quality natural cement. The invention of portland cement is generally credited to Joseph Aspdin, an English mason. In 1824, he obtained a patent for his product, which he named portland cement because when set, it resembled the colour of the natural limestone quarried on the Isle of Portland in the English Channel (Fig. 2-2) (Aspdin 1824). The name has endured and is used throughout the world, with many manufacturers adding their own trade or brand names.

Aspdin was the first to prescribe a formula for portland cement and the first to have his product patented. However, in 1845, I. C. Johnson, of White and Sons, Swanscombe, England, claimed to have "burned the cement raw materials with unusually strong heat until the mass was nearly vitrified," producing a portland cement as we now know it. This cement became the popular choice during the middle of the 19th century and was exported from England to various parts of the world. Production also began in Belgium, France, and Germany about the same time and export of these products from Europe to North America began about 1865.

Natural cement producers in northeastern United States produced the first portland cement in North America in 1871 at a plant near Coplay, Pennsylvania. The first imports of portland cement into Canada date back to 1877. Canadian producers of natural cement recognized the superiority of portland cement and made a determined effort to compete by converting existing facilities to produce the proven material for local consumption. In 1889, C. B. Wright & Sons, Hull, Quebec, became the first producer of portland cement in Canada. Almost at the same time Napanee Cement Works, Napanee, Ontario came on line producing portland cement. This was followed shortly after by the construction of two new plants for the production of portland cement, one at Shallow Lake, near Owen Sound, Ontario, and one at Longue Pointe, east of Montreal, Quebec. These four plants are credited with much of the pioneer work in the development of the portland cement industry in Canada. The first portland cement plant in western Canada was built by the Canadian Pacific Railway Company in Vancouver in 1893.

In 1900, cement production in Canada totaled 73,072 tons (51,122 tons of portland cement and 21,950 tons of natural rock cement (Canada Department of Mines 1920). One hundred years later in 2000, there were 16 plants that produced a total of 12.8 million tonnes of cement in 15 locations across Canada.

In the industry's early days each producer made portland cement according to their own specifications. The first standard was established in 1916 by the Canadian Society for Civil Engineering. Taking over the job for the development of standards, the Canadian Standards Association (CSA) produced its first standard for hydraulic cement in 1922. All subsequent cement standards, including the current standard, A5-98, have been developed by CSA.

MANUFACTURE OF PORTLAND CEMENT

By definition, portland cement is a product obtained by pulverizing clinker consisting essentially of hydraulic calcium silicates to which various forms of calcium sulphate, limestone, water, and processing additions may be added at the option of the manufacturer.

Unlike cement manufactured to meet ASTM specification C 150, *Standard Specification for Portland Cement*, the Canadian Standards Association's Technical Committee on Hydraulic Cement and

Fig. 2-2. Isle of Portland quarry stone (after which portland cement was named) next to a cylinder of modern concrete. (68976)

22

Table 2-1. Sources of Raw Materials Used in Manufacture of Portland Cement

Calcium	Iron	Silica	Alumina	Sulphate
Alkali waste	Blast-furnace flue dust	Calcium silicate	Aluminum-ore refuse*	Anhydrite
Aragonite*	Clay*	Cement rock	Bauxite	Calcium sulphate
Calcite*	Iron ore*	Clay*	Cement rock	Gypsum*
Cement-kiln dust	Mill scale*	Fly ash	Clay*	
Cement rock	Ore washings	Fuller's earth	Copper slag	
Chalk	Pyrite cinders	Limestone	Fly ash*	
Clay	Shale	Loess	Fuller's earth	
Fuller's earth		Marl*	Granodiorite	
Limestone*		Ore washings	Limestone	
Marble		Quartzite	Loess	
Marl*		Rice-hull ash	Ore washings	
Seashells		Sand*	Shale*	
Shale*		Sandstone	Slag	
Slag		Shale^	Staurolite	
		Slag		
		Traprock		

Note: Many industrial byproducts have potential as raw materials for the manufacture of portland cement.
*Most common sources.

Supplementary Materials recognizes carbonate additions for some portland cements. At the option of the manufacturer, a maximum of 5% addition of limestone is permitted for Type 10, Normal portland cement, and for Type 30, High-early-strength portland cement. The limestone addition must be of a quality suitable for the manufacture of portland cement clinker.

Materials used in the manufacture of portland cement must contain appropriate proportions of calcium, silica, alumina, and iron components. During manufacture, analyses of all materials are made frequently to ensure a uniformly high quality cement.

While the operations of all cement plants are basically the same, no flow diagram can adequately illustrate all plants. There is no typical portland cement manufacturing plant; every plant has significant differences in layout, equipment, or general appearance (Fig. 2-3).

Selected raw materials (Table 2-1) are transported from the quarry (Fig. 2-4), crushed (Fig. 2-5), milled, and proportioned so that the resulting mixture has the desired chemical composition. The raw materials are generally a mixture of calcareous (calcium oxide) material, such as limestone, chalk or shells, and an argillaceous (silica and alumina) material such as clay, shale, or blast-furnace slag. Either a dry or a wet process is used (Fig. 2-6). In the dry process, grinding and blending are done with dry materials. In the wet process, the grinding and blending operations are done with the materials mixed with water in a slurry form. In other respects, the dry and wet processes are very much alike. Fig. 2-7 illustrates important technological developments that can improve significantly the productivity and energy efficiency of dry-process plants.

After blending, the ground raw material is fed into the upper end of a kiln (Fig. 2-8). The raw mix passes through the kiln at a rate controlled by the slope and rotational speed of the kiln. Burning fuel (powdered coal, new or

Fig. 2-4. Limestone, a primary raw material providing calcium in making cement, is quarried near the cement plant. (59894)

Fig. 2-5. Quarry rock is trucked to the primary crusher. (59893)

Fig. 2-3. Aerial view of a cement plant. (70000)

23

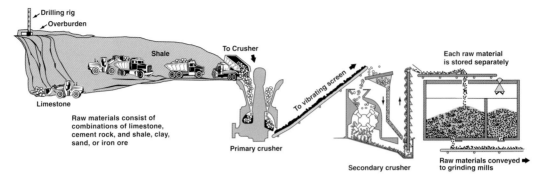

1. Stone is first reduced to 125 mm size, then to 20 mm, and stored.

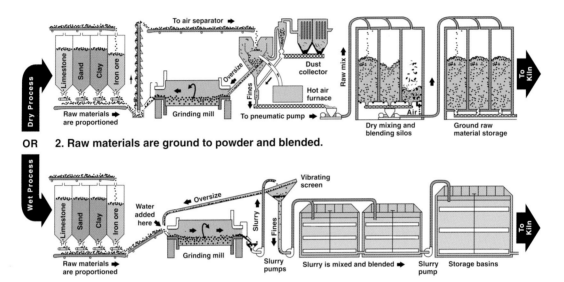

OR 2. Raw materials are ground to powder and blended.

2. Raw materials are ground, mixed with water to form slurry, and blended.

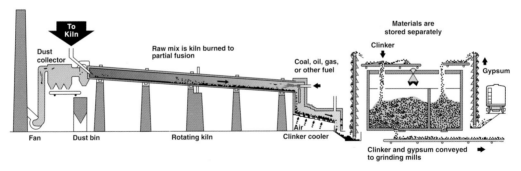

3. Burning changes raw mix chemically into cement clinker.

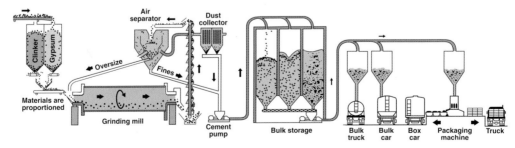

4. Clinker with gypsum is ground into portland cement and shipped.

Fig. 2-6. Steps in the traditional manufacture of portland cement.

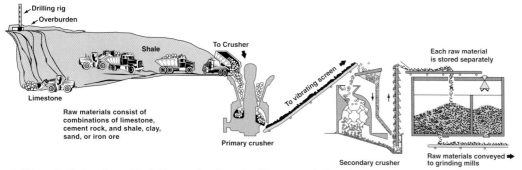

1. Stone is first reduced to 125 mm size, then to 20 mm, and stored.

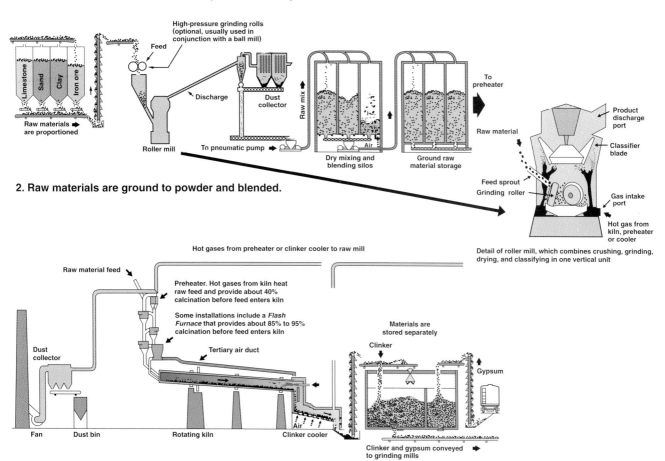

2. Raw materials are ground to powder and blended.

3. Burning changes raw mix chemically into cement clinker. Note four-stage preheater, flash furnaces, and shorter kiln.

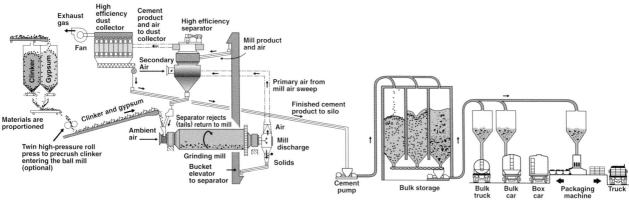

4. Clinker with gypsum is ground into portland cement and shipped.

Fig. 2-7. Steps in the modern dry-process manufacture of portland cement.

25

Fig. 2-8. Rotary kiln (furnace) for manufacturing portland cement clinker. Inset view inside the kiln. (58927, 25139)

recycled oil, gas, rubber tires, and by-product fuel) is forced into the lower end of the kiln where temperatures of 1400°C to 1550°C change the raw material chemically into cement clinker, grayish-black pellets predominantly the size of marbles (Fig. 2-9). Fig. 2-10 shows the clinker production process from raw feed to the final product.

Fig. 2-9. Portland cement clinker is formed by burning calcium and siliceous raw materials in a kiln. This particular clinker is about 20 mm in diameter. (60504)

The clinker is cooled and then pulverized. During this operation a small amount of gypsum (Fig. 2-11) is added to regulate the setting time of the cement and to improve shrinkage and strength development properties (Lerch 1946 and Tang 1992). In the grinding mill clinker is ground so fine that nearly all of it passes through a 45 µm sieve. This extremely fine gray powder is portland cement (Fig. 2-1).

Fig. 2-11. Gypsum, a source of sulphate, is interground with portland clinker to form portland cement. It helps control setting, drying shrinkage properties, and strength development. (60505)

TYPES OF PORTLAND CEMENT

All portland and blended cements are hydraulic cements. Hydraulic cement is the broad term to describe cements that set and harden by reacting chemically with water and are capable of doing so under water. They also stay hard and maintain their stability under water. Portland cements are used in all aspects of concrete construction. Different types of portland cement are manufactured to meet the various physical and chemical requirements needed for specific purposes. In Canada, portland cements are manufactured to meet the specifications of the Canadian Standards Association (CSA) Standard A5. Five types of portland cement are covered in CSA A5 and are identified as follows:

Type 10	Normal portland cement
Type 20	Moderate portland cement*
Type 30	High-early-strength portland cement
Type 40	Low-heat of hydration portland cement
Type 50	Sulphate-resistant portland cement

Portland cements specified and used in the United States normally meet the requirements of ASTM C 150, *Standard Specification for Portland Cement.* ASTM Standards are by far the most widely referenced and used specifications for cement and other concrete related materials. ASTM C 1157, *Performance Specification for Hydraulic Cements,* provides for six types of portland cement, and are discussed under "Other Hydraulic Cements" later in this chapter.

ASTM C 150 provides for eight types of portland cement and uses Roman numeral designations as follows:

Type I	Normal
Type IA	Normal, air-entraining
Type II	Moderate sulphate resistance
Type IIA	Moderate sulphate resistance, air-entraining
Type III	High early strength
Type IIIA	High early strength, air-entraining
Type IV	Low heat of hydration
Type V	High sulphate resistance

ASTM C 150 cement Types I, II, III, IV, and V are essentially the same as CSA A5 Types 10 through 50, respectively except for the allowance of up to 5% limestone addition in Type 10 and Type 30 cements.

*Moderate with respect to the heat of hydration or sulphate resistance.

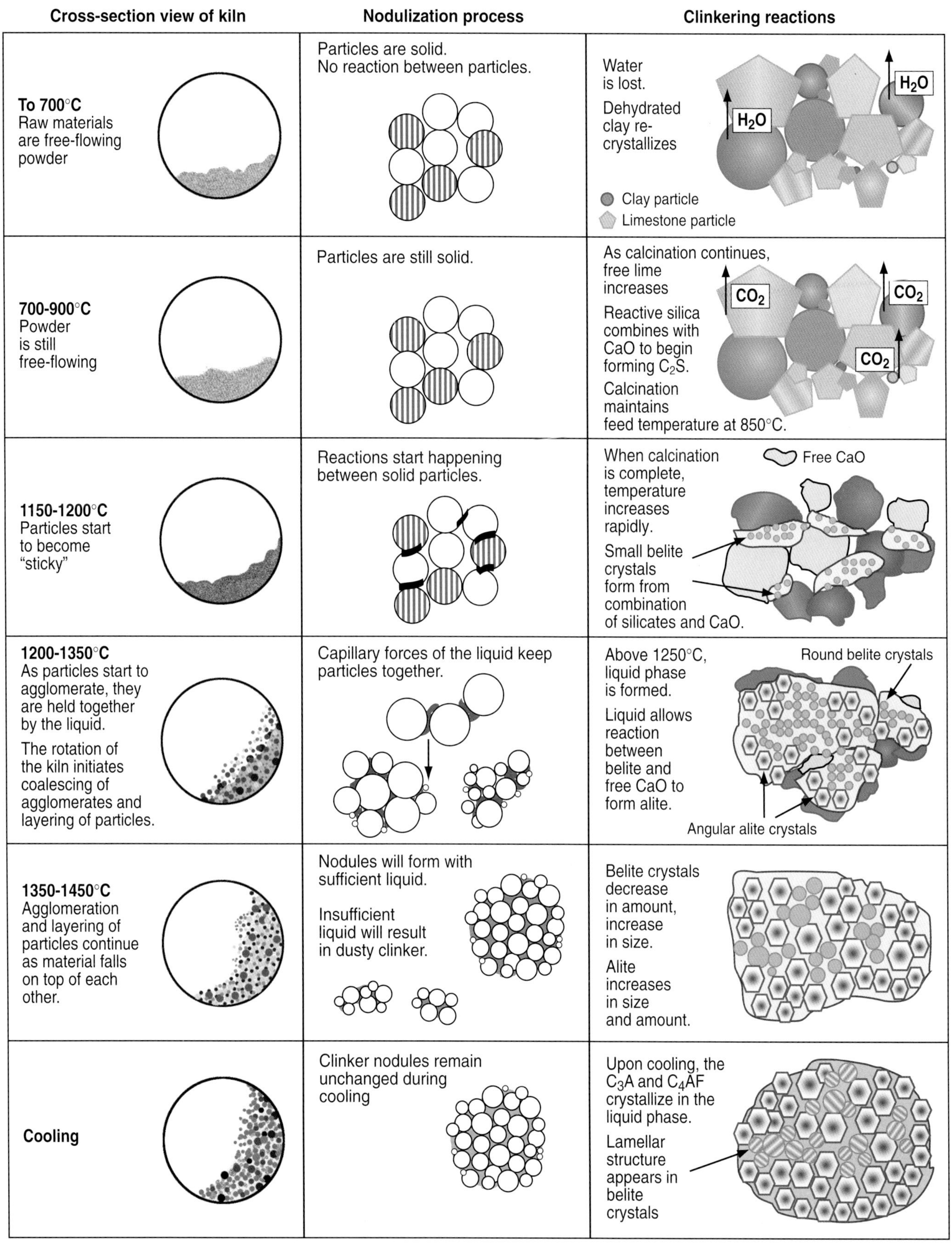

Cross-section view of kiln	Nodulization process	Clinkering reactions
To 700°C Raw materials are free-flowing powder	Particles are solid. No reaction between particles.	Water is lost. Dehydrated clay re-crystallizes ● Clay particle ⬠ Limestone particle
700-900°C Powder is still free-flowing	Particles are still solid.	As calcination continues, free lime increases. Reactive silica combines with CaO to begin forming C_2S. Calcination maintains feed temperature at 850°C.
1150-1200°C Particles start to become "sticky"	Reactions start happening between solid particles.	When calcination is complete, temperature increases rapidly. Small belite crystals form from combination of silicates and CaO. Free CaO
1200-1350°C As particles start to agglomerate, they are held together by the liquid. The rotation of the kiln initiates coalescing of agglomerates and layering of particles.	Capillary forces of the liquid keep particles together.	Above 1250°C, liquid phase is formed. Liquid allows reaction between belite and free CaO to form alite. Round belite crystals. Angular alite crystals
1350-1450°C Agglomeration and layering of particles continue as material falls on top of each other.	Nodules will form with sufficient liquid. Insufficient liquid will result in dusty clinker.	Belite crystals decrease in amount, increase in size. Alite increases in size and amount.
Cooling	Clinker nodules remain unchanged during cooling	Upon cooling, the C_3A and C_4AF crystallize in the liquid phase. Lamellar structure appears in belite crystals

Fig. 2-10. Process of clinker production from raw feed to the final product (Hills 2000).

A detailed review of the five types of portland cements covered in CSA Standard A5, *Portland Cement,* follows:

Type 10

Type 10 portland cement is a general-purpose cement suitable for all uses where the special properties of other types are not required. Its uses in concrete include pavements, floors, reinforced concrete buildings, bridges, tanks, reservoirs, pipe, masonry units, and precast concrete products (Fig. 2-12).

Type 20

Type 20 portland cement is used where precaution against moderate sulphate attack is important. It is used in normal structures or elements exposed to soil or ground waters where sulphate concentrations are higher than normal but not unusually severe (see Table 2-2 and Figs. 2-13 to

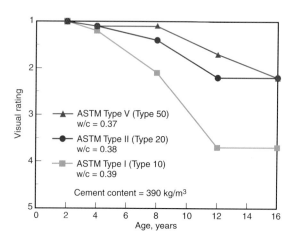

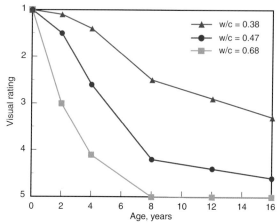

Fig. 2-12. Typical uses for normal or general use cements include (across, top left to bottom right) highway pavements, floors, bridges, and buildings. (68815, 68813, 68809, 63303)

Fig. 2-13. (top) Performance of concretes made with different cements in sulphate soil. Type II (20) and Type V (50) cements have lower C_3A contents that provide improved sulphate resistance. (bottom) Improved sulphate resistance results from low water-to-cementing materials ratios as demonstrated over time for concrete beams exposed to sulphate soils in a wetting and drying environment. Shown are average values for concretes containing a wide range of cementing materials, including cement Types I, II, V (10, 20, 50), blended cements, pozzolans, and slags. See Fig. 2-15 for rating illustration and a description of the concrete beams (Stark 2002).

Table 2.2 Requirements for Concrete Subjected to Sulphate Attack*

Class of exposure	Degree of exposure	Water-soluble sulphate (SO_4) in soil sample, %	Sulphate (SO_4) in groundwater samples, mg/L	Minimum specified 56 day compressive strength, MPa◊	Maximum water-to-cementing materials ratio†	Air content category‡	Cementing materials to be used**◊◊
S-1	Very severe	Over 2.0	Over 10,000	35	0.40	2	50
S-2	Severe	0.20 to 2.0	1500 to 10,000	32	0.45	2	50
S-3	Moderate	0.10 to 0.20	150 to 1500	30	0.50	2	20E††, 40, or 50E

* For seawater exposure refer to CSA A23.1 Clause 15.
◊ Where supplementary cementing materials are used, the owner may specify other test ages.
† The owner shall specify the minimum 28-day compressive strength.
‡ For steel-troweled interior slabs on grade subject to sulphate attack but not freeze thaw, air entrainment is not required.
** When combinations of portland cement and supplementary cementing materials are used, they shall have been proven, to the satisfaction of the owner, to produce concrete resistant to the exposure conditions under consideration.
◊◊ Cementing material combinations with equivalent performance may be used (Refer to CSA A23.1 Clauses 3.2, 3.3, and 3.4).
†† Type 20E cement with moderate sulphate resistance (Refer to CSA A23.1 Clause 3.1.2).

Note: Type 50E cement shall not be used in reinforced concrete exposed to both chlorides and sulphates. Refer to CSA A 23.1 Clause 15.4.
　　　See CSA Test Methods A23.2-2B and A23.2-3B for test methods to determine sulphate ion content.
Source: CSA Standard A23.1

Fig. 2-14. Moderate sulphate resistant cements and high sulphate resistant cements improve the sulphate resistance of concrete elements, such as (left to right) slabs on ground, pipe, and concrete posts exposed to high-sulphate soils. (68985, 52114, 68986)

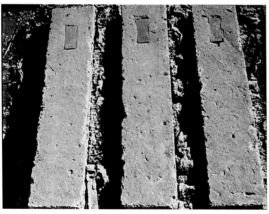

Fig. 2-15. Specimens used in the outdoor sulphate test plot in Sacramento, California, are 150 x 150 x 760-mm beams. A comparison of ratings is illustrated: (top) a rating of 5 for 12-year old concretes made with Type V (50) cement and a water-to-cement ratio of 0.65; and (bottom) a rating of 2 for 16-year old concretes made with Type V (50) cement and a water-to-cement ratio of 0.39 (Stark 2002). (68840, 68841)

2-15). Type 20 cement has moderate sulphate resistant properties because it contains no more than 7.5% tricalcium aluminate (C_3A).

Sulphates in moist soil or water may enter the concrete and develop expansive reactions with the hydrated C_3A, resulting in expansion, scaling, and cracking of concrete. Some sulphate compounds, such as magnesium sulphate, directly attack calcium silicate hydrate. Wetting and drying in a sulphate environment aggravates the formation of sulphate salts or compounds that have sufficient crystallization pressure to disrupt cement paste.

Use of Type 20 cement must be accompanied by the use of a low water to cementing materials ratio and low permeability to control sulphate attack. Fig. 2-13 illustrates the improved sulphate resistance of a Type 20 (Type II) cement over a Type 10 (Type I) cement.

Concrete exposed to seawater is often made with Type 20 cement. Seawater contains significant amounts of sulphates and chlorides. Although sulphates in seawater are capable of attacking concrete, the presence of chlorides inhibits the expansive reaction that is characteristic of sulphate attack. Calcium sulphoaluminate (ettringite) and other sulphate compounds, the reaction products of sulphate attack, are more soluble in a chloride solution and more readily leach out of the concrete, thus resulting in less destructive expansion. This is the major factor explaining observations from a number of sources that the performance of concretes in seawater with portland cements having C_3A contents as high as 10%, and sometimes greater, have shown satisfactory durability, providing the permeability of the concrete is low and the reinforcing steel has adequate cover.

Type 20 cements specially manufactured to meet the moderate heat *option* of CSA A5 will generate less heat at a slower rate than Type 10 cement. The requirement of moderate heat of hydration can be specified at the option of the purchaser. If heat-of-hydration maximums are specified,

Fig. 2-16. Moderate heat and low heat cements minimize heat generation in massive elements or structures such as (left) very thick bridge supports, and (right) dams. Hoover dam, shown here, used a Type 40 cement to control temperature rise. (65258, 68983)

this cement can be used in structures of considerable mass, such as large piers, large foundations, and thick retaining walls (Fig. 2-16). Its use will reduce temperature rise and temperature related cracking, which is especially important when concrete is placed in warm weather.

Type 30

Type 30 portland cement provides high strengths at an early period, usually a week or less. It is chemically and physically similar to Type 10 cement, except that its particles have been ground finer. It is used when forms need to be removed as soon as possible or when the structure must be put into service quickly. In cold weather its use permits a reduction in the length of the curing period (Fig. 2-17). Although higher-cement content mixes of Type 10 cement can be used to gain high early strength, Type 30 may provide it easier and more economically.

Type 40

Type 40 portland cement is used where the rate and amount of heat generated from hydration must be minimized. It develops strength at a slower rate than other cement types. Type 40 cement is intended for use in massive concrete structures, such as large gravity dams, where the temperature rise resulting from heat generated during hardening must be minimized (Fig. 2-16). Type 40 cement is generally only available by specific request for large projects.

Type 50

Type 50 portland cement is used in concrete exposed to severe sulphate action—principally where soils or groundwaters have a high sulphate content (Figs. 2-13 to 2-15). It gains strength more slowly than Type 10 cement. Table 2-2 lists sulphate concentrations requiring the use of Type 50 cement. The high sulphate resistance of Type 50 cement is attributed to a low tricalcium aluminate content, not more than 3.5%. Use of a low water to cementing materials ratio and low permeability are critical to the performance of any concrete exposed to sulphates. Even Type 50 cement concrete cannot withstand a severe sulphate exposure if the concrete has a high water to cementing materials ratio (Fig. 2-15 top). Type 50 cement, like other portland cements, is not resistant to acids and other highly corrosive substances. The chemical and physical requirements for Type 50 cement are given in CSA A5.

White Cement

White portland cement is a true portland cement that differs from gray cement chiefly in colour. It is made to conform to the specifications of CSA A5 (ASTM C 150), usually Type 10 or Type 30; the manufacturing process is controlled

Fig. 2-17. High early strength cements are used where early concrete strength is needed, such as in (left to right) cold weather concreting, fast track paving to minimize traffic congestion, and rapid form removal for precast concrete. (65728, 59950, 68668)

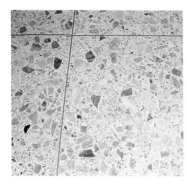  

Fig. 2-18. White portland cement is used in white or light-coloured architectural concrete, ranging from (left to right) terrazzo for floors shown here with white cement and green granite aggregate (68923), to decorative and structural precast and cast-in-place elements (68981), to building exteriors. The far right photograph shows a white precast concrete building housing the ASTM Headquarters in West Conshohocken, Pennsylvania. Photo courtesy of ASTM.

so that the finished product will be white. White portland cement is made of selected raw materials containing negligible amounts of iron and magnesium oxides—the substances that give cement its gray colour. White portland cement is used primarily for architectural purposes such as precast curtain walls and facing panels, terrazzo surfaces, stucco, cement paint, tile grout, and decorative concrete (Fig. 2-18). Its use is recommended wherever white or coloured concrete, grout, or mortar is desired.

BLENDED HYDRAULIC CEMENTS

Blended cements are used in all aspects of concrete construction in the same manner as portland cements. Blended cements can be used as the only cementing material in concrete or they can be used in combination with other supplementary cementing materials added at the concrete plant. Blended cements are often designed to be used in combination with local pozzolans and slags. If a blended cement or portland cement is used alone or in combination with added pozzolans or slags, the concrete should be tested for strength, durability, and other properties required in project specifications (PCA 1995 and Detwiler, Bhatty, and Bhattacharja 1996).

Blended hydraulic cements are produced by intimately and uniformly intergrinding or blending two or more types of fine materials. The primary materials are portland cement, ground granulated blast-furnace slag, fly ash, silica fume, calcined clay, and other pozzolans, hydrated lime, and preblended cement combinations of these materials (Fig. 2-19). Blended hydraulic cements manufactured in Canada conform to the requirements of CSA Standard A362 *Blended Hydraulic Cement.* In United States blended cements are covered under ASTM C 595, *Specification For Blended Hydraulic Cements,* and ASTM C 1157, *Performance Specification for Hydraulic Cements.*

Blended hydraulic cement as defined by CSA A362 is a product consisting of a mixture of portland cement and one or more of granulated blast-furnace slag, fly ash, or silica fume to which no processing additions have been made except as permitted and noted in Clause 3.2 of the Standard.

The four types of blended hydraulic cements addressed in CSA A362 are:

Portland blast-furnace slag cement (S)
Portland fly ash cement (F)
Portland silica fume cement (SF)
Ternary blend cement

These blended products are produced by either of the following methods or a combination of both: (a) by intergrinding portland cement clinker and blast-furnace slag, fly ash, or silica fume; or (b) by blending portland cement and finely ground granulated blast-furnace slag, fly ash, or silica fume.

Blended hydraulic cements may develop lower early compressive strengths than the corresponding portland cement. This effect is more pronounced as the proportion of slag or fly ash is increased. When a portland silica fume cement is used, experience indicates that early strengths

Fig. 2-19. Blended cements CSA A362 (ASTM C 595 and ASTM C 1157) use a combination of portland cement or clinker and gypsum blended or interground with pozzolans, slag, or fly ash. Shown is blended cement (centre) surrounded by (right and clockwise) clinker, gypsum, portland cement, fly ash, slag, silica fume, and calcined clay. (68988)

are frequently higher than the strength of the corresponding portland cement.

The nomenclature and naming practice used for blended cements is as follows:

TE-A/B

where

T = the equivalent performance to Type 10, 20, 30, 40, or 50 portland cement;

E = an indication that the cement has equivalent performance for the physical properties specified in CSA A362, Table 2;

A = the predominant supplementary material; and

B = the secondary supplementary material, only specified in a ternary blend.

Examples:

10E-S is a portland blast furnace slag cement having an equivalent performance to that of a Type 10 portland cement.

40E-F is a portland fly ash cement having an equivalent performance to that of a Type 40 portland cement.

50E-S/SF is a ternary blend cement having an equivalent performance to that of a Type 50 portland cement with slag being the predominant supplementary cementing material and silica fume the secondary supplementary cementing material.

Portland Blast-Furnace Slag Cement

Portland blast-furnace slag cement, Type 10E-S, 20E-S, 30E-S, 40E-S, or 50E-S, is a product consisting of portland cement and finely ground, granulated blast-furnace slag in which the slag content is greater than 0 and less than 70% of the total mass. Blast-furnace slag, also known as iron blast-furnace slag, is the nonmetallic product, consisting essentially of silicates and alumino-silicates of calcium and other bases, that is developed in a molten condition simultaneously with iron in a blast furnace. Granulation of the slag is achieved by immersing the molten slag in water or pelletizing the molten slag to produce a high percentage of glass (a process called vitrification).

Portland Fly Ash Cement

Portland fly ash cement, Type 10E-F, 20E-F, 30E-F, 40E-F, or 50E-F, is a product consisting of portland cement and fly ash, in which the fly ash content is greater than 0 and less than 40% of the total mass. Fly ash is the finely divided residue that results from the combustion of pulverized coal and which is carried from the combustion chamber of a furnace by the exhaust gases.

Portland Silica Fume Cement

Portland silica fume cement, Type 10E-SF, 20E-SF, 30E-SF, 40E-SF, or 50E-SF, is a product consisting of portland cement and silica fume in which the silica fume content does not exceed 10% of the total mass. Silica fume is the finely divided residue resulting from the production of silicon, ferro-silicon, or other silicon-containing alloys that is carried from the burning surface area of an electric-arc furnace by exhaust gases. In some cases purchasers may request silica fume contents in excess of 10% for specific applications such as shotcrete. Cements manufactured to fulfill such requests are beyond the scope of CSA A362, *Blended Hydraulic Cement*.

Ternary Blended Cement

Ternary blended cement, Type 10E-A/B, 20E-A/B, 30E-A/B, 40E-A/B, or 50E-A/B, is a product consisting of portland cement and a combination of two of the following: finely granulated blast-furnace slag, fly ash, or silica fume. The total supplementary cementing material content is greater than 0 and less than or equal to 70% of the total mass; the slag content must be less than 70% of the total mass, the fly ash content less than 40% of the total mass; and the silica fume content less than 10% of the total mass.

OTHER HYDRAULIC CEMENTS

The 1990s saw the creation of the world's first performance specification for hydraulic cements—ASTM C 1157, *Performance Specification for Hydraulic Cements.* This specification is designed generically for hydraulic cement to include portland cement, modified portland cement, and blended hydraulic cement. Cements meeting the requirements of C 1157 meet physical performance test requirements, as opposed to prescriptive restrictions on ingredients or cement chemistry as found in other cement specifications. ASTM C 1157 provides for six types of hydraulic cement as follows:

Type GU	General use
Type HE	High early strength
Type MS	Moderate sulphate resistance
Type HS	High sulphate resistance
Type MH	Moderate heat of hydration
Type LH	Low heat of hydration

In addition, these cements can also have an Option R— Low Reactivity with Alkali-Reactive Aggregates— specified to help control alkali-silica reactivity. For example, Type GU-R would be a general use hydraulic cement with low reactivity with alkali-reactive aggregates.

When specifying a C 1157 cement, the specifier uses the nomenclature of "hydraulic cement," "portland cement," "modified portland cement" or "blended hydraulic cement" along with a type designation. For example, a specification may call for a Hydraulic Cement Type GU, a Blended Hydraulic Cement Type MS, or a Portland Cement Type HS. If a type is not specified, then Type GU is assumed.

ASTM C 1157 defines a blended cement as having more than 15% mineral additive and a modified portland cement as containing up to 15% mineral additive. The mineral additive usually prefixes the modified portland cement nomenclature, for example, slag-modified portland cement.

ASTM C 1157 also allows a strength range to be specified from a table in the standard. If a strength range is not specified, only the minimum strengths apply. Strength ranges are rarely applied in the United States.

A detailed review of ASTM C 1157 cements follows:

Type GU

Type GU is a general-purpose cement suitable for all uses where the special properties of other types are not required. Its uses in concrete include pavements, floors, reinforced concrete buildings, bridges, pipe, precast concrete products, and other applications where Type 10 is used (Fig. 2-12).

Type HE

Type HE cement provides high strengths at an early age, usually a week or less. It is used in the same manner as Type 30 portland cement (Fig. 2-17).

Type MS

Type MS cement is used where precaution against moderate sulphate attack is important, as in drainage structures where sulphate concentrations in ground waters are higher than normal but not unusually severe (see Table 2-2). It is

used in the same manner as Type 20 portland cement (Fig. 2-14). Like Type 20, Type MS cement concrete must be made with a lower water-cementing materials ratio to provide sulphate resistance.

Type HS

Type HS cement is used in concrete exposed to severe sulphate action—principally where soils or ground waters have a high sulphate content (see Table 2-2). It is used in the same manner as Type 50 portland cement (Fig. 2-14).

Type MH

Type MH cement is used where the concrete needs to have a moderate heat of hydration and a controlled temperature rise. Type MH cement is used in the same manner as a moderate heat Type 20 portland cement (Fig. 2-16).

Type LH

Type LH cement is used where the rate and amount of heat generated from hydration must be minimized. It develops strength at a slower rate than other cement types. Type LH cement is intended for use in massive concrete structures where the temperature rise resulting from heat generated during hardening must be minimized. It is used in the same manner as Type 40 portland cement (Fig. 2-16).

Table 2-3 provides a matrix of commonly used cements and where they are used in concrete construction.

Table 2-3. Applications for Commonly Used Cements

Cement specification	Applications*						
	General purpose	Moderate heat of hydration	High early strength	Low heat of hydration	Moderate sulphate resistance	High sulphate resistance	Resistance to alkali-silica reactivity (ASR)**
CSA A5	10	20	30	40	20	50	Low alkali option
ASTM C 150	I	II	III	IV	II	V	
CSA A362	10E-S 10E-F 10E-SF Ternary 10E-A/B	20E-S 20E-F 20E-SF Ternary 20E-A/B	30E-S 30E-F 30E-SF Ternary 30E-A/B	40E-S 40E-F 40E-SF Ternary 40E-A/B	20E-S 20E-F 20E-SF Ternary 20E-A/B	50E-S 50E-F 50E-SF Ternary 50E-A/B	Low reactivity option
ASTM C 595	IS IP I(PM) I(SM) S, P	IS(MH) IP(MH) I(PM)(MH) I(SM)(MH)		P(LH)	IS(MS) IP(MS) P(MS) I(PM)(MS) I(SM)(MS)		
Performance standard ASTM C 1157 hydraulic cements***	GU	MH	HE	LH	MS	HS	Option R

* Check the local availability of specific cements as all cements are not available everywhere.

** The option for low reactivity with ASR susceptible aggregates can be applied to the cement types in the columns to the left.

*** For ASTM C 1157 cements, the nomenclature of hydraulic cement, portland cement, air-entraining portland cement, modified portland cement, or blended hydraulic cement is used with the type designation.

Air-Entraining Portland Cements

Specifications for three types of air-entraining portland cement (Types IA, IIA, and IIIA) are given in ASTM C 150. They correspond in composition to ASTM Types I, II, and III, respectively, except that small quantities of air-entraining material are interground with the clinker during manufacture. These cements produce concrete with improved resistance to freezing and thawing. Such concrete contains minute, well-distributed, and completely separated air bubbles. Air entrainment for most concrete is achieved through the use of an air-entraining admixture, rather than through the use of air-entraining cements. Air-entraining cements are available only in certain areas. Air-entraining cements are not covered by CSA and are not manufactured in Canada.

MODIFIED PORTLAND CEMENTS

The term "modified portland cement" usually refers to a blended cement containing mostly portland cement along with a small amount of mineral additive. However, some local areas have modified portland cements that do not contain a mineral additive. The modification merely refers to a special property that cement has or a cement that has the characteristics of more than one type of portland cement.

HYDRAULIC SLAG CEMENTS (Cementitious Hydraulic Slag)

Hydraulic slag cements are like other cements that set and harden by a chemical interaction with water. Concrete made with hydraulic slag cement is used for the same applications as concrete made with other hydraulic cements. Hydraulic slag cement (defined by CSA as Cementitious Hydraulic Slag) is the product obtained by pulverizing a granulated blast-furnace slag that possesses hydraulic properties, and to which various forms of calcium sulphate and processing additions may be added at the option of the manufacturer. The blast-furnace slag used is the nonmetallic product, consisting essentially of silicates and alumino-silicates of calcium and other bases, which is developed in a molten condition simultaneously with iron in a blast furnace. When the molten blast-furnace slag is chilled rapidly, a glassy granular material is formed which is known as granulated blast-furnace slag. Combining cementitious hydraulic slag with water produces essentially the same binding material (calcium silicate hydrate) that portland cement alone makes when combined with water. Cementitious Hydraulic Slag conforms to CSA Standard A363.

SPECIAL CEMENTS

Special cements are produced for particular applications. A number of the other special cements described, although not covered by CSA, have been used for specific applications throughout the country. Table 2-4 summarizes a number of special cements and their applications.

Masonry Cements

Masonry cement is manufactured in Canada and covered by CSA Standard A8, *Masonry Cement*. Masonry cements are hydraulic cements designed for use in mortar for masonry construction (Fig. 2-20). They are a mixture of portland cement, air entraining materials, and plasticizing materials (such as limestone or hydrated or hydraulic lime), together with other materials introduced to enhance one or more properties such as setting time, workability, water retention, and durability. These materials are proportioned and packed at a cement plant under controlled conditions to assure uniformity of performance.

Masonry cements meet the requirements of CSA A8 (ASTM C 91), which classifies them as Type N masonry cement and Type S masonry cement.

Type N masonry cement is for use in the preparation of Type N masonry mortar, or with the use of portland cement additions it is capable of producing Type S mortar as specified in CSA Standard A179, *Mortar and Grout for Unit Masonry*.

Type S masonry cement is for use in the preparation of Type S masonry mortar as specified in CSA A179, without the use of portland cement additions.

The workability, strength, and colour of masonry cements stay at a high level of uniformity because of manufacturing controls. In addition to mortar for masonry construction, masonry cements are used for parging and portland cement based plaster (stucco) (Fig. 2-21) construction (see ASTM C 926). *Masonry cement must never be used for making concrete.*

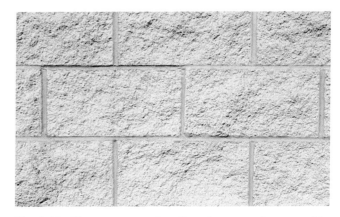

Fig. 2-20. Masonry cement and mortar cement are used to make mortar to bond masonry units together. (68807)

Table 2-4. Applications for Special Cements

Special cements	Type	Application
White portland cements, ASTM C 150 **CSA A5**	I, II, III, V **10**	White or coloured architectural concrete, masonry, mortar, grout, plaster, and stucco
White masonry cements, ASTM C 91 **CSA A8**	M, S, N **N, S**	White or coloured mortar between masonry units
Masonry cements, ASTM C 91 **CSA A8**	M, S, N **N, S**	Mortar between masonry units,* plaster, and stucco
Mortar cements, ASTM C 1329	M, S, N	Mortar between masonry units
Plastic cements, ASTM C 1328	M, S	Plaster and stucco**
Expansive cements, ASTM C 845	E-1(K), E-1(M), E-1(S)	Shrinkage compensating concrete
Oil-well cements, API-10	A, B, C, D, E, F, G, H	Grouting wells
Waterproof cements		Tile grout, paint, and stucco finish coats
Regulated-set cements		Early strength and repair***
Cements with functional additions, see ASTM C 595 and ASTM C 1157		General concrete construction needing special characteristics such as; water-reducing, retarding, air entraining, set control, and accelerating properties
Ultrafine cement		Geotechnical grouting***
Calcium aluminate cement		Repair, chemical resistance, high temperature exposures
Magnesium phosphate cement		Repair and chemical resistance
Geopolymer cement		General construction, repair, waste stabilization
Ettringite cements		Waste stabilization
Sulphur cements		Repair and chemical resistance
Rapid hardening hydraulic cement	VH, MR, GC	General paving where very rapid (about 4 hours) strength development is required

  * CSA portland cement Types 10, 20, 30 and ASTM Types I, II, and III and blended cement Types IS, IP, and I(PM) are also used in making mortar.
 ** CSA portland cement Types 10, 20, 30 and ASTM Types I, II, and III and blended cement Types IP, I(SM) and I(PM) are also used in making plaster.
*** Portland and blended hydraulic cements are also used for these applications.

Fig. 2-21. Masonry cement and plastic cement are used to make plaster or stucco for commercial, institutional, and residential buildings. Shown are a church and home with stucco exteriors. Inset shows a typical stucco texture. (69389, 67878, 68805)

Mortar Cements

Mortar cements are hydraulic cements designed for use in mortar for masonry construction (Fig. 2-20). They consist of a mixture of portland cement or blended hydraulic cement and plasticizing materials together with other materials introduced to enhance one or more properties such as setting time, workability, water retention and durability. These components are proportioned at the cement plant under controlled conditions to assure uniformity of performance.

Mortar cements meet the requirements of ASTM C 1329, which also classifies mortar cements as Type N, Type S, and Type M. A brief description of each type follows:

Type N mortar cements are used in ASTM C 270 Type N and Type O mortars. They may also be used with portland or blended cements to produce Type S and Type M mortars.

Type S mortar cements are used in ASTM C 270 Type S mortar. They may also be used with portland or blended cements to produce Type M mortar.

Type M mortar cements are used in ASTM C 270 Type M mortar without the addition of other cements or hydrated lime.

Types N, S, and M generally have increasing levels of portland cement and higher strength, Type M having the highest strength. Type N is used most commonly.

The increased use of masonry in demanding structural applications and high seismic areas resulted in the recent development of mortar cement. Mortar cement is similar to masonry cement in that it is a factory-prepared cement primarily used to produce masonry mortar. However, ASTM C 1329 places lower maximum air content limits on mortar cement than permitted for masonry cements and ASTM C 1329 is the only ASTM masonry material specification that includes bond strength performance criteria.

The workability, strength, and colour of mortar cements stay at a uniform level because of manufacturing controls. In addition to mortar for masonry construction, mortar cements are also used for parging. *Mortar cement must never be used for making concrete.*

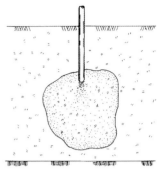

Fig. 2-22. (left) A slurry of finely ground cement and water can be injected into the ground, as shown here, to stabilize in-place materials, to provide strength for foundations, or to chemically retain contaminants in soil. (68810) Illustration (right) of grout penetration in soil.

Plastic Cements

Plastic cement is a hydraulic cement that meets the requirements of ASTM C 1328. It is used to make portland cement-based plaster or stucco (see ASTM C 926) in the Southwest and west coast area of the United States (Fig. 2-21). Plastic cements consist of a mixture of portland and blended hydraulic cement and plasticizing materials (such as limestone, hydrated or hydraulic lime), together with materials introduced to enhance one or more properties such as setting time, workability, water retention, and durability.

ASTM C 1328 defines separate requirements for a Type M and a Type S plastic cement with Type M having higher strength requirements. The Uniform Building Code, UBC 25-1, does not classify plastic cement into different types, but defines just one set of requirements which correspond to those of an ASTM C 1328 Type M plastic cement. When plastic cement is used, no lime or other plasticizer may be added to the plaster at the time of mixing.

The term "plastic" in plastic cement does not refer to the inclusion of any organic compounds in the cement; rather, "plastic" refers to the ability of the cement to impart to the plaster a high degree of workability. Plaster made from this cement must remain workable for a long enough time for it to be reworked to obtain the desired densification and texture. Plastic cement should not be used to make concrete. For more information on the use of plastic cement and plaster, see Melander and Isberner (1996).

Finely-Ground Cements (Ultrafine Cements)

Finely-ground cements, also called ultrafine cements, are hydraulic cements that are ground very fine for use in grouting into fine soil or thin rock fissures (Fig. 2-22). The cement particles are less than 10 μm in diameter with 50% of particles less than 5 μm. Blaine fineness often exceeds 800 m²/kg. These very fine cements consist of portland cement, ground granulated blast-furnace slag, and other mineral additives.

Expansive Cements

Expansive cement is a hydraulic cement that expands slightly during the early hardening period after setting. It must meet the requirements of ASTM C 845 in which it is designated as Type E-1. Currently, three varieties of expansive cement are recognized and have been designated as K, M, and S, which are added as a suffix to the type. Type E-1(K) contains portland cement, anhydrous tetracalcium trialuminosulphate, calcium sulphate, and uncombined calcium oxide (lime). Type E-1(M) contains portland cement, calcium aluminate cement, and calcium sulphate. Type E-1(S) contains portland cement with a high tricalcium aluminate content and calcium sulphate. Type E-1(K) is the most readily available expansive cement in North America.

Expansive cement may also be made of formulations other than those mentioned. The expansive properties of each type can be varied over a considerable range. Type 10 cement may be transformed into expansive cement by the addition of an expansive admixture at the ready mix plant.

When expansion is restrained, for example by reinforcing steel, expansive cement concrete (also called shrinkage compensating concrete) can be used to (1) compensate for the volume decrease due to drying shrinkage, (2) induce tensile stress in reinforcement (post-tensioning), and (3) stabilize the long-term dimensions of post-tensioned concrete structures with respect to original design.

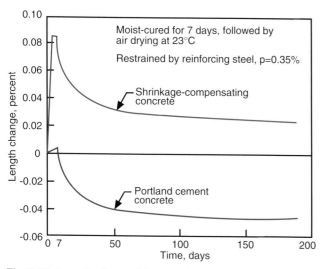

Fig. 2-23. Length-change history of shrinkage compensating concrete containing Type E-1(S) cement and Type I (Type 10) portland cement concrete (Pfeifer and Perenchio 1973).

One of the major advantages of using expansive cement in concrete is noted in (1) above; when you can compensate for volume change due to drying shrinkage you can control and reduce drying shrinkage cracks. Fig. 2-23 illustrates the length change (early expansion and drying shrinkage) history of shrinkage-compensating concrete and conventional portland cement concrete. For more information see Pfeifer and Perenchio (1973), Russell (1978), and ACI (1998).

Oil-Well Cements

Oil-well cements, used for oil-well grouting, often called oil-well cementing, are usually made from portland cement clinker or from blended hydraulic cements. Generally they must be slow setting and resistant to high temperatures and pressures. The American Petroleum Institute *Specification for Cements and Materials for Well Cementing* (API Specification 10A) includes requirements for eight classes of oil-well cements (Classes A through H) and three grades (Grades O—ordinary, MSR—moderate sulphate resistant, and HSR—high sulphate resistant). Each class is applicable for use at a certain range of well depths, temperatures, pressures, and sulphate environments. The petroleum industry also uses conventional types of portland cement with suitable cement-modifiers. Expansive cements have also performed adequately as well cements.

Cements with Functional Additions

Functional additions can be interground with cement clinker to beneficially change the properties of hydraulic cement. These additions must meet the requirements of ASTM C 226 or C 688. ASTM C 226 addresses air-entrain-

ing additions. ASTM C 688 addresses the following types of additions: water-reducing, retarding, accelerating, water reducing and retarding, water reducing and accelerating, and set control additions. Cement specifications ASTM C 595 and C 1157 allow functional additions. These cements can be used for normal or special concrete construction, grouting, and other applications.

Water-Repellent Cements

Water-repellent cement, sometimes called waterproofed cement, is usually made by adding a small amount of water-repellent additive such as stearate (sodium, aluminum, or other) to cement clinker during its final grinding (Lea 1971). Manufactured in either white or gray colour, it reduces capillary water transmission under little to no pressure but does not stop water-vapor transmission. It is used in tile grouts, paint, and stucco finish coats.

Regulated-Set Cements

Regulated-set cement is a calcium fluoroaluminate hydraulic cement that can be formulated and controlled to produce concrete with setting times from a few minutes to an hour and with corresponding rapid early strength development (Greening and others 1971). It is a portland-based cement with functional additions that can be manufactured in the same kiln used to manufacture conventional portland cement. Regulated-set cement incorporates set control and early-strength-development components. Final physical properties of the resulting concrete are in most respects similar to comparable concretes made with portland cement.

Geopolymer Cements

Geopolymer cements are inorganic hydraulic cements that are based on the polymerization of minerals (Davidovits, Davidovits and James 1999). The term more specifically refers to alkali-activated alumino-silicate cements, also called zeolitic cements. They have been used in general construction, high-early strength applications, and waste stabilization. These cements do not contain organic polymers or plastics.

Ettringite Cements

Ettringite cements are calcium sulphoaluminate cements that are specially formulated for particular uses, such as the stabilization of waste materials (Klemm 1998). They can be formulated to form large amounts of ettringite to stabilize particular metallic ions within the ettringite structure. Ettringite cements have also been used in rapid setting applications, including use in coal mines. Also see previous discussion on Expansive Cements.

Calcium Aluminate Cements

Calcium aluminate cement is not portland cement-based. It is used in special applications for early strength gain (design strength in one day), resistance to high temperatures, and resistance to sulphates, weak acids, and seawater. Portland cement and calcium aluminate cement combinations have been used to make rapid setting concretes and mortars. Typical applications for calcium aluminate cement concrete include: chemically resistant, heat resistant and corrosion resistant industrial floors, refractory castables; and repair applications. Standards addressing these cements include British Standard BS 915-2 or French Standard NF P15-315.

The hydrates responsible for the rapid hardening and early strength gain change over time, resulting in a loss of strength. This process, called conversion, always occurs. The process involves the conversion of the less stable hexagonal calcium aluminate hydrate (CAH_{10}) to stable cubic tricalcium aluminate hydrate (C_3AH_6), hydrous alumina (AH_3), and water. With time and particular moisture conditions and temperatures, this conversion causes a 53% decrease in volume of hydrated material. However, this internal volume change occurs without a dramatic alteration of the overall dimensions of a concrete element, resulting in increased paste porosity and decreased compressive strength. At low water-cement ratios, there is insufficient space for all the calcium aluminate to react and form CAH_{10}. The water released from conversion reacts with more calcium aluminate, partially compensating for the effects of conversion. The design of durable concrete using this type of cement must therefore be based on long-term performance, not on the high but transient strengths that can occur initially.

Long-term compressive strengths of 40 MPa are typical for properly designed calcium aluminate cement concrete. Higher strengths can be obtained with limestone coarse aggregate, as compared with some other aggregates, which do not perform as well. For this reason it is recommended that calcium aluminate cement not be used for certain types of construction, such as prestressed concrete. Based on these facts, concrete should be proportioned with a total water (including water absorbed by the aggregate) to cement ratio of not more than 0.40 and with a minimum cement content of 400 kg/m³. Because of eventual conversion, calcium aluminate cement is often used in nonstructural applications and used with caution (or not at all) in structural applications (Taylor 1997).

Magnesium Phosphate Cements

Magnesium phosphate cement is a rapid setting, early strength gain cement. It is usually used for special applications, such as repair of pavements and concrete structures or for resistance to certain aggressive chemicals. It does not contain portland cement.

Sulphur Cements

Sulphur cement is used to make sulphur cement concrete for repairs and chemically resistant applications. Sulphur cement melts at temperatures between 113°C and 121°C. Sulphur concrete is maintained at temperatures around 130°C during mixing and placing. The material gains strength quickly as it cools and is resistant to acids and aggressive chemicals. Sulphur cement does not contain portland or hydraulic cement.

Rapid Hardening Cements

Rapid hardening, high-early strength, hydraulic cement is used in construction applications, such as fast-track paving, where fast strength development is needed (design or load carrying strength in less than three hours). These cements often use calcium sulphoaluminate to obtain early strength. They are classified as Types VH (very high-early strength), MR (middle range high-early strength), and GC (general construction).

SELECTING AND SPECIFYING CEMENTS

When specifying cements for a project, be sure to check the availability of cement types and allow the specifications to be flexible in cement selection. Cements with special properties should not be required unless special characteristics are necessary. In addition, the use of supplementary cementing materials should not inhibit the use of any particular portland or blended cement. The project specifications should focus on the needs of the concrete structure and allow use of a variety of materials to accomplish those needs. A typical specification may call for portland cements meeting CSA A5 (ASTM C 150 or C 1157) or for blended cements meeting CSA A362 (ASTM C 595 or C 1157).

If no special properties (such as low-heat generation or sulphate attack) are required, all general use cements should be allowed. See Tables 2-3 and 2-4 for guidance on using different cements.

Availability of Cements

Some types of cement may not be readily available in all areas of Canada. Before specifying a type of cement, its availability should be determined.

CSA A5 Type 10 (ASTM C 150 Type I) portland cement is usually carried in stock and is furnished when there is not a specific type of cement specified. Type 20 (Type II) cement is usually available where moderate sulphate resistance is needed. Type 30 (Type III) cement and white cement are usually available in most areas. Type 40 (Type IV) cement is manufactured only when specified for particular projects (massive structures like dams) and therefore is usually not readily available. Type 50 (Type V) cement is only readily available in particular parts of Canada where it is needed to resist high sulphate environments. Masonry cement is available in most areas.

If a given type is not available, comparable results frequently can be obtained with one of the available types. For example, high-early-strength concrete can be made by using mixtures with higher Type 10 cement contents and lower water-cementing materials ratios. Also, the effects of heat of hydration can be minimized by using lean mixes, smaller placing lifts, artificial cooling, or by adding a pozzolan to the concrete.

Blended cements are available in most parts of Canada; however, certain types of blended cement may not be available in all areas. When blended cements are required but are not available, similar properties may be obtained by adding pozzolans [CSA A23.5 (ASTM C 618)] or finely ground granulated blast-furnace slag [CSA A23.5 (ASTM C 989)] to the concrete at a ready mix plant using normal portland cement. Like any concrete mixture, these mixes should be tested for time of set, strength gain, durability, and other properties prior to use in construction.

Drinking Water Applications

Concrete has demonstrated decades of safe use in drinking water applications. Materials in contact with drinking water must meet special requirements to control elements entering the water supply. Some localities may require that cement and concrete meet the special requirements of the American National Standards Institute/National Sanitation Foundation standard ANSI/NSF 61, *Drinking Water System Components—Health Effects*. ANSI/NSF 61 is being adopted to assure that products such as pipe, coatings, process media, and plumbing fittings are safe for use in public drinking water systems. Cement is used in drinking water system components such as pipe and tanks, and therefore, is subject to testing under Standard 61. Kanare (1995) outlines the specifications and testing program required to certify cement for drinking water applications. Kanare and West (1993) and PCA (1992) address elements that can leach from cement and concrete. Consult the local building code or local officials to determine if National Sanitation Foundation certified cements are required for water projects such as concrete pipe and concrete tanks (Fig. 2-24).

Fig. 2-24. Concrete has demonstrated decades of safe use in drinking water applications such as concrete tanks. (69082)

European Cement Specifications

In some instances, projects in Canada designed by engineering firms from other countries refer to cement standards other than those in CSA or ASTM. For example, the European cement standard (EN 197-1, -2, 2000) sometimes appears on project specifications. EN 197 cement Types CEM I, II, III, IV, and V do not correspond to the cement types in CSA A5 (ASTM C 150), nor can a CSA cement be simply used in place of an EN specified cement without the designer's approval. CEM I is a portland cement and CEM II through V are blended cements. EN 197 also has strength classes and ranges (32.5, 42.5, and 52.5 MPa). There is no direct equivalency between CSA and other cement standards because of different test methods and different limits on required properties. EN 197 cements are not available in Canada and therefore the best approach is to inform the designer as to what cements are locally available and ask for a change in the project specifications to allow a CSA cement.

CHEMICAL COMPOUNDS AND HYDRATION OF PORTLAND CEMENT

During the burning operation in the manufacture of portland cement clinker, calcium combines with the other components of the raw mix to form four principal compounds that make up 90% of cement by mass. Gypsum (4% to 6%), or other calcium sulphate source, and grinding aids are also added upon grinding. Cement chemists use the following chemical shorthand (abbreviations) to describe compounds:

$A = Al_2O_3$, $C = CaO$, $F = Fe_2O_3$, $H = H_2O$, $M = MgO$, $S = SiO_2$, and $\bar{S} = SO_3$.

The term "phase" can also be used to describe the "compounds" of clinker. Following are the four primary compounds in portland cement, their approximate chemical formulas, and abbreviations:

Tricalcium silicate	$3CaO \cdot SiO_2$	$= C_3S$
Dicalcium silicate	$2CaO \cdot SiO_2$	$= C_2S$
Tricalcium aluminate	$3CaO \cdot Al_2O_3$	$= C_3A$
Tetracalcium aluminoferrite	$4CaO \cdot Al_2O_3 \cdot Fe_2O_3$	$= C_4AF$

Following are the forms of calcium sulphate, their approximate chemical formulas, and abbreviations:

Anhydrous calcium sulphate	$CaSO_4 = CaO \cdot SO_3$	$= C\bar{S}$
Calcium sulphate dihydrate (gypsum)	$CaSO_4 \cdot 2H_2O =$ $CaO \cdot SO_3 \cdot 2H_2O$	$= C\bar{S}H_2$
Calcium sulphate hemihydrate	$CaSO_4 \cdot 1/2\,H_2O =$ $CaO \cdot SO_3 \cdot 1/2\,H_2O$	$= C\bar{S}H_{1/2}$

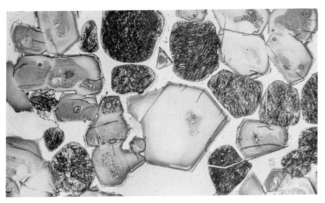

Fig. 2-25. (left) Polished thin-section examination of portland clinker shows alite (C₃S) as light, angular crystals. The darker, rounded crystals are belite (C₂S). Magnification 400X. (right) Scanning electron microscope (SEM) micrograph of alite (C₃S) crystals in portland clinker. Magnification 3000X. (54049, 54068)

Gypsum is the predominant source of sulphate found in cement.

C_3S and C_2S in clinker are also referred to as alite and belite, respectively. Alite constitutes 50% to 70% of the clinker, whereas belite accounts for only 15% to 30%. Aluminate compounds constitute about 5% to 10% of the clinker and ferrite compounds 5% to 15% (Taylor 1997). These and other compounds may be observed and analyzed through the use of microscopical techniques (see ASTM C 1356, Fig. 2-25, and Campbell 1999).

In the presence of water, these compounds hydrate (chemically combined with water) to form new compounds that are the infrastructure of hardened cement paste in concrete (Fig. 2-26). The calcium silicates, C_3S and C_2S, hydrate to form the compounds calcium hydroxide and calcium silicate hydrate (archaically called tobermorite gel). Hydrated portland cement contains 15% to 25% calcium hydroxide and about 50% calcium silicate hydrate by mass. The strength and other properties of hydrated cement are due primarily to calcium silicate hydrate (Fig. 2-27). C_3A reacts with water and calcium hydroxide to form tetracalcium aluminate hydrate. C_4AF reacts with water to form calcium aluminoferrite hydrate. C_3A, sulphates (gypsum, anhydrite, or other sulphate source), and water combine to form ettringite (calcium sulphoaluminate hydrate), monosulphoaluminate, and other related compounds. These basic compound transformations are shown in Table 2-5. Brunauer (1957), Copeland and others (1960), Lea (1971), Powers and Brownyard (1947), Powers (1961), and Taylor (1997) addressed the pore structure and

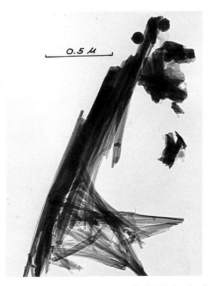

Fig. 2-26. Electron micrographs of (left) dicalcium silicate hydrate, (middle) tricalcium silicate hydrate, and (right) hydrated normal portland cement. Note the fibrous nature of the calcium silicate hydrates. Broken fragments of angular calcium hydroxide crystallites are also present (right). The aggregation of fibres and the adhesion of the hydration particles is responsible for the strength development of portland cement paste. Reference (left and middle) Brunauer 1962 and (right) Copeland and Schulz 1962. (69110, 69112, 69099)

Fig. 2-27. Scanning-electron micrographs of hardened cement paste at (left) 500X, and (right) 1000X. (A7112, A7111)

Table 2-5. Portland Cement Compound Hydration Reactions (Oxide Notation)

$2 (3CaO•SiO_2)$ Tricalcium silicate	$+ 11 H_2O$ Water	$= 3CaO•2SiO_2•8H_2O$ Calcium silicate hydrate (C-S-H)	$+ 3 (CaO•H_2O)$ Calcium hydroxide
$2 (2CaO•SiO_2)$ Dicalcium silicate	$+ 9 H_2O$ Water	$= 3CaO•2SiO_2•8H_2O$ Calcium silicate hydrate (C-S-H)	$+ CaO•H_2O$ Calcium hydroxide
$3CaO•Al_2O_3$ Tricalcium aluminate	$+ 3 (CaO•SO_3•2H_2O)$ Gypsum	$+ 26 H_2O$ Water	$= 6CaO•Al_2O_3•3SO_3•32H_2O$ Ettringite
$2 (3CaO•Al_2O_3)$ Tricalcium aluminate	$+ 6CaO•Al_2O_3•3SO_3•32H_2O$ Ettringite	$+ 4 H_2O$ Water	$= 3 (4CaO•Al_2O_3•SO_3•12H_2O)$ Calcium monosulphoaluminate
$3CaO•Al_2O_3$ Tricalcium aluminate	$+ CaO•H_2O$ Calcium hydroxide	$+ 12 H_2O$ Water	$= 4CaO•Al_2O_3•13H_2O$ Tetracalcium aluminate hydrate
$4CaO• Al_2O_3•Fe_2O_3$ Tetracalcium aluminoferrite	$+ 10 H_2O$ Water	$+ 2 (CaO•H_2O)$ Calcium hydroxide	$= 6CaO•Al_2O_3•Fe_2O_3•12H_2O$ Calcium aluminoferrite hydrate

Note: This table illustrates only primary transformations and not several minor transformations. The composition of calcium silicate hydrate (C-S-H) is not stoichiometric (Tennis and Jennings 2000).

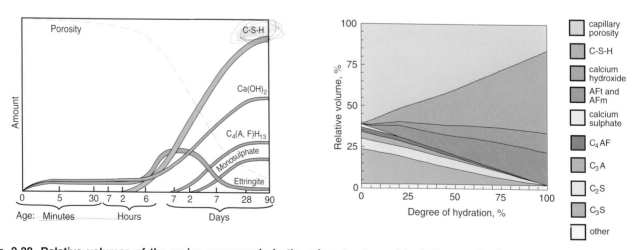

Fig. 2-28. Relative volumes of the major compounds in the microstructure of hydrating portland cement pastes (left) as a function of time (adapted from Locher, Richartz, and Sprung 1976) and (right) as a function of the degree of hydration as estimated by a computer model for a water to cement ratio of 0.50 (adapted from Tennis and Jennings 2000). Values are given for an average Type 10 (Type I) cement composition (Gebhardt 1995): $C_3S=55\%$, $C_2S=18\%$, $C_3A=10\%$ and $C_4AF=8\%$. "AFt and AFm" includes ettringite (AFt) and calcium monosulphoaluminate (AFm) and other hydrated calcium aluminate compounds. See Table 2-5 for compound transformations.

chemistry of cement paste. Fig. 2-28 shows estimates of the relative volumes of the compounds in hydrated portland cement pastes.

A web-based computer model for hydration and microstructure development can be found at NIST (2001) [http://vcctl.cbt.nist.gov].

The approximate percentage of each compound can be calculated from a chemical oxide analysis (ASTM C 114) of the unhydrated cement as per ASTM C 150 (Bogue calculations). Due to the inaccuracies of Bogue calculations, X-ray diffraction techniques can be used to more accurately determine compound percentages (ASTM C 1365). Table 2-6 shows typical elemental and compound composition and fineness for each of the principal types of portland cement.

Although elements are reported as simple oxides for standard consistency, they are usually not found in that oxide form in the cement. For example, sulphur from the gypsum is reported as SO_3 (sulphur trioxide), however, cement does not contain any sulphur trioxide. The amount of calcium, silica, and alumina establish the amounts of the primary compounds in the cement and effectively the properties of hydrated cement. Sulphate is present to control setting, as well as drying shrinkage and strength gain (Tang 1992). Minor and trace elements and their effect on cement properties are discussed by Bhatty (1995) and PCA (1992). Present knowledge of cement chemistry indicates that the primary cement compounds have the following properties:

Tricalcium Silicate, C_3S, hydrates and hardens rapidly and is largely responsible for initial set and early strength (Fig. 2-29). In general, the early strength of portland cement concrete is higher with increased percentages of C_3S.

Dicalcium Silicate, C_2S, hydrates and hardens slowly and contributes largely to strength increase at ages beyond one week.

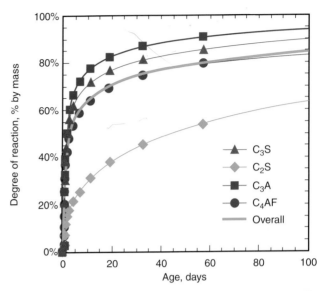

Fig. 2-29. Relative reactivity of cement compounds. The curve labeled "overall" has a composition of 55% C_3S, 18% C_2S, 10% C_3A, and 8% C_4AF, an average Type I (Type 10) cement composition (Tennis and Jennings 2000).

Tricalcium Aluminate, C_3A, liberates a large amount of heat during the first few days of hydration and hardening. It also contributes slightly to early strength development. Cements with low percentages of C_3A are more resistant to soils and waters containing sulphates.

Tetracalcium Aluminoferrite, C_4AF, is the product resulting from the use of iron and aluminum raw materials to reduce the clinkering temperature during cement manufacture. It hydrates rather rapidly but contributes very little to strength. Most colour effects that make cement gray are due to C_4AF and its hydrates.

Table 2-6. Chemical and Compound Composition and Fineness of Cements*

Type of portland cement	Chemical composition, %						Loss on Ignition, %	Na₂O eq	Potential compound composition,%				Blaine fineness, m²/kg
	SiO₂	Al₂O₃	Fe₂O₃	CaO	MgO	SO₃			C₃S	C₂S	C₃A	C₄AF	
I (min-max)	18.7-22.0	4.7-6.3	1.6-4.4	60.6-66.3	0.7-4.2	1.8-4.6	0.6-2.9	0.11-1.20	40-63	9-31	6-14	5-13	300-421
I (mean)	**20.5**	**5.4**	**2.6**	**63.9**	**2.1**	**3.0**	**1.4**	**0.61**	**54**	**18**	**10**	**8**	**369**
II** (min-max)	20.0-23.2	3.4-5.5	2.4-4.8	60.2-65.9	0.6-4.8	2.1-4.0	0.0-3.1	0.05-1.12	37-68	6-32	2-8	7-15	318-480
II** (mean)	**21.2**	**4.6**	**3.5**	**63.8**	**2.1**	**2.7**	**1.2**	**0.51**	**55**	**19**	**6**	**11**	**377**
III (min-max)	18.6-22.2	2.8-6.3	1.3-4.9	60.6-65.9	0.6-4.6	2.5-4.6	0.1-2.3	0.14-1.20	46-71	4-27	0-13	4-14	390-644
III (mean)	**20.6**	**4.9**	**2.8**	**63.4**	**2.2**	**3.5**	**1.3**	**0.56**	**55**	**17**	**9**	**8**	**548**
IV (min-max)	21.5-22.8	3.5-5.3	3.7-5.9	62.0-63.4	1.0-3.8	1.7-2.5	0.9-1.4	0.29-0.42	37-49	27-36	3-4	11-18	319-362
IV (mean)	**22.2**	**4.6**	**5.0**	**62.5**	**1.9**	**2.2**	**1.2**	**0.36**	**42**	**32**	**4**	**15**	**340**
V (min-max)	20.3-23.4	2.4-5.5	3.2-6.1	61.8-66.3	0.6-4.6	1.8-3.6	0.4-1.7	0.24-0.76	43-70	11-31	0-5	10-19	275-430
V (mean)	**21.9**	**3.9**	**4.2**	**63.8**	**2.2**	**2.3**	**1.0**	**0.48**	**54**	**22**	**4**	**13**	**373**
White (min-max)	22.0-24.4	2.2-5.0	0.2-0.6	63.9-68.7	0.3-1.4	2.3-3.1	1.2-2.2	0.09-0.38	51-72	9-25	5-13	1-2	384-564
White (mean)	**22.7**	**4.1**	**0.3**	**66.7**	**0.9**	**2.7**	**1.6**	**0.18**	**63**	**18**	**10**	**1**	**482**

* Values represent a summary of combined statistics. Air-entraining cements are not included. For consistency in reporting, elements are reported in a standard oxide form. This does not mean that the oxide form is present in the cement. For example, sulphur is reported as SO_3, sulphur trioxide, but portland cement does not have sulphur trioxide present. "Potential Compound Composition" refers to ASTM C 150 calculations using the chemical composition of the cement. The actual compound composition may be less due to incomplete or altered chemical reactions.
** Includes fine grind cements. Adapted from PCA (1996) and Gebhardt (1995).

Calcium Sulphate, as anhydrite (anhydrous calcium sulphate—$C\bar{S}$), gypsum (calciumsulphate dihydrate—$C\bar{S}H_2$), or hemihydrate, often called plaster of Paris or bassanite (calcium sulphate hemihydrate—$C\bar{S}H_{1/2}$) is added to cement during final grind to provide sulphate to react with C_3A to form ettringite (calcium sulphoaluminate). This controls the hydration of C_3A. Without sulphate, a cement would set rapidly. In addition to controlling setting and early strength gain, the sulphate also helps control drying shrinkage and can influence strength through 28 days.

In addition to the above primary cement compounds, numerous other lesser compound formulations also exist (PCA 1997, Taylor 1997, and Tennis and Jennings 2000).

Water (Evaporable and Nonevaporable)

Water is a key ingredient of pastes, mortars, and concretes, as the phases in portland cement must chemically react with water to develop strength. The amount of water added to a mixture controls the durability as well. The space initially taken up by water in a cementitious mixture is partially or completely replaced over time as the hydration reactions proceed (Table 2-5). If more than about 35% water by mass of cement—a water to cement ratio of 0.35—is used, the porosity in the hardened material will remain, even after complete hydration. This is called capillary porosity. Fig. 2-30 shows cement pastes with high and low water to cement ratios have equal masses after drying (evaporable water was removed). The cement consumed the same amount of water in both pastes resulting in more bulk volume in the higher water-cement ratio paste. As the water to cement ratio increases, the capillary porosity increases, and the strength decreases. Also, transport properties such as permeability and diffusivity are increased,

allowing detrimental chemicals to more readily attack the concrete or reinforcing steel.

Water is found in cementing materials in several forms. *Free water* is mixing water that has not reacted with the cement pastes. *Bound water* is chemically combined in the solid phases or physically bound to the solid surfaces. A reliable separation of the chemically-combined from the physically-absorbed water is not possible. Therefore, Powers (1948 and 1949) distinguished between *evaporable* and *nonevaporable water.* The non-evaporable water is the amount retained by a sample after it has been subjected to a drying procedure intended to remove all the free water (traditionally, by heating to 105°C). Evaporable water was originally considered to be free water, but it is now recognized that some bound water is also lost upon heating to this temperature. All nonevaporable water is bound water, but the opposite is not true.

For complete hydration of portland cement, only about 40% water (a water-to-cement ratio of 0.40) is needed. If a water to cement ratio greater than about 0.40 is used, the excess water not needed for cement hydration remains in the capillary pores or evaporates. If a water-to-cement ratio less than about 0.40 is used, some cement will remain unhydrated.

To estimate the degree of hydration of a hydrated material, the nonevaporable water content is often used. To convert the measured nonevaporable water into degrees of hydration, it is necessary to know the value of nonevaporable water-to-cement ratio (w_n/c) at complete hydration. Experimentally, this can be determined by preparing a high water-to-cement ratio cement paste (for example, 1.0) and continuously grinding while it hydrates in a roller mill. In this procedure, complete hydration of the cement will typically be achieved after 28 days.

Alternatively, an estimate of the value of nonevaporable water-to-cement ratio (w_n/c) at complete hydration can be obtained from the potential Bogue composition of the cement. Nonevaporable water contents for the major compounds of portland cement are provided in Table 2-7. For a typical Type 10 cement, these coefficients will generally result in a calculated w_n/c for completely hydrated cement somewhere between 0.22 and 0.25.

Fig. 2-30. Cement paste cylinders of equal mass and equal cement content, but mixed with different water-to-cement ratios, after all water has evaporated from the cylinders. (1072)

Table 2-7. Nonevaporable Water Contents for Fully Hydrated Major Compounds of Cement

Hydrated cement compound	Nonevaporable (combined) water content (g water/g cement compound)
C_3S hydrate	0.24
C_2S hydrate	0.21
C_3A hydrate	0.40
C_4AF hydrate	0.37
Free lime (CaO)	0.33

PHYSICAL PROPERTIES OF CEMENT

Specifications for cement place limits on both its physical properties (performance) and often, chemical composition. An understanding of the significance of some of the physical properties is helpful in interpreting results of cement tests. Tests of the physical properties of the cements should be used to evaluate the properties of the cement, rather than the concrete. Cement specifications limit the properties with respect to the type of cement. Cement should be sampled in accordance with CSA A456.2-A1 (ASTM C 183). During manufacture, cement is continuously monitored for its chemistry and the following properties:

Particle Size and Fineness

Portland cement consists of individual angular particles with a range of sizes, the result of pulverizing clinker in the grinding mill (Fig. 2-31 top). Approximately 85% to 95% of cement particles are smaller than 45 μm, with the average particle around 15 μm. Fig. 2-31 bottom illustrates the particle size distribution for a portland cement. The overall particle size distribution of cement is called "fineness". The fineness of cement affects heat released and the rate of hydration. Greater cement fineness (smaller particle size) increases the rate at which cement hydrates and thus accel-

erates strength development. The effects of greater fineness on paste strength are manifested principally during the first seven days.

Cement fineness can be measured by a number of methods. One of the methods by which it is measured is called the Blaine air-permeability test (ASTM C 204—Fig. 2-32); it indirectly measures the surface area of the cement particles per unit mass. Cements with finer particles have more surface area in square metres per kilogram of cement. The older use of square centimetres per gram for reporting fineness is now considered outdated. The Wagner turbidimeter test (ASTM C 115—Fig. 2-32), can also be used.

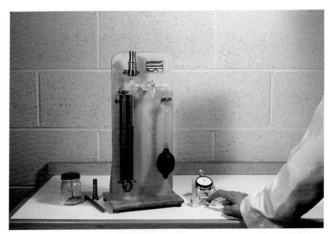

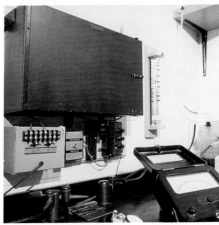

Fig. 2-32. Blaine test apparatus (top) and Wagner turbidimeter (bottom) for determining the fineness of cement. Wagner fineness values are a little more than half of Blaine values. (40262, 43815)

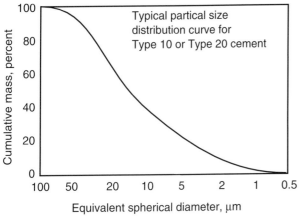

Fig. 2-31. (top) Scanning-electron micrograph of powdered cement at 1000X and (bottom) particle size distribution of a portland cement. (69426)

In Canada fineness is measured by CSA Test Method A 456.2-A3, *Test Method for Determination of Cement Fineness by Wet-Sieving* (ASTM C 430—Fig. 2-33). The fineness is expressed as a percentage retained on the 45-μm sieve. Electronic (x-ray or laser) particle size analyzers can also be used to test fineness (Fig. 2-34). Examples of Blaine fineness data are given in Table 2-6.

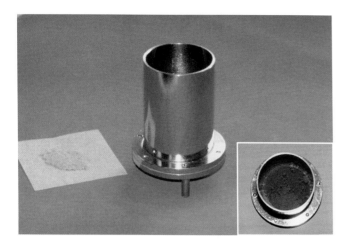

Fig. 2-33. Quick tests, such as washing cement over this 45-μm sieve, help monitor cement fineness during production. Shown is a side view of the sieve holder with an inset top view of a cement sample on the sieve before washing with water. (68818, 68819)

Fig. 2-34. A laser particle analyzer uses laser diffraction to determine the particle size distribution of fine powders. Fig. 2-31 (bottom) illustrates typical results. (69390)

Soundness

Soundness refers to the ability of a hardened paste to retain its volume. Lack of soundness or delayed destructive expansion can be caused by excessive amounts of hard-burned free lime or magnesia. Most specifications for portland cement limit the magnesia (periclase) content and the maximum expansion as measured by the autoclave-expansion test. Since adoption of the autoclave-expansion test, CSA A456.2-B5, *Test Method for Determination of Autoclave Expansion* (ASTM C 151), there have been exceedingly few cases of abnormal expansion attributed to unsound cement (Fig. 2-35) (Gonnermann, Lerch, and Whiteside 1953).

Consistency

Consistency refers to the relative mobility of a freshly mixed cement paste or mortar or to its ability to flow. Dur-

ing cement testing, pastes are mixed to normal consistency as defined by a penetration of 10±1 mm of the Vicat plunger; see CSA A456.2-B1 (ASTM C 187) and Fig. 2-36. Mortars are mixed to obtain either a fixed water-cement ratio or to yield a flow within a prescribed range. The flow is determined on a flow table as described in CSA A456.3 (ASTM C 230), Fig. 2-37. Both the normal consistency

Fig. 2-35. In the soundness test, 25-mm square bars are exposed to high temperature and pressure in the autoclave to determine the volume stability of the cement paste. (23894)

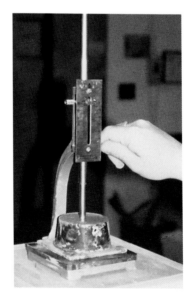

Fig. 2-36. Normal consistency test for paste using the Vicat plunger. (68820)

Fig. 2-37. Consistency test for mortar using the flow table. The mortar is placed in a small brass mould centered on the table (inset). For skin safety the technician wears protective gloves while handling the mortar. After the mould is removed and the table undergoes a succession of drops, the diameter of the pat is measured to determine consistency. (68821, 68822)

method and the flow test are used to regulate water contents of pastes and mortars, respectively, to be used in subsequent tests; both allow comparing dissimilar ingredients with the same penetrability or flow.

Setting Time

The object of the setting time test is to determine (1) the time which elapses from the moment water is added until the paste ceases to be fluid and plastic (called initial set) and (2) the time required for the paste to acquire a certain degree of hardness (called final set).

To determine if a cement sets according to the time limits specified in cement specifications, tests are performed using either the Vicat apparatus [CSA A456.2-B2 (ASTM C 191), Fig. 2-38] or the Gillmore needle [CSA A456.2-B3 (ASTM C 266), Fig. 2-39]. In cases of dispute, the Gillmore setting time test governs. Initial set of cement paste must not occur too early and final set must not occur too late. The setting times indicate that a paste is or is not undergoing normal hydration reactions. Sulphate (from gypsum or other sources) in the cement regulates setting time, but setting time is also affected by cement fineness, water-cement ratio, and any admixtures that may be used. Setting times of concretes do not correlate directly with setting times of pastes because of water loss to the air or substrate, presence of aggregate, and because of temperature differences in the field (as contrasted with the controlled temperature in a testing lab). Fig. 2-40 illustrates mean set times for portland cements.

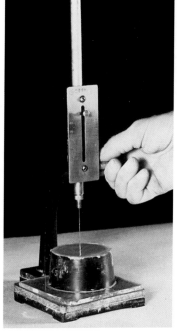

Fig. 2-38. Time of set as determined by the Vicat needle. (23890)

Fig. 2-39. Time of set as determined by the Gillmore needle. (23892)

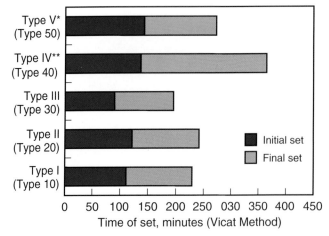

*Average of two values for final set
**Average of two values for initial set; one value for final set

Fig. 2-40. Time of set for portland cements (Gebhardt 1995 and PCA 1996).

Early Stiffening (False Set and Flash Set)

Early stiffening is the early development of stiffness in the working characteristics or plasticity of cement paste, mortar, or concrete. This includes both false set and flash set.

False set is evidenced by a significant loss of plasticity without the evolution of much heat shortly after mixing. From a placing and handling standpoint, false-set tendencies in cement will cause no difficulty if the concrete is mixed for a longer time than usual or if it is remixed without additional water before it is transported or placed. False set occurs when a significant amount of gypsum dehydrates in the cement mill to form plaster. Stiffening is caused by the rapid crystallization of interlocking needle-like secondary gypsum. Additional mixing without added water breaks up these crystals to restore workability. Ettringite precipitation can also contribute to false set.

Flash set or quick set is evidenced by a rapid and early loss of workability in paste, mortar, or concrete, usually with the evolution of considerable heat resulting primarily from the rapid reaction of aluminates. If the proper amount or form of calcium sulphate is not available to control the calcium aluminate hydration, stiffening becomes apparent. Flash set cannot be dispelled, nor can the plasticity be regained by further mixing without the addition of water.

Proper stiffening results from the careful balance of the sulphate and aluminate compounds, as well as the temperature and fineness of the materials (which control the hydration and dissolution rates). The amount of sulphate formed into plaster has a significant effect. For example, with one particular cement, 2% plaster allowed a 5 hour set time, while 1% plaster caused flash set to occur, and 3% allowed false set to occur (Helmuth and others 1995).

CSA Standard A5 treats testing for the early stiffening of cement paste as an Optional Test Requirement of the purchaser as noted in *A5-Appendix A*. When cements are tested for early stiffening, CSA A456.2-B6, *Test Method Determination of Early Stiffening of Cement Paste,* is used. This test method addresses the determination of early stiffening of cement paste by the Vicat needle. (ASTM has two different methods of testing, ASTM C 451 a paste method, and ASTM C 359 a mortar method, which use the penetration techniques of the Vicat apparatus). However, none of these tests address all the mixing, placing, and temperature conditions in the field that can cause early stiffening. They also do not address early stiffening caused by interactions with other concrete ingredients. For example, concretes mixed for very short mixing times, less than a minute, tend to be more susceptible to early stiffening (ACI 225).

Compressive Strength

Compressive strength as specified by CSA cement standards is that obtained from tests of 50-mm mortar cubes tested in accordance with CSA A456.2-C2 (ASTM C 109—Fig. 2-41). CSA requires the cubes to be made and cured in a prescribed manner using a graded sand (ASTM uses a standard sand).

Compressive strength is influenced by the cement type, or more precisely, the compound composition and fineness of the cement. CSA A5 and A362 (ASTM C 150 and C 595) set only a minimum strength requirement. Minimum strength requirements in cement specifications are exceeded comfortably by most manufacturers. Therefore, it should not be assumed that two types of cement, meeting the same minimum requirements, will produce the same strength of mortar or concrete without modification of mix proportions. There is, however, within CSA A5 an option available to purchasers should they choose to invoke, *Appendix A- Optional Test Requirements,* which provides maximum compressive strength requirements at various ages for the cement types within this Standard.

(ASTM C 1157 has both minimum and maximum strength requirements).

In general, cement strengths (based on mortar-cube tests) cannot be used to predict concrete strengths with a great degree of accuracy because of the many variables in aggregate characteristics, concrete mixtures, construction procedures, and environmental conditions in the field. Figs. 2-42 and 2-43 illustrate the strength development for standard mortars made with various types of portland cement. Wood (1992) provides long-term strength properties of mortars and concretes made with portland and blended cements. The strength uniformity of a cement from a single source may be determined by following the procedures outlined in CSA A5, Section 3.4 (ASTM C 917).

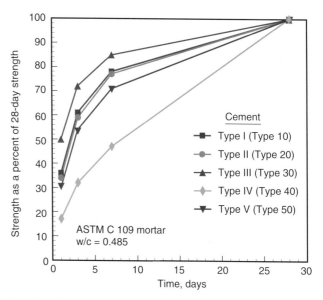

Fig. 2-42. Relative strength development of portland cement mortar cubes as a percentage of 28-day strength. Mean values adapted from Gebhardt 1995.

Fig. 2-41. 50-mm mortar cubes are cast (left) and crushed (right) to determine strength characteristics of cement. (69128, 69124)

Heat of Hydration

Heat of hydration is the heat generated when cement and water react. The amount of heat generated is dependent chiefly upon the chemical composition of the cement, with C_3A and C_3S being the compounds primarily responsible for high heat evolution. The water-cement ratio, fineness of the cement, and temperature of curing also are factors. An increase in the fineness, cement content, and curing temperature increases the heat of hydration. Although portland cement can evolve heat for many years, the rate of heat generation is greatest at early ages. A large amount of heat evolves within the first three days with the greatest rate of heat liberation usually occurring within the first 24

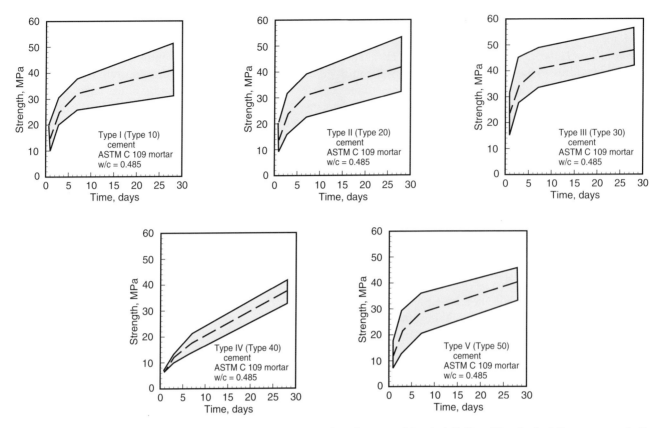

Fig. 2-43. Strength development of portland cement mortar cubes from combined statistics. The dashed line represents the mean value and the shaded area the range of values (adapted from Gebhardt 1995).

hours. (Copeland and others 1960). The heat of hydration is tested in accordance with CSA A456.2-B7, *Test Method for Determination of Heat of Hydration* (ASTM C 186), or by conduction calorimetry (Fig. 2-44).

For most concrete elements, such as slabs, heat generation is not a concern because the heat is quickly dissipated into the environment. However, in structures of considerable mass, greater than a metre thick, the rate and amount of heat generated are important. If this heat is not rapidly dissipated, a significant rise in concrete temperature can occur. This may be undesirable since, after hardening at an elevated temperature, nonuniform cooling of the concrete mass to ambient temperature may create undesirable tensile stresses. On the other hand, a rise in concrete temperature caused by heat of hydration is often beneficial in cold weather since it helps maintain favourable curing temperatures.

Table 2-8 provides heat of hydration values for a variety of portland cements. This limited data show that Type 30 cement has a higher heat of hydration than other types while Type 40 has the lowest. Also note the difference in heat generation between regular Type 20 cements and moderate heat Type 20 cements.

Cements do not generate heat at a constant rate. The heat output during hydration of a typical Type 10 portland cement is illustrated in Fig. 2-45. The first peak shown in

Fig. 2-44. Heat of hydration can be determined by (top) CSA A456.2-B7 (ASTM C 186) or by (bottom) a conduction calorimeter. (68823, 68824)

48

Table 2-8. ASTM C 186 Heat of Hydration for Selected Portland Cements from the 1990s, kJ/kg*

	Type I (Type 10) cement		Type II (Type 20) cement		Type II (Type 20) moderate heat cement	Type III (Type 30) cement		Type IV (Type 40) cement		Type V (Type 50) cement
	7 day	28 day	7 day	28 day	7 day	7 day	28 day	7 day	28 day	7 day
No. of samples	15	7	16	7	4	2	2	3	1	6
Average	349	400	344	398	263	370	406	233	274	310
Maximum	372	444	371	424	283	372	414	251	—	341
Minimum	320	377	308	372	227	368	397	208	—	257
% of Type I (Type 10) (7 day)	100		99		75	106		67		89

*This table is based on very limited data.
1 cal/g = 4.184 kJ/kg. PCA (1997).

the heat profile is due to heat given off by the initial hydration reactions of cement compounds such as tricalcium aluminate. Sometimes called the heat of wetting, this first heat peak is followed by a period of slow activity known as the induction period. After several hours, a broad second heat peak attributed to tricalcium silicate hydration emerges signaling the onset of the paste hardening process. Finally, there is the third peak due to the renewed activity of tricalcium aluminate; its intensity and location normally dependent on the amount of tricalcium aluminate and sulphate in the cement.

In calorimetry testing, the first heat measurements are obtained about 7 minutes after mixing the paste; as a result only the down slope of the first peak is observed (Stage 1, Fig. 2-45). The second peak (C_3S peak) often occurs

between 6 and 12 hours. The third peak (C_3A peak) occurs between 12 and 90 hours. This information can be helpful when trying to control temperature rise in mass concrete (Tang 1992).

When heat generation must be minimized in concrete, designers should choose a lower heat cement, such as a CSA A5 Type 20 (ASTM C 150 Type II) portland cement with the optional moderate heat of hydration requirements. Not all Type 20 (Type II) cements are made for a moderate level of heat development, however, so the moderate heat option must be specifically requested. Type 40 (Type IV) cement can also be used to control temperature rise, but it is not always available. Moderate-heat and low-heat cements are also available under CSA A362 (ASTM C 595 and C 1157). Supplementary cementing materials can also be used to reduce temperature rise.

Physical requirements for heat of hydration are given in Table 2, CSA A5, and Table 2, CSA A362. ASTM C 150 has both a chemical approach and a physical approach to control heat of hydration. Either approach can be specified, but not both. ASTM C 595 and ASTM C 1157 use physical limits. See PCA (1997) for more information.

Loss on Ignition (LOI)

Loss on ignition of portland cement is determined by heating a cement sample of known mass to 1000°C, or 550°C on cements with carbonate additions, until a constant mass of the sample is obtained. The loss of mass of the sample is then determined. Normally, a high loss on ignition is an indication of prehydration and carbonation, which may be caused by improper and prolonged storage or adulteration during transport. The test for loss on ignition (Fig. 2-46) is performed in accordance with CSA A456.1, *Chemical Test Methods for Hydraulic Cement, Supplementary Cementing Materials, and Cementitious Hydraulic Slag*. Requirements for

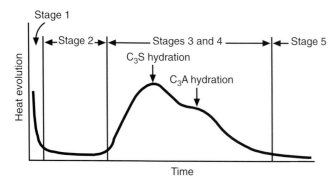

Fig. 2-45. Heat evolution as a function of time for cement paste. Stage 1 is heat of wetting or initial hydrolysis. Stage 2 is a dormant period related to initial set. Stage 3 is an accelerated reaction of the hydration products that determines rate of hardening and final set. Stage 4 decelerates formation of hydration products and determines the rate of early strength gain. Stage 5 is a slow, steady formation of hydration products establishing the rate of later strength gain.

Fig. 2-46. Loss on ignition test of cement. (43814)

Loss on Ignition are given in Table 1 of CSA A5. LOI values range from 0 to 3%.

Density and Relative Density (Specific Gravity)

The density of cement is defined as the mass of a unit volume of the solids or particles, excluding air between particles. It is reported as megagrams per cubic metre or grams per cubic centimetre as done in CSA test reporting (the numeric value is the same for both units). The particle density of portland cement ranges from 3.10 to 3.25, averaging 3.15 g/cm³. Portland blast-furnace slag and port-land-pozzolan cements have densities ranging from 2.90 to 3.15, averaging 3.05 g/cm³. The density of a cement, determined by CSA A456.2-A2, *Test Method for Determination of Density of Cement* (ASTM C 188) (Fig. 2-47), is not an indication of the cement's quality; its principal use is in mixture proportioning calculations.

For mixture proportioning, it may be more useful to express the density as relative density (also called specific

gravity). The relative density is a dimensionless number determined by dividing the cement density by the density of water at 4°C. The density of water at 4°C is 1.0 g/cm³ (Mg/m³).

A relative density of 3.15 is assumed for portland cement in volumetric computations of concrete mix proportioning. However, as mix proportions list quantities of concrete ingredients in kilograms, the relative density must be multiplied by the density of water at 4°C stated as 1000 kg/m³ to determine the particle density in kg/m³. This product is then divided into the mass of cement to determine the absolute volume of cement in cubic metres.

Bulk Density

The bulk density of cement is defined as the mass of cement particles plus air between particles per unit volume. The bulk density of cement can vary considerably depending on how it is handled and stored. Portland cement that is fluffed up may have a mass of only 830 kg/m³, whereas when consolidated by vibration, the same cement can have a mass as much as 1650 kg/m³ (Toler 1963). For this reason, good practice dictates that the mass of cement must be determined for each batch of concrete produced, as opposed to use of a volumetric measure (Fig. 2-48).

Fig. 2-48. Both 500-mL beakers contain 500 grams of dry powdered cement. On the left, cement was simply poured into the beaker. The beaker on the right was slightly vibrated—imitating potential consolidation during transport, or packing in the silo. The 20% difference in bulk volume demonstrates the need to measure cement by mass instead of volume for batching concrete. (68970)

THERMAL ANALYSIS

Thermal analysis techniques have been available for years to analyze hydraulic reactions and the interactions of cement with both mineral and chemical admixtures (Figs. 2-45 and 2-49). Traditionally, thermal analysis was not part of a routine test program for cement. However, thermal analysis has gained recent popularity in analyzing chemical and physical properties of cementing materials and raw materials for cement manufacture (Bhatty 1993, Shkolnik and Miller 1996, and Tennis 1997).

Thermal analysis involves heating a small sample at a controlled rate to high temperature (up to 1000°C or more). As compounds react or decompose, changes that occur as a function of time or temperature are recorded. As the sample's temperature changes, various reactions occur that cause a change in the mass of the sample, temperature,

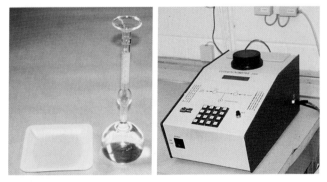

Fig. 2-47. Density of cement can be determined by (left) using a Le Chatelier flask and kerosine or by (right) using a helium pycnometer. (68825, 68826)

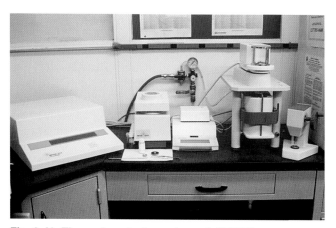

Fig. 2-49. Thermal analysis equipment. (69116)

energy, or state. A solid can melt, vaporize, decompose to a gas with a residual solid, or react with a gas (at elevated temperatures) to form a different solid or a different solid and another gas.

Typical uses for thermal analysis include:
- identifying which hydration products have formed and in what quantities
- solving early stiffening problems
- identifying impurities in raw materials
- determining the amount of weathering in clinker or cement
- estimating the reactivity of pozzolans and slags for use in blended cements
- identifying the organic matter content and variations in a quarry
- quantifying the amount of carbonation that has taken place in an exposed sample
- analyzing durability problems with concrete.

Discussion of specific thermal analysis techniques follows:

Thermogravimetric Analysis (TGA)

Thermogravimetric analysis (TGA) is a technique that measures the mass of a sample as it is being heated (or cooled) at a controlled rate. The change in mass of a sample depends on the composition of the sample, the temperature, the heating rate, and the type of gas in the furnace (air, oxygen, nitrogen, argon, or other gas). A change in mass within a specific temperature range identifies the presence of particular chemical compound. The magnitude of the change in mass indicates the amount of the compound in the sample.

A sample's free water will evaporate, lowering the mass of the sample, as the temperature is raised from room temperature to about 100°C. The sample will also lose some of the water bound in hydration products, particularly some of the water in the aluminate hydrates.

Between about 100°C and 400°C, water bound in the hydration products, primarily the C-S-H gel, will be lost, together with the remainder of water in the aluminate hydrates. Between about 400°C and 500°C, calcium hydroxide provides a relatively distinct loss in mass as it decomposes to calcium oxide (a solid) and water vapor. The amount of this loss in mass can be used to determine how much calcium hydroxide was originally present in the sample. Above 500°C, a small amount of additional water may be lost from the hydration products. Carbonated phases lose carbon dioxide around 800°C.

By quantifying the amount of calcium hydroxide, a TGA gives an indication of the degree of hydration that has taken place in a sample. The reactivity of pozzolans can be assessed by tracking the disappearance of calcium hydroxide due to pozzolanic reaction.

Differential Thermal Analysis (DTA)

Differential thermal analysis (DTA) is an analytical method in which the difference in temperature between a sample and a control is measured as the samples are heated. The control is usually an inert material, like powdered alumina, that does not react over the temperature range being studied. If a sample undergoes a reaction at a certain temperature, then its temperature will increase or decrease relative to the (inert) control as the reaction gives off heat (exothermic) or adsorbs heat (endothermic). A thermocouple measures the temperature of each material, allowing the difference in temperature to be recorded. DTA is ideal for monitoring the transformation of cement compounds during hydration. DTA can be performed simultaneously with TGA.

Differential Scanning Calorimetry (DSC)

In differential scanning calorimetry (DSC), the heat adsorbed or released is measured directly as a function of temperature or time and compared to a reference. An advantage of DTA and DSC methods is that no mass change is required; thus if a sample melts without vaporizing, measurements can still be made.

Like DTA, DSC can be used to determine what compounds are present at different stages of hydration. Fig. 2-50 shows two differential scanning calorimetry curves or thermograms. The upper curve (a) shows a portland cement paste after 15 minutes of reaction. The peaks in the curve between 100°C and 200°C are due to the endothermic (heat adsorbing) decomposition of gypsum and ettringite, while the peak around 270°C is due to syngenite (potassium calcium sulphate hydrate). Near 450°C, a smaller peak can be seen, due to calcium hydroxide.

The lower curve (b) in Fig. 2-50 shows the same cement paste after 24 hours of hydration. Note the disappearance of the peak due to gypsum, the reduction in the size of the peak due to syngenite and the growth of those due to ettringite and calcium hydroxide. The size of the areas under the curves are related to the quantity of the material in the sample.

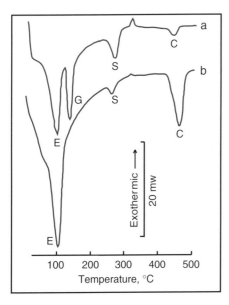

Fig. 2-50. Differential scanning calorimetry thermogram of a portland cement paste after (a) 15 minutes and (b) 24 hours of hydration. C = calcium hydroxide; E = ettringite; G = gypsum; and S = syngenite.

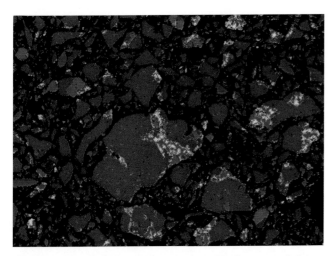

Fig. 2-51. Two-dimensional image of portland cement. Colours are: red–tricalcium silicate, aqua–dicalcium silicate, green–tricalcium aluminate, yellow–tetracalcium aluminoferrite, pale green–gypsum, white–free lime, dark blue–potassium sulphate, and magenta–periclase. The image was obtained by combining a set of SEM backscattered electron and X-ray images (NIST 2001).

VIRTUAL CEMENT TESTING

Computer technology now allows simulation of cement compounds (Fig. 2-51), hydration, microstructure development, and physical properties. Combinations of materials, cement compounds, or particle size distributions can be examined to predict cement performance. Just some of the properties that can be simulated and predicted include heat of hydration, adiabatic heat signature, compressive strength, setting time, rheology (yield stress and viscosity), percolation, porosity, diffusivity, thermal conductivity, electrical conductivity, carbonation, elastic properties, drying profiles, susceptibility to degradation mechanisms, autogenous shrinkage, and volumes of hydration reactants and products as a function of time. The effects of varying sulphate and alkali contents can be examined, along with interaction with supplementary cementing materials and chemical admixtures. Computer modeling predicts performance without the expense and time requirements of physical testing (NIST 2001).

TRANSPORTATION AND PACKAGING

In the 1800's and early 1900's, cement was shipped in barrels, where one barrel contained 171 kg or 376 lbs (four 94-lb bags, each 1 cu ft in size). Today most portland cements are shipped in bulk by rail, truck, barge or ship (Fig. 2-52). Pneumatic loading and unloading of the transport vehicle is the most popular means of bulk cement handling. Bulk cement is measured by the metric ton (tonne); smaller quantities are bagged (Fig. 2-53). In Canada, a bag of portland cement has a mass of 40 kg, though there may be exceptions. The mass of a bag of masonry cement is 30 kg. CSA A5 or A8 does not stipulate the mass of a bag of cement.

In the United States, a bag of portland cement has a mass of 42 kg or 94 lb and has a volume of 28 litres when freshly packed. The mass of masonry cement and mortar cement by the bag is 36 kg for Type M, 34 kg for Type S, and 32 kg for Type N. Plastic cement has a mass of 42 kg for Type M and 35 kg for Type S. The mass of all metric bags are based on a 28 litre volume standard. Information for quantities of blended and specialty cements can be found

Fig. 2-52. Portland cements are shipped from the plant silos to the user in bulk by (left to right) rail, truck, or water. (59899, 59355, 59891)

Fig. 2-53. A small amount of cement is shipped in bags, primarily for mortar applications and for small projects. (59411)

Fig. 2-54. When stored on the job, cement should be protected from moisture. (36052)

on the bag. Partial bag sizes are also available in some areas. In other countries, a bag of cement can have a mass of 25 kg or 50 kg.

Because of the variability of bag sizes and the presence of supplementary cementing materials, cement bag factor terminology, such as "a six bag mix" should not be used to describe the cement content of concrete mixtures.

STORAGE OF CEMENT

Cement is a moisture-sensitive material; if kept dry, it will retain its quality indefinitely. Cement stored in contact with damp air or moisture sets more slowly and has less strength than cement that is kept dry. At the cement plant, and at ready mixed concrete facilities, bulk cement is stored in silos. The relative humidity in a warehouse or shed used to store bagged cement should be as low as possible. All cracks and openings in walls and roofs should be closed. Cement bags should not be stored on damp floors but should rest on pallets. Bags should be stacked close together to reduce air circulation but should never be stacked against outside walls. Bags to be stored for long periods should be covered with tarpaulins or other waterproof covering. Bags should be stored so that the first in are the first out.

On small jobs where a shed is not available, bags should be placed on raised wooden platforms (pallets) above the ground. Waterproof coverings should fit over the pile and extend over the edges of the platform to prevent rain from reaching the cement and the platform (Fig. 2-54). Rain-soaked platforms can damage the bottom bags of cement.

Cement stored for long periods may develop what is called warehouse pack. This can usually be corrected by rolling the bags on the floor. At the time of use, cement should be free-flowing and free of lumps. If lumps do not break up easily, the cement should be tested before it is used in important work. Standard strength tests or loss-on-

ignition tests should be made whenever the quality of a cement is doubtful.

Ordinarily, cement does not remain in storage long, but it can be stored for long periods without deterioration. Bulk cement should be stored in weathertight concrete or steel bins or silos. Dry low-pressure aeration or vibration should be used in bins or silos to keep the cement flowable and avoid bridging. Due to fluffing of cement, silos may hold only about 80% of rated capacity.

HOT CEMENT

When cement clinker is pulverized in the grinding mill, the friction generates heat. Freshly ground cement is therefore hot when placed in storage silos at the cement plant. This heat dissipates slowly, therefore, during summer months when demand is high, cement may still be hot when delivered to a ready mixed concrete plant or jobsite. Tests have shown that the effect of hot cement on the workability and strength development of concrete is not significant (Lerch 1955). The temperatures of the mixing water and aggregates play a much greater role in determining the concrete temperature.

REFERENCES

Abrams, Duff A., "Effect of Hydrated Lime and Other Powdered Admixtures in Concrete," *Proceedings of the American Society for Testing Materials*, Vol. 20, Part 2, 1920. Reprinted with revisions as Bulletin 8, Structural Materials Research Laboratory, Lewis Institute, June 1925, 78 pages. Available through PCA as LS008, http://www.portcement.org/pdf_files/LS008.pdf.

ACI Committee 223, *Standard Practice for the Use of Shrinkage Compensating Concrete*, ACI 223, American Concrete Institute, Farmington Hills, Michigan, 1998, 28 pages.

ACI Committee 225, *Guide to the Selection and Use of Hydraulic Cements*, ACI 225, ACI Committee 225 Report, American Concrete Institute, Farmington Hills, Michigan, 1999.

Aspdin, Joseph, *Artificial Stone*, British Patent No. 5022, December 15, 1824, 2 pages.

Auburn, *Historical Timeline of Concrete*, AU BSC 314, Auburn University, http://www.bsc.auburn.edu/heinmic/bsc314/timeline/timeline.htm, June 2000.

Barger, Gregory S.; Lukkarila, Mark R.; Martin, David L.; Lane, Steven B.; Hansen, Eric R.; Ross, Matt W.; and Thompson, Jimmie L., "Evaluation of a Blended Cement and a Mineral Admixture Containing Calcined Clay Natural Pozzolan for High-Performance Concrete," *Proceedings of the Sixth International Purdue Conference on Concrete Pavement Design and Materials for High Performance*, Purdue University, West Lafayette, Indiana, November 1997, 21 pages.

Bentz, Dale P., and Farney, Glenn P., *User's Guide to the NIST Virtual Cement and Concrete Testing Laboratory, Version 1.0*, NISTIR 6583, November. 2000.

Bhatty, Javed I., "Application of Thermal Analysis to Problems in Cement Chemistry," Chapter 6, *Treatise on Analytical Chemistry*, Part 1, Volume 13, J. D. Winefordner editor, John Wiley & Sons, 1993, pages 355 to 395.

Bhatty, Javed I., *Role of Minor Elements in Cement Manufacture and Use*, Research and Development Bulletin RD109, Portland Cement Association, 1995, 48 pages.

Brown, Gordon E., *Analysis and History of Cement*, Gordon E. Brown Associates, Keswick, Ontario, 1996, 259 pages.

Brunauer, S., *Some Aspects of the Physics and Chemistry of Cement*, Research Department Bulletin RX080, Portland Cement Association, http://www.portcement.org/pdf_files/RX080.pdf, 1957.

Brunauer, Stephen, *Tobermorite Gel—The Heart of Concrete*, Research Department Bulletin RX138, Portland Cement Association, http://www.portcement.org/pdf_files/RX138.pdf, 1962, 20 pages.

Bureau of Reclamation, *Concrete Manual*, 8th ed., U.S. Bureau of Reclamation, Denver, Colorado, revised 1981.

BS 915-2, *Specification for High Alumina Cement*, British Standards Institution, London, 1972.

Campbell, Donald H., *Microscopical Examination and Interpretation of Portland Cement and Clinker*, SP030, Portland Cement Association, 1999.

Canada Department of Mines, *Mineral Production of Canada*, Department of Mines, Mines Branch, 1920.

Canadian Standards Association, CSA Standard A3000-98, *Cementitious Materials Compendium*, Canadian Standards Association, Toronto, March 1998.

Canadian Standards Association, CSA Standard A23.1-00, *Concrete Materials and Methods of Concrete Construction*, CAN/CSA Standard A23.2-00, *Methods of Test for Concrete*, Canadian Standards Association, Toronto, September 2000.

Clausen, C. F., *Cement Materials*, Research Department Report MP-95, Portland Cement Association, 1960.

Copeland, L.E., and Hayes, John C., *The Determination of Non-evaporable Water in Hardened Portland Cement Paste*, Research Department Bulletin RX047, Portland Cement Association, http://www.portcement.org/pdf_files/RX047.pdf, 1953.

Copeland, L. E.; Kantro, D. L.; and Verbeck, George, *Chemistry of Hydration of Portland Cement*, Research Department Bulletin RX153, Portland Cement Association, http://www.portcement.org/pdf_files/RX153.pdf, 1960.

Copeland, L. E., and Schulz, Edith G., *Electron Optical Investigation of the Hydration Products of Calcium Silicates and Portland Cement*, Research Department Bulletin RX135, Portland Cement Association, http://www.portcement.org/pdf_files/RX135.pdf, 1962.

Davidovits, Joseph; Davidovits, Ralph; and James, Claude, editors, *Geopolymer '99*, The Geopolymer Institute, http://www.geopolymer.org, Insset, Université de Picardie, Saint-Quentin, France, July 1999.

DeHayes, Sharon M., "C 109 vs. Concrete Strengths" *Proceedings of the Twelfth International Conference on Cement Microscopy*, International Cement Microscopy Association, Duncanville, Texas, 1990.

Detwiler, Rachel J.; Bhatty, Javed I.; Barger, Gregory; and Hansen, Eric R., "Durability of Concrete Containing Calcined Clay," *Concrete International*, April 2001, pages 43 to 47.

Detwiler, Rachel J.; Bhatty, Javed I.; and Bhattacharja, Sankar, *Supplementary Cementing Materials for Use in Blended Cements*, Research and Development Bulletin RD112, Portland Cement Association, 1996, 108 pages.

EN 197-1, *European Standard for Cement—Part 1: Composition, Specifications and Common Cements,* European Committee for Standardization (CEN), Brussels, 2000.

EN 197-2, *European Standards for Cement—Part 2: Conformity Evaluation,* European Committee for Standardization (CEN), Brussels, 2000.

Farny, James A., *White Cement Concrete,* EB217, Portland Cement Association, 2001, 32 pages.

Gajda, John, *Development of a Cement to Inhibit Alkali-Silica Reactivity,* Research and Development Bulletin RD115, Portland Cement Association, 1996, 58 pages.

Gebhardt, R. F., "Survey of North American Portland Cements: 1994," *Cement, Concrete, and Aggregates,* American Society for Testing and Materials, West Conshohocken, Pennsylvania, December 1995, pages 145 to 189.

Gonnerman, H. F., *Development of Cement Performance Tests and Requirements,* Research Department Bulletin RX093, Portland Cement Association, http://www.portcement org/pdf_files/RX093.pdf, 1958.

Gonnerman, H. F., and Lerch, William, *Changes in Characteristics of Portland Cement As Exhibited by Laboratory Tests Over the Period 1904 to 1950,* Research Department Bulletin RX039, Portland Cement Association, http://www.port cement.org/pdf_files/RX039.pdf, 1952.

Gonnerman, H. F.; Lerch, William; and Whiteside, Thomas M., *Investigations of the Hydration Expansion Characteristics of Portland Cements,* Research Department Bulletin RX045, Portland Cement Association, http://www.portcement. org/pdf_files/RX045.pdf, 1953.

Greening, Nathan R.; Copeland, Liewellyn E.; and Verbeck, George J., *Modified Portland Cement and Process,* United States Patent No. 3,628,973, Issued to Portland Cement Association, December 21, 1971.

Hansen, W. C.; Offutt, J. S.; Roy, D. M.; Grutzeck, M. W.; and Schatzlein, K. J., *Gypsum & Anhydrite in Portland Cement,* United States Gypsum Company, Chicago, 1988, 104 pages.

Helmuth, Richard; Hills, Linda M.; Whiting, David A.; and Bhattacharja, Sankar, *Abnormal Concrete Performance in the Presence of Admixtures,* RP333, Portland Cement Association, 1995, 94 pages.

Hills, Linda M., "Clinker Formation and the Value of Microscopy," *Proceedings of the Twenty-Second International Conference on Cement Microscopy,* Montreal, 2000, pages 1 to 12.

Johansen, Vagn, and Idorn, G. M., "Cement Production and Cement Quality," *Material Science of Concrete I,* Edited by J. Skalny, American Ceramic Society, 1989, pages 27 to 72.

Kanare, Howard M., *Certifying Portland Cement to ANSI/NSF 61 for Use in Drinking Water System Components,* SP117, Portland Cement Association, 1995, 253 pages.

Kanare, Howard M., and West, Presbury B., "Leachability of Selected Chemical Elements from Concrete," *Emerging Technologies Symposium on Cement and Concrete in the Global Environment,* SP114, Portland Cement Association, 1993, 366 pages.

Klemm, Waldemar A., *Ettringite and Oxyanion-Substituted Ettringites—Their Characterization and Applications in the Fixation of Heavy Metals: A Synthesis of the Literature,* Research and Development Bulletin RD116, Portland Cement Association, 1998, 80 pages.

Klieger, Paul, and Isberner, Albert W., *Laboratory Studies of Blended Cements—Portland Blast-Furnace Slag Cements,* Research Department Bulletin RX218, Portland Cement Association, http://www.portcement.org/pdf_files/RX218. pdf, 1967.

Lea, F. M., *The Chemistry of Cement and Concrete,* 3rd ed., Chemical Publishing Co., Inc., New York, 1971.

Lerch, William, *Hot Cement and Hot Weather Concrete Tests,* IS015, Portland Cement Association, http://www.port cement.org/pdf_files/IS015.pdf, 1955.

Lerch, William, *The Influence of Gypsum on the Hydration and Properties of Portland Cement Pastes,* Research Department Bulletin RX012, Portland Cement Association, http:// www.portcement.org/pdf_files/RX012.pdf, 1946.

Locher, F. W.; Richartz, W.; and Sprung, S., "Setting of Cement–Part I: Reaction and Development of Structure;" *ZKG INTERN. 29,* No. 10, 1976, pages 435 to 442.

McMillan, F. R.; Tyler, I. L.; Hansen, W. C.; Lerch, W.; Ford, C. L.; and Brown, L. S., *Long-Time Study of Cement Performance in Concrete,* Research Department Bulletin RX026, Portland Cement Association, 1948.

Melander, John M. and Isberner, Albert W. Jr., *Portland Cement Plaster (Stucco) Manual,* EB049, Portland Cement Association, 1996, 60 pages.

Nehdi, Moncef, "Ternary and Quaternary Cements for Sustainable Development," *Concrete International,* April 2001, pages 35 to 42.

NF P15-315, *Hydraulic Binders. High Alumina Melted Cement,* AFNOR Association française de normalisation, Saint-Denis La Plaine Cedex, France, 1991.

NIST, *Virtual Cement and Concrete Testing Laboratory,* http://vcctl.cbt.nist.gov, 2001.

Odler, Ivan, *Special Inorganic Cements,* E&FN Spon, New York, 2000, 420 pages.

PCA, *An Analysis of Selected Trace Metals in Cement and Kiln Dust,* SP109, Portland Cement Association, 1992, 60 pages.

PCA, *Emerging Technologies Symposium on Cements for the 21st Century,* SP206, Portland Cement Association, 1995, 140 pages.

PCA, "Portland Cement: Past and Present Characteristics," *Concrete Technology Today,* PL962, Portland Cement Association, http://www.portcement.org/pdf_files/PL962.pdf, July 1996, pages 1 to 4.

PCA, "Portland Cement, Concrete, and the Heat of Hydration," *Concrete Technology Today,* PL972, Portland Cement Association, http://www.portcement.org/pdf_files/PL972.pdf, July 1997, pages 1 to 4.

PCA, "What is White Cement?," *Concrete Technology Today,* PL991, Portland Cement Association, http://www.portcement.org/pdf_files/PL991.pdf, April 1999, 4 pages.

PCA, *U.S. Cement Industry Fact Sheet,* Portland Cement Association, 2002.

Perenchio, William F., and Klieger, Paul, *Further Laboratory Studies of Portland-Pozzolan Cements,* Research and Development Bulletin RD041, Portland Cement Association, http://www.portcement.org/pdf_files/RD041.pdf, 1976.

Pfeifer, Donald W., and Perenchio, W. F., *Reinforced Concrete Pipe Made with Expansive Cements,* Research and Development Bulletin RD015, Portland Cement Association, http://www.portcement.org/pdf_files/RD015.pdf, 1973.

Powers, T. C., *Some Physical Aspects of the Hydration of Portland Cement,* Research Department Bulletin RX126, Portland Cement Association, http://www.portcement.org/pdf_files/RX126.pdf, 1961.

Powers, T. C., *The Nonevaporable Water Content of Hardened Portland-Cement Paste–Its Significance for Concrete Research and Its Method of Determination,* Research Department Bulletin RX029, Portland Cement Association, http://www.portcement.org/pdf_files/RX029.pdf, 1949.

Powers, T. C., and Brownyard, T. L., *Studies of the Physical Properties of Hardened Portland Cement Paste,* Research Department Bulletin RX022, Portland Cement Association, http://www.portcement.org/pdf_files/RX022.pdf, 1947.

Russell, H. G., *Performance of Shrinkage-Compensating Concretes in Slabs,* Research and Development Bulletin RD057, Portland Cement Association, http://www.portcement.org/pdf_files/RD057.pdf, 1978.

Shkolnik, E., and Miller, F. M., "Differential Scanning Calorimetry for Determining the Volatility and Combustibility of Cement Raw Meal Organic Matter," *World Cement Research and Development,* May 1996, pages 81 to 87.

Snell, Luke M., and Snell, Billie G., *The Erie Canal—America's First Concrete Classroom,* http://www.sive.edu/~lsnell/erie.htm, 2000.

Stark, David, *Durability of Concrete in Sulfate-Rich Soils,* Research and Development Bulletin RD086, Portland Cement Association, http://www.portcement.org/pdf_files/RD086.pdf, 1989.

Stark, David, *Performance of Concrete in Sulfate Environments,* R&D Serial No. 2248, Portland Cement Association, 2002.

Tang, Fulvio J., *Optimization of Sulfate Form and Content,* Research and Development Bulletin RD105, Portland Cement Association, 1992, 44 pages.

Taylor, H. F. W., *Cement Chemistry,* Thomas Telford Publishing, London, 1997, 477 pages.

Tennis, Paul D., "Laboratory Notes: Thermal Analysis by TGA, DTA, and DSC," *Concrete Technology Today,* PL971, Portland Cement Association, http://www.portcement.org/pdf_files/PL971.pdf, April 1997, pages 6 and 7.

Tennis, Paul D., and Jennings, Hamlin M., "A Model for Two Types of Calcium Silicate Hydrate in the Microstructure of Portland Cement Pastes," *Cement and Concrete Research,* Pergamon, June 2000, pages 855 to 863.

Toler, H. R., *Flowability of Cement,* Research Department Report MP-106, Portland Cement Association, October, 1963.

Verbeck, G. J., *Field and Laboratory Studies of the Sulphate Resistance of Concrete,* Research Department Bulletin RX227, Portland Cement Association, http://www.portcement.org/pdf_files/RX227.pdf, 1967.

Washa, George W., and Wendt, Kurt F., "Fifty Year Properties of Concrete," *ACI Journal,* American Concrete Institute, Farmington Hills, Michigan, January 1975, pages 20 to 28.

Weaver, W. S.; Isabelle, H. L.; and Williamson, F., "A Study of Cement and Concrete Correlation," *Journal of Testing and Evaluation,* American Society for Testing and Materials, West Conshohocken, Pennsylvania, January 1974, pages 260 to 280.

White, Canvass, *Hydraulic Cement,* U. S. patent, 1820.

Wood, Sharon L., *Evaluation of the Long-Term Properties of Concrete,* Research and Development Bulletin RD102, Portland Cement Association, 1992, 99 pages.

CHAPTER 3

Fly Ash, Slag, Silica Fume, and Natural Pozzolans

Fig. 3-1. Supplementary cementing materials. From left to right, fly ash (Class C), metakaolin (calcined clay), silica fume, fly ash (Class F), slag, and calcined shale. (69794)

Fly ash, ground granulated blast-furnace slag, silica fume, and natural pozzolans, such as calcined shale, calcined clay or metakaolin, are materials that, when used in conjunction with portland or blended cement, contribute to the properties of the hardened concrete through hydraulic or pozzolanic activity or both. These materials are also generally catergorized as supplementary cementing materials (SCM's) or mineral admixtures (Fig. 3-1). A pozzolan is a siliceous or aluminosiliceous material that, in finely divided form and in the presence of moisture, chemically reacts with the calcium hydroxide released by the hydration of portland cement to form calcium silicate hydrate and other cementing compounds. The use of these materials in blended cements is discussed in Chapter 2 and by Detwiler, Bhatty, and Bhattacharja (1996).

The practice of using supplementary cementing materials in concrete mixtures has been growing in North America since the 1970s. There are similarities between many of these materials in that most are byproducts of other industrial processes; their judicious use is desirable not only from an environmental and energy conservation standpoint but also for the technical benefits they provide concrete.

Supplementary cementing materials are added to concrete as part of the total cementing system. They may be used in addition to or as a partial replacement of portland cement or blended cement in concrete, depending on the properties of the materials and the desired effect on concrete.

Supplementary cementing materials are used to improve a particular concrete property, such as the mitigation of deleterious alkali-aggregate reactivity. The optimum amount to use should be established by testing. Testing is necessary to determine (1) whether the material is indeed improving the property, and (2) the correct dosage rate. The correct dosage rate is important as an overdose can be harmful while an underdose may not achieve the desired effect. Supplementary cementing materials also react differently with different cements.

Traditionally, fly ash, slag, silica fume and natural pozzolans such as calcined clay and calcined shale were used in concrete individually. Today, due to improved access to these materials, concrete producers can combine two or more of these materials to optimize concrete properties. Mixtures using three cementing materials, called ternary mixtures, are becoming more common.

Supplementary cementing materials are covered under CSA Standard A23.5, *Supplementary Cementing Materials,* which forms part of the CSA A3000, *Cementitious Materials Compendium.* This Standard categorizes supplementary cementing materials into four types: Fly Ashes and Natural Pozzolans; Granulated Slag; Silica Fume; and Blended Supplementary Cementing Materials (ASTM C 618, ASTM C 989, ASTM C 1240). The chemical and physical requirements for supplementary cementing materials are given in Table 1 and Table 2 respectively in CSA A23.5. Optional test requirements that may be specified by the purchaser are given in *Appendix A* of the Standard.

Fly ashes and natural pozzolans are classified as follows:

Type N—natural pozzolans

Type F—fly ash normally produced from burning anthracite or bituminous coal and having a low calcium content (less than 8%).

Type C—Fly ash normally produced from burning lignite or sub-bituminous coal. This type of ash is further classified as follows:

- Type CI—Type C fly ash having an intermediate calcium content (8-20%)

- Type CH—Type C fly ash having a high calcium content (greater than 20%).

Granulated Slag. The slag covered by CSA A23.5 is designated as Type S. Granulated slag is the glassy, granular material formed when molten blast furnace slag is rapidly chilled. This is done by immersing the molten slag in water, by the pelletizing process, or by other methods that will ensure a high percentage of glass or vitrification.

Silica Fume. The type of silica fume covered by CSA A23.5 is designated as Type SF. Silica fume is the finely divided residue resulting from the production of silicon, ferro-silicon, or other silicon-containing alloys that is carried from the burning surface area of an electric-arc furnace by exhaust gases.

Blended Supplementary Cementing Materials. A blended supplementary cementing material is a mixture of any two of the following: fly ash, natural pozzolan, granulated slag, or silica fume. The nomenclature and naming practice for blended supplementary cementing materials as defined by CSA A23.5 is as follows:

P1A/P2B

where

P1 = the percentage of the predominate supplementary component

A = the predominate supplementary component

P2 = the percentage of the secondary supplementary component

B = the secondary supplementary component

The tolerance on these proportions is ± 2.5%, with the exception of Type SF silica fume which is controlled to within ± 1.5%. Where Type SF silica fume is used, the predominant supplementary component will comprise the balance.

Example 1:

80F/20SF indicates 80% Type F fly ash with 20% silica fume; the SF would be in the range of 18.5% to 21.5%, and the other component (F fly ash) would comprise the balance.

Example 2:

60S/40CH indicates 60% Type S slag with 40% Type CH fly ash; the Type S slag content could vary from 57.5% to 62.5%, and the Type CH fly ash content could vary from 37.5% to 42.5%.

When both components of the blend are fly ash, the blend must meet the requirements of CSA A23.5 based on the resultant CaO content. Blended products are produced by either intergrinding, blending, or a combination of both.

Fig. 3-2. Fly ash, a powder resembling cement, has been used in concrete since the 1930s. (69799)

FLY ASH

Fly ash, the most widely used supplementary cementing material in concrete, is a finely divided residue (a powder resembling cement) that results from the combustion of pulverized coal in electric power generating plants (Fig. 3-2). Upon ignition in the furnace, most of the volatile matter and carbon in the coal are burned off. During combustion, the coal's mineral impurities (such as clay, feldspar, quartz, and shale) fuse in suspension and are carried away from the combustion chamber by the exhaust gases. In the process, the fused material cools and solidifies into spherical glassy particles called fly ash (Fig. 3-3). The fly ash is then collected from the exhaust gases by electrostatic precipitators or bag filters.

Most of the fly ash particles are solid spheres and some are hollow cenospheres. Also present are plerospheres, which are spheres containing smaller spheres. Ground materials, such as portland cement, have solid angular particles. The particle sizes in fly ash vary from

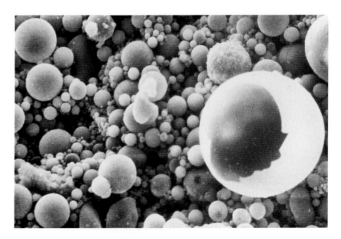

Fig. 3-3. Scanning electron microscope (SEM) micrograph of fly ash particles at 1000X. Although most fly ash spheres are solid, some particles, called cenospheres, are hollow (as shown in the micrograph). (54048)

Fig. 3-4. Fly ash, slag, and calcined clay or calcined shale are used in general purpose construction, such as (left to right) walls for residential buildings, pavements, high-rise towers, and dams. (67279, 48177, 69554, 69555)

less than 1 μm to more than 100 μm with the typical particle size measuring under 20 μm. Only 10% to 30% of the particles by mass are larger than 45 μm. The surface area is typically 300 to 500 m²/kg, although some fly ashes can have surface areas as low as 200 m²/kg and as high as 700 m²/kg. For fly ash without close compaction, the bulk density (mass per unit volume including air between particles) can vary from 540 to 860 kg/m³, whereas with close packed storage or vibration, the range can be 1120 to 1500 kg/m³.

Fly ash is primarily silicate glass containing silica, alumina, iron, and calcium. Minor constituents are magnesium, sulphur, sodium, potassium, and carbon. Crystalline compounds are present in small amounts. The relative density (specific gravity) of fly ash generally ranges between 1.9 and 2.8 and is generally tan or grey in colour.

CSA A23.5 Class F and Class C fly ashes are commonly used as pozzolanic admixtures for general purpose concrete (Fig. 3-4). Class F materials are low-calcium (less than 8% CaO) fly ashes with carbon contents less than 5%, but some may be as high as 10%. Class C materials have higher calcium contents than Class F ashes and are divided into two classifications: Type CI (intermediate calcium content) with a CaO content of 8% to 20%; and Type CH (high calcium content) with a CaO content greater than 20%. Class C ashes generally have carbon contents less than 2%. Many Class C ashes when exposed to water will hydrate and harden in less than 45 minutes. Some fly ashes meet both Class F and Class C classifications.

Class F fly ash is often used at dosages of 15% to 25% by mass of cementing material and Class C fly ash is used at dosages of 15% to 40% by mass of cementing material. However, when concrete is to be deicer-scaling resistant, the maximum amount of fly ash used should be 25% by mass of the cementing material unless testing of the concrete to confirm adequate durability indicates otherwise. Dosage varies with the reactivity of the ash and the desired effects on the concrete (Helmuth 1987 and ACI 232 1996).

SLAG

Ground granulated blast-furnace slag (Fig. 3-5), also called slag cement, is made from iron blast-furnace slag; it is a nonmetallic hydraulic cement consisting essentially of silicates and aluminosilicates of calcium developed in a molten condition simultaneously with iron in a blast furnace. The molten slag, at a temperature of about 1500°C, is rapidly chilled by quenching in water to form a glassy sandlike granulated material. The granulated material, which is ground to less than 45 microns, has a surface area fineness of about 400 to 600 m²/kg Blaine. The relative density (specific gravity) for ground granulated blastfurnace slag is in the range of 2.85 to 2.95. The bulk density varies from 1050 to 1375 kg/m³.

The rough and angular-shaped ground slag (Fig. 3-6) in the presence of water and an activator, NaOH or CaOH, both supplied by portland cement, hydrates and sets in a manner similar to portland cement. However, air-cooled slag does not have the hydraulic properties of water-cooled slag.

Granulated blast furnace slag was first developed in Germany in 1853 (Malhotra 1996). Ground slag has been used as a cementing material in concrete since the beginning of the 1900s (Abrams 1925). Ground granulated blast-furnace slag, when used in general purpose concrete in North

Fig. 3-5. Ground granulated blast-furnace slag. (69800)

Fig. 3-6. Scanning electron microscope micrograph of slag particles at 2100X. (69541)

Fig. 3-8. Scanning electron microscope micrograph of silica-fume particles at 5000X. (54095)

America, commonly constitutes between 30% and 45% of the cementing material in the mix (Fig. 3-4) (PCA 2000). Some slag concretes have a slag component of 70% or more of the cementing material. For concrete that is to be deicer-scaling resistant, the maximum amount of slag used should not exceed 50% by mass of the cementing material. In any case, testing of the combination of materials being used is essential to confirm adequate durability of the concrete. Granulated slag must meet the requirements of CSA Standard A23.5 (ASTM C 989). ACI 233R-95, *Ground Granulated Blast Furnace Slag as a Cementitious Constituent in Concrete,* provides an extensive review of slag.

SILICA FUME

Silica fume, also referred to as microsilica or condensed silica fume, is a byproduct material that is used as a pozzolan (Fig. 3-7). This byproduct is a result of the reduction of high-purity quartz with coal in an electric arc furnace in the manufacture of silicon or ferrosilicon alloy. Silica fume rises as an oxidized vapor from the 2000°C furnaces. When

it cools it condenses and is collected in huge cloth bags. The condensed silica fume is then processed to remove impurities and to control particle size.

Condensed silica fume is essentially silicon dioxide (usually more than 85%) in noncrystalline (amorphorous) form. Since it is an airborne material like fly ash, it has a spherical shape (Fig. 3-8). It is extremely fine with particles less than 1 μm in diameter and with an average diameter of about 0.1 μm, about 100 times smaller than average cement particles.

Condensed silica fume has a surface area of about 20,000 m²/kg (nitrogen adsorption method). For comparison, tobacco smoke's surface area is about 10,000 m²/kg. Type 10 and Type 30 cements have surface areas of about 300 to 400 m²/kg and 500 to 600 m²/kg (Blaine), respectively.

The relative density of silica fume is generally in the range of 2.20 to 2.25, but can be as high as 2.5. Portland cement has a relative density of about 3.15. The bulk density of silica fume varies from 130 to 430 kg/m³. Silica fume is sold in powder form but is more commonly available in a liquid. Silica fume is used in amounts between 5% and 10% by mass of the total cementing material. It is used in applications where a high degree of impermeability is needed (Fig. 3-9) and in high-strength concrete. In cases where the concrete must be deicer-scaling resistant, the maximum amount of silica fume used should be 10% by mass of cementing material. Silica fume must meet CSA Standard A23.5 (ASTM C 1240). ACI 234R-96, *Guide for the Use of Silica Fume in Concrete,* and SFA (2000) provides an extensive review of silica fume.

NATURAL POZZOLANS

Natural pozzolans have been used for centuries. The term "pozzolan" comes from a volcanic ash mined at Pozzuoli, a village near Naples, Italy, following the 79 AD eruption

Fig. 3-7. Silica fume powder. (69801)

Fig. 3-9. Although they can be used in general construction, silica fume and metakaolin are often used in applications such as (left) bridges and (right) parking garages to minimize chloride penetration into concrete. (68681, 69542)

of Mount Vesuvius. However, the use of volcanic ash and calcined clay dates back to 2000 BC and earlier in other cultures. Many of the Roman, Greek, Indian, and Egyptian pozzolan concrete structures can still be seen today, attesting to the durability of these materials.

The North American experience with natural pozzolans dates back to early 20th century public works projects, such as dams, where they were used to control temperature rise in mass concrete and provide cementing material. In addition to controlling heat rise, natural pozzolans were used to improve resistance to sulphate attack and were among the first materials to be found to mitigate alkali-silica reaction.

The most common natural pozzolans used today are the processed materials, which are heat treated in a kiln and then ground to a fine powder (Figs. 3-10, 3-11, and 3-12); they include calcined clay, calcined shale, and metakaolin.

Calcined clays are used in general purpose concrete construction much the same as other pozzolans (Fig. 3-4). They can be used as a partial replacement for the cement, typically in the range of 15% to 35%, and can be used to enhance resistance to sulphate attack, control alkali-silica reactivity, and reduce permeability.

Calcined shale may contain on the order of 5% to 10% calcium, which results in the material having some cementing or hydraulic properties on its own. Because of the amount of residual calcite that is not fully calcined, and the bound water molecules in the clay minerals, calcined shale will have a loss on ignition (LOI) of perhaps 1% to 5%. The LOI value for calcined shale is not a measure or indication of carbon content as would be the case in fly ash.

Metakaolin, a special calcined clay, is produced by low-temperature calcination of high purity kaolin clay. The product is ground to an average particle size of about 1 to 2 μm. Metakaolin is used in special applications where very low permeability or very high strength is required. In these applications, metakaolin is used more as an additive to the concrete rather than a replacement of cement; typical additions are around 10% of the cement mass.

Natural pozzolans are classified by CSA A23.5 as Type N pozzolans (ASTM C 618). ACI 232.1-94, *Use of Natural*

Fig. 3-10. Scanning electron microscope micrograph of calcined shale particles at 5000X. (69543)

Fig. 3-11. Metakaolin, a calcined clay. (69803)

Fig. 3-12. Scanning electron microscope micrograph of calcined clay particles at 2000X. (69544)

Pozzolans in Concrete, provides an extensive review of natural pozzolans. Table 3-1 illustrates typical chemical analysis and selected properties of pozzolans.

EFFECTS ON FRESHLY MIXED CONCRETE

This section provides a brief understanding of the freshly mixed concrete properties that supplementary cementing materials affect and their degree of influence. First it should be noted that these materials vary considerably in their effect on concrete mixtures. The attributes of these materials when added separately to a concrete mixture can

also be found in blended cements using supplementary cementing materials.

Water Requirements

Concrete mixtures containing fly ash generally require less water (about 1% to 10% less at normal dosages) for a given slump than concrete containing only portland cement. Higher dosages can result in greater water reduction. However, some fly ashes can increase water demand up to 5% (Gebler and Klieger 1986). Fly ash reduces water demand in a manner similar to liquid chemical water reducers (Helmuth 1987). Ground slag usually decreases water demand by 1% to 10%, depending on dosage.

The water demand of concrete containing silica fume increases with increasing amounts of silica fume, unless a water reducer or plasticizer is used. Some lean mixes may not experience an increase in water demand when only a small amount (less than 5%) of silica fume is present.

Calcined clays, calcined shales, and metakaolin generally have little effect on water demand at normal dosages; however, other natural pozzolans can significantly increase or decrease water demand.

Workability

Fly ash, slag, and calcined clay and shale generally improve the workability of concretes of equal slump. Silica fume may contribute to stickiness of a concrete mixture; adjustments, including the use of high-range water reducers, may be required to maintain workability and permit proper compaction and finishing.

Table 3-1. Chemical Analysis and Selected Properties of Typical Fly Ash, Slag, Silica Fume, Calcined Clay, Calcined Shale, and Metakaolin

	Class F fly ash	Class CH fly ash	Ground slag	Silica fume	Calcined clay	Calcined shale	Metakaolin
SiO_2, %	52	35	35	90	58	50	53
Al_2O_3, %	23	18	12	0.4	29	20	43
Fe_2O_3, %	11	6	1	0.4	4	8	0.5
CaO, %	5	21	40	1.6	1	8	0.1
SO_3, %	0.8	4.1	9	0.4	0.5	0.4	0.1
Na_2O, %	1.0	5.8	0.3	0.5	0.2	—	0.05
K_2O, %	2.0	0.7	0.4	2.2	2	—	0.4
Total Na eq. alk, %	2.2	6.3	0.6	1.9	1.5	—	0.3
Loss on ignition, %	2.8	0.5	1.0	3.0	1.5	3.0	0.7
Blaine fineness, m^2/kg	420	420	400	20,000	990	730	19,000
Relative density	2.38	2.65	2.94	2.40	2.5	2.63	2.50

Bleeding and Segregation

Concretes using fly ash generally exhibit less bleeding and segregation than plain concretes (Table 3-2). This effect makes the use of fly ash particularly valuable in concrete mixtures made with aggregates that are deficient in fines. The reduction in bleed water is primarily due to the reduced water demand in fly ash concretes. Gebler and Klieger (1986) correlate reduced bleeding of concrete to the reduced water requirement of fly ash mortar.

Concretes containing ground slags of fineness equal to that of the cement tend to show an increased rate and amount of bleeding compared to that of plain concretes, but this appears to have no adverse effect on segregation. Slags ground finer than cement reduce bleeding.

Silica fume is very effective in reducing both bleeding and segregation; as a result, higher slumps may be used. Calcined clays, calcined shales, and metakaolin have little effect on bleeding.

Table 3-2. Effect of Fly Ash on Bleeding of Concrete (ASTM C 232)*

Fly ash mixtures		Bleeding	
Identification	Class of fly ash	Percent	mL/cm²**
A	C	0.22	0.007
B	F	1.11	0.036
C	F	1.61	0.053
D	F	1.88	0.067
E	F	1.18	0.035
F	C	0.13	0.004
G	C	0.89	0.028
H	F	0.58	0.022
I	C	0.12	0.004
J	F	1.48	0.051
Average of: Class C		0.34	0.011
Class F		1.31	0.044
Control mixture		1.75	0.059

* All mixtures had cementing materials contents of 307 kg/m³, a slump of 75 ± 25 mm, and an air content of 6 ± 1%. Fly ash mixtures contain 25% ash by mass of cementing material.
** Volume of bleed water per surface area.
  Gebler and Klieger 1986.

Air Content

The amount of air-entraining admixture required to obtain a specified air content is normally greater when fly ash is used. Class C ash requires less air-entraining admixture than Class F ash and tends to lose less air during mixing (Table 3-3). Ground slags have variable effects on the required dosage rate of air-entraining admixtures. Silica fume has a marked influence on the air-entraining admixture requirement, which in most cases rapidly increases with an increase in the amount of silica fume used in the concrete. The inclusion of both fly ash and silica fume in non-air-entrained concrete will generally reduce the amount of entrapped air.

The amount of air-entraining admixture required for a certain air content in the concrete is a function of the fineness, carbon content, alkali content, organic material content, loss on ignition, and presence of impurities in the fly ash. Increases in alkalies decrease air-entraining agent dosages, while increases in the other properties increase dosage requirements. The Foam Index test provides an indication of the required dosage of air-entraining admixture for various fly ashes (Gebler and Klieger 1983).

The air-entraining dosage and air retention characteristics of concretes containing ground slag or natural pozzolans are similar to mixtures made only with portland cement.

Table 3-3. Effect of Fly Ash on Air-Entraining Admixture Dosage and Air Retention

Fly ash mixtures		Percent of air-entraining admixture relative to control	Air content, %			
			Minutes after initial mixing			
Identification	Class of fly ash		0	30	60	90
A	C	126	7.2	6.0	6.0	5.8
B	F	209	5.3	4.1	3.4	3.1
C	F	553	7.0	4.7	3.8	2.9
D	F	239	6.6	5.4	4.2	4.1
E	F	190	5.6	4.6	4.3	3.8
F	C	173	6.8	6.5	6.3	6.4
G	C	158	5.5	4.8	4.5	4.2
H	F	170	7.6	6.9	6.5	6.6
I	C	149	6.6	6.5	6.5	6.8
J	F	434	5.5	4.2	3.8	3.4
Control mixture		100	6.6	6.0	5.6	5.3

Concretes had a cementing materials content of 307 kg/m³ and slump of 75 ± 25 mm. Fly ash mixtures contained 25% ash by mass of cementing material (Gebler and Klieger 1983).

Heat of Hydration

Fly ash, natural pozzolans, and ground slag have a

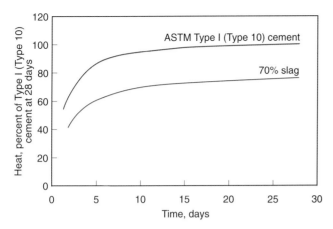

Fig. 3-13. Effect of a slag on heat of hydration at 20°C compared to a Type I (Type 10) cement.

lower heat of hydration than portland cement; consequently their use will reduce the amount of heat built up in a concrete structure (Fig. 3-13). Some pozzolans have a heat of hydration of only 40% that of cement. This reduction in temperature rise is especially beneficial in concrete used for massive structures. Silica fume may or may not reduce the heat of hydration. Detwiler and others (1996) provide a review of the effect of pozzolans and slags on heat generation.

Setting Time

The use of fly ash and ground granulated blast-furnace slag will generally retard the setting time of concrete (Table 3-4). The degree of set retardation depends on factors such as the amount of portland cement, water requirement, the

type and reactivity of the slag or pozzolan dosage, and the temperature of the concrete. Set retardation is an advantage during hot weather, allowing more time to place and finish the concrete. However, during cold weather, pronounced retardation can occur with some materials, significantly delaying finishing operations. Accelerating admixtures can be used to decrease the setting time. Calcined shale and clay have little effect on setting time.

Finishability

Concrete containing supplementary cementing materials will generally have equal or improved finishability compared to similar concrete mixes without them. Mixes that contain high dosages of cementing materials—and especially silica fume—can be sticky and difficult to finish.

Pumpability

The use of supplementary cementing materials generally aids the pumpability of concrete. Silica fume is the most effective, especially in lean mixtures.

Plastic Shrinkage Cracking

Because of its low bleeding characteristics, concrete containing silica fume may exhibit an increase in plastic shrinkage cracking. The problem may be avoided by ensuring that such concrete is protected against drying, both during and after finishing. Other pozzolans and slag generally have little effect on plastic shrinkage cracking. Supplementary cementing materials that significantly increase set time can increase the risk on plastic shrinkage cracking.

Table 3-4. Effect of Fly Ash on Setting Time of Concrete

Fly ash test mixtures		Setting time, hr:min		Retardation relative to control, hr:min	
Identification	Class of fly ash	Initial	Final	Initial	Final
A	C	4:30	5:35	0:15	0:05
B	F	4:40	6:15	0:25	0:45
C	F	4:25	6:15	0:10	0:45
D	F	5:05	7:15	0:50	1:45
E	F	4:25	5:50	0:10	0:20
F	C	4:25	6:00	0:10	0:30
G	C	4:55	6:30	0:40	1:00
H	F	5:10	7:10	0:55	1:40
I	C	5:00	6:50	0:45	1:20
J	F	5:10	7:40	0:55	2:10
<u>Average of:</u> Class C Class F		 4:40 4:50	 6:15 6:45	 0:30 0:35	 0:45 1:15
Control mixture		4:15	5:30	—	—

Concretes had a cementing materials content of 307 kg/m³. Fly ash mixtures contain 25% ash by mass of cementing material. Water to cement plus fly ash ratio = 0.40 to 0.45. Tested at 23°C. (Gebler and Klieger 1986.)

Curing

The effects of temperature and moisture conditions on setting properties and strength development of concretes containing supplementary cementing materials are similar to the effects on concrete made with only portland cement; however, the curing time may need to be longer for certain materials with slow-early-strength gain.

High dosages of silica fume can make concrete highly cohesive with very little aggregate segregation or bleeding. With little or no bleed water available at the concrete surface for evaporation, plastic cracking can readily develop, especially on hot, windy days if special precautions are not taken.

Proper curing of all concrete, especially concrete containing supplementary cementing materials, should commence immediately after finishing. A seven-day moist cure or membrane cure should be adequate for concretes with normal dosages of most supplementary cementing materials. As with portland cement concrete, low curing temperatures can reduce early-strength gain (Gebler and Klieger 1986).

EFFECTS ON HARDENED CONCRETE

Strength

Fly ash, ground granulated blast-furnace slag, calcined clay, metakaolin, calcined shale, and silica fume contribute to the strength gain of concrete. However, the strength of concrete containing these materials can be higher or lower than the strength of concrete using portland cement as the only cementing material. Fig. 3-14 illustrates this for various fly

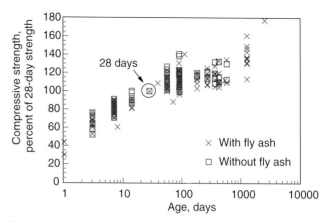

Fig. 3-15. Compressive strength gain as a percentage of 28-day strength of concretes with and without fly ash (Lange 1994).

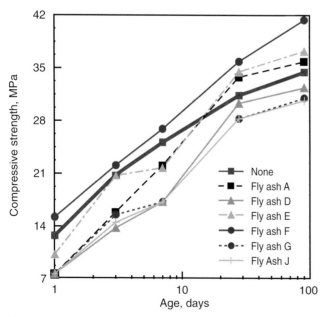

Fig. 3-14. Compressive strength development at 1, 3, 7, 28, and 90 days of concrete mixtures containing 307 kg/m³ of cementing materials with a fly ash dosage of 25% of the cementing materials (Whiting 1989).

ashes. Tensile, flexural, torsional, and bond strength are affected in the same manner as compressive strength.

Because of the slow pozzolanic reaction of some supplementary cementing materials, continuous wet curing and favorable curing temperatures may need to be provided for longer periods than normally required. However, concrete containing silica fume is less affected by this and generally equals or exceeds the one-day strength of a cement-only control mixture. Silica fume contributes to strength development primarily between 3 and 28 days, during which time a silica fume concrete exceeds the strength of a cement-only control concrete. Silica fume also aids the early strength gain of fly ash-cement concretes.

The strength development of concrete with fly ash, ground slag, calcined clay, or calcined shale, is similar to normal concrete when cured around 23°C. Figure 3-15 illustrates that the rate of strength gain of concrete with fly ash, relative to its 28-day strength, is similar to concrete without fly ash. Concretes made with certain highly reactive fly ashes (especially high-calcium Class CH ashes) or ground slags can equal or exceed the control strength in 1 to 28 days. Depending on the mix proportions, some fly ashes and natural pozzolans may require up to 90 days to exceed a 28-day control strength. Concretes containing Class C ashes generally develop higher early-age strength than concretes with Class F ashes.

Strength gain can be increased by: (1) increasing the amount of cementing material in the concrete; (2) adding high-early strength cementing materials; (3) decreasing the water-cementing materials ratio; (4) increasing the curing temperature; or (5) using an accelerating admixture. Figure 3-16 illustrates the benefit of using fly ash as an addition instead of a cement replacement to improve strength development in cold weather. Mass concrete design often takes advantage of the delayed strength gain of pozzolans, as these structures are often not put into full service immediately. Slow early strength gain resulting from the use of some supplementary cementing materials is an advantage in hot weather construction as it allows

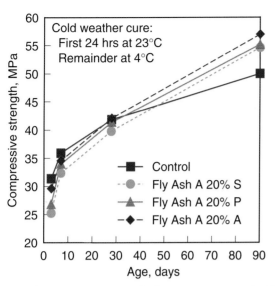

Fig. 3-16. Compressive strengths for concretes cured at 23°C for the first 24 hours and 4°C for the remaining time. Control had a cement content of 332 kg/m³ and water-to-cement ratio of 0.45. The fly ash curves show substitution for cement (S), partial (equal) substitution for cement and sand (P), and addition of fly ash by mass of cement (A). The use of partial cement substitution or addition of fly ash increases strength development comparable to the cement-only control, even in cold weather (Detwiler 2000).

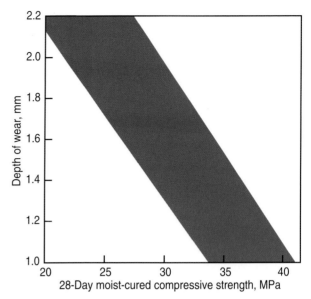

Fig. 3-17. Comparison of abrasion resistance and compressive strength of various concretes with 25% fly ash. Abrasion resistance increases with strength (Gebler and Klieger 1986).

more time to place and finish the concrete. With appropriate mixture adjustments, all supplementary cementing materials can be used in all seasons.

Supplementary cementing materials are often essential to the production of high-strength concrete. Fly ash, especially, has been used in production of concrete with strengths up to 100 MPa. With silica fume, ready mix producers now have the ability to make concrete with strengths up to 140 MPa, when used with high-range water reducers and appropriate aggregates (Burg and Ost 1994).

Impact and Abrasion Resistance

The impact resistance and abrasion resistance of concrete are related to compressive strength and aggregate type. Supplementary cementing materials generally do not affect these properties beyond their influence on strength. Concretes containing fly ash are just as abrasion resistant as portland cement concretes without fly ash (Gebler and Klieger 1986). Fig. 3-17 illustrates that abrasion resistance of fly ash concrete is related to strength.

Freeze-Thaw Resistance

It is imperative for development of resistance to deterioration from cycles of freezing and thawing that concrete has adequate strength and entrained air. For concrete containing supplementary cementing materials to provide the same resistance to freezing and thawing cycles as a con-

crete made using only portland cement as a binder, four conditions for both concretes must be met:

1. They must have the same compressive strength.
2. They must have an adequate entrained air content with proper air-void characteristics.
3. They must be properly cured.
4. They must be air-dried for one month prior to exposure to saturated freezing conditions.

Table 3-5 shows equal frost resistance for concrete with and without fly ash. Fig. 3-18 illustrates the long term durability of concretes with fly ash or slag.

Deicer-Scaling Resistance

Decades of field experience have demonstrated that air-entrained concretes containing normal dosages of fly ash, slag, silica fume, calcined clay, or calcined shale are resistant to scaling caused by the application of deicing salts in a freeze-thaw environment. Laboratory tests also indicate that the deicer-scaling resistance of concrete made without supplementary cementing materials is often equal to concrete made with supplementary cementing materials.

Scaling resistance can decrease as the amount of certain supplementary cementing materials increases. However, concretes that are properly designed, placed, and cured, have demonstrated good scaling resistance even when made with high dosages of some of these materials.

Deicer-scaling resistance of all concrete is significantly improved with the use of a low water-cementing materials ratio, a moderate portland cement content, adequate air entrainment, proper finishing and curing, and an air-drying period prior to exposure of the concrete to salts and freezing temperatures. Lean concrete with about 240 kg or less of

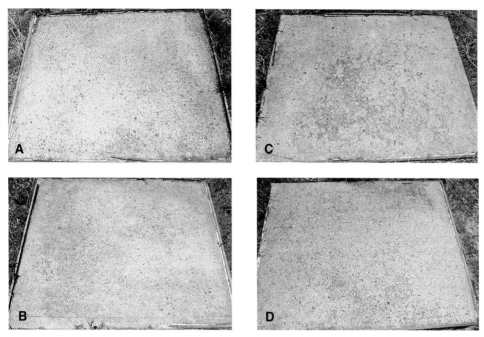

Fig. 3-18. View of concrete slabs containing (A) fly ash, (B) slag, (C) calcined shale, and (D) portland cement after 30 years of deicer exposure. These samples demonstrate the durability of concretes containing fly ash or slag. Source: RX 157, LTS Cement No. 51, slabs with 335 kg/m³ of cementing material and air entrainment. (69714, 69716, 69715, 69717)

cementing material per cubic metre of concrete can be especially vulnerable to deicer scaling. A minimum of 335 kg of cementing material per cubic metre and a maximum water to cementing materials ratio of 0.45 is recommended for non-reinforced concrete and 0.40 for reinforced concrete. A satisfactory air-void system is also critical.

The importance of using a low water-cementing materials ratio for scale resistance is demonstrated in Fig. 3-19. The effect of high fly ash dosages and low cementing material contents is demonstrated in Fig. 3-20 The performance of scale-resistant concretes containing fly ash at a dosage of 25% of cementing material by mass is presented in Table 3-5. The table demonstrates that

Table 3-5. Frost and Deicer Scaling Resistance of Fly Ash Concrete

| Fly ash mixtures* | | Results at 300 cycles | | | | |
| | | Frost resistance in water, ASTM C 666 Method A | | | Deicer scaling resistance, ASTM C 672** | |
Identification	Class of fly ash	Expansion, %	Mass loss, %	Durability factor	Water cure	Curing compound
A	C	0.010	1.8	105	3	2
B	F	0.001	1.2	107	2	2
C	F	0.005	1.0	104	3	3
D	F	0.006	1.3	98	3	3
E	F	0.003	4.8	99	3	2
F	C	0.004	1.8	99	2	2
G	C	0.008	1.0	102	2	2
H	F	0.006	1.2	104	3	2
I	C	0.004	1.7	99	3	2
J	F	0.004	1.0	100	3	2
Average of: Class C		0.006	1.6	101	3	2
Class F		0.004	1.8	102	3	2
Control mixture		0.002	2.5	101	2	2

* Concrete mixtures had a cementing materials content of 307 kg/m³, a water to cementing materials ratio of 0.40 to 0.45, a target air content of 5% to 7%, and a slump of 75 mm to 100 mm. Fly ash dosage was 25% by mass of cementing material. Gebler and Klieger 1986.

** Scale rating (see at right)

    0 = No scaling
    1 = Slight scaling
    2 = Slight to moderate scaling
    3 = Moderate scaling
    4 = Moderate to severe scaling
    5 = Severe scaling

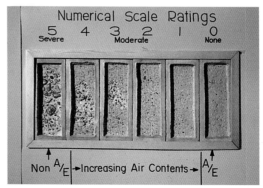

(2718)

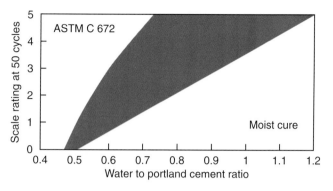

Fig. 3-19. Relationship between deicer-scaling resistance and water to portland cement ratio for several air-entrained concretes with and without fly ash. A scale rating of 0 is no scaling and 5 is severe scaling (Whiting 1989).

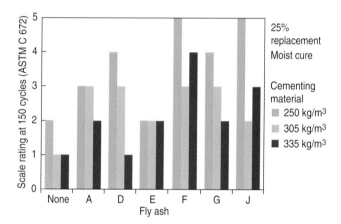

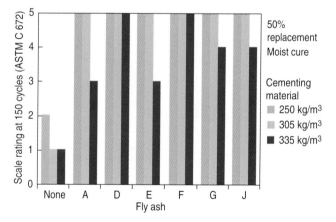

Fig. 3-20. Relationship between deicer-scaling resistance and dosage of fly ash for air-entrained concretes made with moderate to high water-cementing materials ratios. Replacement of portland cement with fly ash: (top) 25% and (bottom) 50%. A scale rating of 0 is no scaling and 5 is severe scaling (Whiting 1989).

well designed, placed and cured concretes with and without fly ash can be equally resistant to deicer scaling.

The ACI 318 building code states that the maximum dosage of fly ash, slag, and silica fume should be 25%, 50%, and 10% by mass of cementing materials, respectively for deicer exposures. Higher dosages of supplementary cementing materials can be used, if adequate durability is demonstrated by laboratory or field performance. The total supplementary cementing materials should not exceed 50% of the cementing material. The selection of materials and dosages should be based on local experience and the durability should be demonstrated by field or laboratory performance.

Drying Shrinkage and Creep

When used in low to moderate amounts, the effect of fly ash, ground granulated blast-furnace slag, calcined clay, calcined shale, and silica fume on the drying shrinkage and creep of concrete is generally small and of little practical significance. Some studies indicate that silica fume may reduce creep (Burg and Ost 1994).

Permeability and Absorption

With adequate curing, fly ash, ground slag, and natural pozzolans generally reduce the permeability and absorption of concrete. Silica fume and metakaolin are especially effective in this regard. Tests show that the permeability of concrete decreases as the quantity of hydrated cementing materials increases and the water-cementing materials ratio decreases. The absorption of fly ash concrete is about the same as concrete without ash, although some ashes can reduce absorption by 20% or more.

Alkali-Aggregate Reactivity

Alkali-silica reactivity can be controlled through the use of certain supplementary cementing materials. Silica fume, fly ash, ground granulated blast-furnace slag, calcined clay, calcined shale, and other pozzolans have been reported to significantly reduce alkali-silica reactivity (Figs. 3-21 and 3-22). Low calcium Class F ashes have reduced reactivity expansion up to 70% or more in some cases. At optimum dosage, Class C ashes can also reduce reactivity but to a lesser degree than most Class F ashes. Supplementary cementing materials provide additional calcium silicate hydrate to chemically tie up the alkalies in the concrete (Bhatty 1985, and Bhatty and Greening 1986). Determination of optimum supplementary cementing materials dosage is important to maximize the reduction in reactivity and to avoid dosages and materials that can aggravate reactivity. Dosage rates should be verified by tests, such as CSA A23.2-25A, *Detection of Alkali-Silica Reactive Aggregate by Accelerated Expansion of Mortar Bars* (ASTM C 1260) and CSA A23.2-14A, *Potential Expansivity of Aggregates (concrete prisms)*(ASTM C 1293).

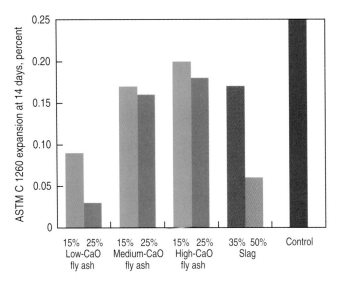

Fig. 3-21. Effect of different fly ashes and slag on alkali-silica reactivity. Note that some ashes are more effective than others in controlling the reaction and that dosage of the ash or slag is critical. A highly reactive natural aggregate was used in this test. A less reactive aggregate would require less ash or slag to control the reaction. A common limit for evaluating the effectiveness of pozzolans or slags is 0.10% expansion using this rapid mortar bar test (Detwiler 2002).

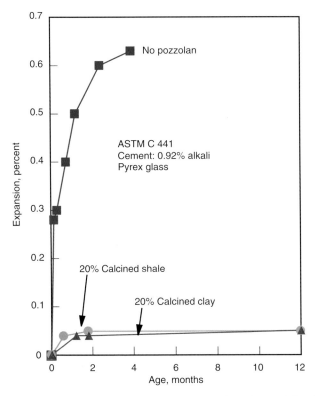

Fig. 3-22. Reduction of alkali-silica reactivity by calcined clay and calcined shale (Lerch 1950).

Supplementary cementing materials that reduce alkali-silica reactions will not reduce alkali-carbonate reactions, a type of reaction involving cement alkalies and certain dolomitic limestones.

Descriptions of aggregate testing and preventive measures to be taken to prevent deleterious alkali-aggregate reaction are discussed in: CSA A23.1 *Appendix B— Alkali-Aggregate Reaction;* CSA A23.2-27A, *Standard Practice to Identify Degree of Alkali-Reactivity of Aggregates and to Identify Measures to Avoid Deleterious Expansion in Concrete;* Farny and Kosmatka (1997), and PCA (1998).

Sulphate Resistance

With proper proportioning and material selection, silica fume, fly ash, calcined shale, and ground slag can improve the resistance of concrete to sulphate or seawater attack. This is done primarily by reducing permeability and by reducing the amount of reactive elements (such as calcium) needed for expansive sulphate reactions. For improved sulphate resistance of lean concrete, one study showed that for a particular Class F ash, an adequate amount was approximately 20% of the cement plus fly ash; this illustrates the need to determine optimum ash contents, as higher ash contents were detrimental (Stark 1989).

The sulphate resistance of high-cement-content, low water-cement ratio concrete made with a sulphate resistant cement is so great that fly ash has little opportunity to improve resistance. Concretes with Class F ashes are generally more sulphate resistant than those made with Class C ashes. Some Class C ashes have been shown to reduce sulphate resistance at normal dosage rates.

Ground slag is generally considered beneficial in sulphate environments. However, one long-term study in a very severe environment showed a slight reduction in sulphate resistance in concrete containing ground slag compared to concrete containing only portland cement as the cementing material (Stark 1986 and 1989). One reason for decreased performance with slag in this study is that the mixtures may not have been optimized for sulphate resistance.

Other studies indicate that concrete with ground slag has a sulphate resistance equal to or greater than concrete made with Type 50 Sulphate-resistant portland cement (ACI 233 and Detwiler, Bhatty, and Bhattacharja 1996). Calcined clay has been demonstrated to provide sulphate resistance greater than high sulphate-resistant Type 50 cement (Barger and others 1997).

Corrosion of Embedded Steel

Some supplementary cementing materials reduce steel corrosion by reducing the permeability of properly cured concrete to water, air, and chloride ions. Fly ash can significantly reduce chloride-ion ingress. Silica fume greatly decreases permeability and chloride-ion ingress and also significantly increases electrical resistivity, thereby reducing the electrochemical reaction of corrosion. Concrete containing silica fume or metakaolin is often used in overlays and full-depth slab placements on bridges and parking garages; these structures are particularly vulnerable to corrosion due to chloride-ion ingress.

Carbonation

Carbonation of concrete is a process by which carbon dioxide from the air penetrates the concrete and reacts with the hydroxides, such as calcium hydroxide, to form carbonates. In the reaction with calcium hydroxide, calcium carbonate is formed. Carbonation lowers the alkalinity of concrete. High alkalinity is needed to protect embedded steel from corrosion; consequently, concrete should be resistant to carbonation to help prevent steel corrosion.

The amount of carbonation is significantly increased in concretes with a high water-cementing materials ratio, low cement content, short curing period, low strength, and highly permeable or porous paste. The depth of carbonation of good quality concrete is generally of little practical significance. At normal dosages, fly ash is reported to slightly increase carbonation, but usually not to a significant amount in concrete with short (normal) moist-curing periods (Campbell, Sturm, and Kosmatka 1991).

Chemical Resistance

Supplementary cementing materials often reduce chemical attack by reducing the permeability of concrete. Although many of these materials may improve chemical resistance, they do not make concrete totally immune to attack. Concrete in severe chemical exposure should be protected using barrier systems. Kerkhoff (2001) contains a good discussion of methods and materials to protect concrete from aggressive chemicals and exposures.

Soundness

Normal dosages of fly ash, slag, silica fume and natural pozzolans do not affect the soundness of concrete. Concrete soundness is protected by soundness requirements, such as autoclave expansion limits, on the materials. Dosages in concrete should not exceed dosages deemed safe in the autoclave test.

Concrete Colour

Supplementary cementing materials may slightly alter the colour of hardened concrete. Colour effects are related to the colour and amount of the material used in concrete. Many supplementary cementing materials resemble the colour of portland cement and therefore have little effect on colour of the hardened concrete. Some silica fumes may give concrete a slightly bluish or dark gray tint and tan fly ash may impart a tan colour to concrete when used in large quantities. Ground slag and metakaolin can make concrete whiter. Ground slag can initially impart a bluish or greenish undertone.

CONCRETE MIX PROPORTIONS

The optimum amounts of supplementary cementing materials used with portland cement or blended cement are determined by testing, by the relative cost and availability of the materials, and by the specified properties of the concrete. When proportioning mixes with supplementary cementing materials, two important concepts have to be considered. One is the blend ratio (R) of supplementary cementing materials to portland cement (SCM/PC), and the other, the total amount of cementing materials needed to produce a concrete with the required properties in the plastic and hardened states (SCM+ PC).

The blend ratio R may be based on regional practice using locally available materials, on job specifications, or on the recommendation of the supplier of the supplementary cementing materials. The blend ratio should be selected to optimize the economy and durability of a particular concrete to meet particular requirements.

Several test mixtures are required to determine the optimum blend ratio R. These mixtures should cover a range of blends to establish the relationship between strength and cementing materials (CM), and between strength and water to cementing materials ratio (W/CM). As a guide to determining mix proportions, refer to Chapter 9 of this publication or to ACI Standards 211.1 and 211.2, taking into account the relative densities of the supplementary cementing materials. These are usually different from the relative density of portland cement. The results of these tests will be a family of strength curves for each age at which the concrete is required to meet certain specified requirements. The dosage of a cementing material is usually stated as a mass percentage of all the cementing material in a concrete mixture.

Typical practice in Canada is to use fly ash, slag, or silica fume, as an addition to portland cement or as a partial replacement for some of the portland cement. Blended cements, which already contain pozzolans or slag, are designed to be used with or without additional supplementary cementing materials.

Concrete mixtures with more than one supplementary cementing material are also used. For example, a concrete mixture may contain portland cement, fly ash, and silica fume. Such mixes are called ternary concretes. When fly ash, slag, silica fume, or natural pozzolans are used in combination with portland or blended cement, the proportioned concrete mixture should be tested to demonstrate that it meets the required concrete properties for the project.

AVAILABILITY

Supplementary cementing materials suitable for use in concrete are not readily available in all areas. Canadian sources of fly ash, for example, are only readily available in those Provinces that have coal burning thermal power generating plants. As all fly ash is not suitable as produced for use in concrete, a number of major centres must import the material from other parts of Canada or the United States. In general, Class F fly ashes are used in Eastern and Central Canada, while Class C ashes are used in Western Canada. Silica fume is available in most locations because only small dosages are used. Ground granulated slags are primarily available and used in Ontario.

STORAGE

In most cases, moisture will not affect the physical performance of supplementary cementing materials. These materials should, however, be kept dry to avoid difficulties in handling and discharge. Class C fly ash and calcined shale must be kept dry as they will set and harden when exposed to moisture. Equipment for handling and storing these materials is similar to that required for cement. Additional modifications may be required where using silica fume, which does not have the same flowing characteristics as other supplementary cementing materials (and may be supplied as a liquid).

These materials are usually kept in bulk storage facilities or silos, although some products are available in bags. Because the materials may resemble portland cement in colour and fineness, the storage facilities should be clearly marked to avoid the possibility of misuse and contamination with other materials at the batch plant. All valves and piping should also be clearly marked and properly sealed to avoid leakage and contamination. Valves sticking open can be the cause of overdosing of SCM's during the batching operation. To help avoid this potential problem, portland cement or blended cement should be measured ahead of any supplementary cementing materials in the batching sequence.

REFERENCES

Abrams, Duff A., "Effect of Hydrated Lime and Other Powdered Admixtures in Concrete," *Proceedings of the American Society for Testing Materials,* Vol. 20, Part 2, 1920. Reprinted with revisions as Bulletin 8, Structural Materials Research Laboratory, Lewis Institute, June 1925, 78 pages. Available through PCA as LS08.

ACAA, http://www.acaa-usa.org, American Coal Ash Association, Alexandria, Virginia, 2001.

ACI, *Concrete Durability, Katherine and Bryant Mather International Conference,* SP100, American Concrete Institute, Farmington Hills, Michigan, 1987.

ACI Committee 211, *Standard Practice for Selecting Proportions for Normal, Heavyweight, and Mass Concrete,* ACI 211.1-91, American Concrete Institute, Farmington Hills, Michigan, 1991, 38 pages.

ACI Committee 211, *Standard Practice for Selecting Proportions for Structural Lightweight Concrete,* ACI 211.2-98, American Concrete Institute, Farmington Hills, Michigan, 1998, 14 pages.

ACI Committee 211, *Guide for Selecting Proportions for High Strength Concrete with Portland Cement and Fly Ash,* ACI 211.4R-93, American Concrete Institute, Farmington Hills, Michigan, 1993, 13 pages.

ACI Committee 232, *Use of Fly Ash in Concrete,* ACI 232.2R-96, American Concrete Institute, Farmington Hills, Michigan, 1996, 34 pages.

ACI Committee 232, *Use of Raw or Processed Natural Pozzolans in Concrete,* ACI 232.1R-00, American Concrete Institute, Farmington Hills, Michigan, 2001, 24 pages.

ACI Committee 233, *Ground Granulated Blast-Furnace Slag as a Cementitious Constituent in Concrete,* ACI 233R-95, American Concrete Institute, Farmington Hills, Michigan, 1995, 18 pages.

ACI Committee 234, *Guide for the Use of Silica Fume in Concrete,* ACI 234R-96, American Concrete Institute, Farmington Hill, Michigan, 1996, 51 pages.

Barger, Gregory S.; Lukkarila, Mark R.; Martin, David L.; Lane, Steven B.; Hansen, Eric R.; Ross, Matt W.; and Thompson, Jimmie L., "Evaluation of a Blended Cement and a Mineral Admixture Containing Calcined Clay Natural Pozzolan for High-Performance Concrete," *Proceedings of the Sixth International Purdue Conference on Concrete Pavement Design and Materials for High Performance,* Purdue University, West Lafayette, Indiana, 1997, 21 pages.

Bhatty, M. S. Y., "Mechanism of Pozzolanic Reactions and Control of Alkali-Aggregate Expansion," *Cement, Concrete, and Aggregates,* American Society for Testing and Materials, West Conshohocken, Pennsylvania. Winter 1985, pages 69 to 77.

Bhatty, M. S. Y., and Greening, N. R., "Some Long Time Studies of Blended Cements with Emphasis on Alkali-Aggregate Reaction," *7th International Conference on Alkali-Aggregate Reaction,* 1986.

Buck, Alan D., and Mather, Katharine, *Methods for Controlling Effects of Alkali-Silica Reaction,* Technical Report SL-87-6, Waterways Experiment Station, U.S. Army Corps of Engineers, Vicksburg, Mississippi, 1987.

Burg, R. G., and Ost, B. W., *Engineering Properties of Commercially Available High-Strength Concretes (Including Three-Year Data),* Research and Development Bulletin RD104, Portland Cement Association, 1994, 62 pages.

Campbell, D. H.; Sturm, R. D.; and Kosmatka, S. H., "Detecting Carbonation," *Concrete Technology Today,* PL911; Portland Cement Association, http://www.portcement.org/pdf_files/PL911.pdf, March 1991, pages 1 to 5.

Canadian Standards Association, CSA Standard A23.5-98, *Supplementary Cementing Materials,* forming part of the CSA A3000-98 *Cementitious Materials Compendium,* Canadian Standards Association, Etobicoke, Ontario, 1998.

Canadian Standards Association, CSA Standard A23.1-00 /A23.2-00, *Concrete Materials and Methods of Concrete Construction/Methods of Test for Concrete,* Canadian Standards Association, Toronto, 2000.

Carette, G. G., and Malhotra, V. M., "Mechanical Properties, Durability, and Drying Shrinkage of Portland Cement Concrete Incorporating Silica Fume," *Cement, Concrete, and Aggregates,* American Society for Testing and Materials, West Conshohocken, Pennsylvania, Summer 1983.

Cohen, Menashi, D., and Bentur, Arnon, "Durability of Portland Cement-Silica Fume Pastes in Magnesium Sulfate and Sodium Sulfate Solutions," *ACI Materials Journal,* American Concrete Institute, Farmington Hills, Michigan, May-June 1988, pages 148 to 157.

Detwiler, Rachel J., "Controlling the Strength Gain of Fly Ash Concrete at Low Temperature," *Concrete Technology Today,* CT003, Portland Cement Association, http://www.portcement.org/pdf_files/CT003.pdf, 2000, pages 3 to 5.

Detwiler, Rachel J., *Documentation of Procedures for PCA's ASR Guide Specification,* SN 2407, Portland Cement Association, 2002.

Detwiler, Rachel J.; Bhatty, Javed I.; and Bhattacharja, Sankar, *Supplementary Cementing Materials for Use in Blended Cements,* Research and Development Bulletin RD112, Portland Cement Association, 1996, 108 pages.

Farny, J. A., and Kosmatka, S. H., *Diagnosis and Control of Alkali-Aggregate Reactivity,* IS413, Portland Cement Association, 1997, 24 pages.

Gaynor, R. D., and Mullarky, J. I., "Survey: Use of Fly Ash in Ready-Mixed Concrete," *NRMCA Technical Information Letter No. 426,* National Ready Mixed Concrete Association, Silver Spring, Maryland, June 1985.

Gebler, S. H., and Klieger, P., *Effect of Fly Ash on the Air-Void Stability of Concrete,* Research and Development Bulletin RD085, Portland Cement Association, http://www.portcement.org/pdf_files/RD085.pdf, 1983, 40 pages.

Gebler, Steven H., and Klieger, Paul, *Effect of Fly Ash on Some of the Physical Properties of Concrete,* Research and Development Bulletin RD089, Portland Cement Association, http://www.portcement.org/pdf_files/RD089.pdf, 1986, 48 pages.

Gebler, Steven H., and Klieger, Paul, *Effect of Fly Ash on the Durability of Air-Entrained Concrete,* Research and Development Bulletin RD090, Portland Cement Association, http://www.portcement.org/pdf_files/RD090.pdf, 1986, 44 pages.

Helmuth, Richard A., *Fly Ash in Cement and Concrete,* SP040T, Portland Cement Association, 1987, 203 pages.

Hogan, F. J., and Meusel, J. W., "Evaluation for Durability and Strength Development of a Ground Granulated Blast Furnace Slag," *Cement, Concrete, and Aggregates,* Vol. 3, No. 1, American Society for Testing and Materials, West Conshohocken, Pennsylvania, Summer 1981, pages 40 to 52.

Kerkhoff, Beatrix, *Effect of Substances on Concrete and Guide to Protective Treatments,* IS001, Portland Cement Association, 2001, 36 pages.

Klieger, Paul, and Isberner, Albert W., *Laboratory Studies of Blended Cements—Portland Blast-Furnace Slag Cements,* Research Department Bulletin RX218, Portland Cement Association, http://www.portcement.org/pdf_files/RX218.pdf, 1967.

Klieger, Paul, and Perenchio, William F., *Laboratory Studies of Blended Cement: Portland-Pozzolan Cements,* Research and Development Bulletin RD013, Portland Cement Association, http://www.portcement.org/pdf_files/RD013.pdf, 1972.

Malhotra, V. M., *Pozzolanic and Cementitious Materials,* Gordon and Breach Publishers, Amsterdam, 1996, 208 pages.

PCA, *Survey of Mineral Admixtures and Blended Cements in Ready Mixed Concrete,* Portland Cement Association, 2000, 16 pages.

PCA Durability Subcommittee, *Guide Specification to Control Alkali-Silica Reactions,* IS415, Portland Cement Association, 1998, 8 pages.

SFA, http://www.silicafume.org, Silica Fume Association, Fairfax, Virginia, 2000.

Stark, David, *Longtime Study of Concrete Durability in Sulfate Soils,* Research and Development Bulletin RD086, Portland Cement Association, http://www.portcement.org/pdf_files/RD086.pdf, 1982.

Stark, David, *Durability of Concrete in Sulfate-Rich Soils,* Research and Development Bulletin RD097, Portland Cement Association, http://www.portcement.org/pdf_files/RD097.pdf, 1989, 14 pages.

Tikalsky, P. J., and Carrasquillo, R. L., *Effect of Fly Ash on the Sulfate Resistance of Concrete Containing Fly Ash,* Research Report 481-1, Center for Transportation Research, Austin, Texas, 1988, 92 pages.

Tikalsky, P. J., and Carrasquillo, R. L., *The Effect of Fly Ash on the Sulfate Resistance of Concrete,* Research Report 481-5, Center for Transportation Research, Austin, Texas, 1989, 317 pages.

Whiting, D., *Strength and Durability of Residential Concretes Containing Fly Ash,* Research and Development Bulletin RD099, Portland Cement Association, http://www.portcement.org/pdf_files/RD099.pdf, 1989, 42 pages.

CHAPTER 4
Mixing Water for Concrete

Almost any natural water that is drinkable and has no pronounced taste or odor can be used as mixing water for making concrete (Fig. 4-1). However, some waters that are not fit for drinking may be suitable for use in concrete.

Six typical analyses of city water supplies and seawater are shown in Table 4-1. These waters approximate the composition of domestic water supplies for most of the cities over 20,000 population in the United States and Canada. Water from any of these sources is suitable for making concrete. A water source comparable in analysis to any of the waters in the table is probably satisfactory for use in concrete (Fig. 4-2).

Water of questionable suitability can be used for making concrete if mortar cubes (CSA A23.2-8A) made with it have 28-day strengths equal to at least 90% of companion specimens made with a known or acceptable water. In addition, it is recommended that ASTM C 191 tests should be made to ensure that impurities in the mixing water do not adversely shorten or extend the setting time of the cement. Acceptable criteria for water to be used in concrete are given in Tables 4-2 and 4-3.

Fig. 4-1. Water that is safe to drink is safe to use in concrete. (44181)

Excessive impurities in mixing water not only may affect setting time and concrete strength, but also may cause efflorescence, staining, corrosion of reinforcement, volume instability, and reduced durability. Therefore, certain optional limits may be set on chlorides, sulphates,

Table 4-1. Typical Analyses of City Water Supplies and Seawater, parts per million

Chemicals	Analysis No.						
	1	2	3	4	5	6	Seawater*
Silica (SiO_2)	2.4	0.0	6.5	9.4	22.0	3.0	—
Iron (Fe)	0.1	0.0	0.0	0.2	0.1	0.0	—
Calcium (Ca)	5.8	15.3	29.5	96.0	3.0	1.3	50 to 480
Magnesium (Mg)	1.4	5.5	7.6	27.0	2.4	0.3	260 to 1410
Sodium (Na)	1.7	16.1	2.3	183.0	215.0	1.4	2190 to 12,200
Potassium (K)	0.7	0.0	1.6	18.0	9.8	0.2	70 to 550
Bicarbonate (HCO_3)	14.0	35.8	122.0	334.0	549.0	4.1	—
Sulphate (SO_4)	9.7	59.9	5.3	121.0	11.0	2.6	580 to 2810
Chloride (Cl)	2.0	3.0	1.4	280.0	22.0	1.0	3960 to 20,000
Nitrate (NO_3)	0.5	0.0	1.6	0.2	0.5	0.0	—
Total dissolved solids	31.0	250.0	125.0	983.0	564.0	19.0	35,000

*Different seas contain different amounts of dissolved salts.

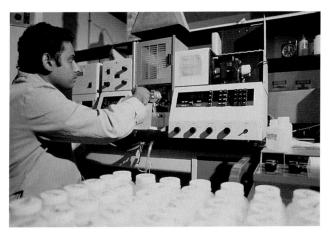

Fig. 4-2. Laboratory analysis of water, in this case using an atomic absorption spectrophotometer to detect concentration of elements. (44182)

(1960) and Abrams (1924). Over 100 different compounds and ions are discussed.

Following is a synopsis of the effects of certain impurities in mixing water on the quality of normal concrete:

ALKALI CARBONATE AND BICARBONATE

Carbonates and bicarbonates of sodium and potassium have different effects on the setting times of different cements. Sodium carbonate can cause very rapid setting, bicarbonates can either accelerate or retard the set. In large concentrations, these salts can materially reduce concrete strength. When the sum of the dissolved salts exceeds 1000 ppm, tests for their effect on setting time and 28-day strength should be made. The possibility of aggravated alkali-aggregate reactions should also be considered.

CHLORIDE

Concern over a high chloride content in mixing water is chiefly due to the possible adverse effect of chloride ions on the corrosion of reinforcing steel or prestressing strands. Chloride ions attack the protective oxide film formed on the steel by the highly alkaline (pH > 12.5) chemical environment present in concrete. The acid-soluble chloride ion level at which steel reinforcement corrosion begins in concrete is about 0.2% to 0.4% by mass of cement (0.15% to

alkalies, and solids in the mixing water or appropriate tests can be performed to determine the effect the impurity has on various properties. Some impurities may have little effect on strength and setting time, yet they can adversely affect durability and other properties.

Water containing less than 2000 parts per million (ppm) of total dissolved solids can generally be used satisfactorily for making concrete. Water containing more than 2000 ppm of dissolved solids should be tested for its effect on strength and time of set. Additional information on the effects of mix water impurities can be found in Steinour

Table 4-2. Acceptance Criteria for Questionable Water Supplies

	Limits	Test method
Compressive strength, minimum percentage of control at 28 days	90	CSA A23-8A*

*Comparisons should be based on fixed proportions and the same volume of test water compared to a control mixture using city water or distilled water.

Table 4-3. Chemical Limits for Wash Water used as Mixing Water

Chemical or type of construction	Maximum concentration, ppm*	Test method**
Chloride, as Cl		ASTM D 512
Prestressed concrete or concrete in bridge decks	500†	
Other reinforced concrete in moist environments or containing aluminum embedments or dissimilar metals or with stay-in-place galvanized metal forms	1,000†	
Sulphate, as SO_4	3,000	ASTM D 516
Alkalies, as (Na_2O + 0.658 K_2O)	600	
Total solids	50,000	AASHTO T 26

* Wash water reused as mixing water in concrete can exceed the listed concentrations of chloride and sulphate if it can be shown that the concentration calculated in the total mixing water, including mixing water on the aggregates and other sources, does not exceed the stated limits.
** Other test methods that have been demonstrated to yield comparable results can be used.
† For conditions allowing use of $CaCl_2$ accelerator as an admixture, the chloride limitation may be waived by the purchaser.

0.3% water soluble). Of the total chloride-ion content in concrete, only about 50% to 85% is water soluble; the rest becomes chemically combined in cement reactions (Whiting 1997, Whiting, Taylor, and Nagi 2000, and Taylor, Whiting, and Nagi 2000).

Chlorides can be introduced into concrete with the separate mixture ingredients—admixtures, aggregates, cement, and mixing water—or through exposure to deicing salts, seawater, or salt-laden air in coastal environments. Placing an acceptable limit on chloride content for any one ingredient, such as mixing water, is difficult considering the several possible sources of chloride ions in concrete. An acceptable limit in the concrete depends primarily upon the type of structure and the environment to which it is exposed during its service life.

A high dissolved solids content of a natural water is sometimes due to a high content of sodium chloride or sodium sulphate. Both can be tolerated in rather large quantities. Concentrations of 20,000 ppm of sodium chloride are generally tolerable in concrete that will be dry in service and has low potential for corrosive reactions. Water used in prestressed concrete or in concrete that is to have aluminum embedments should not contain deleterious amounts of chloride ion. The contribution of chlorides from ingredients other than water should also be considered. Calcium chloride admixtures should be avoided in steel reinforced concrete.

CSA Standard A23.1 limits water soluble chloride ion content in concrete before exposure to the following percentages by mass of cementing materials:

Prestressed concrete	0.06%
Reinforced concrete exposed to a moist environment nor chlorides or both	0.15%
Reinforced concrete exposed to neither a moist environment nor chlorides	1.00%

CSA Standard A23.1 does not limit the amount of chlorides in plain concrete, that is concrete not containing steel. Additional information on limits and tests can be found in ACI 222, *Corrosion of Metals in Concrete*. The water-soluble chloride content of hardened concrete should be determined in accordance with CSA Test Method A23.2-4B on the proposed mix for the job and prior to commencement of work.

SULPHATE

Concern over a high sulphate content in mix water is due to possible expansive reactions and deterioration by sulphate attack, especially in areas where the concrete will be exposed to high sulphate soils or water. Although mixing waters containing 10,000 ppm of sodium sulphate have been used satisfactorily, the limit in Table 4-3 should be considered unless special precautions are taken.

OTHER COMMON SALTS

Carbonates of calcium and magnesium are not very soluble in water and are seldom found in sufficient concentration to affect the strength of concrete. Bicarbonates of calcium and magnesium are present in some municipal waters. Concentrations up to 400 ppm of bicarbonate in these forms are not considered harmful.

Magnesium sulphate and magnesium chloride can be present in high concentrations without harmful effects on strength. Good strengths have been obtained using water with concentrations up to 40,000 ppm of magnesium chloride. Concentrations of magnesium sulphate should be less than 25,000 ppm.

IRON SALTS

Natural ground waters seldom contain more than 20 to 30 ppm of iron; however, acid mine waters may carry rather large quantities. Iron salts in concentrations up to 40,000 ppm do not usually affect concrete strengths adversely.

MISCELLANEOUS INORGANIC SALTS

Salts of manganese, tin, zinc, copper, and lead in mixing water can cause a significant reduction in strength and large variations in setting time. Of these, salts of zinc, copper, and lead are the most active. Salts that are especially active as retarders include sodium iodate, sodium phosphate, sodium arsenate, and sodium borate. All can greatly retard both set and strength development when present in concentrations of a few tenths percent by mass of the cement. Generally, concentrations of these salts up to 500 ppm can be tolerated in mixing water.

Another salt that may be detrimental to concrete is sodium sulphide; even the presence of 100 ppm warrants testing. Additional information on the effects of other salts can be found in the references.

SEAWATER

Seawater containing up to 35,000 ppm of dissolved salts is generally suitable as mixing water for concrete not containing steel. About 78% of the salt is sodium chloride, and 15% is chloride and sulphate of magnesium. Although concrete made with seawater may have higher early strength than normal concrete, strengths at later ages (after 28 days) may be lower. This strength reduction can be compensated for by reducing the water-cementing materials ratio.

Seawater is not suitable for use in making steel reinforced concrete and it should not be used in prestressed concrete due to the risk of corrosion of the reinforcement, particularly in warm and humid environments. If seawater is used in plain concrete (no steel) in marine applications, a moderate sulphate resistant cement, Type 20, should be used along with a low water-cementing materials ratio.

Sodium or potassium in salts present in seawater used for mix water can aggravate alkali-aggregate reactivity. Thus, seawater should not be used as mix water for concrete with potentially alkali-reactive aggregates.

Seawater used for mix water also tends to cause efflorescence and dampness on concrete surfaces exposed to air and water (Steinour 1960). Marine-dredged aggregates are discussed in Chapter 5.

ACID WATERS

Acceptance of acid mixing water should be based on the concentration (in parts per million) of acids in the water. Occasionally, acceptance is based on the pH, which is a measure of the hydrogen-ion concentration on a log scale. The pH value is an intensity index and is not the best measure of potential acid or base reactivity. The pH of neutral water is 7.0; values below 7.0 indicate acidity and those above 7.0 alkalinity (a base).

Generally, mixing waters containing hydrochloric, sulphuric, and other common inorganic acids in concentrations as high as 10,000 ppm have no adverse effect on strength. Acid waters with pH values less than 3.0 may create handling problems and should be avoided if possible. Organic acids, such as tannic acid, can have a significant effect on strength at higher concentrations (Fig. 4-3).

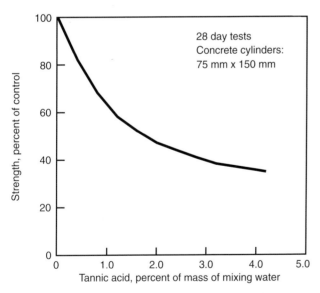

Fig. 4-3. Effect of tannic acid on the strength of concrete (Abrams 1920).

ALKALINE WATERS

Waters with sodium hydroxide concentrations of 0.5% by mass of cement do not greatly affect concrete strength provided quick set is not induced. Higher concentrations, however, may reduce concrete strength.

Potassium hydroxide in concentrations up to 1.2% by mass of cement has little effect on the concrete strength developed by some cements, but the same concentration when used with other cements may substantially reduce the 28-day strength.

The possibility for increased alkali-aggregate reactivity should be considered.

WASH WATER

Environment Canada and some provincial agencies forbid discharging into the nation's waterways untreated wash water used in reclaiming sand and gravel from returned concrete or mixer washout operations. Wash water is commonly used as mixing water in ready mixed concrete (Fig. 4-4) (Yelton 1999). Wash water should meet the limits in Table 4-2. While not a requirement in CSA A23.1, chemical limits for wash water used as mixing water are given in Table 4-3.

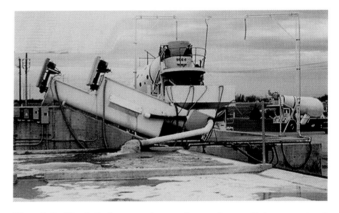

Fig. 4-4. Reclaiming system allows immediate reuse of wash water in batching. (69901)

INDUSTRIAL WASTEWATER

Most waters carrying industrial wastes have less than 4000 ppm of total solids. When such water is used as mixing water in concrete, the reduction in compressive strength is generally not greater than about 10%-15%. Wastewaters such as those from tanneries, paint factories, coke plants, and chemical and galvanizing plants may contain harmful impurities. It is best to test any wastewater that contains even a few hundred parts per million of unusual solids.

WATERS CARRYING SANITARY SEWAGE

A typical sewage may contain about 400 ppm of organic matter. After the sewage is diluted in a good disposal system, the concentration is reduced to about 20 ppm or less. This amount is too low to have any significant effect on concrete strength.

ORGANIC IMPURITIES

The effect of organic substances on the setting time of portland cement or the ultimate strength of concrete is a problem of considerable complexity. Such substances, like surface loams, can be found in natural waters. Highly coloured waters, waters with a noticeable odor, or those in which green or brown algae are visible should be regarded with suspicion and tested accordingly. Organic impurities are often of a humus nature containing tannates or tannic acid (Fig. 4-3).

SUGAR

Small amounts of sucrose, as little as 0.03% to 0.15% by mass of cement, usually retard the setting of cement. The upper limit of this range varies with different cements. The 7-day strength may be reduced while the 28-day strength may be improved. Sugar in quantities of 0.25% or more by mass of cement may cause rapid setting and a substantial reduction in 28-day strength. Each type of sugar influences setting time and strength differently.

Less than 500 ppm of sugar in mix water generally has no adverse effect on strength, but if the concentration exceeds this amount, tests for setting time and strength should be made.

SILT OR SUSPENDED PARTICLES

About 2000 ppm of suspended clay or fine rock particles can be tolerated in mixing water. Higher amounts might not affect strength but may influence other properties of some concrete mixtures. Before use, muddy water should be passed through settling basins or otherwise clarified to reduce the amount of silt and clay added to the mixture by way of the mix water. When cement fines are returned to the concrete in reused wash water, 50,000 ppm can be tolerated.

OILS

Various kinds of oil are occasionally present in mixing water. Mineral oil (petroleum) not mixed with animal or vegetable oils probably has less effect on strength development than other oils. However, mineral oil in concentrations greater than 2.5% by mass of cement may reduce strength by more than 20%.

ALGAE

Water containing algae is unsuited for making concrete because the algae can cause an excessive reduction in strength. Algae in water leads to lower strengths either by influencing cement hydration or by causing a large amount of air to be entrained in the concrete. Algae may also be present on aggregates, in which case the bond between the aggregate and cement paste is reduced. A maximum algae content of 1000 ppm is recommended.

INTERACTION WITH ADMIXTURES

When evaluating waters for their effect on concrete properties, it is important to also test the water with chemical admixtures that will be used in the job concrete. Certain compounds in water can influence the performance and efficiency of certain admixtures. For example, the dosage of air-entraining admixture may need to be increased when used with hard waters containing high concentrations of certain compounds or minerals.

REFERENCES

AASHTO, *Method of Test for Quality of Water to be Used in Concrete*, T 26-79, American Association of State Highway and Transportation Officials, Washington, D.C., 1979.

Abrams, Duff A., *Effect of Tannic Acid on the Strength of Concrete*, Bulletin 7, Structural Materials Research Laboratory, Lewis Institute, Chicago, http://www.port cement.org/pdf_files/LS007.pdf, 1920, 34 pages (available through PCA as LS007).

Abrams, Duff A., *Tests of Impure Waters for Mixing Concrete*, Bulletin 12, Structural Materials Research Laboratory, Lewis Institute, Chicago, http://www.portcement.org/pdf_files/LS012.pdf, 1924, 50 pages (available through PCA as LS012).

ACI Committee 201, *Guide to Durable Concrete*, ACI 201.2R-92, American Concrete Institute, Farmington Hills, Michigan, 1992, 41 pages.

ACI Committee 222, *Corrosion of Metals in Concrete*, ACI 222R-96, American Concrete Institute, Farmington Hills, Michigan, 1997, 30 pages.

ACI Committee 318, *Building Code Requirements for Structural Concrete and Commentary*, ACI 318-99, American Concrete Institute, Farmington Hills, Michigan, 1999, 369 pages. Available through PCA as LT125.

Bhatty, Javed I., *Effects of Minor Elements on Cement Manufacture and Use*, Research and Development Bulletin RD109, Portland Cement Association, 1995, 48 pages.

Bhatty, J.; Miller, F.; West, P.; and Ost, B., *Stabilization of Heavy Metals in Portland Cement, Silica Fume/Portland Cement and Masonry Cement Matrices*, RP348, Portland Cement Association, 1999, 106 pages.

Gaynor, Richard D., "Calculating Chloride Percentages," *Concrete Technology Today,* PL983, Portland Cement Association, http://www.portcement.org/pdf_files/PL983.pdf, December, 1998, 2 pages.

Meininger, Richard C., *Recycling Mixer Wash Water,* National Ready Mixed Concrete Association, Silver Spring, Maryland.

NRMCA, *A System for 100% Recycling of Returned Concrete: Equipment, Procedures, and Effects on Product Quality,* National Ready Mixed Concrete Association, Silver Spring, Maryland, 1975.

Steinour, H. H., *Concrete Mix Water—How Impure Can It Be?,* Research Department Bulletin RX119, Portland Cement Association, http://www.portcement.org/pdf_files/RX119.pdf, 1960, 20 pages.

Taylor, Peter C.; Whiting, David A.; and Nagi, Mohamad A., *Threshold Chloride Content of Steel in Concrete,* R&D Serial No. 2169, Portland Cement Association, http://www.portcement.org/pdf_files/SN2169.pdf, 2000, 32 pages.

Whiting, David A., *Origins of Chloride Limits for Reinforced Concrete,* R&D Serial No. 2153, Portland Cement Association, http://www.portcement.org/pdf_files/SN2153.pdf, 1997, 18 pages.

Whiting, David A.; Taylor, Peter C.; and Nagi, Mohamad A., *Chloride Limits in Reinforced Concrete,* R&D Serial No. 2438, Portland Cement Association, 2000, 96 pages.

Yelton, Rick, "Answering Five Common Questions about Reclaimers," *The Concrete Producer,* Addison, Illinois, September, 1999, pages 17 to 19.

CHAPTER 5
Aggregates for Concrete

The importance of using the right type and quality of aggregates cannot be overemphasized. The fine and coarse aggregates generally occupy 60% to 75% of the concrete volume (70% to 85% by mass) and strongly influence the concrete's freshly mixed and hardened properties, mixture proportions, and economy. Fine aggregates (Fig. 5-1) generally consist of natural sand or crushed stone with most particles smaller than 5 mm. Coarse aggregates (Fig. 5-2) consist of one or a combination of gravels or crushed stone

Fig. 5-1. Closeup of fine aggregate (sand). (69792)

Fig. 5-2. Coarse aggregate. Rounded gravel (left) and crushed stone (right). (69791)

with particles predominantly larger than 5 mm and generally between 10 mm and 40 mm. Some natural aggregate deposits, sometimes called pit-run gravel, consist of gravel and sand that can be readily used in concrete after minimal processing. Natural gravel and sand are usually dug or dredged from a pit, river, lake, or seabed. Crushed stone is produced by crushing quarry rock, boulders, cobbles, or large-size gravel. Crushed air-cooled blast-furnace slag is also used as fine or coarse aggregate.

The aggregates are usually washed and graded at the pit or plant. Some variation in the type, quality, cleanliness, grading, moisture content, and other properties is expected. Close to half of the coarse aggregates used in portland cement concrete in North America are gravels; most of the remainder are crushed stones.

Naturally occurring concrete aggregates are a mixture of rocks and minerals (see Table 5-1). A mineral is a naturally occurring solid substance with an orderly internal structure and a chemical composition that ranges within narrow limits. Rocks, which are classified as igneous, sedimentary, or metamorphic, depending on origin, are generally composed of several minerals. For example, granite contains quartz, feldspar, mica, and a few other minerals; most limestones consist of calcite, dolomite, and minor amounts of quartz, feldspar, and clay. Weathering and erosion of rocks produce particles of stone, gravel, sand, silt, and clay.

Recycled concrete, or crushed waste concrete, is a feasible source of aggregates and an economic reality, especially where good aggregates are scarce. Conventional stone crushing equipment can be used, and new equipment has been developed to reduce noise and dust.

Aggregates must conform to certain standards for optimum engineering use: they must be clean, hard, strong, durable particles free of absorbed chemicals, coatings of clay, and other fine materials in amounts that could affect hydration and bond of the cement paste. Aggregate particles that are friable or capable of being split are undesirable. Aggregates containing any appreciable amounts of shale or other shaly rocks, soft and porous materials, should be avoided; certain types of chert should be espe-

Table 5-1. Rock and Mineral Constituents in Aggregates

Minerals	Igneous rocks	Metamorphic rocks
Silica	Granite	Marble
Quartz	Syenite	Metaquartzite
Opal	Diorite	Slate
Chalcedony	Gabbro	Phyllite
Tridymite	Peridotite	Schist
Cristobalite	Pegmatite	Amphibolite
Silicates	Volcanic glass	Hornfels
Feldspars	Obsidian	Gneiss
Ferromagnesian	Pumice	Serpentinite
Hornblende	Tuff	
Augite	Scoria	
Clay	Perlite	
Illites	Pitchstone	
Kaolins	Felsite	
Chlorites	Basalt	
Montmorillonites	**Sedimentary rocks**	
Mica		
Zeolite	Conglomerate	
Carbonate	Sandstone	
Calcite	Quartzite	
Dolomite	Graywacke	
Sulphate	Subgraywacke	
Gypsum	Arkose	
Anhydrite	Claystone, siltstone,	
Iron sulphide	argillite, and shale	
Pyrite	Carbonates	
Marcasite	Limestone	
Pyrrhotite	Dolomite	
Iron oxide	Marl	
Magnetite	Chalk	
Hematite	Chert	
Goethite		
Imenite		
Limonite		

For brief descriptions, see "Standard Descriptive Nomenclature of Constituents of Natural Mineral Aggregates" (ASTM C 294).

Fig. 5-3. Low-density aggregate. Expanded clay (left) and expanded shale (right). (69793)

cially avoided since they have low resistance to weathering and can cause surface defects such as popouts.

Identification of the constituents of an aggregate cannot alone provide a basis for predicting the behavior of aggregates in service. Visual inspection will often disclose weaknesses in coarse aggregates. Service records are invaluable in evaluating aggregates. In the absence of a performance record, the aggregates should be tested before they are used in concrete. The most commonly used aggregates—sand, gravel, crushed stone, and air-cooled blast-furnace slag—produce freshly mixed normal density concrete of 2200 to 2400 kg/m³. Aggregates of expanded shale, clay, slate, and slag (Fig. 5-3) are used to produce structural low-density concrete with a freshly mixed density ranging from about 1350 to 1850 kg/m³. Structural low-density aggregates must conform to the requirements of ASTM Standard C 330. Other low-density materials such as pumice, scoria, perlite, vermiculite, and diatomite are used to produce insulating low-density con-cretes ranging in density from about 250 to 1450 kg/m³. High-density materials such as barite, limonite, magnetite, ilmenite, hematite, iron, and steel punchings or shot are used to produce high-density concrete and radiation-shielding concrete (CSA Standard N287 Series) (ASTM C 637 and C 638). Only normal density aggregates are dis-cussed in this chapter. See Chapter 18 for special types of aggregates and concretes.

Normal-density aggregates should meet the require-ments of CSA Standard A23.1 (ASTM C 33). This Standard limits the permissible amounts of deleterious substances and provides the requirements for aggregate characteris-tics. Compliance is determined by using one or more of the several standard tests cited in the following sections and tables. However, the fact that aggregates satisfy CSA A23.1 (ASTM C 33) requirements does not necessarily assure defect-free concrete.

For adequate consolidation of concrete, the desirable amount of air, water, cement, and fine aggregate (that is, the mortar fraction) should be about 50% to 65% by absolute volume (45% to 60% by mass). Rounded aggre-gate, such as gravel, requires slightly lower values, while crushed aggregate requires slightly higher values. Fine aggregate content is usually 35% to 45% by mass or volume of the total aggregate content.

CHARACTERISTICS OF AGGREGATES

The important characteristics of aggregates for concrete are listed in Table 5-2 and most are discussed in the following section:

Grading

Grading is the particle-size distribution of an aggregate as determined by a sieve analysis as outlined in CSA Test Method A23.2-2A (ASTM C 136). The range of particle sizes

Table 5-2. Characteristics and Tests of Aggregate

Characteristic	Significance	Test designation*		Requirement or item reported
Resistance to abrasion and degradation	Index of aggregate quality; wear resistance of floors and pavements	CSA A23.2-16A CSA A23.2-17A CSA A23.2-23A	ASTM C 131 ASTM C 535 ASTM C 779	Maximum percentage of loss in mass. Depth of wear and time
Resistance to freezing and thawing	Surface scaling, roughness, and aesthetics	CSA A23.2-24A —	ASTM C 666 ASTM C 682	Maximum number of cycles or period of frost immunity; durability factor
Resistance to disintegration by sulphates	Soundness against weathering action	CSA A23.2-9A	ASTM C 88	Loss in mass, particles exhibiting distress
Particle shape and surface texture	Workability of fresh concrete	CSA A23.2-13A —	ASTM C 295 ASTM D 3398	Maximum percentage of flat and elongated particles
Grading	Workability of fresh concrete; economy	CSA A23.2-2A CSA A23.2-5A	ASTM C 136 ASTM C 117	Minimum and maximum percentage passing standard sieves
Density	Mix design calculations; classification	CSA A23.2-10A	ASTM C 29	Compact density and loose density
Relative density (specific gravity)	Mix design calculations	CSA A23.2-6A CSA A23.2-12A	ASTM C 127 ASTM C 128	—
Absorption and surface moisture	Control of concrete quality (water-cement ratio)	CSA A23.2-6A CSA A23.2-11A CSA A23.2-12A —	ASTM C 128 ASTM C 70 ASTM C 127 ASTM C 566	—
Compressive and flexural strength	Acceptability of fine aggregate failing other tests	CSA A23.2-8C CSA A23.2-9C	ASTM C 78 ASTM C 39	Strength to exceed 95% of strength achieved with purified sand
Definitions of constituents	Clear understanding and communication	ASTM C 125 ASTM C 294	ASTM C 125 ASTM C 294	—
Aggregate constituents	Determine amount of deleterious and organic materials	CSA A23.2-3A CSA A23.2-4A CSA A23.2-5A CSA A23.2-7A CSA A23.2-8A	ASTM C 142 ASTM C 123 ASTM C 117 ASTM C 40 ASTM C 87 ASTM C 295	Maximum percentage allowed of individual constituents
Resistance to alkali reactivity and volume change	Soundness against volume change	CSA A23.2-14A CSA A23.2-25A CSA A23.2-26A ASTM C 227 ASTM C 289 ASTM C 295 ASTM C 586 —	ASTM C 1293 ASTM C 1260 — ASTM C 227 ASTM C 289 ASTM C 295 ASTM C 586 ASTM C 342	Maximum length change, constituents and amount of silica, and alkalinity

* The majority of the tests and characteristics listed are referenced in CSA A23.1 and ASTM C 33. ACI 221-96 presents additional test methods and properties of concrete influenced by aggregate characteristics.

in aggregate is illustrated in Fig. 5-4. The aggregate particle size is determined by using wire-mesh sieves with square openings. The seven standard CSA A23.1 (ASTM C 33) sieves for fine aggregate have openings ranging from 160 µm to 10 mm. The standard sieves for coarse aggregate have openings ranging from 1.25 mm to 112 mm. A typical laboratory "sieve shaker" used in performing a sieve analysis is shown in Fig. 5-5. Sieves must meet the requirements for woven wire cloth testing sieves given in Canadian General Standards Board (CGSB) Standard CAN/CGSB-8.2.

Fig. 5-4. Range of particle sizes found in aggregate for use in concrete. (8985)

Fig. 5-5. Making a sieve analysis test of coarse aggregate in a laboratory. (30175-A)

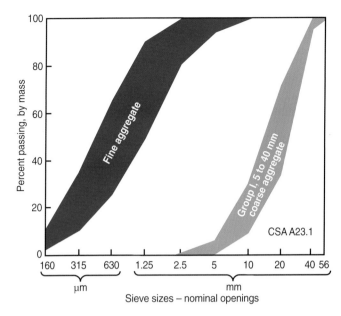

Fig. 5-6. Curves indicate the limits specified in CSA A23.1 (ASTM C 33) for fine aggregate and for one commonly used size number (grading size) of coarse aggregate.

The grading and grading limits are usually expressed as the percentage of material passing each sieve. Fig. 5-6 shows these limits for fine aggregate and for one size of coarse aggregate.

There are several reasons for specifying grading limits and nominal maximum aggregate size; they affect relative aggregate proportions as well as cement and water requirements, workability, pumpability, economy, porosity, shrinkage, and durability of concrete. Variations in grading can seriously affect the uniformity of concrete from batch to batch. Very fine sands are often uneconomical; very coarse sands and coarse aggregate can produce harsh, unworkable mixtures. In general, aggregates that do not have a large deficiency or excess of any size and give a smooth grading curve will produce the most satisfactory results.

The effect of a collection of various sizes in reducing the total volume of voids between aggregates is illustrated by the simple method shown in Fig. 5-7. The beaker on the left is filled with large aggregate particles of uniform size and shape; the middle beaker is filled with an equal volume of small aggregate particles of uniform size and shape; and the beaker on the right is filled with particles of both sizes. Below each beaker is a graduate with the amount of water required to fill the voids in that beaker. Note that when the beakers are filled with one particle size

of equal volume, the void content is constant, regardless of the particle size. When the two aggregate sizes are combined, the void content is decreased. If this operation were repeated with several additional sizes, a further reduction in voids would occur. The cement paste requirement for concrete is related to the void content of the combined aggregates.

During the early years of concrete technology it was sometimes assumed that the smallest percentage of voids (greatest density of aggregates) was the most suitable for concrete. At the same time, limits were placed on the amount and size of the smallest particles. It is now known

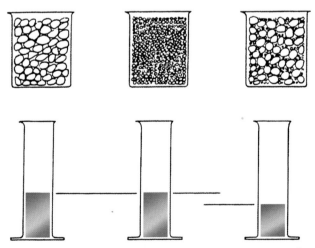

Fig. 5-7. The level of liquid in the graduates, representing voids, is constant for equal absolute volumes of aggregates of uniform but different size. When different sizes are combined, the void-content decreases. The illustration is not to scale.

that, even on this restricted basis, this is not the best target for the mix designer. However, production of satisfactory, economical concrete requires aggregates of low void content, but not the lowest. Voids in aggregates can be tested according to ASTM C 29.

In reality, the amount of cement paste required in concrete is greater than the volume of voids between the aggregates. This is illustrated in Fig. 5-8. Sketch A represents large aggregates alone, with all particles in contact. Sketch B represents the dispersal of aggregates in a matrix of paste. The amount of paste is necessarily greater than the void content of sketch A in order to provide workability to the concrete; the actual amount is influenced by the workability and cohesiveness of the paste.

Fine-Aggregate Grading

Requirements of CSA A23.1 (ASTM C 33) permit a relatively wide range in fine-aggregate gradation, but specifications by other organizations are sometimes more restrictive. The most desirable fine-aggregate grading depends on the type of work, the richness of the mixture, and the maximum size of coarse aggregate. In leaner mixtures, or when small-size coarse aggregates are used, a grading that approaches the maximum recommended percentage passing each sieve is desirable for workability. In general, if the water-cementing materials ratio is kept constant and the ratio of fine-to-coarse aggregate is chosen correctly, a wide range in grading can be used without measurable effect on strength. However, the best economy will sometimes be achieved by adjusting the concrete mixture to suit the gradation of the local aggregates.

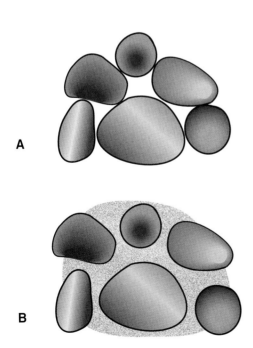

Fig. 5-8. Illustration of the dispersion of aggregates in cohesive concrete mixtures.

Fine-aggregate grading within the limits of CSA A23.1 (ASTM C 33) is generally satisfactory for most concretes. The CSA A23.1 limits with respect to sieve size are shown in Table 5-3.

These specifications permit the minimum percentages (by mass) of material passing the 315-μm and 160-μm sieves to be reduced to 5% and 0% respectively, provided:

1. The aggregate is used in air-entrained concrete containing more than 250 kg/m³ of cementing material.
2. The aggregate is used in non-air-entrained concrete containing more than 300 kg/m³ of cementing material.

Table 5-3. Grading Limits for Fine Aggregate (FA)

Sieve size	Total passing sieve, percentage by mass	
	FA1*	FA2*
10 mm	100	100
5 mm	95 – 100	80 – 90
2.5 mm	80 – 100	60 – 75
1.25 mm	50 – 90	35 – 50
630 μm	25 – 65	15 – 30
315 μm	10 – 35	5 – 15
160 μm	2 – 10	0 – 8

Source: CSA A23.1-00.

The A23.1 Standard also notes that:
- For high-strength concrete, it is desirable to limit the amount of material passing the 160-μm sieve to a maximum of 2%.
- Workability problems have been experienced when the percentage passing the 315-μm sieve is less than 10.
- For fine aggregate, not more than 45% shall pass any sieve and be retained on the next consecutive sieve of those shown in Table 5-3.
- To control the grading, the fineness modulus of any shipment made during the progress of the work should not vary more than ± 0.20 from the initially approved value unless the owner considers the variation acceptable after any changes that were considered necessary have been made to the concrete mix proportions.
- When a fine aggregate with a grading falling outside the limits of those shown in Table 5-3 is proposed for use by the supplier, the supplier must provide the owner with all the necessary test data showing that the material will produce concrete of acceptable quality that meets all relevant requirements of the Standard.

The amounts of fine aggregate passing the 315-μm and 160-μm sieves affect workability, surface texture, air content, and bleeding of concrete. A number of specifications allow 5% to 30% to pass the 315-μm sieve. The lower limit may be sufficient for easy placing conditions or where concrete is mechanically finished, such as in pavements. However, for hand-finished concrete floors, or where a smooth surface texture is desired, fine aggregate with at least 15% passing the 315-μm sieve and 3% or more passing the 160-μm sieve should be used.

Fineness Modulus. The fineness modulus (FM) of either fine or coarse aggregate according to ASTM C 125 is calculated by adding the cumulative percentages by mass retained on each of a specified series of sieves and dividing the sum by 100.

The FM is an index of the fineness of an aggregate—the higher the FM, the coarser the aggregate. Different aggregate grading may have the same FM. The FM of fine aggregate is useful in estimating proportions of fine and coarse aggregates in concrete mixtures. An example of how the FM of a fine aggregate is determined (with an assumed sieve analysis) is shown in Table 5-4.

Degradation of fine aggregate due to friction and abrasion will decrease the FM and increase the amount of materials finer than the 80-μm sieve.

Table 5-4. Determination of Fineness Modulus of Fine Aggregates

Sieve size	Percentage of individual fraction retained, by mass*	Percentage passing, by mass	Percentage retained, by mass
10 mm	0	100	0
5 mm	2	98	2
2.5 mm	13	85	15
1.25 mm	20	65	35
630 μm	20	45	55
315 μm	24	21	79
160 μm	18	3	97
Pan	3	0	—
Total	100		283

Fineness modulus
= 283 ÷ 100 = 2.83

* Percentages of material retained between consecutive sieves. For example, 13% of the material is retained between the 5 mm and the 2.5 mm sieves.

Coarse-Aggregate Grading

The coarse aggregate grading requirements of CSA A23.1 (ASTM C 33) permit a wide range in grading and a variety of grading sizes (see Table 5-5). The grading for a given maximum-size coarse aggregate can be varied over a moderate range without appreciable effect on cement and water requirements of a mixture if the proportion of fine aggregate to total aggregate produces concrete of good workability. Mixture proportions should be changed to produce workable concrete if wide variations occur in the coarse-aggregate grading. Since variations are difficult to anticipate, it is often more economical to maintain unifor-

mity in manufacturing and handling coarse aggregate than to reduce variations in gradation.

The maximum size of coarse aggregate used in concrete has a bearing on the economy of concrete. Usually more paste, water, and cement is required for small-size aggregates than for large sizes, due to an increase in total aggregate surface area. The water and cement required for a slump of approximately 75 mm is shown in Fig. 5-9 for a wide range of coarse-aggregate sizes. Fig. 5-9 shows that, for a given water-cement ratio, the amount of cement required decreases as the maximum size of coarse aggregate increases. The increased cost of obtaining and/or handling aggregates much larger than 50 mm may offset the savings

Table 5-5. Grading Requirements for Coarse Aggregates

	Nominal size of aggregate, mm	Total passing each sieve*, percentage by mass				
		112 mm	80 mm	56 mm	40 mm	
Group I	40–5**	—	—	100	95–100	
	28–5**	—	—	—	100	
	20–5	—	—	—	—	
	14–5	—	—	—	—	
	10–2.5	—	—	—	—	
Group II	80–40	100	90–100	25–60	0–15	
	56–28	—	100	90–100	30–65	
	40–20	—	—	100	90–100	
	28–14	—	—	—	100	
	20–10	—	—	—	—	
	14–10	—	—	—	—	
	10–5	—	—	—	—	
	5–2.5	—	—	—	—	

* Sieves shall meet the requirements for woven wire cloth testing sieves given in CGSB Standard CAN/CGSB-8.2.
** To prevent segregation, aggregates that make up either of these gradings shall be stockpiled and batched in two or more separate sizes selected from Groups I and II.
NOTE: Group I comprises combined aggregate gradings most commonly used in concrete production. Group II provides for special requirements, ie. gap grading, pumping, etc. or for blending two or more sizes to produce Group I gradings. Source: CSA A23.1-00

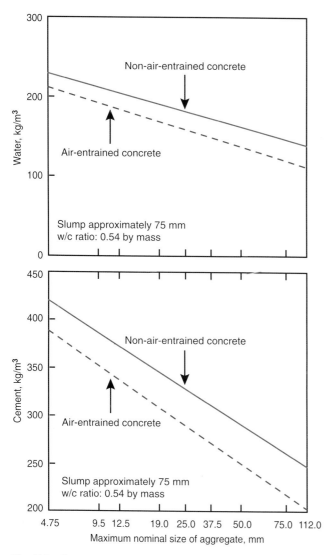

Fig. 5-9. Cement and water contents in relation to maximum size of aggregate, for air-entrained and non-air-entrained concrete. Less cement and water are required in mixtures having large coarse aggregate (Bureau of Reclamation 1981).

in using less cement. Furthermore, aggregates of different maximum sizes may give slightly different concrete strengths for the same water-cementing materials ratio. In some instances, at the same water-cementing materials ratio, concrete with a smaller maximum-size aggregate could have higher compressive strength. This is especially true for high-strength concrete. The optimum maximum size of coarse aggregate for higher strength depends on factors such as relative strength of the paste, cementing materials-aggregate bond, and strength of the aggregate particles.

The terminology used to specify size of coarse aggregate must be chosen carefully. Particle size is determined by size of sieve and applies to the aggregate passing that sieve and not passing the next smaller sieve. When speaking of an assortment of particle sizes, the size number (or grading size) of the gradation is used. The size number applies to the collective amount of aggregate that passes through an assortment of sieves. As shown in Table 5-5, the amount of aggregate passing the respective sieves is given in percentages; it is called a sieve analysis.

Because of past usage, there is sometimes confusion about what is meant by the maximum size of aggregate. ASTM C 125 and ACI 116 define this term and distinguish it from nominal maximum size of aggregate. The maximum size of an aggregate is the smallest sieve that all of a particular aggregate must pass through. The nominal maximum size of an aggregate is the smallest sieve size through which the major portion of the aggregate must pass. The nominal maximum-size sieve may retain 5% to 15% of the aggregate depending on the size number. For example, refer to Group 1 material shown in Table 5-5; an aggregate with a maximum size of 56 mm has a nominal maximum size of 40 mm. Ninety-five to one hundred percent of this aggregate must pass the 40-mm sieve and all of the particles must pass the 56-mm sieve.

28 mm	20 mm	14 mm	10 mm	5 mm	2.5 mm	1.25 mm
—	35–70	—	10–30	0–5	—	—
95–100	—	30–65	—	0–10	0–5	—
100	85–100	60–90	25–60	0–10	0–5	—
—	100	90–100	45–75	0–15	0–5	—
—	—	100	85–100	10–30	0–10	0–5
—	0–5	—	—	—	—	—
0–15	—	0–5	—	—	—	—
25–60	0–15	—	0–5	—	—	—
90–100	30–65	0–15	—	0–5	—	—
100	85–100	—	0–20	0–5	—	—
—	—	85–100	0–45	0–10	—	—
—	—	100	85–100	0–20	0–5	—
—	—	—	100	70–100	10–40	0–10

The nominal maximum size of aggregate that can be used generally depends on the size and shape of the concrete member and the amount and distribution of reinforcing steel. The nominal maximum size of aggregate particles generally should not exceed:

1. One-fifth of the narrowest dimension between sides of forms;
2. Three-quarters of the minimum clear spacing between reinforcing bars and between the reinforcing bars and forms;
3. One-third the depth of slabs;
4. The specified cover for concrete not exposed to earth or weather, refer CSA Standard A23.1, (Clause 12.6.2);
5. Two-thirds of the specified cover for concrete exposed to earth or weather, refer CSA Standard A23.1, (Clause 12.6.2.2 and Table 9);
6. One-half of the specified cover for concrete exposed to chlorides, refer CSA Standard A23.1, (Clause 12.6.2.2 and Table 9); or
7. For concrete that is to be placed by pump, the nominal maximum size of coarse aggregate shall be limited to one-third the smallest internal diameter of the hose or pipe through which the concrete is to be pumped or 40 mm, whichever is smaller.

Except for the limitations of Numbers 6 and 7, the above limitations may be waived if, in the judgement of the owner, workability and methods of consolidation are such that the concrete can be properly placed with a larger nominal maximum size aggregate.

Combined Aggregate Grading

Aggregate is sometimes analyzed using the combined grading of fine and coarse aggregate together, as they exist in a concrete mixture. This provides a more thorough analysis of how the aggregates will perform in concrete. Sometimes mid-sized aggregate, around the 10 mm size, is

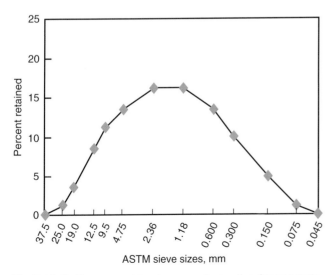

Fig. 5-10. Optimum combined aggregate grading for concrete.

lacking in an aggregate supply, resulting in a concrete with high shrinkage properties, high water demand, poor workability, poor pumpability, and poor placeability. Strength and durability may also be affected.

Fig. 5-10, using standard ASTM C 33 sieve sizes, illustrates an ideal gradation; however, a perfect gradation does not exist in the field—but we can try to approach it. If problems develop due to a poor gradation, alternative aggregates, blending, or special screening of existing aggregates, should be considered. Refer to Shilstone (1990) for options on obtaining optimal grading of aggregate.

The combined gradation can be used to better control workability, pumpability, shrinkage, and other properties of concrete. Abrams (1918) and Shilstone (1990) demonstrate the benefits of a combined aggregate analysis:

- With constant cementing materials content and constant consistency, there is an optimum for every combination of aggregates that will produce the most effective water to cementing materials ratio and highest strength.
- The optimum mixture has the least particle interference and responds best to a high frequency, high amplitude vibrator.

However, this optimum mixture cannot be used for all construction due to variations in placing and finishing needs and availability. Crouch (2000) found in his studies on air-entrained concrete that the water-cementing materials ratio could be reduced by over 8% using combined aggregate gradation. Shilstone (1990) also analyzes aggregate gradation by coarseness and workability factors to improve aggregate gradation.

Gap-Graded Aggregates

In gap-graded aggregates certain particle sizes are intentionally omitted. For cast-in-place concrete, typical gap-graded aggregates consist of only one size of coarse aggregate with all the particles of fine aggregate able to pass through the voids in the compacted coarse aggregate. Gap-graded mixes are used in architectural concrete to obtain uniform textures in exposed-aggregate finishes. They can also used in normal structural concrete because of possible improvements in some concrete properties, and to permit the use of local aggregate gradations (Houston 1962 and Litvin and Pfeifer 1965).

For an aggregate of 20 mm maximum size, the 5 mm to 10 mm particles can be omitted without making the concrete unduly harsh or subject to segregation. In the case of 40 mm aggregate, usually the 5 mm to 20 mm sizes are omitted.

Care must be taken in choosing the percentage of fine aggregate in a gap-graded mixture. A wrong choice can result in concrete that is likely to segregate or honeycomb because of an excess of coarse aggregate. Also, concrete with an excess of fine aggregate could have a high water demand resulting in a low-density concrete. Fine aggregate is usually 25% to 35% by volume of the total aggregate. The lower percentage is used with rounded aggregates and the higher with crushed material. For a smooth off-the-form finish, a somewhat higher percentage of fine aggregate to total aggregate may be used than for an exposed-aggregate finish, but both use a lower fine aggregate content than continuously graded mixes. Fine aggregate content depends upon cementing materials content, type of aggregate, and workability.

Air entrainment is usually required for workability since low-slump, gap-graded mixes use a low fine aggregate percentage and produce harsh mixes without entrained air.

Segregation of gap-graded mixes must be prevented by restricting the slump to the lowest value consistent with good consolidation. This may vary from zero to 75 mm depending on the thickness of the section, amount of reinforcement, and height of casting. Close control of grading and water content is also required because variations might cause segregation. If a stiff mixture is required, gap-graded aggregates may produce higher strengths than normal aggregates used with comparable cementing materials contents. Because of their low fine-aggregate volumes and low water-cementing materials ratios, gap-graded mixtures might be considered unworkable for some cast-in-place construction. When properly proportioned, however, these concretes are readily consolidated with vibration.

Particle Shape and Surface Texture

The particle shape and surface texture of an aggregate influence the properties of freshly mixed concrete more than the properties of hardened concrete. Rough-textured, angular, elongated particles require more water to produce workable concrete than do smooth, rounded, compact aggregates. Hence, aggregate particles that are angular require more cementing material to maintain the same water-cementing materials ratio. However, with satisfactory gradation, both crushed and noncrushed aggregates (of the same rock types) generally give essentially the same strength for the same cementing materials factor. Angular or poorly graded aggregates can also be more difficult to pump.

The bond between the paste and a given aggregate generally increases as particles change from smooth and rounded to rough and angular. This increase in bond is a consideration in selecting aggregates for concrete where flexural strength is important or where high compressive strength is needed.

Void contents of compacted fine or coarse aggregate can be used as an index of differences in the shape and texture of aggregates of the same grading. The mixing water and cement requirement tend to increase as aggregate void content increases. Voids between aggregate particles increase with aggregate angularity.

Aggregate should be relatively free of flat and elongated particles. A particle is called flat when the width or length exceeds the thickness by a ratio of about 3:1. Elongated particles have the shape of a rod (Galloway

Fig. 5-11. Videograder for measuring size and shape of aggregate. (69545)

1994). ASTM D 3398 provides an indirect method of establishing a particle index as an overall measure of particle shape or texture, while ASTM C 295 provides procedures for the petrographic examination of aggregate.

Flat and elongated aggregate particles should be avoided or at least limited to about 15% by mass of the total aggregate. This requirement is equally important for coarse and for crushed fine aggregate, since fine aggregate made by crushing stone often contains flat and elongated particles. Such aggregate particles require an increase in mixing water and thus may affect the strength of concrete, particularly in flexure, if the water-cementing materials ratio is not adjusted.

A number of automated test machines are available for rapid determination of the particle size distribution of aggregate. Designed to provide a faster alternative to the standard sieve analysis test, these machines capture and analyze digital images of the aggregate particles to determine gradation. Fig. 5-11 shows a videograder that measures size and shape of an aggregate by using line-scan cameras wherein two-dimensional images are constructed from a series of line images. Other machines use matrix-scan cameras to capture two-dimensional snapshots of the falling aggregate. Maerz and Lusher (2001) developed a dynamic prototype imaging system that provides particle size and shape information by using a miniconveyor system to parade individual fragments past two orthogonally oriented, synchronized cameras.

Bulk Density and Voids

The bulk density of an aggregate is the mass of the aggregate required to fill a container of a specified unit volume. The volume referred to here is that occupied by both aggregates and the voids between aggregate particles.

The bulk density of aggregate commonly used in normal density concrete ranges from about 1200 to 1750 kg/m³. The void content between particles affects paste requirements in mix design (see preceding sections, "Particle Shape and Surface Texture" and "Grading"). Void contents range from about 30% to 45% for coarse aggregates to about 40% to 50% for fine aggregate. Angularity increases void content while larger sizes of well-graded aggregate and improved grading decreases void content (Fig. 5-7). Methods of determining the density of aggregates and void content are given in CSA A23.2-10A, *Density of Aggregate* (ASTM C 29). Procedures are given for Compact Density determination using rodding for aggregate having a maximum size of 56 mm or less, a jigging procedure for aggregate greater than 56 mm but not more than 112 mm, and for a Loose Density determination using a shovelling procedure for aggregate of 112 mm or less. The measurement of loose uncompacted void content of fine aggregate is described in ASTM C 1252.

Relative Density (Specific Gravity)

The relative density (specific gravity) of an aggregate is the ratio of its mass to the mass of an equal absolute volume of water. It is used in certain computations for mixture proportioning and control, such as the volume occupied by the aggregate in the absolute volume method of mix design. It is not generally used as a measure of aggregate quality, though some porous aggregates that exhibit accelerated freeze-thaw deterioration do have low specific gravities. Most natural aggregates have relative densities between 2.4 and 2.9 with corresponding (mass) densities of 2400 and 2900 kg/m³.

Test methods for determining relative densities for coarse and fine aggregates are described in CSA A23.2-12A and CSA A23.2-6A (ASTM C 127 and ASTM C 128), respectively. The relative density of an aggregate may be determined on an ovendry basis or a saturated surface-dry (SSD) basis. Both the ovendry and saturated surface-dry relative densities may be used in concrete mixture proportioning calculations. Ovendry aggregates do not contain any absorbed or free water. They are dried in an oven to a constant mass. Saturated surface-dry aggregates are those in which the pores in each aggregate particle are filled with water but there is no excess water on the particle surface.

Density

The density of aggregate particles used in mixture proportioning computations (not including voids between particles) is determined by multiplying the relative density of the aggregate times the density of water. A value of 1000 kg/m³ (1 kg/L) is often used for the density of water. The density of aggregate is provided in CSA A23.2-12A and CSA A23.2-6A and more accurate values for water density are given in Table 2 of CSA A23.2-10A (ASTM C 127 and ASTM C 128). Most natural aggregates have mass densities of between 2400 and 2900 kg/m³.

Absorption and Surface Moisture

The absorption and surface moisture of aggregates should be determined according to to CSA Test Methods A23.2-6A, A23.2-11A, CSA A23.2-12A (ASTM C 70, C 127, C 128, and C 566) so that the total water content of the concrete can be controlled and correct batch quantities determined. The internal structure of an aggregate particle is made up of solid matter and voids that may or may not contain water.

The moisture conditions of aggregates are shown in Fig. 5-12. They are designated as:

1. **Ovendry**—fully absorbent
2. **Air dry**—dry at the particle surface but containing some interior moisture, thus still somewhat absorbent
3. **Saturated surface dry (SSD)**—neither absorbing water from nor contributing water to the concrete mixture
4. **Damp or wet**—containing an excess of moisture on the surface (free water)

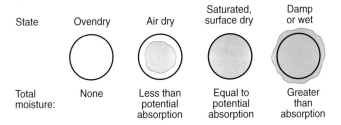

Fig. 5-12. Moisture conditions of aggregate.

The amount of water added at the concrete batch plant must be adjusted for the moisture conditions of the aggregates in order to accurately meet the water requirement of the mix design. If the water content of the concrete mixture is not kept constant, the water-cementing materials ratio will vary from batch to batch causing other properties, such as the compressive strength and workability to vary from batch to batch.

Coarse and fine aggregate will generally have absorption levels (moisture contents at SSD) in the range of 0.2% to 4% and 0.2% to 2%, respectively. Free-water contents will usually range from 0.5% to 2% for coarse aggregate and 2% to 6% for fine aggregate. The maximum water content of drained coarse aggregate is usually less than that of fine aggregate. Most fine aggregates can maintain a maximum drained moisture content of about 3% to 8% whereas coarse aggregates can maintain only about 1% to 6%.

Bulking. Bulking is the increase in total volume of moist fine aggregate over the same mass dry. Surface tension in the moisture holds the particles apart, causing an increase in volume. Bulking of a fine aggregate (such as sand) occurs when it is shoveled or otherwise moved in a damp condition, even though it may have been fully consolidated

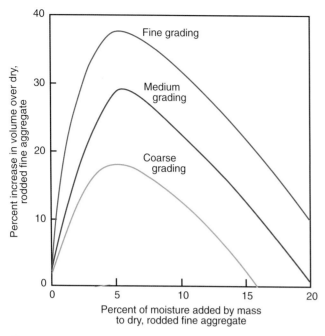

Fig. 5-13. Surface moisture on fine aggregate can cause considerable bulking; the amount varies with the amount of moisture and the aggregate grading (PCA Major Series 172 and PCA ST20).

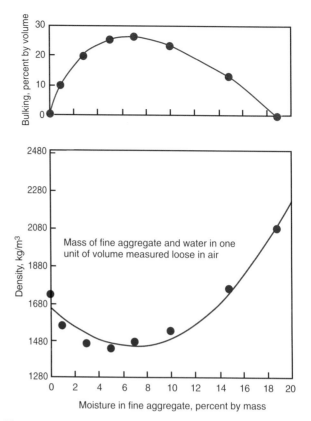

Fig. 5-14. Bulk density and volume increase is compared with the moisture content for a particular sand (PCA Major Series 172).

beforehand. Fig. 5-13 illustrates how the amount of bulking of fine aggregate varies with moisture content and grading; fine gradings bulk more than coarse gradings for a given amount of moisture. Fig. 5-14 shows similar information in terms of mass for a particular fine aggregate. Since most fine aggregates are delivered in a damp condition, wide variations can occur in batch quantities if batching is done by volume. For this reason, good practice has long favoured determining the mass of the aggregate and adjusting for moisture content when proportioning concrete.

Resistance to Freezing and Thawing

The frost resistance of an aggregate, an important characteristic for exterior concrete, is related to its porosity, absorption, permeability, and pore structure. An aggregate particle may absorb so much water (to critical saturation) that it cannot accommodate the expansion and hydraulic pressure that occurs during the freezing of water. If enough of the offending particles are present, the result can be expansion of the aggregate and possible disintegration of the concrete. If a single problem particle is near the surface of the concrete, it can cause a popout. Popouts generally appear as conical fragments that break out of the concrete surface. The offending aggregate particle or a part of it is usually found at the bottom of the void. Generally it is coarse rather than fine aggregate particles with higher porosity values and medium-sized pores (0.1 to 5 μm) that are easily saturated and cause concrete deterioration and popouts. Larger pores do not usually become saturated or

cause concrete distress, and water in very fine pores may not freeze readily.

At any freezing rate, there may be a critical particle size above which a particle will fail if frozen when critically saturated. This critical size is dependent upon the rate of freezing and the porosity, permeability, and tensile strength of the particle. For fine-grained aggregates with low permeability (cherts for example), the critical particle size may be within the range of normal aggregate sizes. It is higher for coarse-grained materials or those with capillary systems interrupted by numerous macropores (voids too large to hold moisture by capillary action). For these aggregates the critical particle size may be sufficiently large to be of no consequence, even though the absorption may be high. If potentially vulnerable aggregates are used in concrete subjected to periodic drying while in service, they may never become sufficiently saturated to cause failure.

Cracking of concrete pavements caused by the freeze-thaw deterioration of the aggregate within concrete is called D-cracking. This type of cracking has been observed in some pavements after three or more years of service. D-cracked concrete resembles frost-damaged concrete caused by paste deterioration. D-cracks are closely spaced crack formations parallel to transverse and longitudinal joints that later multiply outward from the joints toward

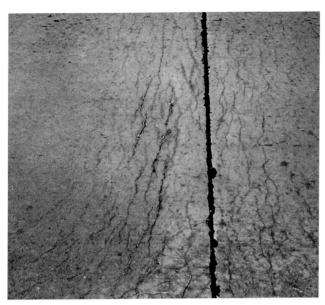

Fig. 5-15. D-cracking along a transverse joint caused by failure of carbonate coarse aggregate (Stark 1976). (30639)

The performance of aggregates under exposure to freezing and thawing can be evaluated in two ways: (1) past performance in the field, and (2) laboratory freeze-thaw tests of concrete specimens. If aggregates from the same source have previously given satisfactory service when used in concrete, they might be considered suitable. Aggregates not having a service record can be considered acceptable if they perform satisfactorily in air-entrained concretes subjected to freeze-thaw tests. CSA Standard A23.1 notes that, testing for D-cracking potential of the aggregates in concrete by a modified version of ASTM C 666, using only two cycles of freezing and thawing per day and a maximum length change measurement of 0.035% after 350 cycles as criteria for failure, has been more successful than absorptive or other tests. It also notes that the change in dynamic modulus has not correlated well with D-cracking deterioration. Deterioration is measured by (1) the reduction in the dynamic modulus of elasticity, (2) linear expansion, and (3) loss in mass of the specimens. Different aggregate types may influence the criteria levels and empirical correlations between laboratory freeze-thaw tests. Field service records should be made to select the proper criterion (Vogler and Grove 1989).

Specifications may require that resistance to weathering be demonstrated by a magnesium sulphate ($MgSO_4$) test (CSA A23.2-9A or ASTM C 88). The test consists of a number of immersion cycles for a sample of the aggregate in a sulphate solution; this creates a pressure through salt-crystal growth in the aggregate pores similar to that produced by freezing water. The sample is then ovendried and the percentage loss in mass calculated. Unfortunately, this test is sometimes misleading. Aggregates behaving satisfactorily in the test might produce concrete with low freeze-thaw resistance; conversely, aggregates performing poorly might produce concrete with adequate resistance. This is attributed, at least in part, to the fact that the aggregates in the test are not confined by paste (as they would be in concrete) and the mechanisms of attack are not the same as in freezing and thawing. The test is most reliable for stratified rocks with porous layers or weak bedding planes.

the center of the pavement panel (Fig. 5-15). D-cracking is a function of the pore properties of certain types of aggregate particles and the environment in which the pavement is placed. Due to the natural accumulation of water under pavements in the base and subbase layers, the aggregate may eventually become saturated. Then with freezing and thawing cycles, cracking of the concrete starts in the saturated aggregate (Fig. 5-16) at the bottom of the slab and progresses upward until it reaches the wearing surface. This problem can be reduced either by selecting aggregates that perform better in freeze-thaw cycles or, where marginal aggregates must be used, by reducing the maximum particle size. Also, installation of effective drainage systems for carrying free water out from under the pavement may be helpful.

Three additional tests are referenced in CSA A23.1. The first two are CSA A23.2-23A and CSA A23.2-29A, Micro-Deval tests for fine and coarse aggregate, respectively. The tests are rapid, have excellent precision and have excellent correlation with the complex and more variable magnesium sulphate ($MgSO_4$) soundness test. The third test is CSA A23.2-24A, an unconfined freeze-thaw test, for coarse aggregate, which has good precision and shows fair correlation with the $MgSO_4$ soundness test.

An additional test that can be used to evaluate aggregates for potential D-cracking is the rapid pressure release method. An aggregate is placed in a pressurized chamber and the pressure is rapidly released causing the aggregate with a questionable pore system to fracture (Janssen and Snyder 1994). The amount of fracturing relates to the potential for D-cracking.

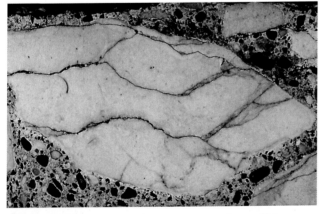

Fig. 5-16. Fractured carbonate aggregate particle as a source of distress in D-cracking (magnification 2.5X) (Stark 1976). (30639-A)

Wetting and Drying Properties

Weathering due to wetting and drying can also affect the durability of aggregates. The expansion and contraction coefficients of rocks vary with temperature and moisture content. If alternate wetting and drying occurs, severe strain can develop in some aggregates, and with certain types of rock this can cause a permanent increase in volume of the concrete and eventual breakdown. Clay lumps and other friable particles can degrade rapidly with repeated wetting and drying. Popouts can also develop due to the moisture-swelling characteristics of certain aggregates, especially clay balls and shales. While no specific tests are available to determine this tendency, an experienced petrographer can often be of assistance in determining this potential for distress.

Abrasion and Skid Resistance

The abrasion resistance of an aggregate is often used as a general index of its quality. Abrasion resistance is essential when the aggregate is to be used in concrete subject to abrasion, as in heavy-duty floors or pavements. Low abrasion resistance of an aggregate may increase the quantity of fines in the concrete during mixing; consequently, this may increase the water requirement and require an adjustment in the water-cementing materials ratio.

The most common test for abrasion resistance is the Los Angeles abrasion test (rattler method) performed in accordance with CSA A23.2-16A for coarse aggregate smaller than 40 mm and CSA A23.2-17A for coarse aggregate larger than 40 mm (ASTM C 131 or C 535). In this test a specified quantity of aggregate is placed in a steel drum containing steel balls, the drum is rotated, and the percentage of material worn away is measured. Specifications often set an upper limit on this loss of mass. However, a comparison of the results of aggregate abrasion tests with the abrasion resistance of concrete made with the same aggregate do not generally show a clear correlation. Mass loss due to impact in the rattler is often as much as that due to abrasion. The wear resistance of concrete is determined more accurately by abrasion tests of the concrete itself (see Chapter 1).

To provide good skid resistance on pavements, the siliceous particle content of the fine aggregate should be at least 25%. For specification purposes, the siliceous particle content is considered equal to the insoluble residue content after treatment in hydrochloric acid under standardized conditions (ASTM D 3042). Certain manufactured sands produce slippery pavement surfaces and should be investigated for acceptance before use.

Strength and Shrinkage

The strength of an aggregate is rarely tested and generally does not influence the strength of conventional concrete as

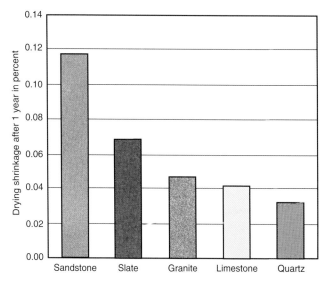

Fig. 5-17. Concretes containing sandstone or slate produce a high shrinkage concrete. Granite, limestone and quartz, are low shrinkage-producing aggregates (ACI 221R).

much as the strength of the paste and the paste-aggregate bond. However, aggregate strength does become important in high-strength concrete. Aggregate stress levels in concrete are often much higher than the average stress over the entire cross section of the concrete. Aggregate tensile strengths range from 2 to 15 MPa and compressive strengths from 65 to 270 MPa.

Different aggregate types have different compressibility, modulus of elasticity, and moisture-related shrinkage characteristics that can influence the same properties in concrete. Aggregates with high absorption may have high shrinkage on drying. Quartz and feldspar aggregates, along with limestone, dolomite, and granite, are considered low shrinkage aggregates; while aggregates with sandstone, shale, slate, hornblende, and graywacke are often associated with high shrinkage in concrete (Fig. 5-17).

Resistance to Acid and Other Corrosive Substances

Portland cement concrete is durable in most natural environments; however, concrete in service is occasionally exposed to substances that will attack it.

Most acidic solutions will slowly or rapidly disintegrate portland cement concrete depending on the type and concentration of acid. Certain acids, such as oxalic acid, are harmless. Weak solutions of some acids have insignificant effects. Although acids generally attack and leach away the calcium compounds of the paste, they may not readily attack certain aggregates, such as siliceous aggregates. Calcareous aggregates often react readily with acids. However, the sacrificial effect of calcareous aggregates is often a benefit over siliceous aggregate in mild acid exposures or in areas where water is not flowing. With calcareous aggregate, the acid attacks the entire exposed concrete surface uniformly, reducing the rate of attack on the paste and preventing loss of aggregate particles at the surface. Calcareous aggregates also tend to neutralize the acid, especially in stagnant locations. Acids can also discolour concrete. Siliceous aggregate should be avoided when strong solutions of sodium hydroxide are present, as these solutions attack this type of aggregate.

Acid rain (often with a pH of 4 to 4.5) can slightly etch concrete surfaces, usually without affecting the performance of exposed concrete structures. Extreme acid rain or strong acid water conditions may warrant special concrete designs or precautions, especially in submerged areas. Continuous replenishment in acid with a pH of less than 4 is considered highly aggressive to buried concrete, such as pipe (Scanlon 1987). Concrete continuously exposed to liquid with a pH of less than 3 should be protected in a manner similar to concrete exposed to dilute acid solutions (ACI 515.1R).

Natural waters usually have a pH of more than 7 and seldom less than 6. Waters with a pH greater than 6.5 may be aggressive if they contain bicarbonates. Carbonic acid solutions with concentrations between 0.9 and 3 parts per million are considered to be destructive to concrete (ACI 515.1R and Kerkhoff 2001).

A low water-cementing materials ratio, low permeability, and low-to-moderate cement content can increase the acid or corrosion resistance of concrete. A low permeability resulting from a low water-cementing materials ratio or the use of silica fume or other pozzolans, helps keep the corrosive agent from penetrating into the concrete. Low-to-moderate cement contents result in less available paste to attack. The use of sacrificial calcareous aggregates should be considered where indicated.

Certain acids, gases, salts, and other substances that are not mentioned here also can disintegrate concrete. Acids and other chemicals that severely attack portland cement concrete should be prevented from coming in contact with the concrete by using protective coatings (Kerkhoff 2001).

Fire Resistance and Thermal Properties

The fire resistance and thermal properties of concrete—conductivity, diffusivity, and coefficient of thermal expansion—depend to some extent on the mineral constituents of the aggregates used. Manufactured and some naturally occurring low-density aggregates are more fire resistant than normal density aggregates due to their insulating properties and high-temperature stability. Concrete containing a calcareous coarse aggregate performs better under fire exposure than a concrete containing quartz or siliceous aggregate such as granite or quartzite. At about 590°C, quartz expands 0.85% causing disruptive expansion (ACI 216 and ACI 221). The coefficient of thermal expansion of aggregates ranges from 0.55×10^{-6} per °C to 5×10^{-6}

per °C. For more information refer to Chapter 15 for temperature-induced volume changes and to Chapter 17 for thermal conductivity and mass concrete considerations.

POTENTIALLY HARMFUL MATERIALS

Harmful substances that may be present in aggregates include organic impurities, silt, clay, shale, iron oxide, coal, lignite, and certain low-density and soft particles (Table 5-6). In addition, rocks and minerals such as some cherts, strained quartz (Buck and Mather 1984), and certain dolomitic limestones are alkali reactive (see Table 5-7). Gypsum and anhydrite may cause sulphate attack. Certain aggregates, such as some shales, will cause popouts by swelling (simply by absorbing water) or by freezing of absorbed water (Fig. 5-18). Most specifications limit the permissible amounts of these substances. The performance history of an aggregate should be a determining factor in setting the limits for harmful substances. Test methods for detecting harmful substances qualitatively or quantitatively are listed in Table 5-6.

Aggregates are potentially harmful if they contain compounds known to react chemically with portland cement concrete and produce any of the following: (1) significant volume changes of the paste, aggregates, or both; (2) interference with the normal hydration of cement; and (3) otherwise harmful byproducts.

Organic impurities may delay setting and hardening of concrete, may reduce strength gain, and in unusual cases may cause deterioration. Organic impurities such as peat, humus, and organic loam may not be as detrimental but should be avoided.

Materials finer than an 80-μm sieve, especially silt and clay, may be present as loose dust and may form a coating on the aggregate particles. Even thin coatings of silt or clay

Fig. 5-18. A popout is the breaking away of a small fragment of concrete surface due to internal pressure that leaves a shallow, typically conical depression. (113)

Table 5-6. Harmful Materials in Aggregates

Substances	Effect on concrete	Test designation	
Organic impurities	Affects setting and hardening, may cause deterioration	CSA A23.2-7A CSA A23.2-8A	ASTM C 40 ASTM C 87
Materials finer than 80-µm sieve	Affects bond, increases water requirement	CSA A23.2-5A	ASTM C 117
Coal, lignite, or other low-density materials	Affects durability, may cause stains and popouts	CSA A23.2-4A	ASTM C 123
Soft particles	Affects durability	ASTM C 235	ASTM C 235
Clay lumps and friable particles	Affects workability and durability, may cause popouts	CSA A23.2-3A	ASTM C 142
Chert of less than 2.40 relative density	Affects durability, may cause popouts	CSA A23.2-4A ASTM C 295	ASTM C 123 ASTM C 295
Alkali-reactive aggregates	Causes abnormal expansion, map cracking, and popouts	CSA A23.2-14A CSA A23.2-25A CSA A23.2-26A ASTM C 227 ASTM C 289 ASTM C 295 ASTM C 586 —	ASTM C 1293 ASTM C 1260 — ASTM C 227 ASTM C 289 ASTM C 295 ASTM C 586 ASTM C 342

Table 5-7. Some Potentially Harmful Reactive Minerals, Rock, and Synthetic Materials

Alkali-silica reactive substances*		Alkali-carbonate reactive substances**
Andesites	Opal	Calcitic dolomites
Argillites	Opaline shales	Dolomitic limestones
Certain siliceous	Phylites	Fine-grained dolomites
limestones	Quartzites	
and dolomites	Quartzoses	
Chalcedonic cherts	Cherts	
Chalcedony	Rhyolites	
Cristobalite	Schists	
Dacites	Siliceous shales	
Glassy or	Strained quartz	
cryptocrystalline	and certain	
volcanics	other forms	
Granite gneiss	of quartz	
Graywackes	Synthetic and	
Metagraywackes	natural silicious	
	glass	
	Tridymite	

* Several of the rocks listed (granite gneiss and certain quartz formations for example) react very slowly and may not show evidence of any harmful degree of reactivity until the concrete is over 20 years old.
** Only certain sources of these materials have shown reactivity.

on gravel particles can be harmful because they may weaken the bond between the cement paste and aggregate. If certain types of silt or clay are present in excessive amounts, water requirements may increase significantly.

There is a tendency for some fine aggregates to degrade from the grinding action in a concrete mixer; this effect, which is measured using ASTM C 1137 may alter mixing water, entrained air and slump requirements.

Coal or lignite, or other low-density materials such as wood or fibrous materials, in excessive amounts will affect the durability of concrete. If these impurities occur at or near the surface, they might disintegrate, pop out, or cause stains. Potentially harmful chert in coarse aggregate can be identified by using CSA A23.2-4A (ASTM C 123).

Soft particles in coarse aggregate are especially objectionable because they cause popouts and can affect durability and wear resistance of concrete. If friable, they could break up during mixing and thereby increase the amount of water required. Where abrasion resistance is critical, such as in heavy-duty industrial floors, testing may indicate that further investigation or another aggregate source is warranted.

Clay lumps present in concrete may absorb some of the mixing water, cause popouts in hardened concrete, and affect durability and wear resistance. They can also break up during mixing and thereby increase the mixing-water demand.

Aggregates can occasionally contain particles of iron oxide and iron sulphide that result in unsightly stains on exposed concrete surfaces (Fig. 5-19). The aggregate should meet the staining requirements of ASTM C 330 when tested according to ASTM C 641; the quarry face and aggregate stockpiles should not show evidence of staining.

As an additional aid in identifying staining particles, the aggregate can be immersed in a lime slurry. If staining particles are present, a blue-green gelatinous precipitate will form within 5 to 10 minutes; this will rapidly change to a brown colour on exposure to air and light. The reaction should be complete within 30 minutes. If no brown gelatinous precipitate is formed when a suspect aggregate is placed in the lime slurry, there is little likelihood of any reaction taking place in concrete. These tests should be required when aggregates with no record of successful prior use are used in architectural concrete.

ALKALI-AGGREGATE REACTIVITY

Aggregates containing certain constituents can react with alkali hydroxides in concrete.

Fig. 5-19. Iron oxide stain caused by impurities in the coarse aggregate. (70024)

The reactivity is potentially harmful only when it produces significant expansion (Mather 1975). This alkali-aggregate reactivity (AAR) has two forms—alkali-silica reaction (ASR) and alkali-carbonate reaction (ACR). ASR is of more concern than ACR because the occurrence of aggregates containing reactive silica minerals is more common. Alkali-reactive carbonate aggregates have a specific composition that is not very common.

Alkali-silica reactivity has been recognized as a potential source of distress in concrete since the late 1930s (Stanton 1940 and PCA 1940). Even though potentially reactive aggregates exist throughout North America, ASR distress in structural concrete is not common. There are a number of reasons for this:

• Most aggregates are chemically stable in hydraulic-cement concrete.
• Aggregates with good service records are abundant in many areas.
• Most concrete in service is dry enough to inhibit ASR.
• Use of certain pozzolans or slags can control ASR.
• In many concrete mixtures, the alkali content of the concrete is low enough to control harmful ASR.
• Some forms of ASR do not produce significant deleterious expansion.

To reduce ASR potential requires understanding the ASR mechanism; properly using tests to identify potentially reactive aggregates; and, if needed, taking steps to minimize the potential for expansion and related cracking.

A thorough discussion of alkali-aggregate reaction is presented in CSA Standard A23.1-*Appendix B*. This part of the Standard provides advice on strategies, test methods, and selection criteria to avoid AAR problems in concrete. Recommendations for the management of structures affected by AAR are given in CSA A864-00, *Guide to the Evaluation and Management of Concrete Structures Affected by Alkali-Aggregate Reaction.*

Alkali-Silica Reaction

Aggregates exhibiting this type of reactivity contain various forms of reactive silica. For convenience, CSA Standard A23.1-*Appendix B*, divides alkali-silica reaction into two categories according to the type of reactive silica involved.

a. Alkali-silica reaction that occurs with poorly crystalline or metastable silica minerals and volcanic or artificial glasses: Aggregates containing such materials (see Table 5-7 and CSA A23.1-*Appendix B*) may cause deterioration of concrete when the reactive component is present in amounts as small as 1%. Cracking of concrete containing these aggregates and a high alkali content is usually seen within 10 years of construction.

b. Alkali-silica reaction that occurs with various varieties of quartz such as chalcedony, cryptocrystalline, and macrogranular quartz: Aggregates containing such forms of quartz may cause deterioration of the con-

crete when the reactive component is present in amounts as small as 5% by mass of the aggregate and there is a high alkali content. Cracking of the concrete may be seen within 10 years of construction. Canadian experience has been that this category also includes several slowly expanding aggregates in which microcrystalline quartz is thought to be the reactive component. Rocks such as greywacke, argillite, quartz-wacke, quartzite, hornfels, granite and granite gneiss are some. See Table 5-7 and CSA A23.1-*Appendix B* for a more complete list. Such rock types may not show cracking and deterioration for up to 20 years. In other cases however, particularly when exposed to deicing salts, cracking may occur in 5 years or less.

Visual Symptoms of Expansive ASR. Typical indicators of ASR might be any of the following: a network of cracks (Fig. 5-20); closed or spalled joints; relative displacements of different parts of a structure; or fragments breaking out of the surface of the concrete (popouts) (Fig. 5-21). Because ASR deterioration is slow, the risk of catastrophic failure is low. However, ASR can cause serviceability problems and can exacerbate other deterioration mechanisms such as those that occur in frost, deicer, or sulphate exposures.

Mechanism of ASR. The alkali-silica reaction forms a gel that swells as it draws water from the surrounding cement paste. Reaction products from ASR have a great affinity for moisture. In absorbing water, these gels can induce pressure, expansion, and cracking of the aggregate and surrounding paste. The reaction can be visualized as a two-step process:

1. Alkali hydroxide + reactive silica gel → reaction product (alkali-silica gel)
2. Gel reaction product + moisture → expansion

The amount of gel formed in the concrete depends on the amount and type of silica and alkali hydroxide concentration. The presence of gel does not always coincide with distress, and thus, gel presence does not necessarily indicate destructive ASR.

Factors Affecting ASR. For alkali-silica reaction to occur, the following three conditions must be present:

1. reactive forms of silica in the aggregate,
2. high-alkali (pH) pore solution, and
3. sufficient moisture.

If one of these conditions is absent, ASR cannot occur.

Test Methods for Identifying ASR Distress. It is important to distinguish between the reaction and damage resulting from the reaction. In the diagnosis of concrete deterioration, it is most likely that a gel product will be identified. But, in some cases significant amounts of gel are formed without causing damage to concrete. To pinpoint ASR as the cause of damage, the presence of deleterious ASR gel must be verified. A site of expansive reaction can be defined as an aggregate particle that is recognizably

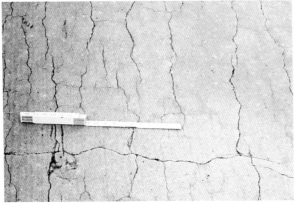

Fig. 5-20 (top and bottom). Deterioration of concrete from alkali-silica reactivity. (69549, 58352)

reactive or potentially reactive and is at least partially replaced by gel. Gel can be present in cracks and voids and may also be present in a ring surrounding an aggregate particle at its edges. A network of internal cracks connecting reacted aggregate particles is an almost certain indication that ASR is responsible for cracking. A petrographic examination (ASTM C 856) is the most positive method for identifying ASR gel in concrete (Powers 1999). Petrography, when used to study a known reacted concrete, can confirm the presence of reaction products and verify ASR as an underlying cause of deterioration (Fig. 5-22).

Control of ASR in New Concrete. The best way to avoid ASR is to take appropriate precautions before concrete is placed. Standard concrete specifications may require modifications to address ASR. These modifications should be carefully tailored to avoid limiting the concrete producer's options. This permits careful analysis of cementing materials and aggregates and choosing a control strategy that optimizes effectiveness and the economic selection of materials. If the aggregate is not reactive by historical identification or testing, no special requirements are needed.

Identification of Potentially Reactive Aggregates. Field performance history is the best method of evaluating the susceptibility of an aggregate to ASR. For the most definitive evaluation, the existing concrete should have been in service for at least 15 years. Comparisons should be made between the existing and proposed concrete's mix proportions, ingredients, and service environments. This process should tell whether special requirements are needed, are not needed, or whether testing of the aggregate or job concrete is required. The use of newer, faster test methods can be utilized for initial screening. Where uncertainties arise, lengthier tests can be used to confirm results. Table 5-8 describes different test methods used to evaluate potential alkali-silica reactivity. These tests should not be used to disqualify use of potentially reactive aggregates, as reactive

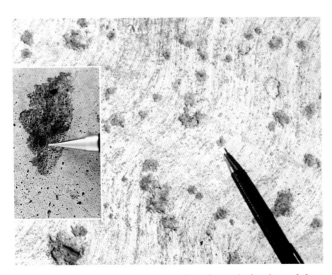

Fig. 5-21. Popouts caused by ASR of sand-sized particles. Inset shows closeup of a popout. (51117, 51118)

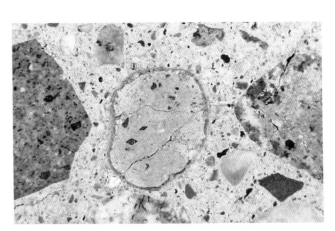

Fig. 5-22. Polished section view of an alkali reactive aggregate in concrete. Observe the alkali-silica reaction rim around the reactive aggregate and the crack formation. (43090)

Table 5-8. Test Methods for Alkali-Silica Reactivity (Farny and Kosmatka 1997)

Test name	Purpose	Type of test	Type of sample
CSA A23.2-14A (ASTM C 1293), Potential expansivity of aggregates (procedure for length change due to alkali-aggregate reaction in concrete prisms)	To determine the potential ASR expansion of cement-aggregate combinations	Concrete prisms stored at 38°C and 100% humidity	Three prisms per cement-aggregate combination of dimensions: not less than 75 x 75 x 275 mm nor more than 75 x 75 x 405 mm
CSA A23.2-25A (ASTM C 1260), Detection of alkali-silica reactive aggregate by accelerated expansion of mortar bars	To test the potential for deleterious alkali-silica reaction of aggregate in mortar bars	Immersion of mortar bars in alkaline solution at 80°C	At least 3 mortar bars
Modified CSA A23.2-14A (ASTM C 1293), Potential expansivity of aggregates (procedure for length change due to alkali-aggregate reaction in concrete prisms)	To determine the potential ASR expansion of cement-aggregate combinations	Concrete prisms stored at 60°C and 100% humidity	Three prisms per cement-aggregate combination of dimensions: not less than 75 x 75 x 275 mm nor more than 75 x 75 x 405 mm
Other Test Methods			
ASTM C 227, Potential alkali-reactivity of cement-aggregate combinations (mortar-bar method)	To test the susceptibility of cement-aggregate combinations to expansive reactions involving alkalies	Mortar bars stored over water at 37.8°C and high relative humidity	At least 4 mortar bars of standard dimensions: 25 x 25 x 285 mm
ASTM C 289, Potential alkali-silica reactivity of aggregates (chemical method)	To determine potential reactivity of siliceous aggregates	Sample reacted with alkaline solution at 80°C	Three 25-grams samples of crushed and sieved aggregates
ASTM C 294, Constituents of natural mineral aggregates	To give descriptive nomenclature for the more common or important natural minerals—an aid in determining their performance	Visual identification	Varies, but should be representative of entire source
ASTM C 295, Petrographic examination of aggregates for concrete	To outline petrographic examination procedures for aggregates—an aid in determining their performance	Visual and microscopic examination of prepared samples—sieve analysis, microscopy, scratch or acid tests	Varies with knowledge of quarry: cores 53 to 100 mm in diameter 45 kg or 300 pieces, or 2 kg
ASTM C 342, Potential volume change or cement-aggregation combinations	To determine the potential ASR expansion of cement-aggregate combinations	Mortar bars stored in water at 23°C	Three mortar bars per cement-aggregate combination of standard dimensions: 25 x 25 x 285 mm
ASTM C 441 Effectiveness of mineral admixtures or GBFS in preventing excessive expansion of concrete due to alkali-silica reaction	To determine effectiveness of supplementary cementing materials in controlling expansion from ASR	Mortar bars—using Pyrex glass as aggregate—stored over water at 37.8°C and high relative humidity	At least 3 mortar bars and also 3 mortar bars of control mixture
ASTM C 856, Petrographic examination of hardened concrete	To outline petrographic examination procedures for hardened concrete—useful in determining condition or performance	Visual (unmagnified) and microscopic examination of prepared samples	At least one 150 mm diameter by 300 mm long core
ASTM C 856, Annex Uranyl-acetate treatment procedure	To identify products of ASR in hardened concrete	Staining of a freshly-exposed concrete surface and viewing under UV light	Varies: core with lapped surface, broken surface
Los Alamos staining method (Powers 1999)	To identify products of ASR in hardened concrete	Staining of a freshly-exposed concrete surface with two different reagents	Varies: core with lapped surface, broken surface

Duration of test	Measurement	Criteria	Comments
Measure for length change after 1, 2, 4, 8, 13, 18, 26, 39, and 52 weeks and approx. every 6 months after that	Length change	Potentially deleteriously reactive if expansion exceeds 0.04% at one year	Requires long test duration for meaningful results.
16 days total—14 days from zero measurement	Length change	Greater than 0.15% at 14 days indicative of potential deleterious expansion	Very fast, useful for slowly reacting aggregates or those that produce expansion late in the reaction.
3 month (91 days)	Length change	Potentially deleteriously reactive if expansion exceeds 0.04% at 91 days	Fast alternative to CSA A23.2-14A (ASTM C 1293) Good correlation to CSA A23.2-14A for carbonate and sedimentary rocks.
Varies: first measurement at 14 days, then 1, 2, 3, 4, 6, 9, and 12 months; every 6 months after that as necessary	Length change	Per ASTM C 33, maximum 0.10% expansion at 6 months, or if not available for a 6-month period, maximum of 0.05% at 3 months	Test may not produce significant expansion, especially for carbonate aggregate. Long test duration. Expansions may not be from AAR
24 hours	Drop in alkalinity and amount of silica solubilized	Point plotted on graph falls in deleterious or potentially deleterious area	Quick results. Some aggregates give low expansions even though they have high silica content. Not reliable
Short duration—as long as it takes to visually examine the sample	Description of type and proportion of minerals in aggregate	Not applicable	These descriptions are used to characterize naturally-occurring minerals that make up common aggregate sources
Short duration–visual examination does not involve long test periods	Particle characteristics, such as shape, size, texture, colour, mineral composition, physical condition	Not applicable	Usually includes optical microscopy. Also may include XRD analysis, differential thermal analysis, or infrared spectroscopy—see ASTM C 294 for descriptive nomenclature
52 weeks	Length change	Per ASTM C 33, unsatisfactory aggregate if expansion equals or exceeds 0.200% at 1 year	Primarily used for aggregates from Oklahoma, Kansas, Nebraska, and Iowa
Varies: first measurement at 14 days, then 1, 2, 3, 4, 5, 9, and 12 months; every 6 months after that as necessary	Length change	Per ASTM C 989, minimum 75% reduction in expansion or 0.02% maximum expansion or per ASTM C 618 comparison against low-alkali control	Highly reactive artificial aggregate may not represent real aggregate conditions Pyrex contains alkalies
Short duration—includes preparation of samples and visual and microscopic examination	Is the aggregate known to be reactive? Orientation and geometry of cracks. Is there any gel present?	See measurements—this examination determines if ASR reactions have taken place and their effects upon the concrete. Used in conjunction with other tests.	Specimens can be examined with stereomicroscopes, polarizing microscopes, metallographic microscopes, and scanning electron microscope
Immediate results	Intensity of fluorescence	Lack of fluorescence	Identifies small amounts of ASR gel whether they cause expansion or not
Immediate results	Colour of stain	Dark pink stain corresponds to ASR gel and indicates an advanced state of degradation	Opal, a natural aggregate, and Carbonated paste can glow—interpret results accordingly. Tests must be supplemented by petrographic examination and physical tests for determining concrete expansion.

aggregates can be safely used with the careful selection of cementing materials.

Materials and Methods to Control ASR. The most effective way of controlling expansion due to ASR is to design mixtures specifically to control ASR, preferably using locally available materials. The following options are not listed in priority order and, although usually not necessary, they can be used in combination with one another.

In North America, current practices include the use of a supplementary cementing material or blended cement proven by testing to control ASR or limiting the alkali content of the concrete. Supplementary cementing materials include fly ash, ground granulated blast-furnace slag, silica fume, and natural pozzolans. Blended cements use slag, fly ash, silica fume, and natural pozzolans to control ASR. Low-alkali portland cement CSA A-5 (ASTM C 150) with an alkali content of not more than 0.60% (equivalent sodium oxide) can be used to control ASR.

In Canada, the procedures to be followed in assessing the suitability of aggregate for concrete are given in CSA Standard Practice A23.2-27A-*Standard Practice to Identify Degree of Alkali-Reactivity of Aggregates and to Identify Measures to Avoid Deleterious Expansions in Concrete.* This Standard Practice provides requirements for the determination of the degree of alkali-silica reactivity of aggregates; it provides a table to help determine the risk level of having poor performance of concrete considering the size of the concrete element, the humidity of the environment, and the degree of reactivity of the aggregates. Information is also provided to determine the need for and level of preventive measures required by taking into account the service life of the concrete structure or element.

A flow chart is provided to aid in the process of determining the potential alkali-aggregate reactivity of concrete aggregate and to recommend limits of linear expansion under tests that indicate when expansion should be considered deleterious. A table identifying possible measures to prevent deleterious alkali-silica reaction as it relates to a recommended level of prevention is given. Where supplementary cementing materials are used, an additional table is provided giving guidance on cement replacement levels to counteract alkali-silica reaction considering the alkali content and the chemical composition requirement of SCM's, and the prevention level and respective possible measures to prevent a deleterious reaction.

A second Standard Practice is also provided in CSA Standard A23.2 and is identified as CSA A23.2-28A-*Standard Practice for Laboratory Testing to Demonstrate the Effectiveness of Supplementary Cementing Materials and Chemical Admixtures to Prevent Alkali-Silica Reaction in Concrete.* Standard Practice CSA A23.2-27A provides the necessary information to identify the degree of alkali-reactivity of aggregates and identify measures to avoid deleterious expansion. Standard Practice CSA A23.2-28A on the other hand describes the procedure to be followed to demonstrate the effectiveness of supplementary cementing

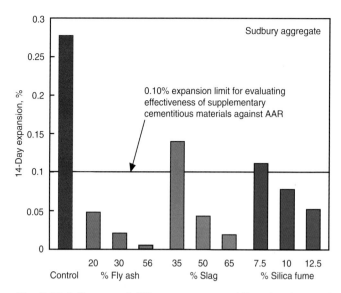

Fig. 5-23. Influence of different amounts of fly ash, slag, and silica fume by mass of cementing material on mortar bar expansion (CSA A23.2-25A or ASTM 1260) after 14 days with a highly reactive aggregate (Fournier 1997).

materials and chemical admixtures in preventing excessive expansion caused by alkali-reaction.

When pozzolans, slags, or blended cements are used to control ASR, expansion usually decreases as the dosage of the pozzolan or slag increases (see Fig. 5-23). Lithium-based admixtures are also available to control ASR. Limestone sweetening (the popular term for replacing approximately 30% of the reactive sand-gravel aggregate with crushed limestone) is effective in controlling deterioration in some sand-gravel aggregate concretes. See AASHTO (2001), Farny and Kosmatka (1997), and PCA (1998) for more information on tests to demonstrate the effectiveness of the above control measures.

Alkali-Carbonate Reaction

Mechanism of ACR. Reactions observed with certain dolomitic rocks are associated with alkali-carbonate reaction (ACR). Reactive rocks usually contain large crystals of dolomite scattered in and surrounded by a fine-grained matrix of calcite and clay. Calcite is one of the mineral forms of calcium carbonate; dolomite is the common name for calcium-magnesium carbonate. ACR is relatively rare because aggregates susceptible to this reaction are usually unsuitable for use in concrete for other reasons, such as strength potential. Argillaceous dolomitic limestone contains calcite and dolomite with appreciable amounts of clay and can contain small amounts of reactive silica (see Table 5-7). Alkali reactivity of carbonate rocks is not usually dependent upon its clay mineral composition (Hadley 1961). Aggregates have potential for expansive ACR if the following lithological characteristics exist (Ozol 1994 and Swenson 1967):

- clay content, or insoluble residue content, in the range of 5% to 25%;
- calcite-to-dolomite ratio of approximately 1:1;
- increase in the dolomite volume up to a point at which interlocking texture becomes a restraining factor; and
- small size of the discrete dolomite crystals (rhombs) suspended in a clay matrix.

Dedolomitization. Dedolomitization, or the breaking down of dolomite, is normally associated with expansive ACR (Hadley 1961). Concrete that contains dolomite and has expanded also contains brucite (magnesium hydroxide, $Mg(OH)_2$), which is formed by dedolomitization. Dedolomitization proceeds according to the following equation (Ozol 1994):

$CaMgCO_3$ (dolomite) + alkali hydroxide solution → $MgOH_2$ (brucite) + $CaCO3$ (calcium carbonate) + K_2CO_3 (potassium carbonate) + alkali hydroxide

The dedolomitization reaction and subsequent crystallization of brucite may cause considerable expansion. Whether dedolomitization causes expansion directly or indirectly, it's usually a prerequisite to other expansive processes (Tang, Deny, and Lon 1994).

Test Methods for Identifying ACR Distress. The test methods commonly used to identify potentially alkali-carbonate reactive aggregate are:

- petrographic examination (ASTM C 295);
- rock cylinder method (ASTM C 586);
- concrete prism expansion test CSA A23.2-14A (ASTM C 1105); and
- chemical composition test – CSA A23.2-26A.

See CSA Standard Practice A23.2-27A and Farny and Kosmatka (1997) for more detailed information.

Materials and Methods to Control ACR. ACR-susceptible aggregate has a specific composition that is readily identified by petrographic testing. If a rock indicates ACR-susceptibility, one of the following preventive measures should be taken:

- selective quarrying to completely avoid reactive aggregate;
- blend aggregate according to Appendix in ASTM C 1105; or
- limit aggregate size to smallest practical.

Low-alkali cement and pozzolans are generally not very effective in controlling expansive ACR.

AGGREGATE BENEFICIATION

Aggregate processing consists of: (1) basic processing—crushing, screening, and washing—to obtain proper gradation and cleanliness; and (2) beneficiation—upgrading quality by processing methods such as heavy media separation, jigging, rising-current classification, and crushing.

In heavy media separation, aggregates are passed through a high-density liquid comprised of finely ground high-density minerals and water proportioned to have a relative density (specific gravity) less than that of the desirable aggregate particles but greater than that of the deleterious particles. The higher density particles sink to the bottom the lower density particles float to the surface. This process can be used when acceptable and harmful particles have distinguishable relative densities.

Jigging separates particles with small differences in density by pulsating water current. Upward pulsations of water through a jig (a box with a perforated bottom) move the lower density material into a layer on top of the higher density material; the top layer is then removed.

Rising-current classification separates particles with large differences in density. Low-density materials, such as wood and lignite, are floated away in a rapidly upward moving stream of water.

Crushing is also used to remove soft and friable particles from coarse aggregates. This process is sometimes the only means of making material suitable for use. Unfortunately, with any process some acceptable material is always lost and removal of all harmful particles may be difficult or expensive.

HANDLING AND STORING AGGREGATES

Aggregates should be handled and stored in a way that minimizes segregation and degradation and prevents contamination by deleterious substances (Fig. 5-24). Stockpiles

Fig. 5-24. Stockpile of aggregate at a ready mix plant. (69552)

should be built up in thin layers of uniform thickness to minimize segregation. The most economical and acceptable method of forming aggregate stockpiles is the truck-dump method, which discharges the loads in a way that keeps them tightly joined. The aggregate is then reclaimed with a front-end loader. The loader should remove slices from the edges of the pile from bottom to top so that every slice will contain a portion of each horizontal layer.

When aggregates are not delivered by truck, acceptable and inexpensive results can be obtained by forming the stockpile in layers with a clamshell bucket (cast-and-spread method); in the case of aggregates not subject to degradation, spreading the aggregates with a rubber-tire dozer and reclaiming with a front-end loader can be used. By spreading the material in thin layers, segregation is minimized. Whether aggregates are handled by truck, bucket loader, clamshell, or conveyor belt, stockpiles should not be built up in high, cone-shaped piles since this results in segregation. However, if circumstances necessitate construction of a conical pile, or if a stockpile has segregated, gradation variations can be minimized when the pile is reclaimed; in such cases aggregates should be loaded by continually moving around the circumference of the pile to blend sizes rather than by starting on one side and working straight through the pile.

Crushed aggregates segregate less than rounded (gravel) aggregates and larger-size aggregates segregate more than smaller sizes. To avoid segregation of coarse aggregates, size fractions can be stockpiled and batched separately. Proper stockpiling procedures, however, should eliminate the need for this. Specifications provide a range in the amount of material permitted in any size fraction partly because of segregation in stockpiling and batching operations.

Washed aggregates should be stockpiled in sufficient time before use so that they can drain to a uniform moisture content. Damp fine material has less tendency to segregate than dry material. When dry fine aggregate is dropped from buckets or conveyors, wind can blow away the fines; this should be avoided if possible.

Bulkheads or dividers should be used to avoid contamination of aggregate stockpiles. Partitions between stockpiles should be high enough to prevent intermingling of materials. Storage bins should be circular or nearly square. Their bottoms should slope not less than 50 degrees from the horizontal on all sides to a center outlet. When loading the bin, the material should fall vertically over the outlet into the bin. Chuting the material into a bin at an angle and against the bin sides will cause segregation. Baffle plates or dividers will help minimize segregation. Bins should be kept as full as possible since this reduces breakage of aggregate particles and the tendency to segregate. Recommended methods of handling and storing aggregates are discussed at length in Matthews (1965 to 1967), NCHRP (1967), and Bureau of Reclamation (1981).

MARINE-DREDGED AGGREGATE

Marine-dredged aggregate from tidal estuaries and sand and gravel from the seashore can be used with caution in some concrete applications when other aggregate sources are not available. Aggregates obtained from seabeds have two problems: (1) seashells and (2) salt.

Seashells may be present in the aggregate source. These shells are a hard material that can produce good quality concrete, however, a higher cement content may be required. Also, due to the angularity of the shells, additional cement paste is required to obtain the desired workability. Aggregate containing complete shells (uncrushed) should be avoided as their presence may result in voids in the concrete and lower the compressive strength.

Marine-dredged aggregates often contain salt from the seawater. The primary salts are sodium chloride and magnesium sulphate and the amount of salt on the aggregate is often not more than about 1% of the mass of the mixing water. The highest salt content occurs in sands located just above the high-tide level. Use of these aggregates with drinkable mix water often contributes less salt to the mixture than the use of seawater (as mix water) with salt-free aggregates.

Marine aggregates can be an appreciable source of chlorides. The presence of these chlorides may affect the concrete by (1) altering the time of set, (2) increasing drying shrinkage, (3) significantly increasing the risk of corrosion of steel reinforcement, and (4) causing efflorescence. Generally, marine aggregates containing large amounts of chloride should not be used in reinforced concrete.

Marine-dredged aggregates can be washed with fresh water to reduce the salt content. There is no maximum limit on the salt content of coarse or fine aggregate; however, the chloride limits presented in Chapter 9 should be followed.

RECYCLED-CONCRETE AGGREGATE

In recent years, the concept of using old concrete pavements, buildings, and other structures as a source of aggregate has been demonstrated on several projects, resulting in both material and energy savings (ECCO 1999). The procedure involves (1) breaking up and removing the old concrete, (2) crushing in primary and secondary crushers (Fig. 5-25), (3) removing reinforcing steel and other embedded items, (4) grading and washing, and (5) finally stockpiling the resulting coarse and fine aggregate (Fig. 5-26). Dirt, gypsum board, wood, and other foreign materials should be prevented from contaminating the final product.

Recycled concrete is simply old concrete that has been crushed to produce aggregate. Recycled-concrete aggregate is primarily used in pavement reconstruction. It has been satisfactorily used as an aggregate in granular sub-

Fig. 5-25. Heavily reinforced concrete is crushed with a beamcrusher. (69779)

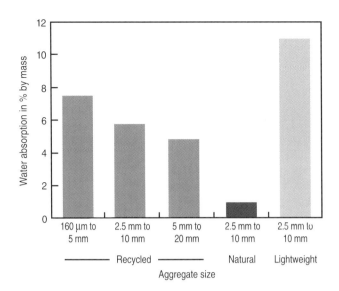

Fig. 5-27. Comparison of water absorption of three different recycled aggregate particle sizes and natural coarse and low-density coarse aggregate (Kerkhoff and Siebel 2001).

Fig. 5-26. Stockpile of recycled-concrete aggregate. (69813)

bases, lean-concrete subbases, soil-cement, and in new concrete as the only source of aggregate or as a partial replacement of new aggregate.

Recycled-concrete aggregate generally has a higher absorption and a lower specific gravity than conventional aggregate. This results from the high absorption of porous mortar and hardened cement paste within the recycled concrete aggregate. Absorption values typically range from 3% to 10%, depending on the concrete being recycled; this absorption lies between those values for natural and low-density aggregate. The specific gravity decreases progressively as particle size decreases. The values increase as coarse particle size decreases (Fig. 5-27). The high absorption of the recycled aggregate makes it necessary to add more water to achieve the same workability and slump than for concrete with conventional aggregates. Dry recycled aggregate absorbs water during and after mixing. To avoid this, recycled aggregate should be prewetted or stockpiles should be kept moist.

The particle shape is similar to crushed rock as shown in Fig. 5-28. The relative density decreases progressively as particle size decreases.

The sulphate content of recycled-concrete aggregate should be determined to assess the possibility of deleterious sulphate reactivity. The chloride content should also be determined where applicable (Stark 1996).

New concrete made from recycled concrete aggregate generally has good durability. Carbonation, permeability, and resistance to freeze-thaw action have been found to be the same or even better than concrete with conventional aggregates. Concrete made with recycled coarse aggregates and conventional fine aggregate can obtain an adequate compressive strength. The use of recycled fine aggregate can result in minor compressive strength reductions. However, drying shrinkage and creep of concrete

Fig. 5-28. Recycled-concrete aggregate. (69812)

made with recycled aggregates is up to 100% higher than concrete with a corresponding conventional aggregate. This is due to the large amount of old cement paste and mortar especially in the fine aggregate. Therefore, considerably lower values of drying shrinkage can be achieved using recycled coarse aggregate with natural sand (Kerkhoff and Siebel 2001). As with any new aggregate source, recycled concrete aggregate should be tested for durability, gradation, and other properties.

Recycled concrete used as coarse aggregate in new concrete possesses some potential for alkali-silica-reaction if the old concrete contained alkali-reactive aggregate. The alkali content of the cement used in the old concrete has little effect on expansion due to alkali-silica-reaction. For highly reactive aggregates made from recycled concrete special measures discussed under "Alkali-Silica Reaction" should be used to control ASR. Also, even if expansive ASR did not develop in the original concrete it can not be assumed that it will not develop in the new concrete if special control measures are not taken. Petrographic examination and expansion tests are recommended to make this judgment.

Concrete trial mixtures should be made to check the new concrete's quality and to determine the proper mixture proportions. One potential problem with using recycled concrete is the variability in the properties of the old concrete that may in turn affect the properties of the new concrete. This can partially be avoided by frequent monitoring of the properties of the old concrete that is being recycled. Adjustments in the mixture proportions may then be needed.

REFERENCES

AASHTO, *Guide Specification For Highway Construction SECTION 56X Portland Cement Concrete Resistant to Excessive Expansion Caused by Alkali-Silica Reaction* (Appendix F to ASR Transition Plan), http://leadstates.tamu.edu /ASR/library/gspec.stm, 2001.

ACI Committee 201, *Guide to Durable Concrete,* ACI 201.2R-92, reapproved 1997, American Concrete Institute, Farmington Hills, Michigan, 1992.

ACI Committee 216, *Guide for Determining the Fire Endurance of Concrete Elements,* ACI 216R-89, reapproved 1994, American Concrete Institute, Farmington Hills, Michigan, 1989.

ACI Committee 221, *Guide for Use of Normal Weight Aggregates in Concrete,* ACI 221R-96, American Concrete Institute, Farmington Hills, Michigan, 1996.

ACI Committee 221, *Guide to Alkali Aggregate Reactions,* ACI 221.1-98, American Concrete Institute, Farmington Hills, Michigan, 1998.

ACI Committee 515, *A Guide to the Use of Waterproofing, Dampproofing, Protective, and Decorative Barrier Systems for Concrete,* ACI 515.1R-79, revised 1985, American Concrete Institute, Farmington Hills, Michigan, 1979.

Barksdale, Richard D., *The Aggregate Handbook,* National Stone Association, Washington D.C., 1991.

Bérubé, M. A.; Fournier, B.; and Durant, B., *Alkali-Aggregate Reaction in Concrete,* 11th International Conference, Québec City, Canada, June 2000.

Bhatty, Muhammad S. Y., "Mechanism of Pozzolanic Reactions and Control of Alkali-Aggregate Expansion," *Cement, Concrete, and Aggregates,* American Society for Testing and Materials, West Conshohocken, Pennsylvania, Winter 1985.

Brown, L. S., *Some Observations on the Mechanics of Alkali-Aggregate Reaction,* Research Department Bulletin RX054, Portland Cement Association, http://www.portcement. org/pdf_files/RX054.pdf, 1955.

Buck, Alan D., "Recycled Concrete as a Source of Aggregate," *ACI Journal,* American Concrete Institute, Farmington Hills, Michigan, May 1977, pages 212 to 219.

Buck, Alan D., and Mather, Katharine, *Reactivity of Quartz at Normal Temperatures,* Technical Report SL-84-12, Structures Laboratory, Waterways Experiment Station, U.S. Army Corps of Engineers, Vicksburg, Mississippi, July 1984.

Buck, Alan D.; Mather, Katharine; and Mather, Bryant, *Cement Composition and Concrete Durability in Sea Water,* Technical Report SL-84-21, Structures Laboratory, Waterways Experiment Station, U.S. Army Corps of Engineers, Vicksburg, Mississippi, December 1984.

Bureau of Reclamation, *Concrete Manual,* 8th ed., U.S. Bureau of Reclamation, Denver, 1981.

Crouch, L. K.; Sauter, Heather J.; and Williams, Jakob A., "92-Mpa Air-entrained HPC," TRB-Record 1698, *Concrete 2000,* page 24.

CSA A23.1-00/A23.2-00, *Concrete Materials and Methods of Concrete Construction/Methods of Test for Concrete,* Canadian Standards Association, Toronto, 2000.

CSA A864-00, *Guide to the Evaluation and Management of Concrete Structures Affected by Alkali-Aggregate Reaction,* Canadian Standards Association, Toronto, 2000.

CSA A3000-98, *Cementitious Materials Compendium,* Canadian Standards Association, Toronto, 1998.

ECCO (Environmental Council of Concrete Organizations), *"Recycling Concrete and Masonry,"* EV 22, Skokie, Illinois, http://www.ecco.org/pdfs/ev22.pdf, 1999, 12 pages.

EPA, *Acid Rain, Research Summary,* EPA-600/8-79-028, U.S. Environmental Protection Agency, Washington, D.C., October 1979.

EPA, *Final Report, U.S. EPA Workshop on Acid Deposition Effects on Portland Cement Concrete and Related Materials,* Atmospheric Sciences Research Laboratory, U.S. Environmental Protection Agency, Research Triangle Park, North Carolina, February 1986.

Farny, James A. and Kosmatka, Steven H., *Diagnosis and Control of Alkali-Aggregate Reactions,* IS413, Portland Cement Association, 1997, 24 pages.

Fournier, B., *CANMET/Industry Joint Research Program on Alkali-Aggregate Reaction—Fourth Progress Report,* Canada Centre for Mineral and Energy Technology, Ottawa, 1997.

Galloway, Joseph E., Jr., "Grading, Shape and Surface Properties," *Significance of Tests and Properties of Concrete and Concrete-Making Materials,* ASTM STP 169C, edited by Klieger, Paul and Lamond, Joseph F., American Society for Testing and Materials, West Conshohocken, Pennsylvania, 1994, pages 401 to 410.

Hadley, D. W., *"Alkali Reactivity of Carbonate Rocks—Expansion and Dedolomitization,"* Research Department Bulletin RX139, Portland Cement Association, http://www.portcement.org/pdf_files/RX139.pdf, 1961.

Helmuth, Richard, *Alkali-Silica Reactivity: An Overview of Research,* SHRP-C-342, Strategic Highway Research Program, Washington, D. C., 1993. Also PCA Publication LT177, 105 pages.

Houston, B. J., *Investigation of Gap-Grading of Concrete Aggregates; Review of Available Information,* Technical Report No. 6-593, Report 1, Waterways Experiment Station, U.S. Army Corps of Engineers, Vicksburg, Mississippi, February 1962.

Janssen, Donald J., and Snyder Mark B., "Resistance of Concrete to Freezing and Thawing," SHRP-C-391, Strategic Highway Research Program, Washington, D.C., 1994, 201 pages.

Kerkhoff, Beatrix, *Effects of Substances on Concrete and Guide to Protective Treatments,* IS001, Portland Cement Association, 2001, 24 pages.

Kerkhoff, Beatrix and Siebel, Eberhard, "Properties of Concrete with Recycled Aggregates (Part 2)," *Beton 2/2001,* Verlag Bau + Technik, 2001, pages 105 to 108.

Kong, Hendrik, and Orbison, James G., "Concrete Deterioration Due to Acid Precipitation," *ACI Materials Journal,* American Concrete Institute, Farmington Hills, Michigan, March-April 1987.

Litvin, Albert, and Pfeifer, Donald W., *Gap-Graded Mixes for Cast-in-Place Exposed Aggregate Concrete,* Development Department Bulletin DX090, Portland Cement Association, http://www.portcement.org/pdf_files/DX090.pdf, 1965.

Maerz, Norbert H., and Lusher, Mike, "Measurement of flat and elongation of coarse aggregate using digital image processing," *80th Annual Meeting, Transportation Research Board,* Washington D.C., 2001, pages 2 to 14.

Mather, Bryant, *New Concern over Alkali-Aggregate Reaction,* Joint Technical Paper by National Aggregates Association and National Ready Mixed Concrete Association, NAA Circular No. 122 and NRMCA Publication No. 149, Silver Spring, Maryland, 1975.

Matthews, C. W., "Stockpiling of Materials," *Rock Products,* series of 21 articles, Maclean Hunter Publishing Company, Chicago, August 1965 through August 1967.

National Cooperative Highway Research Program (NCHRP), *Effects of Different Methods of Stockpiling and Handling Aggregates,* NCHRP Report 46, Transportation Research Board, Washington, D.C., 1967.

Ozol, Michael A., "Alkali-Carbonate Rock Reaction," *Significance of Tests and Properties of Concrete and Concrete-Making Materials,* ASTM STP 169C, edited by Klieger, Paul and Lamond, Joseph F., American Society for Testing and Materials, Philadelphia, 1994, pages 372 to 387.

PCA, *Tests of Concrete Road Materials from California,* Major Series 285, Research Reports, Portland Cement Association, April 1940.

PCA, *Recycling D-Cracked Pavement in Minnesota,* PL146, Portland Cement Association, 1980.

PCA, "Popouts: Causes, Prevention, Repair," *Concrete Technology Today*, PL852, Portland Cement Association, http://www.portcement.org/pdf_files/PL852.pdf, June 1985.

PCA Durability Subcommittee, *Guide Specification for Concrete Subject to Alkali-Silica Reactions*, IS415, Portland Cement Association, 1998.

PCA, "Controlling ASR," *Concrete Technology Today*, PL971, Portland Cement Association, http://www.portcement.org/pdf_files/PL971.pdf, April 1997.

Powers, Laura J., "Developments in Alkali-Silica Gel Detection," *Concrete Technology Today*, PL991, Portland Cement Association, http://www.portcement.org/pdf_files/PL991.pdf, April 1999.

Scanlon, John M., *Concrete Durability, Proceedings of the Katherine and Bryant Mathes International Conference*, SP100, American Concrete Institute, Farmington Hills, Michigan, 1987.

Shilstone, James M., "Changes in Concrete Aggregate Standards," *The Construction Specifier*, Alexandria, Virginia, July 1994, pages 119 to 128.

Shilstone, James M., Sr., "Concrete Mixture Optimization," *Concrete International*, American Concrete Institute, Farmington Hills, Michigan, June 1990, pages 33 to 39.

Stanton, Thomas E., "Expansion of Concrete through Reaction between Cement and Aggregate," *Proceedings, American Society of Civil Engineers*, Vol. 66, New York, 1940, pages 1781 to 1811.

Stark, David, *Characteristics and Utilization of Coarse Aggregates Associated with D-Cracking*, Research and Development Bulletin RD047, Portland Cement Association, http://www.portcement.org/pdf_files/RD047.pdf, 1976.

Stark, D. C., *Alkali-Silica Reactivity: Some Reconsiderations*, Research and Development Bulletin RD076, Portland Cement Association, http://www.portcement.org/pdf_files/RD076.pdf, 1981.

Stark, David, *"The Use of Recycled-Concrete Aggregate from Concrete Exhibiting Alkali-Silica Reactivity,"* Research and Development Bulletin RD114, Portland Cement Association, 1996.

Stark, David, and Klieger, Paul, *Effect of Maximum Size of Coarse Aggregate on D-Cracking in Concrete Pavements*, Research and Development Bulletin RD023, Portland Cement Association, http://www.portcement.org/pdf_files/RD023.pdf, 1974.

Stark, David, *Eliminating or Minimizing Alkali-Silica Reactivity*, SHRP-C-343, Strategic Highway Research Program, Washington, D. C., 1993. Also PCA Publication LT178, 266 pages.

Swenson, E. G., and Gillott, J. E., "Alkali Reactivity of Dolomitic Limestone Aggregate," *Magazine of Concrete Research*, Vol. 19, No. 59, Cement and Concrete Association, London, June 1967, pages 95 to 104.

Tang, Mingshu; Deng, Min; Lon, Xianghui; and Han, Sufeng, "Studies on Alkali-Carbonate Reaction," *ACI Materials Journal*, American Concrete Institute, Farmington Hills, Michigan, January-February 1994, pages 26 to 29.

Thomas, M. D. A.; Hooton, R. Doug; and Rogers, C. A., "Prevention of Damage Due to Alkali-Aggregate Reaction (AAR) in Concrete Construction—Canadian Approach," *Cement, Concrete, and Aggregates*, American Society for Testing and Materials, West Conshohocken, Pennsylvania, 1997, pages 26 to 30.

Thomas, M. D. A., and Innis, F. A., "Effect of Slag on Expansion Due to Alkali-Aggregate Reaction in Concrete," *ACI Materials Journal*, American Concrete Institute, Farmington Hills, Michigan, November-December 1998.

Touma, W. E.; Fowler, D. W.; and Carrasquillo, R. L., *Alkali-Silica Reaction in Portland Cement Concrete: Testing Methods and Mitigation Alternatives*, Research Report ICAR 301-1F, University of Texas, Austin, 2001, 520 pages.

Verbeck, George, and Landgren, Robert, *Influence of Physical Characteristics of Aggregates on Frost Resistance of Concrete*, Research Department Bulletin RX126, Portland Cement Association, http://www.portcement.org/pdf_files/RX126.pdf, 1960.

Vogler, R. H., and Grove, G. H., "Freeze-thaw testing of coarse aggregate in concrete: Procedures used by Michigan Department of Transportation and other agencies," *Cement, Concrete, and Aggregates*, American Society for Testing and Materials, West Conshohocken, Pennsylvania, Vol. 11, No. 1, Summer 1989, pages 57 to 66.

CHAPTER 6
Admixtures for Concrete

Admixtures are those ingredients in concrete other than portland cement, water, and aggregates that are added to the mixture immediately before or during mixing (Fig. 6-1). Admixtures can be classified by function as follows:

1. Air-entraining admixtures
2. Water-reducing admixtures
3. Plasticizers
4. Accelerating admixtures
5. Retarding admixtures
6. Hydration-control admixtures
7. Corrosion inhibitors
8. Shrinkage reducers
9. Alkali-silica reactivity inhibitors
10. Colouring admixtures
11. Miscellaneous admixtures such as workability, bonding, dampproofing, permeability reducing, grouting, gas-forming, antiwashout, foaming, and pumping admixtures

Table 6-1 provides a much more extensive classification of admixtures.

Note: Conformance requirements for air-entraining and chemical admixtures within CSA Standard A23.1-00, *Concrete Materials and Methods of Concrete Construction*, references ASTM Standards. There are no separate CSA standards for admixtures.

Concrete should be workable, finishable, strong, durable, watertight, and wear resistant. These qualities can often be obtained easily and economically by the selection of suitable materials rather than by resorting to admixtures (except air-entraining admixtures when needed).

The major reasons for using admixtures are:

1. To reduce the cost of concrete construction
2. To achieve certain properties in concrete more effectively than by other means

Fig. 6-1. Liquid admixtures, from left to right: antiwashout admixture, shrinkage reducer, water reducer, foaming agent, corrosion inhibitor, and air-entraining admixture. (69795)

3. To maintain the quality of concrete during the stages of mixing, transporting, placing, and curing in adverse weather conditions
4. To overcome certain emergencies during concreting operations

Despite these considerations, it should be borne in mind that no admixture of any type or amount can be considered a substitute for good concreting practice.

The effectiveness of an admixture depends upon factors such as type and amount of cementing materials; water content; aggregate shape, gradation, and proportions; mixing time; slump; and temperature of the concrete.

Admixtures being considered for use in concrete should meet applicable specifications as presented in Table 6-1. Trial mixtures should be made with the admixture and the job materials at temperatures and humidities anticipated on the job. In this way the compatibility of the admixture with other admixtures and job materials, as

Table 6-1. Concrete Admixtures by Classification

Type of admixture	Desired effect	Material
Accelerators (ASTM C 494, Type C)	Accelerate setting and early-strength development	Calcium chloride (ASTM D 98) Triethanolamine, sodium thiocyanate, calcium formate, calcium nitrite, calcium nitrate
Air detrainers	Decrease air content	Tributyl phosphate, dibutyl phthalate, octyl alcohol, water-insoluble esters of carbonic and boric acid, silicones
Air-entraining admixtures (ASTM C 260)	Improve durability in freeze-thaw, deicer, sulphate, and alkali-reactive environments Improve workability	Salts of wood resins (Vinsol resin), some synthetic detergents, salts of sulphonated lignin, salts of petroleum acids, salts of proteinaceous material, fatty and resinous acids and their salts, alkylbenzene sulphonates, salts of sulphonated hydrocarbons
Alkali-aggregate reactivity inhibitors	Reduce alkali-aggregate reactivity expansion	Barium salts, lithium nitrate, lithium carbonate, lithium hydroxide
Antiwashout admixtures	Cohesive concrete for underwater placements	Cellulose, acrylic polymer
Bonding admixtures	Increase bond strength	Polyvinyl chloride, polyvinyl acetate, acrylics, butadiene-styrene copolymers
Colouring admixtures (ASTM C 979)	Coloured concrete	Modified carbon black, iron oxide, phthalocyanine, umber, chromium oxide, titanium oxide, cobalt blue
Corrosion inhibitors	Reduce steel corrosion activity in a chloride-laden environment	Calcium nitrite, sodium nitrite, sodium benzoate, certain phosphates or fluosilicates, fluoaluminates, ester amines
Dampproofing admixtures	Retard moisture penetration into dry concrete	Soaps of calcium or ammonium stearate or oleate Butyl stearate Petroleum products
Foaming agents	Produce foamed concrete with low density	Cationic and anionic surfactants Hydrolized protein
Fungicides, germicides, and insecticides	Inhibit or control bacterial and fungal growth	Polyhalogenated phenols Dieldrin emulsions Copper compounds
Gas formers	Cause expansion before setting	Aluminum powder
Grouting admixtures	Adjust grout properties for specific applications	See Air-entraining admixtures, Accelerators, Retarders, and Water reducers
Hydration control admixtures	Suspend and reactivate cement hydration with stabilizer and activator	Carboxylic acids Phosphorus-containing organic acid salts
Permeability reducers	Decrease permeability	Latex Calcium stearate
Pumping aids	Improve pumpability	Organic and synthetic polymers Organic flocculents Organic emulsions of paraffin, coal tar, asphalt, acrylics Bentonite and pyrogenic silicas Hydrated lime (ASTM C 141)
Retarders (ASTM C 494, Type B)	Retard setting time	Lignin Borax Sugars Tartaric acid and salts
Shrinkage reducers	Reduce drying shrinkage	Polyoxyalkylene alkyl ether Propylene glycol
Superplasticizers* (ASTM C 1017, Type 1)	Increase flowability of concrete Reduce water-cementing materials ratio	Sulphonated melamine formaldehyde condensates Sulphonated naphthalene formaldehyde condensates Lignosulphonates Polycarboxylates

Table 6-1. Concrete Admixtures by Classification (Continued)

Type of admixture	Desired effect	Material
Superplasticizer* and retarder (ASTM C 1017, Type 2)	Increase flowability with retarded set Reduce water–cementing materials ratio	See superplasticizers and also water reducers
Water reducer (ASTM C 494, Type A)	Reduce water content at least 5%	Lignosulphonates Hydroxylated carboxylic acids Carbohydrates (Also tend to retard set so accelerator is often added)
Water reducer and accelerator (ASTM C 494, Type E)	Reduce water content (minimum 5%) and accelerate set	See water reducer, Type A (accelerator is added)
Water reducer and retarder (ASTM C 494, Type D)	Reduce water content (minimum 5%) and retard set	See water reducer, Type A (retarder is added)
Water reducer—high range (ASTM C 494, Type F)	Reduce water content (minimum 12%)	See superplasticizers
Water reducer—high range—and retarder (ASTM C 494, Type G)	Reduce water content (minimum 12%) and retard set	See superplasticizers and also water reducers
Water reducer—mid range	Reduce water content (between 6 and 12%) without retarding	Lignosulphonates Polycarboxylates

* Superplasticizers are also referred to as high-range water reducers or plasticizers. These admixtures often meet both ASTM C 494 and ASTM C 1017 specifications.

well as the effects of the admixture on the properties of the fresh and hardened concrete, can be observed. The amount of admixture recommended by the manufacturer or the optimum amount determined by laboratory tests should be used.

AIR-ENTRAINING ADMIXTURES

Air-entraining admixtures are used to purposely introduce and stabilize microscopic air bubbles in concrete. Air-entrainment will dramatically improve the durability of concrete exposed to cycles of freezing and thawing (Fig. 6-2). Entrained air greatly improves concrete's resistance to surface scaling caused by chemical deicers (Fig. 6-3). Furthermore, the workability of fresh concrete is improved significantly, and segregation and bleeding are reduced or eliminated.

Air-entrained concrete contains minute air bubbles that are distributed uniformly throughout the cement paste. In Canada, entrained air is produced in concrete by the introduction of an air-entraining admixture. However in the United States, in addition to air-entraining admixtures, air-entraining cements are also available which

allow air entrained concrete to be produced by the use of an admixture, an air-entrained cement, or by a combination of both methods. An air-entraining cement is a portland cement with an air-entraining addition interground with the clinker during manufacture. An air-entraining admixture, on the other hand, is added directly to the concrete materials either before or during mixing.

The primary ingredients used in air-entraining admixtures are listed in Table 6-1. Specifications and methods of testing air-entraining admixtures are given in ASTM C 260 and C 233. See Chapter 8, Air-Entrained Concrete, Klieger (1966), and Whiting and Nagi (1998) for more information.

WATER-REDUCING ADMIXTURES

Water-reducing admixtures are used to reduce the quantity of mixing water required to produce concrete of a certain slump, reduce water-cement ratio, reduce cement content, or increase slump. Typical water reducers reduce the water content by approximately 5% to 10%. Adding a water-reducing admixture to concrete without reducing the water content can produce a mixture with a higher slump.

Fig. 6-3. Scaled concrete surface resulting from lack of air entrainment, use of deicers, and poor finishing and curing practices. (52742)

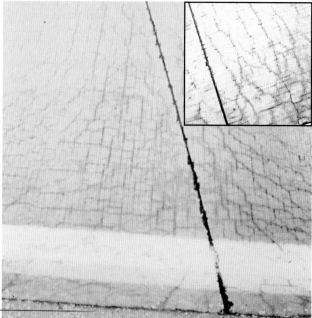

Fig. 6-2. Frost damage (crumbling) at joints of a pavement (top), frost induced cracking near joints (bottom), and enlarged view of cracks (inset). (61621, 67834, 67835)

The rate of slump loss, however, is not reduced and in most cases is increased (Fig. 6-4). Rapid slump loss results in reduced workability and less time to place concrete.

An increase in strength is generally obtained with water-reducing admixtures as the water-cement ratio is reduced. For concretes of equal cement content, air content, and slump, the 28-day strength of a water-reduced concrete containing a water reducer can be 10% to 25%

greater than concrete without the admixture. Despite reduction in water content, water-reducing admixtures may cause increases in drying shrinkage. Usually the effect of the water reducer on drying shrinkage is small compared to other more significant factors that cause shrinkage cracks in concrete. Using a water reducer to reduce the cement and water content of a concrete mixture—while maintaining a constant water-cement ratio—can result in equal or reduced compressive strength, and can increase slump loss by a factor of two or more (Whiting and Dziedzic 1992).

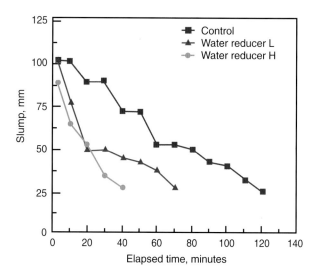

Fig. 6-4. Slump loss at 23°C in mixtures containing conventional water reducers (ASTM C 494, Type D) compared with a control mixture (Whiting and Dziedzic 1992).

Water reducers decrease, increase, or have no effect on bleeding, depending on the chemical composition of the admixture. A reduction of bleeding can result in finishing difficulties on flat surfaces when rapid drying conditions are present. Water reducers can be modified to give varying degrees of retardation while others do not significantly affect the setting time. ASTM C 494 Type A water reducers can have little effect on setting, while Type D admixtures provide water reduction with retardation, and Type E admixtures provide water reduction with accelerated setting. Type D water-reducing admixtures usually retard the setting time of concrete by one to three hours (Fig. 6-5). Some water-reducing admixtures may also entrain some air in concrete. Lignin-based admixtures can increase air contents by 1 to 2 percentage points. Concretes with water reducers generally have good air retention (Table 6-2).

The effectiveness of water reducers on concrete is a function of their chemical composition, concrete temperature, cement composition and fineness, cement content, and the presence of other admixtures. The classifications and components of water reducers are listed in Table 6-1. See Whiting and Dziedzic (1992) for more information on the effects of water reducers on concrete properties.

MID-RANGE WATER REDUCING ADMIXTURES

Mid-range water reducers were first introduced in 1984. These admixtures provide significant water reduction (between 6 and 12%) for concretes with slumps of 125 to 200 mm without the retardation associated with high dosages of conventional (normal) water reducers. Normal water reducers are intended for concretes with slumps of 100 to 125 mm. Mid-range water reducers can be used to reduce stickiness and improve finishability, pumpability, and placeability of concretes containing silica fume and other supplementary cementing materials. Some can also entrain air and be used in low slump concretes (Nmai, Schlagbaum, and Violetta 1998).

HIGH-RANGE WATER REDUCING ADMIXTURES

High-range water reducers, ASTM C 494 Types F (water reducing) and G (water reducing and retarding), can be used to impart properties induced by regular water reducers, only much more efficiently. They can greatly reduce water demand and cementing material contents and make low water-cementing materials ratio, high-strength concrete with normal or enhanced workability. A water reduction of 12% to 30% can be obtained through the use of these admixtures. The reduced water content and water-cement-

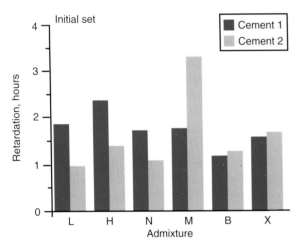

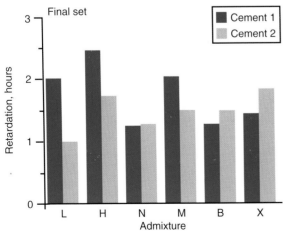

Fig. 6-5. Retardation of set in cement-reduced mixtures relative to control mixture. Concretes L and H contain conventional water reducer, concretes N, M, B, and X contain high-range water reducer (Whiting and Dziedzic 1992).

Fig. 6-6. Low water-to-cement ratio concrete with low chloride permeability—easily made with high-range water reducers—is ideal for bridge decks. (69924)

ing materials ratio can produce concretes with (1) ultimate compressive strengths in excess of 70 MPa, (2) increased early strength gain, (3) reduced chloride-ion penetration, and (4) other beneficial properties associated with low water-cementing materials ratio concrete (Fig. 6-6).

High-range water reducers are generally more effective than regular water-reducing admixtures in producing workable concrete. A significant reduction of bleeding can result with large reductions of water content; this can result in finishing difficulties on flat surfaces when rapid drying conditions are present. Some of these admixtures can cause significant slump loss (Fig. 6-7). Significant retardation is also possible, but can aggravate plastic shrinkage cracking without proper protection and curing (Fig. 6-5). Drying shrinkage, chloride permeability, air retention (Table 6-2), and strength development of concretes with high-range water reducers are comparable to concretes without them when compared at constant water-cementing materials ratios (reduced cementing materials and water contents) (Fig. 6-8).

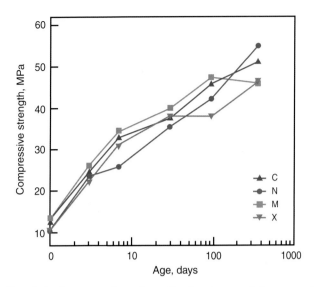

Fig. 6-8. Compressive strength development in cement-reduced concretes: control mixture (C) and concretes containing high-range water reducers (N, M, and X) (Whiting and Dziedzic 1992).

Concretes with high-range water reducers can have larger entrained air voids and higher void-spacing factors than normal air-entrained concrete. This would generally indicate a reduced resistance to freezing and thawing; however, laboratory tests have shown that concretes with a moderate slump using high-range water reducers have good freeze-thaw durability, even with slightly higher void-spacing factors. This may be the result of lower water-cement ratios often associated with these concretes.

When the same chemicals used for high-range water reducers are used to make flowing concrete, they are often called plasticizers or superplasticizers (see discussion that follows).

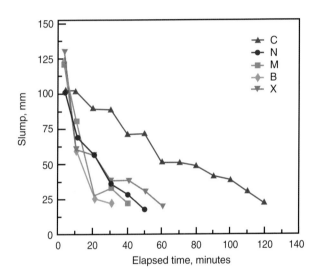

Fig. 6-7. Slump loss at 23°C in mixtures containing high-range water reducers (N, M, B, and X) compared with control mixture (C) (Whiting and Dziedzic 1992).

Table 6-2. Loss of Air from Cement Reduced Concrete Mixtures

Mixture		Initial air content, %*	Final air content, %†	Percent air retained	Rate of air loss, %/minute
C	Control	5.4	3.0	56	0.020
L	} Water	7.0	4.7	67	0.038
H	} reducer	6.2	4.6	74	0.040
N		6.8	4.8	71	0.040
M	} High-range	6.4	3.8	59	0.065
B	} water	6.8	5.6	82	0.048
X	} reducer	6.6	5.0	76	0.027

* Represents air content measured after addition of admixture.
† Represents air content taken at point where slump falls below 25 mm.
Whiting and Dziedzic 1992

Fig. 6-9. Flowable concrete with a high slump (top) is easily placed (middle), even in areas of heavy reinforcing steel congestion (bottom). (47343, 69900, 47344)

PLASTICIZERS FOR FLOWING CONCRETE

Plasticizers, often called superplasticizers, are essentially high-range water reducers meeting ASTM C 1017; these admixtures are added to concrete with a low-to-normal slump and water-cementing materials ratio to make high-slump flowing concrete (Fig. 6-9). Flowing concrete is a highly fluid but workable concrete that can be placed with

Fig. 6-10. Plasticized, flowing concrete is easily placed in thin sections such as this bonded overlay that is not much thicker than 1½ diameters of a quarter. (69874)

little or no vibration or compaction while still remaining essentially free of excessive bleeding or segregation. Following are a few of the applications where flowing concrete is used: (1) thin-section placements (Fig. 6-10), (2) areas of closely spaced and congested reinforcing steel, (3) tremie pipe (underwater) placements, (4) pumped concrete to reduce pump pressure, thereby increasing lift and distance capacity, (5) areas where conventional consolidation methods are impractical or can not be used, and (6) for reducing handling costs. The addition of a plasticizer to a 75-mm slump concrete can easily produce a concrete with a 225-mm slump. Flowing concrete is defined by ASTM C 1017 as a concrete having a slump greater than 190 mm, yet maintaining cohesive properties.

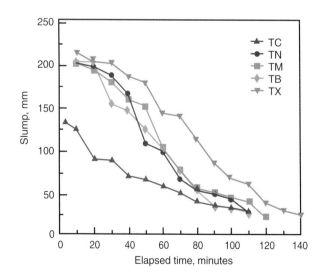

Fig. 6-11. Slump loss at 32°C in flowing concretes (TN, TM, TB, and TX) compared with control mixture (TC) (Whiting and Dziedzic 1992).

ASTM C 1017 has provisions for two types of admixtures: Type 1—plasticizing, and Type 2—plasticizing and retarding. Plasticizers are generally more effective than regular or mid-range water-reducing admixtures in producing flowing concrete. The effect of certain plasticizers in increasing workability or making flowing concrete is short-lived, 30 to 60 minutes; this period is followed by a rapid loss in workability or slump loss (Fig. 6-11). High temperatures can also aggravate slump loss. Due to their propensity for slump loss, these admixtures are sometimes added to the concrete mixer at the jobsite. They are available in liquid and powder form. Extended-slump-life plasticizers added at the batch plant help reduce slump-loss problems. Setting time may be accelerated or retarded based on the admixture's chemistry, dosage rate, and interaction with other admixtures and cementing materials in the concrete mixture. Some plasticizers can retard final set by one to almost four hours (Fig. 6-12). Strength development of flowing concrete is comparable to normal concrete (Fig. 6-13).

While it was previously noted that flowing concretes are essentially free of excessive bleeding, tests have shown that some plasticized concretes bleed more than control concretes of equal water-cement ratio (Fig. 6-14); but plasticized concretes bleed significantly less than control concretes of equally high slump and higher water content. High-slump, low-water-content, plasticized concrete has less drying shrinkage than a high-slump, high-water-content conventional concrete; however this concrete has similar or higher drying shrinkage than conventional low-slump, low-water-content concrete (Whiting 1979, Gebler 1982, and Whiting and Dziedzic 1992).

The effectiveness of the plasticizer is increased with an increasing amount of cement and fines in the concrete. It is also affected by the initial slump of the concrete.

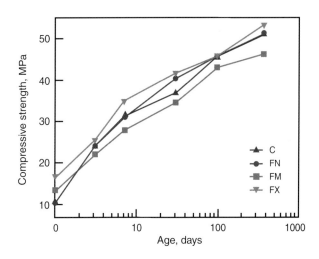

Fig. 6-13. Compressive strength development in flowing concretes. C is the control mixture. Mixtures FN, FM, and FX contain plasticizers (Whiting and Dziedzic 1992).

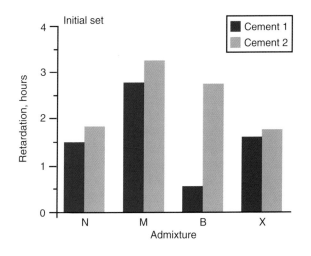

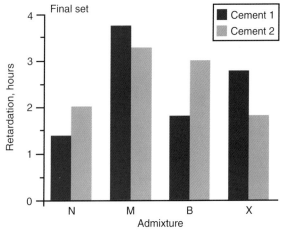

Fig. 6-12. Retardation of set in flowing concrete with plasticizers (N, M, B, and X) relative to control mixture (Whiting and Dziedzic 1992).

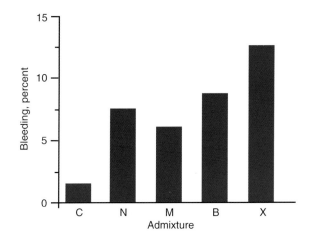

Fig. 6-14. Bleeding of flowing concretes with plasticizers (N, M, B, and X) compared to control (C) (Whiting and Dziedzic 1992).

Plasticized flowing concrete can have larger entrained air voids and greater void-spacing factors than conventional concrete. Air loss can also be significant. Some research has indicated poor frost- and deicer-scaling resistance for some flowing concretes when exposed to a continuously moist environment without the benefit of a drying period (Whiting and Dziedzic 1992). However, field performance of flowing concretes with low water to portland cement ratios has been good in most frost environments.

Table 6-1 lists the primary components and specifications for plasticizing (superplasticizer) admixtures.

RETARDING ADMIXTURES

Retarding admixtures are used to delay the rate of setting of concrete. High temperatures of fresh concrete (30°C) are often the cause of an increased rate of hardening that makes placing and finishing difficult. One of the most practical methods of counteracting this effect is to reduce the temperature of the concrete by cooling the mixing water and/or the aggregates. Retarders do not decrease the initial temperature of concrete. The bleeding rate and bleeding capacity of concrete is increased with retarders.

Retarding admixtures are useful in extending the setting time of concrete, but they are often also used in attempts to decrease slump loss and extend workability, especially prior to placement at elevated temperatures. The fallacy of this approach is shown in Fig. 6-15, where the addition of a retarder resulted in an increased rate of slump loss compared to the control mixtures (Whiting and Dziedzic 1992).

Retarders are sometimes used to: (1) offset the accelerating effect of hot weather on the setting of concrete; (2) delay the initial set of concrete or grout when difficult or unusual conditions of placement occur, such as placing concrete in large piers and foundations, cementing oil wells, or pumping grout or concrete over considerable distances; or (3) delay the set for special finishing techniques, such as an exposed aggregate surface.

The amount of water reduction for an ASTM C 494 Type B retarding admixture is normally less than that obtained with a Type A water reducer. Type D admixtures are designated to provide both water reduction and retardation.

In general, some reduction in strength at early ages (one to three days) accompanies the use of retarders. The effects of these materials on the other properties of concrete, such as shrinkage, may not be predictable. Therefore, acceptance tests of retarders should be made with actual job materials under anticipated job conditions. The classifications and components of retarders are listed in Table 6-1.

HYDRATION-CONTROL ADMIXTURES

Hydration controlling admixtures became available in the late 1980s. They consist of a two-part chemical system: (1) a stabilizer or retarder that essentially stops the hydration of cementing materials, and (2) an activator that reestablishes normal hydration and setting when added to the stabilized concrete. The stabilizer can suspend hydration for 72 hours and the activator is added to the mixture just before the concrete is used. These admixtures make it possible to reuse concrete returned in a ready-mix truck by suspending setting overnight. The admixture is also useful in maintaining concrete in a stabilized non-hardened state during long hauls. The concrete is reactivated when it arrives at the project. This admixture presently does not have a standard specification (Kinney 1989).

ACCELERATING ADMIXTURES

An accelerating admixture is used to accelerate the rate of hydration (setting) and strength development of concrete at an early age. The strength development of concrete can also be accelerated by other methods: (1) using Type 30 (Type III) cement, (2) lowering the water-cementing materials ratio by adding 60 to 120 kg/m³ of additional cement to the concrete, (3) using a water reducer, or (4) curing at higher temperatures. Accelerators are designated as Type C admixtures under ASTM C 494.

Calcium chloride ($CaCl_2$) is the chemical most commonly used in accelerating admixtures, especially for non-reinforced concrete. It should conform to the requirements of ASTM D 98 and should be sampled and tested in accordance with ASTM D 345.

The widespread use of calcium chloride as an accelerating admixtures has provided much data and experience

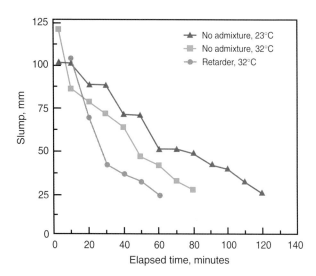

Fig. 6-15. Slump loss at various temperatures for conventional concretes prepared with and without set-retarding admixture (Whiting and Dziedzic 1992).

on the effect of this chemical on the properties of concrete. Besides accelerating strength gain, calcium chloride causes an increase in drying shrinkage, potential reinforcement corrosion, discolouration (a darkening of concrete), and an increase in the potential for scaling.

Calcium chloride is not an antifreeze agent. When used in allowable amounts, it will not reduce the freezing point of concrete by more than a few degrees. Attempts to protect concrete from freezing by this method are foolhardy. Instead, proven reliable precautions should be taken during cold weather (see Chapter 14, Cold-Weather Concreting).

When used, calcium chloride should be added to the concrete mixture in solution form as part of the mixing water. If added to the concrete in dry flake form, all of the dry particles may not be completely dissolved during mixing. Undissolved lumps in the mix can cause popouts or dark spots in hardened concrete.

The amount of calcium chloride added to concrete should be no more than is necessary to produce the desired results and in no case exceed 2% by mass of cementing material. When calculating the chloride content of commercially available calcium chloride, it can be assumed that:

1. Regular flake contains a minimum of 77% $CaCl_2$
2. Concentrated flake, pellet, or granular forms contain a minimum of 94% $CaCl_2$

An overdose can result in placement problems and can be detrimental to concrete. It may cause: rapid stiffening, a large increase in drying shrinkage, corrosion of reinforcement, and loss of strength at later ages (Abrams 1924 and Lackey 1992).

Applications where calcium chloride should be used with caution:

1. Concrete subjected to steam curing
2. Concrete containing embedded dissimilar metals, especially if electrically connected to steel reinforcement
3. Concrete slabs supported on permanent galvanized-steel forms
4. Architectural and coloured concrete

Calcium chloride or admixtures containing soluble chlorides *should not be used* in the following:

1. Construction of parking garages
2. Prestressed concrete because of possible steel corrosion hazards
3. Concrete containing embedded aluminum (for example, conduit) since serious corrosion of the aluminum can result, especially if the aluminum is in contact with embedded steel and the concrete is in a humid environment
4. Concrete containing aggregates that, under standard test conditions, have been shown to be potentially deleteriously reactive
5. Concrete exposed to soil or water containing sulphates
6. Floor slabs intended to receive dry-shake metallic finishes

Table 6-3. Maximum Chloride-Ion Content for Corrosion Protection of Reinforcement*

Type of member	Maximum water soluble chloride ion (Cl⁻) in concrete, percent by mass of cement
Prestressed concrete	0.06
Reinforced concrete exposed to a moist environment or chlorides or both	0.15
Reinforced concrete exposed to neither a moist environment nor chlorides	1.00

* Requirements from CSA A23.1–Section 15.1.6.

7. Hot weather generally
8. Massive concrete placements

The maximum chloride-ion content for corrosion protection of prestressed and reinforced concrete as recommended by the CSA A23.1 is presented in Table 6-3. Resistance to the corrosion of embedded steel is further improved with an increase in the depth of concrete cover over reinforcing steel, and a lower water-cement ratio. Stark (1989) demonstrated that concretes made with 1% $CaCl_2 \cdot 2H_2O$ by mass of cement developed active steel corrosion when stored continuously in fog. When 2% $CaCl_2 \cdot 2H_2O$ was used, active corrosion was detected in concrete stored in a fog room at 100% relative humidity. Risk of corrosion was greatly reduced at lower relative humidities (50%). Gaynor (1998) demonstrates how to calculate the chloride content of fresh concrete and compare it with recommended limits.

Several nonchloride, noncorrosive accelerators are available for use in concrete where chlorides are not recommended (Table 6-1). However, some nonchloride accelerators are not as effective as calcium chloride. Certain nonchloride accelerators are specially formulated for use in cold weather applications with ambient temperatures down to -7°C.

CORROSION INHIBITORS

Corrosion inhibitors are used in concrete for parking structures, marine structures, and bridges where chloride salts are present. The chlorides can cause corrosion of steel reinforcement in concrete (Fig. 6-16). Ferrous oxide and ferric oxide form on the surface of reinforcing steel in concrete. Ferrous oxide, though stable in concrete's alkaline environment, reacts with chlorides to form complexes that move away from the steel to form rust. The chloride ions continue to attack the steel until the passivating oxide layer is destroyed. Corrosion-inhibiting admixtures chemically arrest the corrosion reaction.

Fig. 6-16. The damage to this concrete parking structure resulted from chloride-induced corrosion of steel reinforcement. (50051)

Commercially available corrosion inhibitors include: calcium nitrite, sodium nitrite, dimethyl ethanolamine, amines, phosphates, and ester amines. Anodic inhibitors, such as nitrites, block the corrosion reaction of the chloride-ions by chemically reinforcing and stabilizing the passive protective film on the steel; this ferric oxide film is created by the high pH environment in concrete. The nitrite-ions cause the ferric oxide to become more stable. In effect, the chloride-ions are prevented from penetrating the passive film and making contact with the steel.

A certain amount of nitrite can stop corrosion up to some level of chloride-ion. Therefore, increased chloride levels require increased levels of nitrite to stop corrosion.

Cathodic inhibitors react with the steel surface to interfere with the reduction of oxygen. The reduction of oxygen is the principal cathodic reaction in alkaline environments (Berke and Weil 1994).

SHRINKAGE-REDUCING ADMIXTURES

Shrinkage-reducing admixtures, introduced in the 1980s, have potential uses in bridge decks, critical floor slabs, and buildings where cracks and curling must be minimized for durability or aesthetic reasons (Fig. 6-17). Propylene glycol and polyoxyalkylene alkyl ether have been used as shrinkage reducers. Drying shrinkage reductions of between 25% and 50% have been demonstrated in laboratory tests. These admixtures have negligible effects on slump and air loss, but can delay setting. They are generally compatible with other admixtures (Nmai, Tomita, Hondo and Buffenbarger 1998 and Shah, Weiss and Yang 1998).

CHEMICAL ADMIXTURES TO REDUCE ALKALI-AGGREGATE REACTIVITY (ASR INHIBITORS)

Chemical admixtures to control alkali-silica reactivity (alkali-aggregate expansion) were introduced in the 1990s (Fig. 6-18). Lithium nitrate, lithium carbonate, lithium hydroxide, lithium aluminum silicate (decrepitated spodumene), and barium salts have shown reductions of alkali-silica reaction (ASR) in laboratory tests (Thomas and Stokes 1999 and AASHTO 2001). Some of these materials have potential for use as an additive to cement (Gajda 1996). There is little long-term field experience available on the effectiveness of these materials.

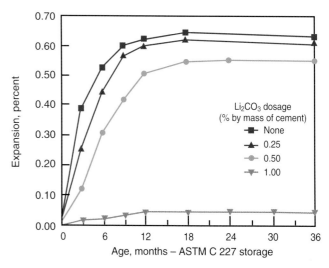

Fig. 6-18. Expansion of specimens made with lithium carbonate admixture (Stark 1992).

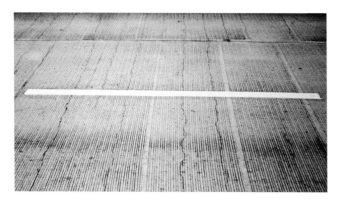

Fig. 6-17. Shrinkage cracks, such as shown on this bridge deck, can be reduced with the use of good concreting practices and shrinkage reducing admixtures. (69883)

Fig. 6-19. Red and blue pigments were used to colour this terrazzo floor. (69873)

COLOURING ADMIXTURES (PIGMENTS)

Natural and synthetic materials are used to colour concrete for aesthetic and safety reasons (Fig. 6-19). Red concrete is used around buried electrical or gas lines as a warning to anyone near these facilities. Yellow concrete safety curbs are used in paving applications. Generally, the amount of pigments used in concrete should not exceed 10% by mass of the cement. Pigments used in amounts less than 6% generally do not affect concrete properties.

Unmodified carbon black substantially reduces air content. Most carbon black for colouring concrete contains an admixture to offset this effect on air. Before a colouring admixture is used on a project, it should be tested for colour fastness in sunlight and autoclaving, chemical stability in cement, and effects on concrete properties. Calcium chloride should not be used with pigments to avoid colour distortions. Pigments should conform to ASTM C 979.

DAMPPROOFING ADMIXTURES

The passage of water through concrete can usually be traced to the existence of cracks or areas of incomplete consolidation. Sound, dense concrete made with a water-cementing materials ratio of less than 0.50 by mass will be watertight if it is properly placed and cured.

Admixtures known as dampproofing agents include certain soaps, stearates, and petroleum products. They may, but generally do not, reduce the permeability of concretes that have low cementing material contents, high water-cementing materials ratios, or a deficiency of fines in the aggregate. Their use in well-proportioned mixes, may increase the mixing water required and actually result in increased rather than reduced permeability.

Dampproofing admixtures are sometimes used to reduce the transmission of moisture through concrete that is in contact with water or damp earth. Many so-called dampproofers are not effective, especially when used in concretes that are in contact with water under pressure.

PERMEABILITY-REDUCING ADMIXTURES

Permeability-reducing admixtures reduce the rate at which water under pressure is transmitted through concrete. One of the best methods of decreasing permeability in concrete is to increase the moist-curing period and reduce the water-cementing materials ratio to less than 0.5. Most admixtures that reduce the water-cementing materials ratio consequently reduce permeability.

Some supplementary cementing materials, especially silica fume, reduce permeability through the hydration and pozzolanic-reaction process. Other admixtures that act to block the capillaries in concrete have been shown to be effective in reducing concrete corrosion in chemically aggressive environments. Such admixtures, designed for use in high-cementing materials content/low-water-cementing materials ratio concretes, contain aliphatic fatty acid and an aqueous emulsion of polymeric and aromatic globules (Aldred 1988).

PUMPING AIDS

Pumping aids are added to concrete mixtures to improve pumpability. Pumping aids cannot cure all unpumpable concrete problems; they are best used to make marginally pumpable concrete more pumpable. These admixtures increase viscosity or cohesion in concrete to reduce dewatering of the paste while under pressure from the pump.

Some pumping aids may increase water demand, reduce compressive strength, increase air content, or retard setting time. These side effects can be corrected by adjusting the mix proportions or adding another admixture to offset the side effect.

A partial list of materials used in pumping aids is given in Table 6-1. Some admixtures that serve other primary purposes but also improve pumpability are air-entraining agents, and some water-reducing and retarding admixtures.

BONDING ADMIXTURES AND BONDING AGENTS

Bonding admixtures are usually water emulsions of organic materials including rubber, polyvinyl chloride, polyvinyl acetate, acrylics, styrene butadiene copolymers, and other polymers. They are added to portland cement mixtures to increase the bond strength between old and new concrete. Flexural strength and resistance to chloride-ion ingress are also improved. They are added in proportions equivalent to 5% to 20% by mass of the cementing materials; the actual quantity depending on job conditions and type of admixture used. Some bonding admixtures may increase the air content of mixtures. Nonreemulsifiable types are resistant to water, better suited to exterior application, and used in places where moisture is present.

The ultimate result obtained with a bonding admixture will be only as good as the surface to which the concrete is applied. The surface must be dry, clean, sound, free of dirt, dust, paint, and grease, and at the proper temperature. Organic or polymer modified concretes are acceptable for patching and thin-bonded overlays, particularly where feather-edged patches are desired.

Bonding agents should not be confused with bonding admixtures. Admixtures are an ingredient in the concrete; bonding agents are applied to existing concrete surfaces immediately before the new concrete is placed. Bonding agents help "glue" the existing and the new materials together. Bonding agents are often used in restoration and repair work; they consist of portland cement or latex-modified portland cement grout or polymers such as epoxy resins (ASTM C 881) or latex (ASTM C 1059).

GROUTING ADMIXTURES

Portland cement grouts are used for a variety of purposes: to stabilize foundations, set machine bases, fill cracks and joints in concrete work, cement oil wells, fill cores of masonry walls, grout prestressing tendons and anchor bolts, and fill the voids in preplaced aggregate concrete. To alter the properties of grout for specific applications, various air-entraining admixtures, accelerators, retarders, and nonshrink admixtures are often used.

GAS-FORMING ADMIXTURES

Aluminum powder and other gas-forming materials are sometimes added to concrete and grout in very small quantities to cause a slight expansion of the mixture prior to hardening. This may be of benefit where the complete grouting of a confined space is essential, such as under machine bases or in post-tensioning ducts of prestressed concrete. These materials are also used in larger quantities to produce autoclaved cellular concretes. The amount of expansion that occurs is dependent upon the amount of gas-forming material used, the temperature of the fresh mixture, the alkali content of the cement, and other variables. Where the amount of expansion is critical, careful control of mixtures and temperatures must be exercised. Gas-forming agents will not overcome shrinkage after hardening caused by drying or carbonation.

AIR DETRAINERS

Air-detraining admixtures reduce the air content in concrete. They are used when the air content cannot be reduced by adjusting the mix proportions or by changing the dosage of the air-entraining agent and other admixtures. However, air-detrainers are rarely used and their effectiveness and dosage rate should be established on trial mixes prior to use on actual job mixes. Materials used in air-detraining agents are listed in Table 6-1.

FUNGICIDAL, GERMICIDAL, AND INSECTICIDAL ADMIXTURES

Bacteria and fungal growth on or in hardened concrete may be partially controlled through the use of fungicidal, germicidal, and insecticidal admixtures. The most effective materials are polyhalogenated phenols, dieldrin emulsions, and copper compounds. The effectiveness of these materials is generally temporary, and in high dosages they may reduce the compressive strength of concrete.

ANTIWASHOUT ADMIXTURES

Antiwashout admixtures increase the cohesiveness of concrete to a level that allows limited exposure to water with little loss of cement. This allows placement of concrete in water and under water without the use of tremies. The admixtures increase the viscosity of water in the mixture resulting in a mix with increased thixotropy and resistance to segregation. They usually consist of water soluble cellulose ether or acrylic polymers.

COMPATIBILITY OF ADMIXTURES AND CEMENTING MATERIALS

Fresh concrete problems of varying severity are encountered due to cement-admixture incompatibility and incompatibility between admixtures. Incompatibility between supplementary cementing materials and admixtures or cements can also occur. Slump loss, air loss, early stiffening, and other factors affecting fresh concrete properties can result from incompatibilities. While these problems primarily affect the plastic-state performance of concrete, long-term hardened concrete performance may also be adversely affected. For example, early stiffening can cause difficulties with consolidation of concrete, therefore compromising strength.

Reliable test methods are not available to adequately address incompatibility issues due to variations in materials, mixing equipment, mixing time, and environmental factors. Tests run in a laboratory do not reflect the conditions experienced by concrete in the field. When incompatibility is discovered in the field, a common solution is to simply change admixtures or cementing materials

(Helmuth, Hills, Whiting, and Bhattacharja 1995, Tagni-Hamou and Aitcin 1993, and Tang and Bhattacharja 1997).

STORING AND DISPENSING CHEMICAL ADMIXTURES

Liquid admixtures can be stored in barrels or bulk tankers. Powdered admixtures can be placed in special storage bins and some are available in premeasured plastic bags. Admixtures added to a truck mixer at the jobsite are often in plastic jugs or bags. Powdered admixtures, such as certain plasticizers, or a barrel of admixture may be stored at the project site.

Dispenser tanks at concrete plants should be properly labeled for specific admixtures to avoid contamination and avoid dosing the wrong admixture. Most liquid chemical admixtures should not be allowed to freeze; therefore, they should be stored in heated environments. Consult the admixture manufacturer for proper storage temperatures. Powdered admixtures are usually less sensitive to temperature restrictions, but may be sensitive to moisture.

Liquid chemical admixtures are usually dispensed individually in the batch water by volumetric means (Fig. 6-20). Liquid and powdered admixtures can be measured by mass, but powdered admixtures should not be measured by volume. Care should be taken to not combine certain admixtures prior to their dispensing into the batch as some combinations may neutralize the desired effect of the admixtures. Consult the admixture manufacturer concerning compatible admixture combinations or perform laboratory tests to document performance.

Fig. 6-20. Liquid admixture dispenser at a ready mix plant provides accurate volumetric measurement of admixtures. (44220)

REFERENCES

AASHTO, "Portland Cement Concrete Resistant to Excessive Expansion Caused by Alkali-Silica Reaction," Section 56X, *Guide Specification For Highway Construction,* http://leadstates.tamu.edu/ASR/library/gspec.stm, American Association of State Highway and Transportation Officials, Washington, D.C., 2001.

Abrams, Duff A., *Calcium Chloride as an Admixture in Concrete,* Bulletin 13 (PCA LS013), Structural Materials Research Laboratory, Lewis Institute, Chicago, http://www.portcement.org/pdf_files/LS013.pdf, 1924.

ACI Committee 212, *Chemical Admixtures for Concrete,* ACI 212.3R-91, American Concrete Institute, Farmington Hills, Michigan, 1991.

ACI Committee 212, *Guide for the Use of High-Range Water-Reducing Admixtures (Superplasticizers) in Concrete,* ACI 212.4R-93 (Reapproved 1998), American Concrete Institute, Farmington Hills, Michigan, 1998.

ACI Committee 222, *Corrosion of Metals in Concrete,* ACI 222R-96, American Concrete Institute, Farmington Hills, Michigan, 1996.

ACI Committee 318, *Building Code Requirements for Structural Concrete and Commentary,* ACI 318-99, American Concrete Institute, Farmington Hills, Michigan, 1999.

ACI E4, *Chemical and Air-Entraining Admixtures for Concrete,* ACI Education Bulletin No. E4-96, American Concrete Institute, Farmington Hills, Michigan, 1999, 16 pages.

Aldred, James M., "HPI Concrete," *Concrete International,* American Concrete Institute, Farmington Hills, Michigan, November 1988.

Berke, N. S., and Weil, T. G., "World Wide Review of Corrosion Inhibitors in Concrete," *Advances in Concrete Technology,* CANMET, Ottawa, 1994, pages 891 to 914.

Chou, Gee Kin, "Cathodic Protection: An Emerging Solution to the Rebar Corrosion Problem," *Concrete Construction,* Addison, Illinois, June 1984.

Gajda, John, *Development of a Cement to Inhibit Alkali-Silica Reactivity,* Research and Development Bulletin RD115, Portland Cement Association, 1996, 58 pages.

Gaynor, Richard D., "Calculating Chloride Percentages," *Concrete Technology Today,* PL983, Portland Cement Association, http://www.portcement.org/pdf_files/PL983.pdf, 1998, pages 4 to 5.

Gebler, S. H., *The Effects of High-Range Water Reducers on the Properties of Freshly Mixed and Hardened Flowing Concrete,* Research and Development Bulletin RD081, Portland Cement Association, http://www.portcement.org/pdf_files/RD081.pdf, 1982.

Helmuth, Richard; Hills, Linda M.; Whiting, David A.; and Bhattacharja, Sankar, *Abnormal Concrete Performance in the Presence of Admixtures,* RP333, Portland Cement Association, http://www.portcement.org/pdf_files/RP333.pdf, 1995.

Hester, Weston T., *Superplasticizers in Ready Mixed Concrete (A Practical Treatment for Everyday Operations),* National Ready Mixed Concrete Association, Publication No. 158, Silver Spring, Maryland, 1979.

Kinney, F. D., "Reuse of Returned Concrete by Hydration Control: Characterization of a New Concept," *Superplasticizers and Other Chemical Admixtures in Concrete,* SP119, American Concrete Institute, Farmington Hills, Michigan, 1989, pages 19 to 40.

Klieger, Paul, *Air-Entraining Admixtures,* Research Department Bulletin RX199, Portland Cement Association, http://www.portcement.org/pdf_files/RX199.pdf, 1966, 12 pages.

Kosmatka, Steven H., "Discoloration of Concrete—Causes and Remedies," *Concrete Technology Today,* PL861, Portland Cement Association, http://www.portcement.org/pdf_files/PL861.pdf, 1986.

Lackey, Homer B., "Factors Affecting Use of Calcium Chloride in Concrete," *Cement, Concrete, and Aggregates,* American Society for Testing and Materials, West Conshohocken, Pennsylvania, Winter 1992, pages 97 to 100.

Nmai, Charles K.; Schlagbaum, Tony; and Violetta, Brad, "A History of Mid-Range Water-Reducing Admixtures," *Concrete International,* American Concrete Institute, Farmington Hills, Michigan, April 1998, pages 45 to 50.

Nmai, Charles K.; Tomita, Rokuro; Hondo, Fumiaki; and Buffenbarger, Julie, "Shrinkage-Reducing Admixtures," *Concrete International,* American Concrete Institute, Farmington Hills, Michigan, April 1998, pages 31 to 37.

Ramachandran, V. S., *Concrete Admixtures Handbook,* Noyes Publications, Park Ridge, New Jersey, 1995.

Rixom, M. R., and Mailvaganam, N. P., *Chemical Admixtures for Concrete,* E. & F. N. Spon, New York, 1986.

Shah, Surendra P.; Weiss, W. Jason; and Yang, Wei, "Shrinkage Cracking—Can it be Prevented?," *Concrete International,* American Concrete Institute, Farmington Hills, Michigan, April 1998, pages 51 to 55.

Stark, David, *Influence of Design and Materials on Corrosion Resistance of Steel in Concrete,* Research and Development Bulletin RD098, Portland Cement Association, http://www.portcement.org/pdf_files/RD098.pdf, 1989, 44 pages.

Stark, David C., *Lithium Salt Admixtures—An Alternative Method to Prevent Expansive Alkali-Silica Reactivity,* RP307, Portland Cement Association, 1992, 10 pages.

Tagnit-Hamou, Arezki, and Aitcin, Pierre-Claude, "Cement and Superplasticizer Compatibility," *World Cement,* Palladian Publications Limited, Farnham, Surrey, England, August 1993, pages 38 to 42.

Tang, Fulvio J., and Bhattacharja, Sankar, *Development of an Early Stiffening Test,* RP346, Portland Cement Association, 1997, 36 pages.

Thomas, Michael D. A., and Stokes, David B., "Use of a Lithium-Bearing Admixture to Suppress Expansion in Concrete Due to Alkali-Silica Reaction," *Transportation Research Record No. 1668,* Transportation Research Board, Washington, D.C., 1999, pages 54 to 59.

Whiting, David, *Effects of High-Range Water Reducers on Some Properties of Fresh and Hardened Concretes,* Research and Development Bulletin RD061, Portland Cement Association, http://www.portcement.org/pdf_files/RD061.pdf, 1979.

Whiting, David A., *Evaluation of Super-Water Reducers for Highway Applications,* Research and Development Bulletin RD078, Portland Cement Association, http://www.portcement.org/pdf_files/RD078.pdf, 1981, 169 pages.

Whiting, D., and Dziedzic, W., *Effects of Conventional and High-Range Water Reducers on Concrete Properties,* Research and Development Bulletin RD107, Portland Cement Association, 1992, 25 pages.

Whiting, David A., and Nagi, Mohamad A., *Manual on the Control of Air Content in Concrete,* EB116, National Ready Mixed Concrete Association and Portland Cement Association, 1998, 42 pages.

CHAPTER 7
Fibres

Fibres have been used in construction materials for many centuries. The last three decades have seen a growing interest in the use of fibres in ready-mixed concrete, precast concrete, and shotcrete. Fibres made from steel, plastic, glass, and natural materials (such as wood cellulose) are available in a variety of shapes, sizes, and thicknesses; they may be round, flat, crimped, and deformed with typical lengths of 6 mm to 150 mm and thicknesses ranging from 0.005 mm to 0.75 mm (Fig. 7-1). They are added to concrete during mixing. The main factors that control the performance of the composite material are:

1. Physical properties of fibres and matrix
2. Strength of bond between fibres and matrix

Although the basic governing principles are the same, there are several characteristic differences between conventional reinforcement and fibre systems:

1. Fibres are generally distributed throughout a given cross section whereas reinforcing bars or wires are placed only where required
2. Most fibres are relatively short and closely spaced as compared with continuous reinforcing bars or wires

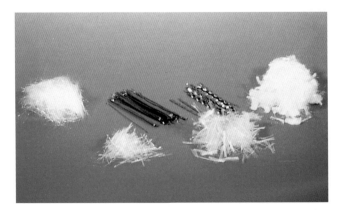

Fig. 7-1. Steel, glass, synthetic and natural fibres with different lengths and shapes can be used in concrete. (69965)

3. It is generally not possible to achieve the same area of reinforcement to area of concrete using fibres as compared to using a network of reinforcing bars or wires

Fibres are typically added to concrete in low volume dosages (often less than 1%), and have been shown to be effective in reducing plastic shrinkage cracking.

Fibres typically do not significantly alter free shrinkage of concrete, however at high enough dosages they can increase the resistance to cracking and decrease crack width (Shah, Weiss, and Yang 1998).

ADVANTAGES AND DISADVANTAGES OF USING FIBRES

Fibres are generally distributed throughout the concrete cross section. Therefore, many fibres are inefficiently located for resisting tensile stresses resulting from applied loads. Depending on fabrication method, random orientation of fibres may be either two-dimensional (2-D) or three-dimensional (3-D). Typically, the spray-up fabrication method has a 2-D random fibre orientation where as the premix (or batch) fabrication method typically has a 3-D random fibre orientation. Also, many fibres are observed to extend across cracks at angles other than 90° or may have less than the required embedment length for development of adequate bond. Therefore, only a small percentage of the fibre content may be efficient in resisting tensile or flexural stresses. "Efficiency factors" can be as low as 0.4 for 2-D random orientation and 0.25 for 3-D random orientation. The efficiency factor depends on fibre length and critical embedment length. From a conceptual point of view, reinforcing with fibres is not a highly efficient method of obtaining composite strength.

Fibre concretes are best suited for thin section shapes where correct placement of conventional reinforcement would be extremely difficult. In addition, spraying of fibre concrete accommodates the fabrication of irregularly shaped products. A substantial reduction in mass can be realized using relatively thin fibre concrete sections having the equivalent strength of thicker conventionally reinforced concrete sections.

TYPES AND PROPERTIES OF FIBRES AND THEIR EFFECT ON CONCRETE

Steel Fibres

Steel fibres are short, discrete lengths of steel with an aspect ratio (ratio of length to diameter) from about 20 to 100, and with any of several cross sections. Some steel fibres have hooked ends to improve resistance to pullout from a cement-based matrix (Fig. 7-2).

ASTM A 820 classifies four different types based on their manufacture. Type I – Cold-drawn wire fibres are the most commercially available, manufactured from drawn steel wire. Type II – Cut sheet fibres are manufactured as the name implies: steel fibres are laterally sheared off steel sheets. Type III – Melt-extracted fibres are manufactured with a relatively complicated technique where a rotating wheel is used to lift liquid metal from a molten metal surface by capillary action. The extracted molten metal is then rapidly frozen into fibres and thrown off the wheel by centrifugal force. The resulting fibres have a crescent-shaped cross section. Type IV – Other fibres. For tolerances for length, diameter, and aspect ratio, as well as minimum tensile strength and bending requirement, see ASTM A 820.

Steel-fibre volumes used in concrete typically range from 0.25% to 2%. Volumes of more than 2% generally reduce workability and fibre dispersion and require special mix design or concrete placement techniques.

The compressive strength of concrete is only slightly affected by the presence of fibres. The addition of 1.5% by volume of steel fibres can increase the direct tensile strength by up to 40% and the flexural strength up to 150%.

Steel fibres do not affect free shrinkage. Steel fibres delay the fracture of restrained concrete during shrinkage and they improve stress relaxation by creep mechanisms (Altoubat and Lange 2001).

The durability of steel-fibre concrete is contingent on the same factors as conventional concrete. Freeze-thaw durability is not diminished by the addition of steel fibres provided the mix is adjusted to accommodate the fibres, the concrete is properly consolidated during placement, and is air-entrained. With properly designed and placed concrete, little or no corrosion of the fibres occurs. Any surface corrosion of fibres is cosmetic as opposed to a structural condition.

Steel fibres have a relatively high modulus of elasticity (Table 7-1). Their bond to the cement matrix can be enhanced by mechanical anchorage or surface roughness and they are protected from corrosion by the alkaline environment in the cement matrix (ACI 544.1R-96).

Steel fibres are most commonly used in airport pavements and runway/taxi overlays. They are also used in bridge decks (Fig. 7-3), industrial floors, and highway pavements. Structures exposed to high-velocity water flow have been shown to last about three times longer than conventional concrete alternatives. Steel fibre concrete is also used for many precast concrete applications that make use of the improved impact resistance or toughness imparted by the fibres. In utility boxes and septic tanks, steel fibres replace conventional reinforcement.

Steel fibres are also widely used with shotcrete in thin-layer applications, especially rock-slope stabilization and tunnel linings. Silica fume and accelerators have enabled shotcrete to be placed in thicker layers. Silica fume also reduces the permeability of the shotcrete material (Morgan 1987). Steel-fibre shotcrete has been successfully applied with fibre volumes up to 2%.

Slurry-infiltrated concrete (SIFCON) with fibre volumes up to 20% has been used since the late 1970s. Slurry-

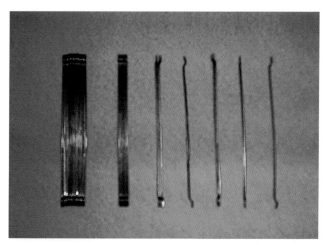

Fig. 7-2. Steel fibres with hooked ends are collated into bundles to facilitate handling and mixing. During mixing the bundles separate into individual fibres. (69992)

Fig. 7-3. Bridge deck with steel fibres. (70007)

Table 7.1. Properties of Selected Fibre Types

Fibre type	Relative density	Diameter, μm	Tensile strength, MPa	Modulus of elasticity, MPa	Strain at failure, %
Steel	7.80	100-1000	500-2600	210,000	0.5-3.5
Glass					
E	2.54	8-15	2000-4000	72,000	3.0-4.8
AR	2.70	12-20	1500-3700	80,000	2.5-3.6
Synthetic					
Acrylic	1.18	5-17	200-1000	17,000-19,000	28-50
Aramid	1.44	10-12	2000-3100	62,000-120,000	2-3.5
Carbon	1.90	8-0	1800-2600	230,000-380,000	0.5-1.5
Nylon	1.14	23	1000	5,200	20
Polyester	1.38	10-80	280-1200	10,000-18,000	10-50
Polyethylene	0.96	25-1000	80-600	5,000	12-100
Polypropylene	0.90	20-200	450-700	3,500-5,200	6-15
Natural					
Wood cellulose	1.50	25-125	350-2000	10,000-40,000	
Sisal			280-600	13,000-25,000	3.5
Coconut	1.12-1.15	100-400	120-200	19,000-25,000	10-25
Bamboo	1.50	50-400	350-500	33,000-40,000	
Jute	1.02-1.04	100-200	250-350	25,000-32,000	1.5-1.9
Elephant grass		425	180	4,900	3.6

Adapted from PCA (1991) and ACI 544.1R-96.

infiltrated concrete can be used to produce a component or structure with strength and ductility that far exceeds that of conventionally mixed or sprayed fibre concrete. SIFCON is not inexpensive and needs fine-tuning, but it holds potential for applications exposed to severe conditions and requiring very high strength and toughness. These applications include impact and blast-resistant structures, refractories, protective revetment, and taxiway and pavement repairs (Fig. 7-4). Table 7-2 shows a SIFCON mix design.

Table 7-2. SIFCON Mix Design.

Cement	1000 kg/m³
Water	330 kg/m³
Siliceous Sand ≤ 0.7 mm	860 kg/m³
Silica Slurry	13 kg/m³
High-Range Water Reducer	35 kg/m³
Steel Fibres (about 10 Vol.-%)	800 kg/m³

Glass Fibres

The first research on glass fibres in the early 1960s used conventional borosilicate glass (E-glass) (Table 7-1) and soda-lime-silica glass fibres (A-glass). The test results showed that alkali reactivity between the E-glass fibres and the cement-paste reduced the strength of the concrete. Continued research resulted in alkali-resistant glass fibres (AR-glass) (Table 7-1), that improved long-term durability, but sources of other strength-loss trends were observed. One acknowledged source was fibre embrittlement stemming from infiltration of calcium hydroxide particles, byproducts of cement hydration, into fibre bundles. Alkali reactivity and cement hydration are the basis for the following two widely held theories explaining strength and ductility loss, particularly in exterior glass fibre concrete:

- Alkali attack on glass-fibre surfaces reduces fibre tensile strength and, subsequently, lowers compressive strength.
- Ongoing cement hydration causes calcium hydroxide particle penetration of fibre bundles, thereby in-

Fig. 7-4. Tightly bunched steel fibres are placed in a form, before cement slurry is poured into this application of slurry-infiltrated steel-fibre concrete (SIFCON). (60672)

creasing fibre-to-matrix bond strength and embrittlement; the latter lowers tensile strength by inhibiting fibre pullout.

Fibre modifications to improve long-term durability involve (1) specially formulated chemical coatings to help combat hydration-induced embrittlement, and (2) employment of a dispersed microsilica slurry to adequately fill fibre voids, thereby reducing potential for calcium hydroxide infiltration.

A low-alkaline cement has been developed in Japan that produces no calcium hydroxide during hydration. Accelerated tests with the cement in alkali-resistant glass-fibre-reinforced concrete samples have shown greater long-term durability than previously achieved.

Metakaolin can be used in glass-fibre-reinforced concrete without significantly affecting flexural strength, strain, modulus of elasticity, and toughness. (Marikunte, Aldea, Shah 1997).

The single largest application of glass-fibre concrete has been the manufacture of exterior building façade panels (Fig. 7-5). Other applications are listed in PCA (1991).

Synthetic Fibres

Synthetic fibres are man-made fibres resulting from research and development in the petrochemical and textile industries. Fibre types that are used in portland cement concrete are: acrylic, aramid, carbon, nylon, polyester, polyethylene, and polypropylene. Table 7-1 summarizes the range of physical properties of these fibres.

Synthetic fibres can reduce plastic shrinkage and subsidence cracking and may help concrete after it is fractured. Ultra-thin whitetopping often uses synthetic fibres for potential containment properties to delay pothole development. Problems associated with synthetic fibres include: (1) low fibre-to-matrix bonding; (2) inconclusive performance testing for low fibre-volume usage with polypropylene, polyethylene, polyester and nylon; (3) a low modulus of elasticity for polypropylene and polyethylene; and (4) the high cost of carbon and aramid fibres.

Fig. 7-5. (top) Glass-fibre-reinforced concrete panels are light and strong enough to reduce this building's structural requirements. (bottom) Spray-up fabrication made it easy to create their contoured profiles. (60671, 46228)

Polypropylene fibres (Fig. 7-6), the most popular of the synthetics, are chemically inert, hydrophobic, and lightweight. They are produced as continuous cylindrical monofilaments that can be chopped to specified lengths or cut as films and tapes and formed into fine fibrils of rectangular cross section (Fig. 7-7).

Used at a rate of at least 0.1 percent by volume of concrete, polypropylene fibres reduce plastic shrinkage cracking and subsidence cracking over steel reinforcement (Suprenant and Malisch 1999). The presence of polypropylene fibres in concrete may reduce settlement of aggregate particles, thus reducing capillary bleed channels. Polypropylene fibres can also help reduce spalling of a high moisture content, low-permeability concrete when it is exposed to fire.

New developments show that monofilament fibres are able to fibrillate during mixing if produced with both, polypropylene and polyethylene resins. The two polymers are incompatible and tend to separate when manipulated. Therefore, during the mixing process each fibre turns into a unit with several fibrils at its end. The fibrils provide better mechanical bonding than conventional monofilaments. The high number of fine fibrils also reduces plastic shrinkage cracking and may increase the ductility and toughness of the concrete (Trottier and Mahoney 2001).

Acrylic fibres have been found to be the most promising replacement for asbestos fibres. They are used in cement board and roof-shingle production, where fibre volumes of up to 3% can produce a composite with mechanical properties similar to that of an asbestos-cement composite. Acrylic-fibre concrete composites exhibit high postcracking toughness and ductility. Although lower than that of asbestos-cement composites, acrylic-fibre-reinforced concrete's flexural strength is ample for many building applications.

Fig. 7-7. Polypropylene fibres are produced either as (left) fine fibrils with rectangular cross section or (right) cylindrical monofilament. (69993)

Aramid fibres have high tensile strength and a high tensile modulus. Aramid fibres are two and a half times as strong as E-glass fibres and five times as strong as steel fibres. A comparison of mechanical properties of different aramid fibres is provided in PCA (1991). In addition to excellent strength characteristics, aramid fibres also have excellent strength retention up to 160°C, dimensional stability up to 200°C, static and dynamic fatigue resistance, and creep resistance. Aramid strand is available in a wide range of diameters.

Carbon fibres were developed primarily for their high strength and elastic modulus and stiffness properties for applications within the aerospace industry. Compared with most other synthetic fibres, the manufacture of carbon fibres is expensive and this has limited commercial development. Carbon fibres have high tensile strength and modulus of elasticity (Table 7-1). They are also inert to most chemicals. Carbon fibre is typically produced in strands that may contain up to 12,000 individual filaments. The strands are commonly prespread prior to incorporation in concrete to facilitate cement matrix penetration and to maximize fibre effectiveness.

Nylon fibres exist in various types in the marketplace for use in apparel, home furnishing, industrial, and textile applications. Only two types of nylon fibre are currently marketed for use in concrete, nylon 6 and nylon 66. Nylon fibres are spun from nylon polymer and transformed through extrusion, stretching, and heating to form an oriented, crystalline, fibre structure. For concrete applications, high tenacity (high tensile strength) heat and light stable yarn is spun and subsequently cut into shorter length. Nylon fibres exhibit good tenacity, toughness, and elastic recovery. Nylon is hydrophilic, with moisture retention of 4.5 percent, which increases the water demand of concrete. However, this does not affect con-

Fig. 7-6. Polypropylene fibres. (69796)

crete hydration or workability at low prescribed contents ranging from 0.1 to 0.2 percent by volume, but should be considered at higher fibre volume contents. This comparatively small dosage has potentially greater reinforcing value than low volumes of polypropylene or polyester fibre. Nylon is relatively inert and resistant to a wide variety of organic and inorganic materials including strong alkalis.

Synthetic fibres are also used in stucco and mortar. For this use the fibres are shorter than synthetic fibres used in concrete. Usually small amounts of 13-mm long alkali-resistant fibres are added to base coat plaster mixtures. They can be used in small line stucco and mortar pumps and spray guns. They should be added to the mix in accordance with manufacturer's recommendation.

For further details about chemical and physical properties of synthetic fibres and properties of synthetic fibre concrete, see ACI 544.1R-96. ASTM C 1116 classifies Steel, Glass, and Synthetic Fibre Concrete or Shotcrete.

The technology of interground fibre cement takes advantage of the fact that some synthetic fibres are not destroyed or pulverized in the cement finishing mill. The fibres are mixed with dry cement during grinding where they are uniformly distributed; the surface of the fibres is roughened during grinding, which offers a better mechanical bond to the cement paste (Vondran 1995).

Natural Fibres

Natural fibres were used as a form of reinforcement long before the advent of conventional reinforced concrete. Mud bricks reinforced with straw and mortars reinforced with horsehair are just a few examples of how natural fibres were used long ago as a form of reinforcement. Many natural reinforcing materials can be obtained at low levels of cost and energy using locally available manpower and technical know-how. Such fibres are used in the manufacture of low-fibre-content concrete and occasionally have been used in thin-sheet concrete with high-fibre content. For typical properties of natural fibres see Table 7-1.

Unprocessed Natural Fibres. In the late 1960s, research on the engineering properties of natural fibres, and concrete made with these fibres was undertaken; the result was these fibres can be used successfully to make thin sheets for walls and roofs. Products were made with portland cement and unprocessed natural fibres such as coconut coir, sisal, bamboo, jute, wood, and vegetable fibres. Although the concretes made with unprocessed natural fibres shows good mechanical properties, they have some deficiencies in durability. Many of the natural fibres are highly susceptible to volume changes due to variations in fibre moisture content. Fibre volumetric changes that accompany variations in fibre moisture content can drastically affect the bond strength between the fibre and cement matrix.

Wood Fibres (Processed Natural Fibres). The properties of wood cellulose fibres are greatly influenced by the method by which the fibres are extracted and the refining processes involved. The process by which wood is reduced to a fibrous mass is called pulping. The kraft process is the one most commonly used for producing wood cellulose fibres. This process involves cooking wood chips in a solution of sodium hydroxide, sodium carbonate, and sodium sulphide. Wood cellulose fibres have relatively good mechanical properties compared to many manmade fibres such as polypropylene, polyethylene, polyester, and acrylic. Delignified cellulose fibres (lignin removed) can be produced with a tensile strength of up to approximately 2000 MPa for selected grades of wood and pulping processes. Fibre tensile strength of approximately 500 MPa can be routinely achieved using a chemical pulping process and the more common, less expensive grades of wood.

MULTIPLE FIBRE SYSTEMS

For a multiple fibre system, two or more fibres are blended into one system. The hybrid-fibre concrete combines macro- and microsteel fibres. A common macrofibre blended with a newly developed microfibre, which is less than 10 mm long and less than 100 μm in diameter, leads to a closer fibre-to-fibre spacing, which reduces microcracking and increases tensile strength. The intended applications include thin repairs and patching (Banthia and Bindiganavile 2001). A blend of steel and polypropylene fibres has also been used for some applications. This system is supposed to combine the toughness and impact-resistance of steel fibre concrete with the reduced plastic cracking of polypropylene fibre concrete. For a project in the Chicago area (Wojtysiak and others 2001), a blend of 30 kg/m³ of steel fibres and 0.9 kg/m³ of fibrillated polypropylene fibres were used for slabs on grade. The concrete with blended fibres had a lower slump compared to plain concrete but seemed to have enhanced elastic and post-elastic strength.

REFERENCES

ACI Committee 544, *State-of-the-Art Report on Fiber Reinforced Concrete,* ACI 544.1R-96, American Concrete Institute, Farmington Hills, Michigan, 1997.

Altoubat, Salah A., and Lange, David A., "Creep, Shrinkage, and Cracking of Restrained Concrete at Early Age," *ACI Materials Journal,* American Concrete Institute, Farmington Hills, Michigan, July-August 2001, pages 323 to 331.

Banthia, Nemkumar, and Bindiganavile, Vivek, "Repairing with Hybrid-Fiber-Reinforced Concrete," *Concrete International,* American Concrete Institute, Farmington Hills, Michigan, June 2001, pages 29 to 32.

Bijen, J., "Durability of Some Glass Fiber Reinforced Cement Composites," *ACI Journal,* American Concrete Institute, Farmington Hills, Michigan, July-August 1983, pages 305 to 311.

CSA Standard A23.1-00 *Appendix H- Fiber-Reinforced Concrete,* Canadian Standards Association, Toronto, 2000.

Hanna, Amir N., *Steel Fiber Reinforced Concrete Properties and Resurfacing Applications,* Research and Development Bulletin RD049, Portland Cement Association, http://www.portcement.org/pdf_files/RD049.pdf, 1977, 18 pages.

Johnston, Colin D., *Fiber Reinforced Cement and Concretes,* LT249, Gordon & Breach, Amsterdam, 2000, 368 pages.

Marikunte, S.; Aldea, C.; and Shah, S., "Durability of Glass Fiber Reinforced Cement Composites: Effect of Silica Fume and Metakaolin," *Advanced Cement Based Materials,* Volume 5, Numbers 3/4, April/May 1997, pages 100 to 108.

Morgan, D. R., "Evaluation of Silica Fume Shotcrete," *Proceedings, CANMET/CSCE International Workshop on Silica Fume in Concrete,* Montreal, May 1987.

Monfore, G. E., *A Review of Fiber Reinforcement of Portland Cement Paste, Mortar and Concrete,* Research Department Bulletin RX226, Portland Cement Association, http://www.portcement.org/pdf_files/RX226.pdf, 1968, 7 pages.

PCA, *Fiber Reinforced Concrete,* SP039, Portland Cement Association, 1991, 54 pages.

PCA, "Steel Fiber Reinforced Concrete," *Concrete Technology Today,* PL931 Portland Cement Association, http://www.portcement.org/pdf_files/PL931.pdf, March 1993, pages 1 to 4.

Panarese, William C., "Fiber: Good for the Concrete Diet?," *Civil Engineering,* American Society of Civil Engineers, New York, May 1992, pages 44 to 47.

Shah, S. P.; Weiss, W. J.; and Yang, W., "Shrinkage Cracking – Can it be prevented?," *Concrete International,* American Concrete Institute, Farmington Hills, Michigan, April 1998, pages 51 to 55.

Suprenant, Bruce A., and Malisch, Ward R., "The fiber factor," *Concrete Construction,* Addison, Illinois, October 1999, pages 43 to 46.

Trottier, Jean-Francois, and Mahoney, Michael, "Innovative Synthetic Fibers," *Concrete International,* American Concrete Institute, Farmington Hills, Michigan, June 2001, pages 23 to 28

Vondran, Gary L., "Interground Fiber Cement in the Year 2000," *Emerging Technologies Symposium on Cements for the 21st Century,* SP206, Portland Cement Association, March 1995, pages 116 to 134.

Wojtysiak, R.; Borden, K. K.; and Harrison P., *Evaluation of Fiber Reinforced Concrete for the Chicago Area – A Case Study,* 2001.

CHAPTER 8
Air-Entrained Concrete

One of the greatest advances in concrete technology was the development of air-entrained concrete in the mid-1930s. Today air entrainment is recommended for nearly all concretes, principally to improve freeze-thaw resistance when exposed to water and deicing chemicals. However, there are other important benefits of entrained air in both freshly mixed and hardened concrete.

Air-entrained concrete is produced by adding an air-entraining admixture during batching (air-entraining cements are also available in the United States). The air-entraining admixture stabilizes bubbles formed during the mixing process, enhances the incorporation of bubbles of various sizes by lowering the surface tension of the mixing water, impedes bubble coalescence, and anchors bubbles to cement and aggregate particles.

Anionic air-entraining admixtures are hydrophobic (repel water) and are electrically charged (nonionic or no-charge admixtures are also available). The negative electric charge is attracted to positively charged cement grains, which aids in stabilizing bubbles. The air-entraining admixture forms a tough, water-repelling film, similar to a soap film, with sufficient strength and elasticity to contain and stabilize the air bubbles and prevent them from coalescing. The hydrophobic film also keeps water out of the bubbles. The stirring and kneading action of mechanical mixing disperses the air bubbles. The fine aggregate particles also act as a three-dimensional grid to help hold the bubbles in the mixture.

Entrained air bubbles are not like entrapped air voids, which occur in all concretes as a result of mixing, handling, and placing and are largely a function of aggregate characteristics. Intentionally entrained air bubbles are extremely small in size, between 10 to 1000 µm in diameter, while entrapped voids are usually 1000 µm (1 mm) or larger. The majority of the entrained air voids in normal concrete are between 10 µm and 100 µm in diameter. As shown in Fig. 8-1, the bubbles are not interconnected; they are well dispersed and randomly distributed. Non-air-entrained concrete with a 25-mm maximum-size aggregate has an air content of approximately 1.5%. This same mixture air entrained for severe frost exposure would require a total air content of about 6%, made up of both the coarser "entrapped" air voids and the finer "entrained" air voids.

PROPERTIES OF AIR-ENTRAINED CONCRETE

The primary concrete properties influenced by air entrainment are presented in the following sections. A brief summary of other properties is presented in Table 8-1.

Freeze-Thaw Resistance

The resistance of hardened concrete to freezing and thawing in a moist condition is significantly improved by the use of intentionally entrained air, even when various deicers are involved. Convincing proof of the improvement in durability effected by air entrainment is shown in Figs. 8-2 and 8-3.

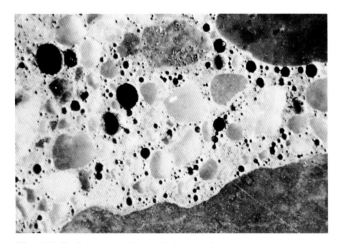

Fig. 8-1. Polished section of air-entrained concrete as seen through a microscope. (67840)

Table 8-1. Effect of Entrained Air on Concrete Properties

Properties	Effect
Abrasion	Little effect; increased strength increases abrasion resistance
Absorption	Little effect
Alkali-silica reactivity	Expansion decreases with increased air
Bleeding	Reduced significantly
Bond to steel	Decreased
Compressive strength	Reduced approximately 2% to 6% per percentage point increase in air; harsh or lean mixes may gain strength
Creep	Little effect
Deicer scaling	Significantly reduced
Density	Decreases with increased air
Fatigue	Little effect
Finishability	Reduced due to increased cohesion (stickiness)
Flexural strength	Reduced approximately 2% to 4% per percentage point increase in air
Freeze-thaw resistance	Significantly improved resistance to water-saturated freeze-thaw deterioration
Heat of hydration	No significant effect
Modulus of elasticity (static)	Decreases with increased air approximately 720 to 1380 MPa per percentage point of air
Permeability	Little effect; reduced water-cementing materials ratio reduces permeability
Scaling	Significantly reduced
Shrinkage (drying)	Little effect
Slump	Increases with increased air approximately 25 mm per ½ to 1 percentage point increase in air
Specific heat	No effect
Sulphate resistance	Significantly improved
Stickiness	Increased cohesion—harder to finish
Temperature of wet concrete	No effect
Thermal conductivity	Decreases 1% to 3% per percentage point increase in air
Thermal diffusivity	Decreases about 1.6% per percentage point increase in air
Water demand of wet concrete for equal slump	Decreases with increased air; approximately 3 to 6 kg/m³ per percentage point of air
Watertightness	Increases slightly; reduced water-cementing materials ratio increases watertightness
Workability	Increases with increased air

Note: The table information may not apply to all situations.

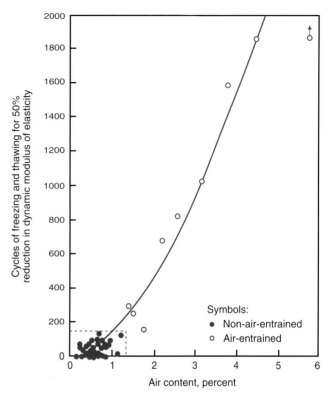

Fig. 8-2. **Effect of entrained air on the resistance of concrete to freezing and thawing in laboratory tests. Concretes were made with cements of different fineness and composition and with various cement contents and water-cementing materials ratios (Bates and others 1952, and Mielenz and Lerch 1960).**

As the water in moist concrete freezes, it produces osmotic and hydraulic pressures in the capillaries and pores of the cement paste and aggregate. If the pressure exceeds the tensile strength of the paste or aggregate, the cavity will dilate and rupture. The accumulative effect of successive freeze-thaw cycles and disruption of paste and aggregate eventually cause significant expansion and deterioration of the concrete. Deterioration is visible in the form of cracking, scaling, and crumbling (Fig. 8-3). Powers (1965) and Pigeon and Pleau (1995) extensively review the mechanisms of frost action.

Hydraulic pressures are caused by the 9% expansion of water upon freezing; in this process growing ice crystals displace unfrozen water. If a capillary is above critical saturation (91.7% filled with water), hydraulic pressures result as freezing progresses. At lower water contents, no hydraulic pressure should exist.

Osmotic pressures develop from differential concentrations of alkali solutions in the paste (Powers 1965a). As pure water freezes, the alkali concentration increases in the adjacent unfrozen water. A high-alkali solution, through the mechanism of osmosis, draws water from lower-alkali solutions in the pores. This osmotic transfer of water continues until equilibrium in the fluids' alkali concentration is achieved. Osmotic pressure is considered a minor factor, if present at all, in aggregate frost action, whereas it may be dominant in certain cement pastes. Osmotic pressures, as described above, are considered to be a major factor in "salt scaling."

Capillary ice (or any ice in large voids or cracks) draws water from pores to advance its growth. Also, since most pores in cement paste and some aggregates are too small for ice crystals to form, water attempts to migrate to locations where it can freeze.

Entrained air voids act as empty chambers in the paste where freezing and migrating water can enter, thus

Fig. 8-3. Effect of weathering on boxes and slabs on ground at the Long-Time Study outdoor test plot, Project 10, PCA, Skokie, Illinois. Specimens at top are air-entrained, specimens at bottom exhibiting severe crumbling and scaling are non-air-entrained. All concretes were made with 335 kg of Type 10 (Type I) portland cement per cubic metre. Periodically, calcium chloride deicer was applied to the slabs. Specimens were 40 years old when photographed (see Klieger 1963 for concrete mixture information). (69977, 69853, 69978, 69854)

relieving the pressures described above and preventing damage to the concrete. Upon thawing, most of the water returns to the capillaries due to capillary action and pressure from air compressed in the bubbles. Thus the bubbles are ready to protect the concrete from the next cycle of freezing and thawing (Powers 1955, Lerch 1960, and Powers 1965).

The pressure developed by water as it expands during freezing depends largely upon the distance the water must travel to the nearest air void for relief. Therefore, the voids must be spaced close enough to reduce the pressure below that which would exceed the tensile strength of the concrete. The amount of hydraulic pressure is also related to the rate of freezing and the permeability of the paste.

The spacing and size of air voids are important factors contributing to the effectiveness of air entrainment in concrete. CSA Standard A23.1 requires that when concrete will be subjected to frequent cycles of freezing and thawing in the presence of moisture or deicing chemicals (Class C-1, C-2, or F-1, refer to Table 8-2), the air-void spacing factor ($\bar{L}$) of the air-void system is to be determined in accordance with ASTM Standard C 457, using a magnification factor between 100 and 125.

The concrete is considered to have a satisfactory air-void system if the average of all tests shows a spacing factor ($\bar{L}$) not exceeding 230 μm, and no single test is greater than 260 μm, and there is an air content greater than or equal to 3.0% in the hardened concrete. The 3% air content value is a lower bound specified by CSA A23.1 for C-3 exposure class concrete with 40-mm aggregate. C-1 or C-2 exposure class concrete with a 20-mm aggregate would require a minimum of 5% air to be considered satisfactory. For high-performance concrete, with a water-cementing materials ratio of 0.36 or less, the average spacing factor shall not exceed 250 μm with no single value greater than 300 μm (See *CSA A23.1-00-Appendix-J*).

Another requirement of CSA Standard A23.1 is that, for concrete subjected to Class C-1, C-2, or F-1 exposure (see Table 8-2), the air-void system must be proven satisfactory prior to the start of construction. The spacing factor is to be determined by data from tests performed on concrete cylinders made with the same material, mix proportions, and mixing procedures as planned for the project. If the owner deems it necessary to check the air-void system during construction, testing is to be carried out on cylinders made from the same concrete as delivered to the jobsite or on cores drilled from hardened concrete in the structure. In this latter case, this requirement must be clearly stated in the project specifications.

Field control practice involves only the measurement of air volume in freshly mixed concrete. Fig. 8-4 illustrates the relationship between spacing factor and total air content.

Table 8-2. Definitions of C, F, and N Exposure Classes

C-1	Structurally reinforced concrete exposed to chlorides with or without freezing and thawing conditions. *Examples:* bridge decks, parking decks and ramps, portions of marine structures located in tidal and splash zone
C-2	Non-structurally reinforced (plain) concrete exposed to chlorides and freezing and thawing. *Examples:* garage floors, porches, steps, pavements, sidewalks, curbs and gutters
C-3	Continuously submerged concrete exposed to chlorides but not to freezing and thawing. *Examples:* underwater portions of marine structures
C-4	Non-structurally reinforced concrete exposed to chlorides but not to freezing and thawing. *Examples:* underground parking slabs on grade
F-1	Concrete exposed to freezing and thawing in a saturated condition but not to chlorides. *Examples:* pool decks, patios, tennis courts, freshwater pools, and fresh water control structures
F-2	Concrete in an unsaturated condition exposed to freezing and thawing but not to chlorides. *Examples:* exterior walls and columns
N	Concrete not exposed to chlorides nor to freezing and thawing. *Examples:* footings and interior slabs, walls and columns

Notes:
(1) "C" classes pertain to chloride exposure.
(2) "F" classes pertain to freezing and thawing exposure only.
(3) "N" class pertains to nonexposure to either chlorides or freezing and thawing.
Source: CSA Standard A23.1

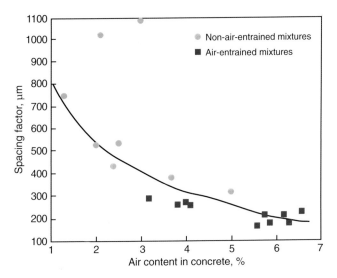

Fig. 8-4. Spacing factor as a function of total air content in concrete (Pinto and Hover 2001).

Measurement of air volume alone does not permit full evaluation of the important characteristics of the air-void system; however, air-entrainment is generally considered effective for freeze-thaw resistance when the volume of air in the mortar fraction of the concrete—material passing the 5-mm sieve—is about 9 ± 1% (Klieger 1952) or about 18% by paste volume. For equal admixture dosage rates per unit of cement, the air content of ASTM C 185 mortar would be about 19% due to the standard aggregate's properties. The air content of concrete with 20-mm maximum-size aggregate would be about 6% for effective freeze-thaw resistance.

The relationship between air content of standard mortar and concrete is illustrated by Taylor (1948). Pinto and Hover (2001) address paste air content versus frost resistance. The total required concrete air content for durability increases as the maximum-size aggregate is reduced (due to greater paste volume) and the exposure conditions become more severe (see "Recommended Air Contents" later in this chapter).

CSA Standard A23.1 requirements for concrete subjected to various exposure conditions (freeze/thaw, chlorides) are given in Tables 8-2 and 8-3.

Freeze-thaw resistance of a non-structurally reinforced (plain) concrete (that is, pavements, driveways, sidewalks, curbs, and gutters, etc.) is also significantly increased with the use of the following: (1) a good quality aggregate, (2) a low water to cementing materials ratio (maximum 0.45), (3) a minimum cementitious materials content of 335 kg/m³, (4) a minimum 28-day compressive strength of 32 MPa, and (5) proper finishing and curing techniques. Even non-air-entrained concretes will be more freeze-thaw resistant with a low water-cementing materials ratio. Fig. 8-5 illustrates the effect of water-cementing materials ratio on the durability of non-air-entrained concrete.

Concrete elements should be properly drained and kept as dry as possible as greater degrees of saturation increase the likelihood of distress due to freeze-thaw cycles. Concrete that is dry or contains only a small amount of moisture in service is essentially not affected by even a large number of cycles of freezing and thawing. Refer to the sections on "Deicer-Scaling Resistance" and "Recommended Air Contents" in this chapter and to Chapter 9 for mixture design considerations.

Deicer-Scaling Resistance

Deicing chemicals used for snow and ice removal can cause and aggravate surface scaling. The damage is primarily a physical action. Deicer scaling of inadequately air-entrained or non-air-entrained concrete during freezing is believed to be caused by a buildup of osmotic and hydraulic pressures in excess of the normal hydraulic pressures produced when water in concrete freezes. These pressures become critical and result in scaling unless entrained air voids are present at the surface and throughout the sample to relieve the pressure. The hygroscopic (mois-

Table 8-3. Requirements for C, F, and N Classes of Exposure

Requirements for specifying concrete	Requirements for concrete		
Class of Exposure*	Maximum water-to-cementing materials ratio	Minimum, specified 28-day compressive strength, MPa	Air content category
C-1	0.40	35	**
C-2	0.45	32	1
C-3	0.50	30	2
C-4	0.55	25	2
F-1	0.50	30	1***
F-2	0.55	25	2***
N	For structural design	For structural design	

* See Table 8-2 for a description of classes of exposure
** Use Category 1 for concrete exposed to freezing and thawing.
 Use category 2 for concrete not exposed to freezing and thawing.
*** Interior ice rink slabs and freezer slabs with a steel troweled finish have been found to perform satisfactorily without entrained air.
Source: CSA Standard A23.1

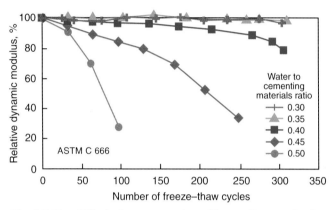

Fig. 8-5. Durability factors vs. number of freeze-thaw cycles for selected non-air-entrained concretes (Pinto and Hover 2001).

ture absorbing) properties of deicing salts also attract water and keep the concrete more saturated, increasing the potential for freeze-thaw deterioration. However, properly designed and placed air-entrained concrete will withstand deicers for many years.

Studies have also shown that, in the absence of freezing, the formation of salt crystals in concrete (from external sources of chloride, sulphate, and other salts) may contribute to concrete scaling and deterioration similar to the crumbling of rocks by salt weathering. The entrained air voids in concrete allow space for salt crystals to grow; this relieves internal stress similar to the way the voids relieve stress from freezing water in concrete (ASCE 1982 and Sayward 1984).

Deicers can have many effects on concrete and the immediate environment. All deicers can aggravate scaling of concrete that is not properly air entrained. Sodium chloride (rock salt) (ASTM D 632), calcium chloride (ASTM D 98), and urea are the most frequently used deicers. In the absence of freezing, sodium chloride has little to no chemical effect on concrete but can damage plants and corrode metal. Calcium chloride in weak solutions generally has little chemical effect on concrete and vegetation but does corrode metal.

Studies have shown that concentrated calcium chloride solutions can chemically attack concrete (Brown and Cady 1975). Urea does not chemically damage concrete, vegetation, or metal. Nonchloride deicers are used to minimize corrosion of reinforcing steel and minimize groundwater chloride contamination. *The use of deicers containing ammonium nitrate and ammonium sulphate should be strictly prohibited as they rapidly attack and disintegrate concrete.* Magnesium chloride deicers have come under recent criticism for aggravating scaling. One study found that magnesium chloride, magnesium acetate, magnesium nitrate, and calcium chloride are more damaging to concrete than sodium chloride (Cody, Cody, Spry, and Gan 1996).

The extent of scaling depends upon the amount of deicer used and the frequency of application. Relatively low concentrations of deicer (on the order of 2% to 4% by mass) produce more surface scaling than higher concentrations or the absence of deicer (Verbeck and Klieger 1956).

Deicers can reach concrete surfaces in ways other than direct application, such as splashing by vehicles and dripping from the undersides of vehicles. Scaling is more severe in poorly drained areas because more of the deicer solution remains on the concrete surface during freezing and thawing. Air entrainment is effective in preventing surface scaling and is recommended for all concretes that may come in contact with deicing chemicals (Fig. 8-6).

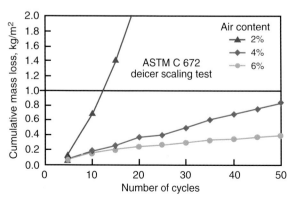

Fig. 8-6. Cumulative mass loss for mixtures with a water to cementing materials ratio of 0.45 and on-time finishing (Pinto and Hover 2001).

A good air-void system with a low spacing factor, maximum of 200 μm (230 μm maximum per CSA A23.1), is perhaps more important to deicer environments than saturated frost environments without deicers. The relationship between spacing factor and deicer scaling is illustrated in Fig. 8-7. A low water to cementing materials ratio helps minimize scaling, but is not sufficient to control scaling at normal water-cementing materials ratios. Fig. 8-8 illustrates the overriding impact of air content over water-cementing materials ratio in controlling scaling.

To provide adequate durability and scale resistance in severe exposures with deicers present (C-1 exposure), air-entrained concrete should be composed of durable materials and have the following: (1) a low water to cementing materials ratio (maximum 0.40), (2) a slump of 100 mm or less unless a plasticizer is used, (3) a cementitious materials content of 335 kg/m³, (4) proper finishing after bleed water has evaporated from the surface, (5) adequate drainage, (6) a minimum of 7 days moist curing at or above 10°C, or for the time necessary to attain 70% of the specified 28-day strength, (7) a compressive strength of 35 MPa when exposed to repeated freeze-thaw cycling, and (8) a minimum 30-day drying period after moist curing if concrete will be exposed to freeze-thaw cycles and deicers when saturated. Recommended air contents are discussed at the end of this chapter.

Normal dosages of supplementary cementing materials should not effect scaling resistance of properly designed, placed, and cured concrete (Table 8-4). The ACI 318 building code allows up to 10% silica fume, 25% fly ash, and 50% slag as part of the cementing materials for deicer exposures. Higher dosages of supplementary cementing materials can be used, if adequate durability is demonstrated by laboratory or field performance. Abuse of these materials along with poor placing and curing practices can aggravate scaling. Consult local guidelines on allowable dosages, practices, and field performance for using these materials in deicer environments as they can vary from ACI 318 requirements.

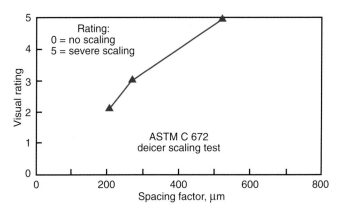

Fig. 8-7. Visual rating as a function of spacing factor, for a concrete mixture with a water to cementing materials ratio of 0.45 (Pinto and Hover 2001).

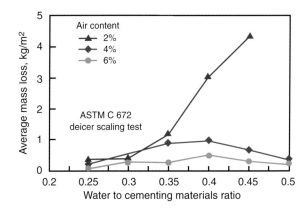

Fig. 8-8. Measured mass loss of concrete after 40 cycles of deicer and frost exposure at various water to cementing materials ratios (Pinto and Hover 2001).

Air Drying. The resistance of air-entrained concrete to freeze-thaw cycles and deicers is greatly increased by an air drying period after initial moist curing. Air drying removes excess moisture from the concrete which in turn reduces the internal stress caused by freeze-thaw conditions and deicers. Water-saturated concrete will deteriorate faster than an air-dried concrete when exposed to

Table 8-4. Deicer Scaling Resistance (Visual Ratings) of Concrete with Selected Supplementary Cementing Materials

Mixture	Control	Fly ash (Class F)	Slag	Calcined shale	Calcined shale
Mass replacement of cement, %	0	15	40	15	25
Scale rating at 25 cycles	1	1	1	1	1
Scale rating at 50 cycles	2	2	1	2	1

Concrete had 335 kg of cementing materials per cubic metre, a Type 10 (Type I) portland cement, a water to cementing materials ratio of 0.50, a nominal slump of 75 mm, and a nominal air content of 6%. Test method: ASTM C 672. Results are for specific materials tested in 2000 and may not be representative of other materials. Scale rating: 1 = very slight scaling (3 mm depth maximum) with no coarse aggregate visible, 2 = slight to moderate scaling.

moist freeze-thaw cycling and deicers. Concrete placed in the spring or summer has an adequate drying period. Concrete placed in the fall season, however, often does not dry out enough before deicers are used. This is especially true of fall paving cured by membrane-forming compounds. These membranes remain intact until worn off by traffic; thus, adequate drying may not occur before the onset of winter. Curing methods, such as use of plastic sheets, that allow drying at the completion of the curing period are preferable for fall paving on all projects where deicers will be used. Concrete placed in the fall should be allowed at least 30 days for air drying after the moist-curing period. The exact length of time for sufficient drying to take place may vary with climate and weather conditions.

Treatment of Scaled Surfaces. If surface scaling (an indication of an inadequate air-void system or poor finishing practices) should develop during the first frost season, or if the concrete is of poor quality, a breathable surface treatment can be applied to the dry concrete to help protect it against further damage. Treatment often consists of a penetrating sealer made with boiled linseed oil (ACPA 1996), breathable methacrylate, or other materials. Non-breathable formulations should be avoided as they can cause delamination.

The effect of mix design, surface treatment, curing, or other variables on resistance to surface scaling can be evaluated by ASTM C 672.

Sulphate Resistance

Sulphate resistance of concrete is improved by air entrainment, as shown in Figs. 8-9 and 8-10, when advantage is taken of the reduction in water-cementing materials ratio possible with air entrainment. Air-entrained concrete made with a low water-cementing materials ratio, an adequate cement content and a sulphate-resistant cement will be resistant to attack from sulphate soils and waters.

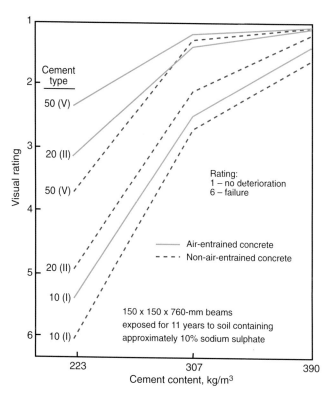

Fig. 8-10. Performance of a variety of air-entrained and non-air-entrained concretes exposed to sulphate soil. Sulphate resistance is increased with the use of Types 20 (II) and 50 (V) cements, a higher cement content, lower water-cementing materials ratio, and air entrainment (Stark 1984).

Resistance to Alkali-Silica Reactivity

The expansive disruption caused by alkali-silica reactivity is reduced through the use of air-entrainment (Kretsinger 1949). Alkali hydroxides react with the silica of reactive aggregates to form expansive reaction products, causing the concrete to expand. Excessive expansion will disrupt and deteriorate concrete. As shown in Fig. 8-11, the expansion of mortar bars made with reactive materials is reduced as the air content is increased.

Strength

When the air content is maintained constant, strength varies inversely with the water-cement ratio. Fig. 8-12 shows a typical relationship between 28-day compressive strength and water-cementing materials ratio for concrete that has the recommended percentages of entrained air. As air content is increased, a given strength generally can be maintained by holding the voids (air + water) to

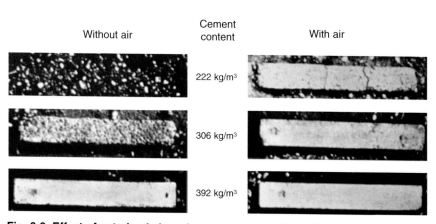

Fig. 8-9. Effect of entrained air and cement content (Type 20) on performance of concrete specimens exposed to a sulphate soil. Without entrained air the specimens made with lesser amounts of cement deteriorated badly. Specimens made with the most cement and the lowest water-cementing materials ratio were further improved by air entrainment. Specimens were 5 years old when photographed (Stanton 1948 and Lerch 1960).

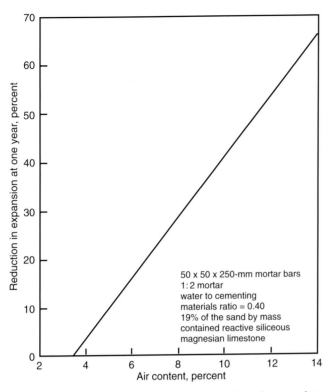

Fig. 8-11. Effect of air content on the reduction of expansion due to alkali-silica reaction (Kretsinger 1949).

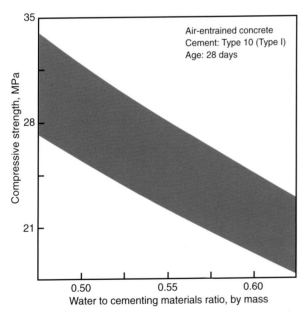

Fig. 8-12. Typical relationship between 28-day compressive strength and water to cementing materials ratio for a wide variety of air-entrained concretes using Type 10 (I) cement.

cementing materials ratio constant; this may, however, necessitate some increase in cementing materials content.

Air-entrained as well as non-air-entrained concrete can readily be proportioned to provide similar moderate strengths. Both generally must contain the same amount of coarse aggregate. When the cementing materials content and slump are held constant, air entrainment reduces the sand and water requirements as illustrated in Fig. 8-13. Thus, air-entrained concretes can have lower water-cementing materials ratios than non-air-entrained concretes; this minimizes the reductions in strength that generally accompany air entrainment. At constant water-cementing materials ratios, increases in air will proportionally reduce strength (Fig. 8-14). Pinto and Hover (2001) found that for a decrease in strength of 10 MPa, resulting from a four percentage point decrease in air, the water-cementing materials ratio had to be decreased by 0.14 to maintain strength. Some reductions in strength may be tolerable in view of other benefits of air, such as improved workability. Reductions in strength become more signifi-

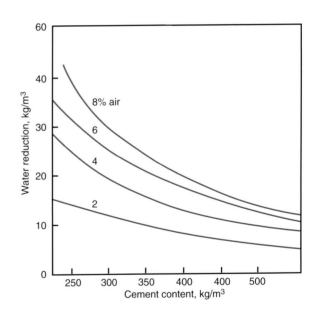

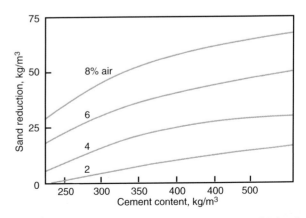

Fig. 8-13. Reduction of water and sand content obtained at various levels of air and cement contents (Gilkey 1958).

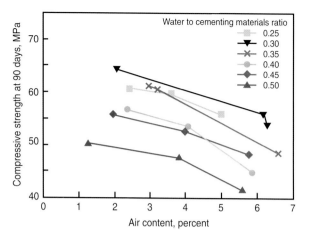

Fig. 8-14. Relationship between compressive strength at 90 days and air content (Pinto and Hover 2001).

cant in higher-strength mixes, as illustrated in Fig. 8-15. In lower-cementing-materials-content, harsh mixes, strength is generally increased by entrainment of air in proper amounts due to the reduced water-cement ratio and improved workability. For moderate to high-strength concrete, each percentile of increase in entrained air reduces the compressive strength about 2% to 9% (Cordon 1946, Klieger 1952, Klieger 1956, Whiting and Nagi 1998, and Pinto and Hover 2001). Actual strength varies and is affected by the cement source, admixtures, and other concrete ingredients.

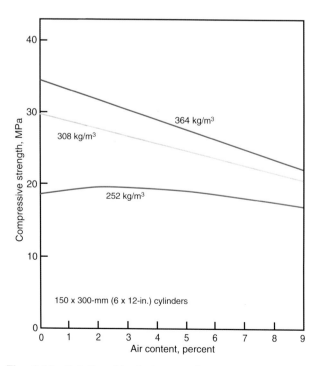

Fig. 8-15. Relationship between air content and 28-day compressive strength for concrete at three cement contents. Water content was reduced with increased air content to maintain a constant slump (Cordon 1946).

Attainment of high strength with air-entrained concrete may be difficult at times. Even though a reduction in mixing water is associated with air entrainment, mixtures with high cementing materials contents require more mixing water than lower-cementing-materials-content mixtures; hence, the increase in strength expected from the additional cementing materials is offset somewhat by the additional water. Water reducers can offset this effect.

Workability

Entrained air improves the workability of concrete. It is particularly effective in lean (low cementing materials content) mixes that otherwise might be harsh and difficult to work. In one study, an air-entrained mixture made with natural aggregate, 3% air, and a 40-mm slump had about the same workability as a non-air-entrained concrete with 1% air and a 75-mm slump, even though less cementing materials was required for the air-entrained mix (Cordon 1946). Workability of mixes with angular and poorly graded aggregates is similarly improved.

Because of improved workability with entrained air, water and sand content can be reduced significantly (Fig. 8-13). A volume of air-entrained concrete requires less water than an equal volume of non-air-entrained concrete of the same consistency and maximum size aggregate. Freshly mixed concrete containing entrained air is cohesive, looks and feels fatty or workable, and can usually be handled with ease; on the other hand, high air contents can make a mixture sticky and more difficult to finish. Entrained air also reduces segregation and bleeding in freshly mixed and placed concrete.

AIR-ENTRAINING MATERIALS

The entrainment of air in concrete is accomplished by adding an air-entraining admixture at the mixer. Adequate control and monitoring is required to ensure the proper air content at all times.

Numerous commercial air-entraining admixtures, manufactured from a variety of materials are available. Most air-entraining admixtures consist of one or more of the following materials: wood resin (Vinsol resin), sulphonated hydrocarbons, fatty and resinous acids, and synthetic materials. Chemical descriptions and performance characteristics of common air-entraining agents are shown in Table 8-5. Air-entraining admixtures are usually liquids and should not be allowed to freeze. Admixtures added at the mixer should conform to ASTM C 260.

Variations in air content can be expected with variations in aggregate proportions and gradation, mixing time, temperature, and slump. The order of batching and mixing concrete ingredients when using an air-entraining admixture has a significant influence on the amount of air entrained; therefore, consistency in batching is needed to maintain adequate control.

When entrained air is excessive, it can be reduced by using one of the following defoaming (air-detraining) agents: tributyl phosphate, dibutyl phthalate, octyl alcohol, water-insoluble esters of carbonic acid and boric acid, and silicones. Only the smallest possible dosage of defoaming agent should be used to reduce the air content to the specified limit. An excessive amount might have adverse effects on concrete properties (Whiting and Stark 1983).

FACTORS AFFECTING AIR CONTENT

Cement

As cement content increases, the air content decreases for a set dosage of air-entraining admixture per unit of cement within the normal range of cement contents (Fig. 8-16). In going from 240 to 360 kilograms of cement per cubic metre, the dosage rate may have to be doubled to maintain a constant air content. However, studies indicate that when this is done the air-void spacing factor generally decreases with an increase in cement content; and for a given air content the specific surface increases, thus improving durability.

An increase in cement fineness will result in a decrease in the amount of air entrained. Type 30 (Type III) cement, a very finely ground material, may require twice as much air-entraining agent as a Type 10 (Type I) cement of normal fineness.

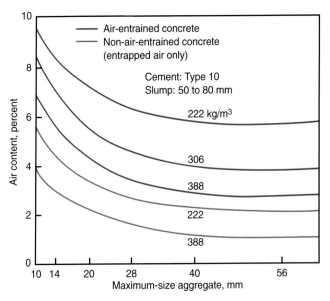

Fig. 8-16. Relationship between aggregate size, cement content, and air content of concrete. The air-entraining admixture dosage per unit of cement was constant for air-entrained concrete. PCA Major Series 336.

Table 8-5. Classification and Performance Characteristics of Common Air-Entraining Admixtures (Whiting and Nagi 1998)

Classification	Chemical description	Notes and performance characteristics
Wood derived acid salts Vinsol® resin	Alkali or alkanolamine salt of: A mixture of tricyclic acids, phenolics, and terpenes.	Quick air generation. Minor air gain with initial mixing. Air loss with prolonged mixing. Mid-sized air bubbles formed. Compatible with most other admixtures.
Wood rosin	Tricyclic acids-major component. Tricyclic acids-minor component.	Same as above.
Tall oil	Fatty acids-major component. Tricyclic acids-minor component.	Slower air generation. Air may increase with prolonged mixing. Smallest air bubbles of all agents. Compatible with most other admixtures.
Vegetable oil acids	Coconut fatty acids, alkanolamine salt.	Slower air generation than wood rosins. Moderate air loss with mixing. Coarser air bubbles relative to wood rosins. Compatible with most other admixtures.
Synthetic detergents	Alkyl-aryl sulphonates and sulphates (e.g., sodium dodecylbenzenesulphonate).	Quick air generation. Minor air loss with mixing. Coarser bubbles. May be incompatible with some HRWR. Also applicable to cellular concretes.
Synthetic workability aids	Alkyl-aryl ethoxylates.	Primarily used in masonry mortars.
Miscellaneous	Alkali-alkanolamine acid salts of lignosulphonate. Oxygenated petroleum residues. Proteinaceous materials. Animal tallows.	All these are rarely used as concrete air-entraining agents in current practice.

High-alkali cements may entrain more air than low alkali cements with the same amount of air-entraining material. A low-alkali cement may require 20% to 40% (occasionally up to 70%) more air-entraining agent than a high-alkali cement to achieve an equivalent air content. Precautions are therefore necessary when using more than one cement source in a batch plant to ensure that proper admixture requirements are determined for each cement (Greening 1967).

Coarse Aggregate

The size of coarse aggregate has a pronounced effect on the air content of both air-entrained and non-air-entrained concrete, as shown in Fig. 8-16. There is little change in air content when the size of aggregate is increased above 40 mm.

Fine Aggregate

The fine-aggregate content of a mixture affects the percentage of entrained air. As shown in Fig. 8-17, increasing the amount of fine aggregate causes more air to be entrained for a given amount of admixture (more air is also entrapped in non-air-entrained concrete).

Fine-aggregate particles passing the 630 μm to 160 μm sieves entrap more air than either very fine or coarser particles. Appreciable amounts of material passing the 160 μm sieve will result in a significant reduction of entrained air.

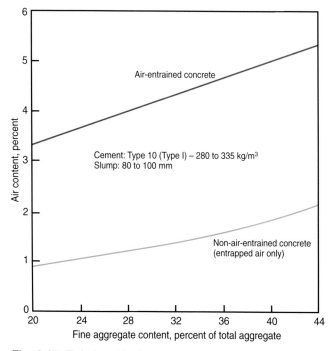

Fig. 8-17. Relationship between percentage of fine aggregate and air content of concrete. PCA Major Series 336.

Fine aggregates from different sources may entrap different amounts of air even though they have identical gradations. This may be due to differences in shape and surface texture or contamination by organic materials.

Mixing Water and Slump

An increase in the mixing water makes more water available for the generation of air bubbles, thereby increasing the air content as slumps increase up to about 150 or 175 mm. An increase in the water-cementing materials ratio from 0.4 to 1.0 can increase the air content by four percentage points. A portion of the air increase is due to the relationship between slump and air content. Air content increases with slump even when the water-cementing materials ratio is held constant. The spacing factor, ($\bar{L}$), of the air-void system also increases, that is, the voids become coarser at higher water-cementing materials ratios, thereby reducing concrete freeze-thaw durability (Stark 1986).

The addition of 5 kg of water per cubic metre of concrete can increase the slump by 25 mm. A 25-mm increase in slump increases the air content by approximately one-half to one percentage point for concretes with a low-to-moderate slump and constant air-entraining admixture dosage. However, this approximation is greatly affected by concrete temperature, slump, and the type and amount of cementing materials and admixtures present in the concrete. A low-slump concrete with a high dosage of water-reducing and air-entraining admixtures can undergo large increases in slump and air content with a small addition of water. On the other hand, a very fluid concrete mixture with a 200 to 250-mm slump may lose air with the addition of water. Refer to Tables 8-6 and 8-7 for more information.

The mixing water used may also affect air content. Algae-contaminated water increases air content. Highly alkaline wash water from truck mixers can affect air contents. The effect of water hardness in most municipal water supplies is generally insignificant; however, very hard water from wells, as used in rural communities, may decrease the air content in concrete.

Slump and Vibration

The effect of slump and vibration on the air content of concrete is shown in Fig. 8-18. For a constant amount of air-entraining admixture, air content increases as slump increases up to about 150 or 175 mm; then it begins to decrease with further increases in slump. At all slumps, however, even 15 seconds of vibration will cause a considerable reduction in air content. Prolonged vibration of concrete should be avoided.

The greater the slump, air content, and vibration time, the larger the percentage of reduction in air content during vibration (Fig. 8-18). However, if vibration is properly applied, little of the intentionally entrained air is lost. The air lost during handling and moderate vibration consists mostly of the larger bubbles that are usually undesirable

Table 8-6. Effect of Mixture Design and Concrete Constituents on Control of Air Content in Concrete

	Characteristic/Material	Effects	Guidance
Portland cement	Alkali content	Air content increases with increase in cement alkali level. Less air-entraining agent dosage needed for high-alkali cements. Air-void system may be more unstable with some combinations of alkali level and air-entraining agent used.	Changes in alkali content or cement source require that air-entraining agent dosage be adjusted. Decrease dosage as much as 40% for high-alkali cements.
	Fineness	Decrease in air content with increased fineness of cement.	Use up to 100% more air-entraining admixture for very fine Type 30 (Type III) cements. Adjust admixture if cement source or fineness changes.
	Cement content in mixture	Decrease in air content with increase in cement content. Smaller and greater number of voids with increased cement content.	Increase air-entraining admixture dosage rate as cement content increases.
	Contaminants	Air content may be altered by contamination of cement with finish mill oil.	Verify that cement meets CSA A.5 (ASTM C 150) requirements on air content of test mortar.
Supplementary cementing materials	Fly ash	Air content decreases with increase in loss on ignition (carbon content). Air-void system may be more unstable with some combinations of fly ash/cement/air-entraining agents.	Changes in LOI or fly ash source require that air-entraining admixture dosage be adjusted. Perform "foam index" test to estimate increase in dosage. Prepare trial mixes and evaluate air-void systems.
	Ground granulated blast-furnace slag	Decrease in air content with increased fineness of GGBFS.	Use up to 100% more air-entraining admixture for finely ground slags.
	Silica fume	Decrease in air content with increase in silica fume content.	Increase air-entraining admixture dosage up to 100% for fume contents up to 10%.
	Metakaolin	No apparent effect.	Adjust air-entraining admixture dosage if needed.
Chemical admixtures	Water reducers	Air content increases with increases in dosage of lignin-based materials. Spacing factors may increase when water-reducers used.	Reduce dosage of air-entraining admixture. Select formulations containing air-detraining agents. Prepare trial mixes and evaluate air-void systems.
	Retarders	Effects similar to water-reducers.	Adjust air-entraining admixture dosage.
	Accelerators	Minor effects on air content.	No adjustments normally needed.
	High-range water reducers (Plasticizers)	Moderate increase in air content when formulated with lignosulphonate. Spacing factors increase.	Only slight adjustments needed. No significant effect on durability.
Aggregate	Maximum size	Air content requirement decreases with increase in maximum size. Little increase over 40 mm maximum size aggregate.	Decrease air content.
	Sand-to-total aggregate ratio	Air content increases with increased sand content.	Decrease air-entraining admixture dosage for mixtures having higher sand contents.
	Sand grading	Middle fractions of sand promote air-entrainment.	Monitor gradation and adjust air-entraining admixture dosage accordingly.

Table 8-6. Effect of Mixture Design and Concrete Constituents on Control of Air Content in Concrete (Continued)

	Characteristic/Material	Effects	Guidance
Mix water and slump	Water chemistry	Very hard water reduces air content.	Increase air entrainer dosage.
		Batching of admixture into concrete wash water decreases air.	Avoid batching into wash water.
		Algae growth may increase air.	
	Water-to-cementing materials ratio	Air content increases with increased water to cementing materials ratio.	Decrease air-entraining admixture dosage as water to cementing materials ratio increases.
	Slump	Air increases with slumps up to about 150 mm.	Adjust air-entraining admixture dosages for slump.
		Air decreases with very high slumps.	Avoid addition of water to achieve high-slump concrete.
		Difficult to entrain air in low-slump concretes.	Use additional air-entraining admixture; up to ten times normal dosage.

Table 8-7. Effect of Production Procedures, Construction Practices, and Environment on Control of Air Content in Concrete

	Procedure/Variable	Effects	Guidance
Production procedures	Batching sequence	Simultaneous batching lowers air content.	Add air-entraining admixture with initial water or on sand.
		Cement-first raises air content.	
	Mixer capacity	Air increases as capacity is approached.	Run mixer close to full capacity. Avoid overloading.
	Mixing time	Central mixers: air content increases up to 90 sec. of mixing.	Establish optimum mixing time for particular mixer.
		Truck mixers: air content increases with mixing.	Avoid overmixing.
		Short mixing periods (30 seconds) reduces air content and adversely affects air-void system.	Establish optimum mixing time (about 60 seconds).
	Mixing speed	Air content gradually increases up to approx. 20 rpm.	Follow truck mixer manufacturer recommendations.
		Air may decrease at higher mixing speeds.	Maintain blades and clean truck mixer.
	Admixture metering	Accuracy and reliability of metering system will affect uniformity of air content.	Avoid manual-dispensing or gravity-feed systems and timers. Positive-displacement pumps interlocked with batching system are preferred.
Transport and delivery	Transport and delivery	Some air (1% to 2%) normally lost during transport.	Normal retempering with water to restore slump will restore air.
		Loss of air in nonagitating equipment is slightly higher.	If necessary, retemper with air-entraining admixture to restore air.
			Dramatic loss in air may be due to factors other than transport.
	Haul time and agitation	Long hauls, even without agitation, reduce air, especially in hot weather.	Optimize delivery schedules. Maintain concrete temperature in recommended range.
	Retempering	Regains some of the lost air.	Retemper only enough to restore workability. Avoid addition of excess water.
		Does not usually affect the air-void system.	Higher admixture dosage is needed for jobsite admixture additions.
		Retempering with air-entraining admixtures restores the air-void system.	

Table 8-7. Effect of Production Procedures, Construction Practices, and Environment on Control of Air Content in Concrete (Continued)

	Procedure/Variable	Effects	Guidance
Placement techniques	Belt conveyors	Reduces air content by an average of 1%.	Avoid long conveyed distance if possible. Reduce the free-falling effect at the end of conveyor.
	Pumping	Reduction in air content ranges from 2% to 3%. Does not significantly affect air-void system. Minimum effect on freeze-thaw resistance.	Use of proper mix design provides a stable air-void system. Avoid high slump, high air content concrete. Keep pumping pressure as low as possible. Use loop in descending pump line.
	Shotcrete	Generally reduces air content in wet-process shotcrete.	Air content of mix should be at high end of target zone.
Finishing and environment	Internal vibration	Air content decreases under prolonged vibration or at high frequencies. Proper vibration does not influence the air-void system.	Do not overvibrate. Avoid high-frequency vibrators (greater than 10,000 vpm). Avoid multiple passes of vibratory screeds. Closely spaced vibrator insertion is recommended for better consolidation.
	Finishing	Air content reduced in surface layer by excessive finishing.	Avoid finishing with bleed water still on surface. Avoid overfinishing. Do not sprinkle water on surface prior to finishing. Do not steel trowel exterior slabs.
	Temperature	Air content decreases with increase in temperature. Changes in temperature do not significantly affect spacing factors.	Increase air-entraining admixture dosage as temperature increases.

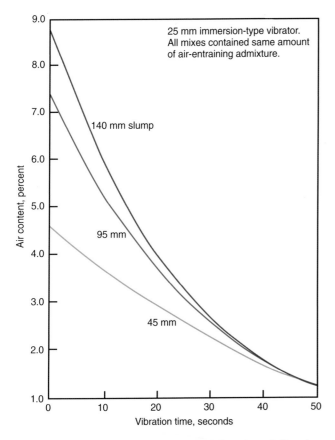

Fig. 8-18. Relationship between slump, duration of vibration, and air content of concrete (Brewster 1949).

from the standpoint of strength. While the average size of the air voids is reduced, the air-void spacing factor remains relatively constant.

Internal vibrators reduce air content more than external vibrators. The air loss due to vibration increases as the volume of concrete is reduced or the vibration frequency is significantly increased. Lower vibration frequencies (8000 vpm) have less effect on spacing factors and air contents than high vibration frequencies (14,000 vpm). High frequencies can significantly increase spacing factors and decrease air contents after 20 seconds of vibration (Brewster 1949 and Stark 1986).

For pavements, specified air contents and uniform air void distributions can be achieved by operating within paving machine speeds of 1.22 to 1.88 metres per minute and with vibrator frequencies of 5,000 to 8,000 vibrations per minute. The most uniform distribution of air voids throughout the depth of concrete, in and out of the vibrator trails, is obtained with the combination of a vibrator frequency of approximately 5,000 vibrations per minute and a slipform paving machine forward track speeds of 1.22 metres per minute. Higher frequencies of speeds singularly or in combination can result in discontinuities and lack of required air content in the upper portion of the concrete pavement. This in turn provides a greater opportunity for water and salt to enter the pavement and reduce the durability and life of the pavement (Cable, McDaniel, Schlorholtz, Redmond, and Rabe 2000).

Concrete Temperature

Temperature of the concrete affects air content, as shown in Fig. 8-19. Less air is entrained as the temperature of the concrete increases, particularly as slump is increased. This effect is especially important during hot-weather concreting when the concrete might be quite warm. A decrease in air content can be offset when necessary by increasing the quantity of air-entraining admixture.

In cold-weather concreting, the air-entraining admixture may lose some of its effectiveness if hot mix water is used during batching. To offset this loss, such admixtures should be added to the batch after the temperature of the concrete ingredients have equalized.

Although increased concrete temperature during mixing generally reduces air volume, the spacing factor and specific surface are only slightly affected.

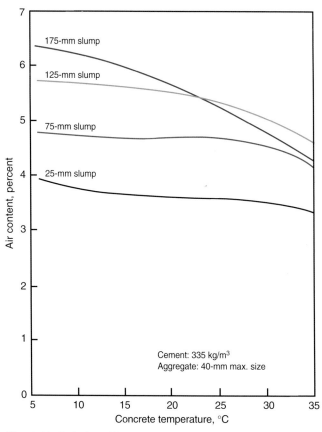

Fig. 8-19. Relationship between temperature, slump, and air content of concrete. PCA Major Series 336 and Lerch 1960.

Supplementary Cementing Materials

The effect of fly ash on the required dosage of air-entraining admixtures can range from no effect to an increase in dosage of up to five times the normal amount (Gebler and Klieger 1986). Large quantities of slag and silica fume can double the dosage of air-entraining admixtures (Whiting and Nagi 1998).

Admixtures and Colouring Agents

Colouring agents such as carbon black usually decrease the amount of air entrained for a given amount of admixture. This is especially true for colouring materials with increasing percentages of carbon (Taylor 1948).

Water-reducing and set-retarding admixtures generally increase the efficiency of air-entraining admixtures by 50% to 100%; therefore, when these are used, less air-entraining admixture will usually give the desired air content. Also, the time of addition of these admixtures into the mix affects the amount of entrained air; delayed additions generally increasing air content.

Set retarders may increase the air-void spacing in concrete. Some water-reducing or set-retarding admixtures are not compatible with some air-entraining admixtures. If they are added together to the mixing water before being dispensed into the mixer, a precipitate may form. This will settle out and result in large reductions in entrained air. The fact that some individual admixtures interact in this manner does not mean that they will not be fully effective if dispensed separately into a batch of concrete.

Superplasticizers (high-range water reducers) may increase or decrease the air content of a concrete mixture based on the admixture's chemical formulation and the slump of the concrete. Naphthalene-based superplasticizers tend to increase the air content while melamine-based materials may decrease or have little effect on air content. The normal air loss in flowing concrete during mixing and transport is about 2 to 4 percentage points (Whiting and Dziedzic 1992).

Superplasticizers also affect the air-void system of hardened concrete by increasing the general size of the entrained air voids. This results in a higher-than-normal spacing factor, occasionally higher than what may be considered desirable for freeze-thaw durability. However, tests on superplasticized concrete with slightly higher spacing factors have indicated that superplasticized concretes have good freeze-thaw durability. This may be due to the reduced water-cementing materials ratio often associated with superplasticized concretes.

A small quantity of calcium chloride is sometimes used in cold weather to accelerate the hardening of concrete. It can be used successfully with air-entraining admixtures if it is added separately in solution form to the mix water. Calcium chloride will slightly increase air content. However, if calcium chloride comes in direct contact with some air-entraining admixtures, a chemical reaction can take place that makes the admixture less effective. Nonchloride accelerators may increase or decrease air content, depending upon the individual chemistry of the admixture, but they generally have little effect on air content.

Mixing Action

Mixing action is one of the most important factors in the production of entrained air in concrete. Uniform distribution of entrained air voids is essential to produce scale-resistant concrete; nonuniformity might result from inadequate dispersion of the entrained air during short mixing periods. In production of ready mixed concrete, it is especially important that adequate and consistent mixing be maintained at all times.

The amount of entrained air varies with the type and condition of the mixer, the amount of concrete being mixed, and the rate and duration of mixing. The amount of air entrained in a given mixture will decrease appreciably as the mixer blades become worn, or if hardened concrete is allowed to accumulate in the drum or on the blades. Because of differences in mixing action and time, concretes made in a stationary mixer and those made in a transit mixer may differ significantly in amounts of air entrained. The air content may increase or decrease when the size of the batch departs significantly from the rated capacity of the mixer. Little air is entrained in very small batches in a large mixer; however, the air content increases as the mixer capacity is approached.

Fig. 8-20 shows the effect of mixing speed and duration of mixing on the air content of freshly mixed concretes made in a transit mixer. Generally, more air is entrained as the speed of mixing is increased up to about 20 rpm, beyond which air entrainment decreases. In the tests from which the data in Fig. 8-20 were derived, the air content reached an upper limit during mixing and a gradual decrease in air content occurred with prolonged mixing. Mixing time and speed will have different effects on the air content of different mixes. Significant amounts of air can be lost during mixing with certain mixtures and types of mixing equipment.

Fig. 8-21 shows the effect of continued mixer agitation on air content. The changes in air content with prolonged agitation can be explained by the relationship between slump and air content. For high-slump concretes, the air content increases with continued agitation as the slump decreases to about 150 or 175 mm. Prolonged agitation will decrease slump further and decrease air content. For initial slumps lower than 150 mm, both the air content and slump decrease with continued agitation. When concrete is retempered (the addition of water and remixing to restore original slump), the air content is increased; however, after 4 hours, retempering is ineffective in increasing air content. Prolonged mixing or agitation of concrete is accompanied by a progressive reduction in slump.

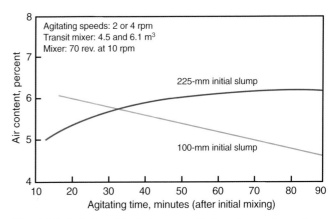

Fig. 8-21. Relationship between agitating time, air content, and slump of concrete. PCA Major Series 336.

Transporting and Handling

Generally, some air—approximately 1 to 2 percentage points—is lost during transportation of concrete from the mixer to the jobsite. The stability of the air content during transport is influenced by several variables including concrete ingredients, haul time, amount of agitation or vibration during transport, temperature, slump, and amount of retempering.

Once at the jobsite, the concrete air content remains essentially constant during handling by chute discharge, wheelbarrow, power buggy, and shovel. However, concrete pumping, crane and bucket, and conveyor-belt handling can cause some loss of air, especially with high-air-content mixtures. Pumping concrete can cause a loss of up to 3 percentage points of air (Whiting and Nagi 1998).

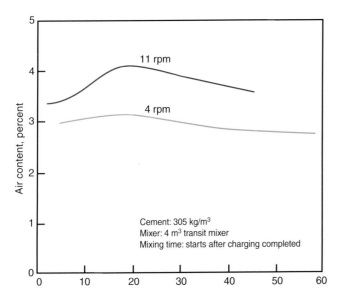

Fig. 8-20. Relationship between mixing time and air content of concrete. PCA Major Series 336.

Finishing

Proper screeding, floating, and general finishing practices should not affect the air content. McNeal and Gay (1996) and Falconi (1996) demonstrated that the sequence and timing of finishing and curing operations are critical to surface durability. Overfinishing (excessive finishing) may reduce the amount of entrained air in the surface region of slabs—thus making the concrete surface vulnerable to scaling. However, as shown in Fig. 8-22, early finishing does not necessarily affect scale resistance unless bleed water is present (Pinto and Hover 2001). *Concrete to be exposed to deicers should not be steel troweled.*

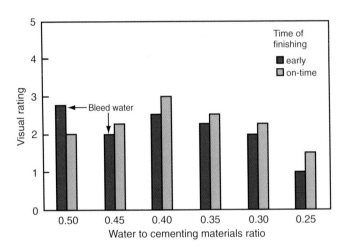

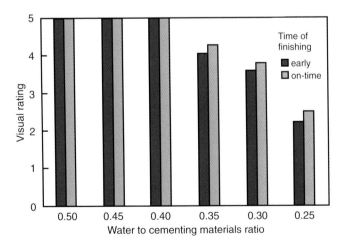

Fig. 8-22. Effect of early finishing—magnesium floating 20 minutes after casting—on scale resistance for: (top) 6% air-entrained concrete; (bottom) non-air-entrained concrete.

TESTS FOR AIR CONTENT

Samples of plastic concrete for the determination of air content should be obtained in accordance with CSA Test Method A23.2-1C, *Sampling Plastic Concrete.*

CSA Standard A23.1 requires that where concrete will be subjected to frequent cycles of freezing and thawing in the presence of moisture or deicing chemicals (Class F-1 and Class C-1 when exposed to freezing and thawing and Class C-2 when exposed to both freezing and thawing and chlorides), every load or batch of concrete shall be tested for air content until satisfactory control is established. An air content determination is also to be made with each strength test.

When concrete will be subjected to Class C-1, C-2, or F-1 exposure, CSA A23.1 also requires that the spacing factor ($\overline{L}$) of the air-void system be determined in accordance with ASTM Standard C 457.

CSA A23.1 requires that air content determinations are to be made in accordance with one of the following CSA Test Methods:

- Pressure Method (CSA Test Method A23.2-4C, *Air Content of Plastic Concrete by the Pressure Method*)—practical for field testing all concretes except those made with highly porous and low-density aggregates (ASTM C 231)
- Volumetric Method (CSA Test Method A23.2-7C, *Air Content of Plastic Concrete by the Volumetric Method*)—practical for field testing of all concretes containing any type of aggregate (ASTM C 173)

Although they measure only air volume and not air-void characteristics, it has been shown by laboratory tests that these methods are generally indicative of the adequacy of the air-void system.

Additional methods for determining the air content in freshly mixed concrete:

- Gravimetric method (ASTM C 138, *Standard Test Method for Unit Weight, Yield, and Air Content [Gravimetric] of Concrete*)—requires accurate knowledge of relative density and absolute volumes of concrete ingredients.
- Chace air indicator (AASHTO T 199, *Standard Method of Test for Air Content of Freshly Mixed Concrete by the Chace Indicator*)—is a very simple and inexpensive way to check the approximate air content of freshly mixed concrete. It is a pocket-size device that tests a mortar sample from the concrete. This test is not a substitute, however, for the more accurate pressure, volumetric, and gravimetric methods.

The foam-index test can be used to measure the relative air-entraining admixture requirement for concretes containing fly ash-cement combinations (Gebler and Klieger 1983).

Air-Void Analysis of Fresh Concrete

The conventional methods for analyzing air in fresh concrete, such as the pressure method noted previously, measure the total air content only; consequently, they provide no information about the parameters that determine the quality of the air-void system. These parameters—the size and number of voids and spacing between them—can be measured on polished samples of hardened concrete (ASTM C 457); but the result of such analysis will only be available several days after the concrete has hardened. Therefore, test equipment called an air-void analyzer (AVA) has been developed to determine the standard ASTM C 457 air-void parameters in fresh samples of air-entrained concrete (Fig. 8-23). The test apparatus determines the volume and size distributions of entrained air voids; thus an estimation of the spacing factor, specific surface, and total amount of entrained air can be made.

In this test method, air bubbles from a sample of fresh concrete rise through a viscous liquid, enter a column of water above it, then rise through the water and collect under a submerged buoyancy recorder (Fig. 8-24). The viscous liquid retains the original bubble sizes. Large bubbles rise faster than small ones through the liquids. The change in buoyancy is recorded as a function of time and can be related to the number of bubbles of different size.

Fresh concrete samples can be taken at the ready mix plant and on the jobsite. Testing concrete before and after placement into forms can verify how the applied methods of transporting, placing, and consolidation affect the air-void system. Since the samples are taken on fresh concrete, the air content and air-void system can be adjusted during production.

Currently, no standard exists for this method. The AVA was not developed for measuring the total air-content of concrete, and because of the small sample size, may not give accurate results for this quantity. However, this does not mean the AVA is not useful as a method for assessing the quality of the air-void system; it gives good results in conjunction with traditional methods for measuring air content (Aarre 1998).

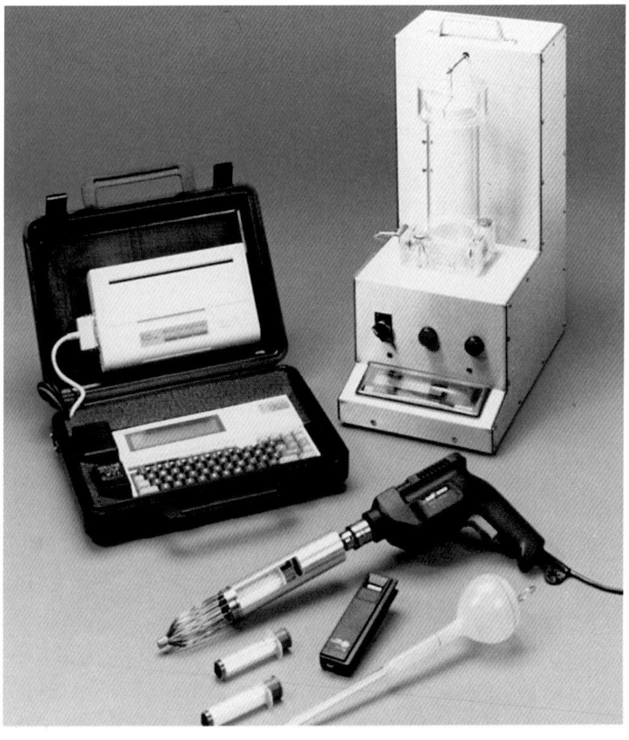

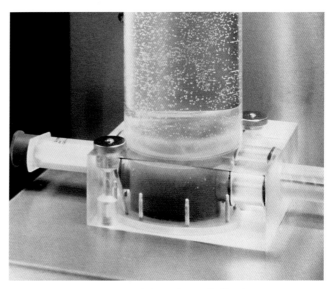

Fig. 8-24. Air bubbles rising through liquids in column. (67962)

RECOMMENDED AIR CONTENTS

CSA Standard A23.1 provides guidance on the required air contents for concretes in various exposure conditions. Table 8-8 gives the required air content for various classes of exposure and respective air content categories described in Tables 8-2 and 8-3. Fig. 8-25 illustrates how deicer-scaling resistance is impacted by air content and low water to cementing materials ratios (strength ranging from 40 to 50 MPa).

CSA A23.1 notes that air contents less than those shown in Table 8-8 may not give the required resistance to freezing and thawing or deicing salts which is the primary purpose of air entrainment. Air contents higher than those levels shown may reduce strength without contributing further improvement to durability. Higher air contents however can be used as long as the design strength is

Fig. 8-23. Equipment for the air-void analyzer. (67961)

achieved. The entrained air, regardless of exposure conditions, helps reduce bleeding and segregation and can improve the workability of the concrete. Fig. 8-26 illustrates the effect of increased air content with respect to aggregate size on reducing expansion due to saturated freezing and thawing.

More information on air-entrained concrete can be found in Whiting and Nagi (1998).

Table 8-8. Recommended Air Content* Relative to Maximum Aggregate Size, Exposure Conditions, and Air

	Content categories defined in Table 8-3		
Air content category	Range in air content for concretes with indicated nominal maximum sizes of coarse aggregates, %		
	10 mm	14-20 mm	28-40 mm
1	6 to 9	5 to 8	4 to 7
2	5 to 8	4 to 7	3 to 6

*At point of discharge from the delivery equipment unless otherwise specified.
Source: CSA Standard A23.1

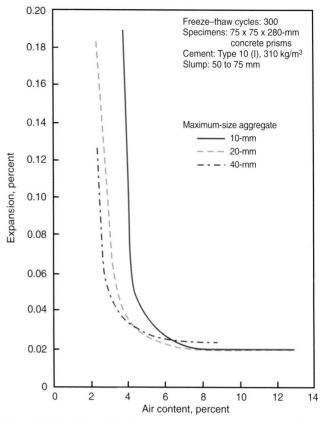

Fig. 8-26. Relationship between air content and expansion of concrete test specimens during 300 cycles of freezing and thawing for various maximum aggregate sizes. Smaller aggregate sizes require more air (Klieger 1952).

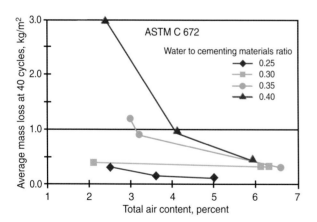

Fig. 8-25. Measured mass loss after 40 cycles of deicers and freeze-thaw exposures at various air contents.

REFERENCES

Aarre, Tine, "Air-Void Analyzer," *Concrete Technology Today,* PL981, Portland Cement Association, http://www.portcement.org/pdf_files/PL981.pdf, April 1998, page 4.

ACI Committee 201, *Guide to Durable Concrete,* ACI 201.2R-92, reapproved 1997, ACI Committee 201 Report, American Concrete Institute, Farmington Hills, Michigan, 1992.

ACI Committee 308, *Standard Practice for Curing Concrete,* ACI 308-92, reapproved 1997, ACI Committee 308 Report, American Concrete Institute, Farmington Hills, Michigan, 1992.

ACI Committee 309, *Guide for Consolidation of Concrete,* ACI 309R-96, ACI Committee 309 Report, American Concrete Institute, Farmington Hills, Michigan, 1996.

ACI Committee 318, *Building Code Requirements for Structural Concrete and Commentary,* ACI 318-99, ACI Committee 318 Report, American Concrete Institute, Farmington Hills, Michigan, 1999.

ACPA, *Scale-Resistant Concrete Pavements,* IS117, American Concrete Pavement Association, Skokie, Illinois, 1996.

Bates, A. A.; Woods, H.; Tyler, I. L.; Verbeck, G.; and Powers, T. C., "Rigid-Type Pavement," Association of Highway Officials of the North Atlantic States, *28th Annual Convention Proceedings,* pages 164 to 200, March 1952.

Bloem, D. L., *Air-Entrainment in Concrete,* National Sand and Gravel Association and National Ready Mixed Concrete Association, Silver Spring, Maryland, 1950.

Brewster, R. S., *Effect of Vibration Time upon Loss of Entrained Air from Concrete Mixes,* Materials Laboratories Report No. C-461, Research and Geology Division, Bureau of Reclamation, Denver, November 25, 1949.

Brown, F. P., and Cady, P. D., "Deicer Scaling Mechanisms in Concrete," *Durability of Concrete,* ACI SP-47, American Concrete Institute, Farmington Hills, Michigan, 1975, pages 101 to 119.

Cable, J. K.; McDaniel, L.; Schlorholtz, S.; Redmond, D.; and Rabe, K., *Evaluation of Vibrator Performance vs. Concrete Consolidation and Air Void System,* Serial No. 2398, Portland Cement Association, http://www.portcement.org/pdf_files/SN2398.pdf, 2000, 60 pages.

Cody, Rober D.; Cody, Anita M.; Spry, Paul G.; and Gan, Guo-Liang, "Concrete Deterioration by Deicing Salts: An Experimental Study," http://www.ctre.iastate.edu/pubs/semisesq/index.htm, *Semisequicentennial Transportation Conference Proceedings,* Center for Transportation Research and Education, Ames, Iowa, 1996.

Cordon, W. A., *Entrained Air—A Factor in the Design of Concrete Mixes,* Materials Laboratories Report No. C-310, Research and Geology Division, Bureau of Reclamation, Denver, March 15, 1946.

CSA Standard A23.1-00/A23.2-00, *Concrete Materials and Methods of Concrete Construction/Methods of Test for Concrete,* Canadian Standards Association, Toronto, 2000.

Elfert, R. J., *Investigation of the Effect of Vibration Time on the Bleeding Property of Concrete With and Without Entrained Air,* Materials Laboratories Report No. C-375, Research and Geology Division, Bureau of Reclamation, Denver, January 26, 1948.

"Entrained Air Voids in Concrete Help Prevent Salt Damage," *Civil Engineering,* American Society of Civil Engineers, New York, May 1982.

Falconi, M. I., *Durability of Slag Cement Concretes Exposed to Freezing and Thawing in the Presence of Deicers,* Masters of Science Thesis, Cornell University, Ithaca, New York, 306 pages.

Gebler, S. H., and Klieger, P., *Effect of Fly Ash on the Air-Void Stability of Concrete,* Research and Development Bulletin RD085, Portland Cement Association, http://www.portcement.org/pdf_files/RD085.pdf, 1983.

Gebler, Steven H., and Klieger, Paul, *Effect of Fly Ash on Durability of Air-Entrained Concrete,* Research and Development Bulletin RD090, Portland Cement Association, http://www.portcement.org/pdf_files/RD090.pdf, 1986

Gilkey, H. J., "Re-Proportioning of Concrete Mixtures for Air Entrainment," *Journal of the American Concrete Institute, Proceedings,* vol. 29, no. 8, Farmington Hills, Michigan, February 1958, pages 633 to 645.

Gonnerman, H. F., "Durability of Concrete in Engineering Structures," *Building Research Congress 1951,* collected papers, Division No. 2, Section D, Building Research Congress, London, 1951, pages 92 to 104.

Gonnerman, H. F., *Tests of Concretes Containing Air-Entraining Portland Cements or Air-Entraining Materials Added to Batch at Mixer,* Research Department Bulletin RX013, Portland Cement Association, http://www.portcement.org/pdf_files/RX013.pdf, 1944.

Greening, Nathan R., *Some Causes for Variation in Required Amount of Air-Entraining Agent in Portland Cement Mortars,* Research Department Bulletin RX213, Portland Cement Association, http://www.portcement.org/pdf_files/RX213.pdf, 1967.

Klieger, Paul, *Air-Entraining Admixtures,* Research Department Bulletin RX199, Portland Cement Association, http://www.portcement.org/pdf_files/RX199.pdf, 1966.

Klieger, Paul, *Extensions to the Long-Time Study of Cement Performance in Concrete,* Research Department Bulletin RX157, Portland Cement Association, 1963.

Klieger, Paul, *Further Studies on the Effect of Entrained Air on Strength and Durability of Concrete with Various Sizes of Aggregate,* Research Department Bulletin RX077, Portland Cement Association, http://www.portcement.org/pdf_files/RX077.pdf, 1956.

Klieger, Paul, *Studies of the Effect of Entrained Air on the Strength and Durability of Concretes Made with Various Maximum Sizes of Aggregates,* Research Department Bulletin RX040, Portland Cement Association, http://www.portcement.org/pdf_files/RX040.pdf, 1952.

Kretsinger, D. G., *Effect of Entrained Air on Expansion of Mortar Due to Alkali-Aggregate Reaction,* Materials Laboratories Report No. C-425, Research and Geology Division, U. S. Bureau of Reclamation, Denver, February 16, 1949.

Lerch, William, *Basic Principles of Air-Entrained Concrete,* T-101, Portland Cement Association, 1960.

McNeal, F., and Gay, F., "Solutions to Scaling Concrete," *Concrete Construction,* Addison, Illinois, March 1996, pages 250 to 255.

Menzel, Carl A., and Woods, William M., *An Investigation of Bond, Anchorage and Related Factors in Reinforced Concrete Beams,* Research Department Bulletin RX042, Portland Cement Association, 1952.

Mielenz, R. C.; Wokodoff, V. E.; Backstrom, J. E.; and Flack, H. L., "Origin, Evolution, and Effects of the Air-Void System in Concrete. Part 1-Entrained Air in Unhardened Concrete," July 1958, "Part 2-Influence of Type and Amount of Air-Entraining Agent," August 1958, "Part 3-Influence of Water-Cement Ratio and Compaction," September 1958, and "Part 4-The Air-Void System in Job Concrete," October 1958, *Journal of the American Concrete Institute,* Farmington Hills, Michigan, 1958.

Pigeon, M. and Pleau, R., *Durability of Concrete in Cold Climates,* E&FN Spon, New York, 1995, 244 pages.

Pinto, Roberto C. A., and Hover, Kenneth C., *Frost and Scaling Resistance of High-Strength Concrete,* Research and Development Bulletin RD122, Portland Cement Association, 2001, 70 pages.

Powers, T. C., "The Mechanism of Frost Action in Concrete," *Stanton Walker Lecture Series on the Materials Sciences,* Lecture No. 3, National Sand and Gravel Association and National Ready Mixed Concrete Association, Silver Spring, Maryland, 1965a.

Powers, T. C., and Helmuth, R. A., *Theory of Volume Changes in Hardened Portland Cement Paste During Freezing,* Research Department Bulletin RX046, Portland Cement Association, http://www.portcement.org/pdf_files/RX046.pdf, 1953.

Powers, T. C., *Basic Considerations Pertaining to Freezing and Thawing Tests,* Research Department Bulletin RX058, Portland Cement Association, http://www.portcement. org/pdf_files/RX058.pdf, 1955.

Powers, T. C., *The Air Requirements of Frost-Resistant Concrete,* Research Department Bulletin RX033, Portland Cement Association, http://www.portcement.org/pdf_ files/RX033.pdf, 1949.

Powers, T. C., *Topics in Concrete Technology:... (3) Mixtures Containing Intentionally Entrained Air; (4) Characteristics of Air-Void Systems,* Research Department Bulletin RX174, Portland Cement Association, http://www.portcement. org/pdf_files/RX174.pdf, 1965.

Powers, T. C., *Void Spacing as a Basis for Producing Air-Entrained Concrete,* Research Department Bulletin RX049, Portland Cement Association, http://www.portcement. org/pdf_files/RX049.pdf, 1954

Sayward, John M., *Salt Action on Concrete,* Special Report 84-25, U.S. Army Cold Regions Research and Engineering Laboratory, Hanover, New Hampshire, August 1984.

South Dakota Department of Transportation, *Investigation of Low Compressive Strength of Concrete in Paving, Precast and Structural Concrete,* Study SD 1998-03, PCA Serial No. 52002-17, Pierre, South Dakota, 2000.

Stanton, Thomas E., "Durability of Concrete Exposed to Sea Water and Alkali Soils-California Experience," *Journal of the American Concrete institute,* Farmington Hills, Michigan, May 1948.

Stark, David C., *Effect of Vibration on the Air-System and Freeze-Thaw Durability of Concrete,* Research and Development Bulletin RD092, Portland Cement Association, http://www.portcement.org/pdf_files/RD092.pdf, 1986.

Stark, David, *Longtime Study of Concrete Durability in Sulfate Soils,* Research and Development Bulletin RD086, Portland Cement Association, http://www.portcement. org/pdf_files/RD086.pdf, 1984.

Taylor, Thomas G., *Effect of Carbon Black and Black Iron Oxide on Air Content and Durability of Concrete,* Research Department Bulletin RX023, Portland Cement Association, http://www.portcement.org/pdf_files/RX023.pdf, 1948.

Verbeck, G. J., *Field and Laboratory Studies of the Sulphate Resistance of Concrete,* Research Department Bulletin RX227, Portland Cement Association, http://www.port cement.org/pdf_files/RX227.pdf, 1967.

Verbeck, George, and Klieger, Paul, *Studies of "Salt" Scaling of Concrete,* Research Department Bulletin RX083, Portland Cement Association, http://www.portcement.org/pdf_ files/RX083.pdf, 1956.

Walker, S., and Bloem, D. L., *Design and Control of Air-Entrained Concrete,* Publication No. 60, National Ready Mixed Concrete Association, Silver Spring, Maryland, 1955.

Wang, Kejin; Monteiro, Paulo J. M.; Rubinsky, Boris; and Arav, Amir, "Microscopic Study of Ice Propagation in Concrete," *ACI Materials Journal,* July-August 1996, American Concrete Institute, Farmington Hills, Michigan, pages 370 to 376.

Whiting, D., and Stark, D., *Control of Air Content in Concrete,* National Cooperative Highway Research Program Report No. 258 and Addendum, Transportation Research Board and National Research Council, Washington, D.C., May 1983.

Whiting, David A., and Nagi, Mohamad A., *Manual on Control of Air Content in Concrete,* EB116, National Ready Mixed Concrete Association and Portland Cement Association, 1998, 42 pages.

Whiting, D., and Dziedzic, D., *Effects of Conventional and High-Range Water Reducers on Concrete Properties,* Research and Development Bulletin RD107, Portland Cement Association, 1992, 25 pages.

Woods, Hubert, *Observations on the Resistance of Concrete to Freezing and Thawing,* Research Department Bulletin RX067, Portland Cement Association, http://www.port cement.org/pdf_files/RX067.pdf, 1954.

CHAPTER 9
Designing and Proportioning Normal Concrete Mixtures

The process of determining required and specifiable characteristics of a concrete mixture is called mix design. Characteristics can include: (1) fresh concrete properties; (2) required mechanical properties of hardened concrete such as strength, and durability requirements; and (3) the inclusion, exclusion, or limits on specific ingredients. Mix design leads to the development of a concrete specification.

Mixture proportioning refers to the process of determining the quantities of concrete ingredients, using local materials, to achieve the specified characteristics of the concrete. A properly proportioned concrete mix should possess these qualities:

1. Acceptable workability of the freshly mixed concrete
2. Durability, strength, and uniform appearance of the hardened concrete
3. Economy

Understanding the basic principles of mixture design is as important as the actual calculations used to establish mix proportions. Only with proper selection of materials and mixture characteristics can the above qualities be

Fig. 9-1. Trial batching (inset) verifies that a concrete mixture meets design requirements prior to use in construction. (69899, 70008).

obtained in concrete construction (Fig. 9-1) (Abrams 1918, Hover 1998, and Shilstone 1990).

SELECTING MIX CHARACTERISTICS

Before a concrete mixture can be proportioned, mixture characteristics are selected based on the intended use of the concrete, the exposure conditions, the size and shape of building elements, and the physical properties of the concrete (such as frost resistance and strength) required for the structure. The characteristics should reflect the needs of the structure; for example, resistance to chloride ions should be verifiable and the appropriate test methods specified.

Once the characteristics are selected, the mixture can be proportioned from field or laboratory data. Since most of the desirable properties of hardened concrete depend primarily upon the quality of the cementing paste, the first step in proportioning a concrete mixture is the selection of the appropriate water-cementing materials ratio for the durability and strength needed. Concrete mixtures should be kept as simple as possible, as an excessive number of ingredients often make a concrete mixture difficult to control. The concrete technologist should not, however, overlook the opportunities provided by modern concrete technology.

Water-Cementing Materials Ratio and Strength Relationship

Strength (compressive or flexural) is the most universally used measure for concrete quality. Although it is an important characteristic, other properties such as durability, permeability, and wear resistance are now recognized as being equal and in some cases more important, especially when considering life-cycle design of structures.

Within the normal range of strengths used in concrete construction, the compressive strength is inversely related to the water-cement ratio or water-cementing materials ratio. For fully compacted concrete made with clean, sound aggregates, the strength and other desirable properties of concrete under given job conditions are governed by the quantity of mixing water used per unit of cement or cementing materials (Abrams 1918).

The strength of the cementing paste binder in concrete depends on the quality and quantity of the reacting paste components and on the degree to which the hydration reaction has progressed. Concrete becomes stronger with time as long as there is moisture and a favourable temperature available. Therefore, the strength at any particular age is both a function of the original water-cementing materials ratio and the degree to which the cementing materials have hydrated. The importance of prompt and thorough curing is easily recognized.

Differences in concrete strength for a given water-cementing materials ratio may result from: (1) changes in the aggregate size, grading, surface texture, shape, strength, and stiffness; (2) differences in types and sources of cementing materials; (3) entrained-air content; (4) the presence of admixtures; and (5) the length of curing time.

Strength

The specified compressive strength, f'_c, at 28 days is the strength that is expected to be equal to or exceeded by the average of any set of three consecutive strength tests. No individual test (average of two cylinders) can be more than 3.5 MPa below the specified strength. Specimens must be cured under laboratory conditions for an individual class of concrete (CSA Standard A23.1). Some specifications allow alternative ranges.

The average strength should equal the specified strength plus an allowance to account for variations in materials; variations in methods of mixing, transporting, and placing the concrete; and variations in making, curing, and testing concrete cylinder specimens. The average strength, which is greater than f'_c, is called f'_{cr}; it is the strength required in the mix design. Requirements for f'_{cr} are discussed in detail under "Proportioning" later in this chapter. Tables 9-1 and 9-2 show strength requirements for various exposure conditions.

Flexural strength is sometimes used on paving projects instead of compressive strength; however, flexural strength is avoided due to its greater variability. For more information on flexural strength, see "Strength" in Chapter 1 and "Strength Specimens" in Chapter 16.

Water-Cementing Materials Ratio

The water-cementing materials ratio is simply the mass of water divided by the mass of the cementing material (portland cement, blended cement, fly ash, slag, silica fume, and natural pozzolans). The water-cementing materials ratio selected for mix design must be the lowest value required to meet anticipated exposure conditions. Tables 9-1 and 9-2 show requirements for various exposure conditions.

When durability does not control, the water-cementing materials ratio should be selected on the basis of concrete compressive strength. In such cases the water-cementing materials ratio and mixture proportions for the required strength should be based on adequate field data or trial mixtures made with actual job materials to determine the relationship between the ratio and strength. Fig. 9-2 or Table 9-3 can be used to select a water-cementing materials ratio with respect to the required average strength, f'_{cr}, for trial mixtures when no other data are available.

In mix design, the water to cementing materials ratio, W/CM, is often used synonymously with water to cement ratio (W/C); however, some specifications differentiate between the two ratios. Traditionally, the water to cement ratio referred to the ratio of water to portland cement or water to blended cement.

Table 9-1. Maximum Water-Cementing Materials Ratios and Minimum Design Strengths for Various Exposure Conditions*

Requirements for specifying concrete	Requirements for concrete		
Class of Exposure*	Maximum water-to-cementing materials ratio	Minimum, specified 28-day compressive strength, MPa	Air content category
C-1	0.40	35	**
C-2	0.45	32	1
C-3	0.50	30	2
C-4	0.55	25	2
F-1	0.50	30	1***
F-2	0.55	25	2***
N	For structural design	For structural design	

  * See Table 8-2 or this Chapter for a description of classes of exposure.
 ** Use Category 1 for concrete exposed to freezing and thawing.
    Use Category 2 for concrete not exposed to freezing and thawing.
*** Interior ice rink slabs and freezer slabs with a steel-troweled finish have been found to perform satisfactorily without entrained air.
Source: CSA Standard A23.1.

Table 9-2. Requirements for Concrete Subjected to Sulphate Attack*

Class of exposure	Degree of exposure	Water-soluble sulphate (SO₄) in soil sample, %	Sulphate (SO₄) in groundwater samples, mg/L	Minimum specified 56-day compressive strength, MPa†	Maximum water-to-cementing materials ratio‡	Air content category▲	Cementing materials to be used**††
S-1	Very severe	Over 2.0	Over 10,000	35	0.40	2	50
S-2	Severe	0.20 – 2.0	1500 – 10,000	32	0.45	2	50
S-3	Moderate	0.10 – 0.20	150 – 1500	30	0.50	2	20E‡‡, 40, or 50E

* For seawater exposure refer to CSA A23.1, Clause 15.

† Where supplementary cementing materials are used, the owner may specify other test ages.

‡ The owner shall specify the minimum 28-day compressive strength.

** When combinations of portland cement and supplementary cementing materials are used, they shall have been proven, to the satisfaction of the owner, to produce concrete resistant to the exposure conditions under consideration.

▲ For steel-troweled interior slabs on grade, subject to sulphate attack but not freeze thaw, air entrainment is not required.

†† Cementing material combinations with equivalent performance may be used. (Refer to CSA A23.1, Clauses 3.2, 3.3, and 3.4).

‡‡ Type 20E cement with moderate sulphate resistance (Refer to CSA A23.1, Clause 3.1.2).

Note: Type 50E cement shall not be used in reinforced concrete exposed to both chlorides and sulphates.
Refer to CSA A 23.1, Clause 15.4.
See CSA Test Methods A23.2-2B and A23.2-3B for test methods to determine sulphate ion content.

Source: CSA Standard A23.1

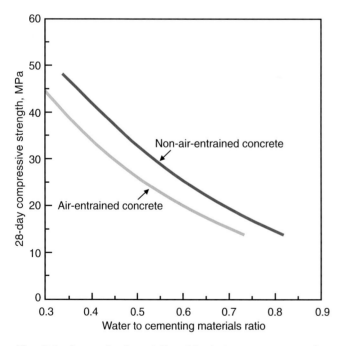

Fig. 9-2. Approximate relationship between compressive strength and water to cementing materials ratio for concrete using 20-mm to 28-mm nominal maximum size coarse aggregate. Strength is based on cylinders moist cured 28 days per CSA A23.2-3C (ASTM C 31). Adapted from Table 9-3, ACI 211.1, ACI 211.3, and Hover 1995.

Table 9-3. Relationship Between Water to Cementing Materials Ratio and Compressive Strength of Concrete

Compressive strength at 28 days, MPa	Water-cementing materials ratio by mass	
	Non-air-entrained concrete	Air-entrained concrete
45	0.38	0.30
40	0.42	0.34
35	0.47	0.39
30	0.54	0.45
25	0.61	0.52
20	0.69	0.60
15	0.79	0.70

Strength is based on cylinders moist-cured 28 days in accordance with CSA A23.2-3C (ASTM C 31). Relationship assumes nominal maximum size aggregate of about 20 to 28 mm.
Adapted from ACI 211.1 and ACI 211.3.

Aggregates

Two characteristics of aggregates have an important influence on proportioning concrete mixtures because they affect the workability of the fresh concrete. They are:

1. Grading (particle size and distribution)
2. Nature of particles (shape, porosity, surface texture)

Grading is important for attaining an economical mixture because it affects the amount of concrete that can be made with a given amount of cementing materials and water. Coarse aggregates should be graded up to the largest size practical under job conditions. The maximum size that can be used depends on factors such as the size and shape of the concrete member to be cast, the amount and distribution of reinforcing steel in the member, and the thickness of slabs. Grading also influences the workability and placeability of the concrete. Sometimes mid-sized aggregate, around the 10 mm size, is lacking in an aggregate supply; this can result in a concrete with high shrinkage properties, high water demand, and poor workability and placeability. Durability may also be affected. Various options are available for obtaining optimal grading of aggregate (Shilstone 1990).

The maximum size of coarse aggregate should not exceed one-fifth the narrowest dimension between sides of forms nor three-fourths the clear space between individual reinforcing bars or wire, bundles of bars, or prestressing tendons or ducts. It is also good practice to limit aggregate size to not more than three-fourths the clear space between reinforcement and the forms. For unreinforced slabs on ground, the maximum size should not exceed one third the slab thickness. Smaller sizes can be used when availability or economic consideration require them. Other requirements of CSA Standard A23.1 limit the nominal maximum size of aggregate to; the specified cover for concrete not exposed to earth or weather, and to two-thirds the specified cover for concrete exposed to earth or weather, and to one-half the specified cover for concrete exposed to chlorides. For pumped concrete, the limit is one-third the smallest internal diameter of the hose or pipe through which the concrete is to be pumped or 40 mm, whichever is smaller.

The amount of mixing water required to produce a cubic metre of concrete of a given slump is dependent on the shape and the maximum size and amount of coarse aggregate. Larger sizes minimize the water requirement and thus allow the cement content to be reduced. Also, rounded aggregate requires less mixing water than a crushed aggregate in concretes of equal slump (see "Water Content").

The maximum size of coarse aggregate that will produce concrete of maximum strength for a given cementing materials content depends upon the aggregate source as well as its shape and grading. For high compressive-strength concrete (greater than 70 MPa in 28 days), the maximum size is about 20 mm. Higher strengths can also sometimes be achieved through the use of crushed stone aggregate rather than rounded-gravel aggregate.

The most desirable fine-aggregate grading will depend upon the type of work, the paste content of the mixture, and the size of the coarse aggregate. For leaner mixtures, a fine grading (lower fineness modulus) is desirable for workability. For richer mixtures, a coarse grading (higher fineness modulus) is used for greater economy.

In some areas, the chemically bound chloride in aggregate may make it difficult for concrete to pass chloride limits set by CSA Standard A23.1 (ACI 318) or other standards or specifications. However, some or all of the chloride in the aggregate may not be available for participation in corrosion of reinforcing steel, thus that chloride may be ignored.

The bulk volume of coarse aggregate can be determined from Fig. 9-3 or Table 9-4. These bulk volumes are based on aggregates in a dry-rodded condition as described in CSA Test Method A23.2-10A (ASTM C 29); they

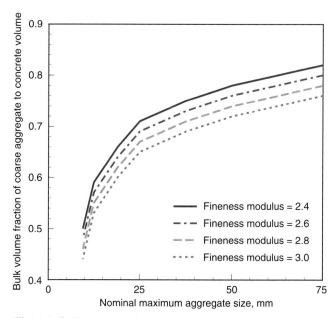

Fig. 9-3. Bulk volume of coarse aggregate per unit volume of concrete. Bulk volumes are based on aggregates in a dry-rodded condition as described in CSA A23.2-10A (ASTM C 29). For more workable concrete, such as may be required when placement is by pump, they may be reduced up to 10%. Adapted from Table 9-4, ACI 211.1 and Hover (1995 and 1998).

Table 9-4. Bulk Volume of Coarse Aggregate Per Unit Volume of Concrete

Nominal maximum size of aggregate, mm	Bulk volume of dry-rodded coarse aggregate per unit volume of concrete for different fineness moduli of fine aggregate*			
	2.40	2.60	2.80	3.00
10	0.50	0.48	0.46	0.44
14	0.59	0.57	0.55	0.53
20	0.66	0.64	0.62	0.60
28	0.71	0.69	0.67	0.65
40	0.75	0.73	0.71	0.69
56	0.78	0.76	0.74	0.72
80	0.82	0.80	0.78	0.76
150	0.87	0.85	0.83	0.81

*Bulk volumes are based on aggregates in dry-rodded condition as described in CSA A23.2-10A (ASTM C 29). Adapted from ACI 211.1.

are selected from empirical relationships to produce concrete with a degree of workability suitable for general reinforced concrete construction. For less workable concrete, such as required for concrete pavement construction, they may be increased about 10%. For more workable concrete, such as may be required when placement is by pump, they may be reduced up to 10%.

Air Content

Entrained air must be used in all concrete that will be exposed to freezing and thawing and deicing chemicals and can be used to improve workability even where not required.

Air entrainment is accomplished by adding an air-entraining admixture at the mixer. The amount of admixture should be adjusted to meet variations in concrete ingredients and job conditions. The amount recommended by the admixture manufacturer will, in most cases, produce the desired air content.

Recommended air contents for air-entrained concrete are shown in Table 9-5. Note that the amount of air required to provide adequate durability (resistance to freeze-thaw, chlorides, and sulphate attack) is dependent upon the nominal maximum size of aggregate and the air category which correlates to the class of exposure. In properly proportioned mixes, the mortar content decreases as maximum aggregate size increases, thus decreasing the

required concrete air content. The Class of Exposure and corresponding recommended Air Content Category as defined by CSA Standard A23.1 are as follows:

C-1: Structurally reinforced concrete exposed to chlorides with or without freezing and thawing conditions. Examples: bridge decks, parking decks and ramps, portions of marine structures located in tidal and splash zones. **Air Content Category:** Category **1** when exposed to freeze/thaw and Category **2** when not exposed to freeze/thaw.

C-2: Non-structurally reinforced (plain) concrete exposed to chlorides and freezing and thawing. Examples: garage floors, porches, steps, pavements, sidewalks, curbs and gutters. **Air Content Category: 1**

C-3: Continuously submerged concrete exposed to chlorides but not to freezing and thawing. Examples: underwater portions of marine structures. **Air Content Category: 2**

C-4: Non-structurally reinforced concrete exposed to chlorides but not to freezing and thawing. Examples: underground parking slabs on grade. **Air Content Category: 2**

F-1: Concrete exposed to freezing and thawing in a saturated condition but not to chlorides. Examples: pool decks, patios, tennis courts, freshwater control structures. **Air Content Category: 1**

Table 9-5. Approximate Mixing Water and Air Content Requirements for Different Slumps and Nominal Maximum Sizes of Aggregate

Slump, mm	Water, kilograms per cubic metre of concrete, for indicated sizes of aggregate*							
	10 mm	14 mm	20 mm	28 mm	40 mm	56 mm**	80 mm**	150 mm**
Non-air-entrained concrete								
25 to 50	207	199	190	179	166	154	130	113
75 to 100	228	216	205	193	181	169	145	124
150 to 175	243	228	216	202	190	178	160	—
Approximate amount of entrapped air in non-air-entrained concrete, percent	3	2.5	2	1.5	1	0.5	0.3	0.2
Air-entrained concrete								
25 to 50	181	175	168	160	150	142	122	107
75 to 100	202	193	184	175	165	157	133	119
150 to 175	216	205	197	184	174	166	154	—
CSA A23.1 Recommended total air content percent†								
Category 1	6 to 9	5 to 8		4 to 7		—	—	—
Category 2	5 to 8	4 to 7		3 to 6		—	—	—

* These quantities of mixing water are for use in computing cementing material contents for trial batches. They are maximums for reasonably well-shaped angular coarse aggregates graded within limits of accepted specifications.

** The slump values for concrete containing aggregates larger than 40 mm are based on slump tests made after removal of particles larger than 40 mm by wet screening.

† See Tables 9-1 and 9-2 for class of exposure and corresponding air content category.

Adapted from CSA Standard A23.1, ACI 211.1, and ACI 318. Hover (1995) presents this information in graphical form.

F-2: Concrete in an unsaturated condition exposed to freezing and thawing but not to chlorides. Examples: exterior walls and columns. **Air Content Category: 2**

N: Concrete not exposed to chlorides nor to freezing or thawing. Examples: footings and interior slabs, walls and columns. **Air Content Category:** air not required for durability.

S-1: Concrete subjected to sulphate attack—Very Severe. **Air Content Category: 2**

S-2: Concrete subjected to sulphate attack—Severe. **Air Content Category: 2**

S-3: Concrete subjected to sulphate attack—Moderate. **Air Content Category: 2**

When mixing water is held constant, the entrainment of air will increase slump. When the cementing material content and slump are held constant, the entrainment of air results in the need for less mixing water, particularly in leaner concrete mixtures. In batch adjustments, in order to maintain a constant slump while changing the air content, the water should be decreased by about 3 kg/m^3 for each percentage point increase in air content or increased 3 kg/m^3 for each percentage point decrease.

A specific air content cannot be readily or repeatedly achieved because of the many variables affecting air content; therefore, a permissible range of air contents is provided in CSA A23.1 and should be used in project specifications.

Slump

Concrete must always be made with a workability, consistency, and plasticity suitable for job conditions. Workability is a measure of how easy or difficult it is to place, consolidate, and finish concrete. Consistency is the ability of freshly mixed concrete to flow. Plasticity determines concrete's ease of molding. If more aggregate is used in a concrete mixture, or if less water is added, the mixture becomes stiff (less plastic and less workable) and difficult to mold. Neither very dry, crumbly mixtures nor very watery, fluid mixtures can be regarded as having plasticity.

The slump test is used to measure concrete consistency. For a given proportion of cementing materials and aggregate without admixtures, the higher the slump, the more fluid the mixture. Slump is indicative of workability when assessing similar mixtures. However, slump should not be used to compare mixtures of totally different proportions. When used with different batches of the same mix design, a change in slump indicates a change in consistency. While in many instances it is the result of a change in the water content of the mix, it could also be the result of a change in the characteristics of materials being used in the mix, mixture proportions, mixing, time of test, or the testing itself.

Different slumps are needed for various types of concrete construction. Slump is usually indicated in the job specifications as a range, such as 50 to 100 mm, or as a maximum value not to be exceeded. CSA Standard A23.1 (ASTM C 94) addresses slump tolerances in detail. When the specified slump is 80 mm or less, the allowable variation is ±20 mm; when the specified slump is between 80 mm and 170 mm, the allowable variation is ±30 mm; and when the specified slump is 180 mm or greater, the allowable variation is ±40 mm. When it is not specified, an approximate value can be selected from Table 9-6 for concrete consolidated by mechanical vibration. For batch adjustments, the slump can be increased by about 10 mm by adding 2 kilograms of water per cubic metre of concrete.

Table 9-6. Recommended Slumps for Various Types of Construction

Concrete construction	Slump, mm	
	Maximum*	Minimum
Reinforced foundation walls and footings	75	25
Plain footings, caissons, and substructure walls	75	25
Beams and reinforced walls	100	25
Building columns	100	25
Pavements and slabs	75	25
Mass concrete	75	25

*May be increased 25 mm for consolidation by hand methods, such as rodding and spading.
Plasticizers can safely provide higher slumps.
Adapted from ACI 211.1.

Water Content

The water content of concrete is influenced by a number of factors: aggregate size, aggregate shape, aggregate texture, slump, water to cementing materials ratio, air content, cementing materials type and content, admixtures, and environmental conditions. An increase in air content and aggregate size, a reduction in water-cementing materials ratio and slump, and the use of rounded aggregates, water-reducing admixtures or fly ash will reduce water demand. On the other hand, increased temperatures, cementing materials contents, slump, water-cementing materials ratio, aggregate angularity, and a decrease in the proportion of coarse aggregate to fine aggregate increase water demand.

The approximate water contents in Table 9-5 and Fig. 9-4, used in proportioning, are for angular coarse aggregates (crushed stone). For some concretes and aggregates,

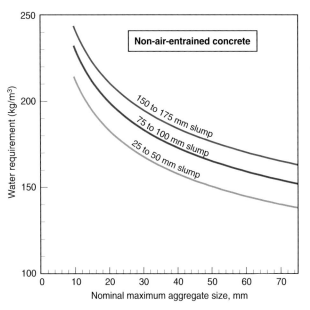
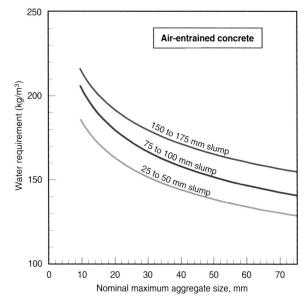

Fig. 9-4. Approximate water requirement for various slumps and crushed aggregate sizes for (left) non-air-entrained concrete and (right) air-entrained concrete. Adapted from Table 9-5, ACI 211.1 and Hover (1995 and 1998).

the water estimates in Table 9-5 and Fig. 9-4 can be reduced by approximately 10 kg for subangular aggregate, 20 kg for gravel with some crushed particles, and 25 kg for a rounded gravel to produce the slumps shown. This illustrates the need for trial batch testing of local materials, as each aggregate source is different and can influence concrete properties differently.

It should be kept in mind that changing the amount of any single ingredient in a concrete mixture normally effects the proportions of other ingredients as well as alter the properties of the mixture. For example, the addition of 2 kg of water per cubic metre will increase the slump by approximately 10 mm; it will also increase the air content and paste volume, decrease the aggregate volume, and lower the density of the concrete. In mixture adjustments, for the same slump, a decrease in air content by 1 percentage point will increase the water demand by about 3 kg per cubic metre of concrete.

Cementing Materials Content and Type

The cementing materials content is usually determined from the selected water-cementing materials ratio and water content, although a minimum cement content frequently is included in specifications in addition to a maximum water-cementing materials ratio. Minimum cementing materials requirements serve to ensure satisfactory durability and finishability, to improve wear resistance of slabs, and guarantee a suitable appearance of vertical surfaces. This is important even though strength requirements may be met at lower cementing materials contents. However, exces-

sively large amounts of cementing materials should be avoided to maintain economy in the mixture and to not adversely affect workability and other properties.

For severe freeze-thaw, deicer, and sulphate exposures, it is desirable to specify: (1) a minimum cementing materials content of 335 kg per cubic metre, and (2) only enough mixing water to achieve the desired consistency without exceeding the maximum water-cementing materials ratios shown in Tables 9-1 and 9-2. For placing concrete underwater, usually not less than 390 kg of cementing materials per cubic metre should be used with a water to cementing materials ratio not exceeding 0.45. For workability, finishability, abrasion resistance, and durability in flatwork, the quantity of cementing materials to be used should be not less than shown in Table 9-7.

Table 9-7. Minimum Requirements of Cementing Materials for Concrete Used in Flatwork

Nominal maximum size of aggregate, mm	Cementing materials, kg/m³*
40	280
28	310
20	320
14	350
10	360

* Cementing materials quantities may need to be greater for severe exposure. For example, for deicer exposures, concrete should contain at least 335 kg/m³ of cementing materials.
Adapted from ACI 302.

To obtain economy, proportioning should minimize the amount of cementing materials required without sacrificing concrete quality. Since quality depends primarily on water-cementing materials ratio, the water content should be held to a minimum to reduce the cement requirement. Steps to minimize water and cementing materials requirements include use of (1) the stiffest practical mixture, (2) the largest practical maximum size of aggregate, and (3) the optimum ratio of fine-to-coarse aggregate.

Concrete that will be exposed to sulphate conditions should be made with the type of cement shown in Table 9-2.

Seawater contains significant amounts of sulphates and chlorides. Although sulphates in seawater are capable of attacking concrete, the presence of chlorides in seawater inhibits the expansive reaction that is characteristic of sulphate attack. This is the major factor explaining observations from a number of sources that the performance of concretes in seawater have shown satisfactory durability; this is despite the fact these concretes were made with portland cements having tricalcium aluminate (C_3A) contents as high as 10%, and sometimes greater. However, the permeability of these concretes is low, and the reinforcing steel had adequate cover. Portland cements meeting a C_3A requirement of not more than 10% or less than 4% (to ensure durability of reinforcement) are acceptable (CSA A23.1, ACI 357R).

Supplementary cementing materials have varied effects on water demand and air contents. The addition of fly ash will generally reduce water demand and decrease the air content if no adjustment in the amount of air-entraining admixture is made. Silica fume increases water demand and decreases air content. Slag and metakaolin have a minimal effect at normal dosages.

Table 9-8 shows limits on the amount of supplementary cementing materials in concrete to be exposed to deicers. When combinations of portland cement and supplementary cementing materials are used, the concrete should be tested to confirm adequate durability. Local practices should also be consulted as dosages smaller or larger than those shown in Table 9-8 can be used without jeopardizing scale-resistance, depending on the exposure severity.

Admixtures

Water-reducing admixtures are added to concrete to reduce the water-cementing materials ratio, reduce cementing materials content, reduce water content, reduce paste content, or to improve the workability of a concrete without changing the water-cementing materials ratio. Water reducers will usually decrease water contents by 5% to 10% and some will also increase air contents by ½ to 1 percentage point. Retarding admixtures may also increase the air content.

High-range water reducers (plasticizers) reduce water contents between 12% and 30% and some can simultaneously increase the air content up to 1 percentage point; others can reduce or not affect the air content.

Calcium chloride-based admixtures reduce water contents by about 3% and increase the air content by about ½ percentage point.

When using a chloride-based admixture, the risks of reinforcing steel corrosion should be considered. Table 9-9 provides recommended limits on the water-soluble chloride-ion content in reinforced and prestressed concrete for various conditions.

When using more than one admixture in concrete, the compatibility of intermixing admixtures should be assured by the admixture manufacturer or the combination of admixtures should be tested in trial batches. The water contained in admixtures should be considered part of the mixing water if the admixture's water content is sufficient to affect the water-cementing materials ratio by 0.01 or more.

Table 9-8. Cementing Materials Requirements for Concrete Exposed to Deicing Chemicals

Cementing materials*	Maximum percent of total cementing materials by mass**
Fly ash and natural pozzolans	25
Slag	50
Silica fume	10
Total of fly ash, slag, silica fume and natural pozzolans	50†
Total of natural pozzolans and silica fume	35†

* Includes portion of supplementary cementing materials in blended cements.
** Total cementing materials include the summation of portland cements, blended cements, fly ash, slag, silica fume and other pozzolans.
† Silica fume should not constitute more than 10% of total cementing materials and fly ash or other pozzolans shall not constitute more than 25% of cementing materials.
Adapted from ACI 318.

Table 9-9. Maximum Chloride-Ion Content for Corrosion Protection

Type of member	Maximum water-soluble chloride ion (Cl⁻) in concrete, percent by mass of cementing material
Prestressed concrete	0.06
Reinforced concrete exposed to a moist environment or chlorides or both	0.15
Reinforced concrete exposed to neither a moist environment nor chlorides	1.00

Source: CSA Standard A23.1

An excessive use of multiple admixtures should be minimized to allow better control of the concrete mixture in production and to reduce the risk of admixture incompatibility.

PROPORTIONING

The design of concrete mixtures involves the following: (1) the establishment of specific concrete characteristics, and (2) the selection of proportions of available materials to produce concrete of required properties, with the greatest economy. Proportioning methods have evolved from the arbitrary volumetric method (1:2:3—cement:sand: coarse aggregate) of the early 1900s (Abrams 1918) to the present-day mass and absolute-volume methods described in the ACI's Committee 211 *Standard Practice for Selecting Proportions for Normal, Heavyweight and Mass Concrete* (ACI 211.1).

Mass proportioning methods are fairly simple and quick for estimating mixture proportions using an assumed or known mass of concrete per unit volume. A more accurate method, absolute volume, involves use of relative density values for all the ingredients to calculate the absolute volume each will occupy in a unit volume of concrete. The absolute volume method will be illustrated. A concrete mixture also can be proportioned from field experience (statistical data) or from trial mixtures.

Other valuable documents to help proportion concrete mixtures include the *Standard Practice for Selecting Proportions for Structural Lightweight Concrete* (ACI 211.2); *Guide for Selecting Proportions for No-Slump Concrete* (ACI 211.3); *Guide for Selecting Proportions for High-Strength Concrete with Portland Cement and Fly Ash* (ACI 211.4R); and *Guide for Submittal of Concrete Proportions* (ACI 211.5). Hover (1995 and 1998) provides a graphical process for designing concrete mixtures in accordance with ACI 211.1.

Proportioning from Field Data

A presently or previously used concrete mixture design can be used for a new project if strength-test data and standard deviations show that the mixture is acceptable. Durability aspects previously presented must also be met. Standard deviation computations are outlined in ACI 318. The statistical data should essentially represent the same materials, proportions, and concreting conditions to be used in the new project. The data used for proportioning should also be from a concrete with an f'_c that is within 7 MPa of the strength required for the proposed work. Also, the data should represent at least 30 consecutive tests or two groups of consecutive tests totaling at least 30 tests (one test is the average strength of two cylinders from the same sample). If only 15 to 29 consecutive tests are available, an adjusted standard deviation can be obtained by multiplying the standard deviation (S) for the 15 to 29 tests

Table 9-10. Modification Factor for Standard Deviation When Less Than 30 Tests Are Available

Number of tests*	Modification factor for standard deviation**
Less than 15	Use Table 9-11
15	1.16
20	1.08
25	1.03
30 or more	1.00

* Interpolate for intermediate numbers of tests.
** Modified standard deviation to be used to determine required average strength, f'_{cr}.
Adapted from ACI 318.

and a modification factor from Table 9-10. The data must represent 45 or more days of tests.

The standard or modified deviation is then used in Equations 9-1 and 9-2, or Equations 9-3 and 9-4, depending on the acceptance criteria being used. The average compressive strength from the test record must equal or exceed the required average compressive strength, f'_{cr}, in order for the concrete proportions to be acceptable. The f'_{cr} for the selected mixture proportions is equal to the larger of Equations 9-1 and 9-2 for CSA acceptance evaluation or Equations 9-3 and 9-4 for ACI 318 acceptance evaluation.

CSA Standard A23.1 states that the 28-day strength criteria can be expected to be met 99% of the time if the concrete mix is proportioned to produce an average strength as follows:

$$f'_{cr} = f'_c + 1.4S \qquad \text{Eq. 9-1}$$
$$f'_{cr} = f'_c + (2.4S - 3.5 \text{ MPa}) \qquad \text{Eq. 9-2}$$

ACI 318 uses an average strength criteria very similar to that of CSA Standard A23.1 and is as follows:

$$f'_{cr} = f'_c + 1.34S \qquad \text{Eq. 9-3}$$
$$f'_{cr} = f'_c + (2.33S - 3.45 \text{ MPa}) \qquad \text{Eq. 9-4}$$

where

f'_{cr} = required average compressive strength of concrete (MPa) used as the basis for selection of concrete proportions

f'_c = specified compressive strength of concrete, (MPa)

S = standard deviation, (MPa)

When field strength test records do not meet the previously discussed requirements, f'_{cr} can be obtained from Table 9-11. A field strength record, several strength test records, or tests from trial mixtures must be used for documentation showing that the average strength of the mixture is equal to or greater than f'_{cr}.

If less than 30, but not less than 10 tests are available, the tests may be used for average strength documentation if the time period is not less than 45 days. Mixture proportions may also be established by interpolating between two or more test records if each meets the above and project

Table 9-11. Required Average Compressive Strength When Data Are Not Available to Establish a Standard Deviation

Specified compressive strength, f'_c, MPa	Required average compressive strength, f'_{cr}, MPa
Less than 21	$f'_c + 7.0$
21 to 35	$f'_c + 8.5$
Over 35	$f'_c + 10.0$

Adapted from ACI 318.

requirements. If a significant difference exists between the mixtures that are used in the interpolation, a trial mixture should be considered to check strength gain. If the test records meet the above requirements and limitations of CSA Standard A23.1 (ACI 318), the proportions for the mixture may then be considered acceptable for the proposed work.

If the average strength of the mixtures with the statistical data is less than f'_{cr}, or statistical data or test records are insufficient or not available, the mixture should be proportioned by the trial-mixture method. The approved mixture must have a compressive strength that meets or exceeds f'_{cr}. Three trial mixtures using three different water to cementing materials ratios or cementing materials contents should be tested. A water to cementing materials ratio to strength curve (similar to Fig. 9-2) can then be plotted and the proportions interpolated from the data. It is also good practice to test the properties of the newly proportioned mixture in a trial batch.

ACI 214 provides statistical analysis methods for monitoring the strength of the concrete in the field to ensure that the mix properly meets or exceeds the design strength, f'_c.

Proportioning by Trial Mixtures

When field test records are not available or are insufficient for proportioning by field experience methods, the concrete proportions selected should be based on trial mixtures. The trial mixtures should use the same materials proposed for the work. Three mixtures with three different water-cementing materials ratios or cementing materials contents should be made to produce a range of strengths that encompass f'_{cr}. The trial mixtures should have a slump and air content within ±20 mm (±0.75 in.) and ± 0.5%, respectively, of the maximum permitted. Three cylinders for each water-cementing materials ratio should be made and cured according to CSA Test Method A23.2-3C (ASTM C 192). At 28 days, or the designated test age, the compressive strength of the concrete should be determined by testing the cylinders in compression. The test results should be plotted to produce a strength versus water-

cementing materials ratio curve (similar to Fig. 9-2) that is used to proportion a mixture.

A number of different methods of proportioning concrete ingredients have been used at one time or another, including:

Arbitrary assignment (1:2:3), volumetric
Void ratio
Fineness modulus
Surface area of aggregates
Cementing materials content

Any one of these methods can produce approximately the same final mixture after adjustments are made in the field. The best approach, however, is to select proportions based on past experience and reliable test data with an established relationship between strength and water to cementing materials ratio for the materials to be used in the concrete. The trial mixtures can be relatively small batches made with laboratory precision or job-size batches made during the course of normal concrete production. Use of both is often necessary to reach a satisfactory job mixture.

The following parameters must be selected first: (1) required strength, (2) minimum cementing materials content or maximum water-cementing materials ratio, (3) nominal maximum size of aggregate, (4) air content, and (5) desired slump. Trial batches are then made varying the relative amounts of fine and coarse aggregates as well as other ingredients. Based on considerations of workability and economy, the proper mixture proportions are selected.

When the quality of the concrete mixture is specified by water-cementing materials ratio, the trial-batch procedure consists essentially of combining a paste (water, cementing materials, and, generally, a chemical admixture) of the correct proportions with the necessary amounts of fine and coarse aggregates to produce the required slump and workability. Representative samples of the cementing materials, water, aggregates, and admixtures must be used. Quantities per cubic metre are then calculated.

To simplify calculations and eliminate error caused by variations in aggregate moisture content, the aggregates should be prewetted then dried to a saturated surface-dry (SSD) condition; place the aggregates in covered containers to keep them in this SSD condition until they are used. The moisture content of the aggregates should be determined and the batch quantities corrected accordingly.

The size of the trial batch is dependent on the equipment available and on the number and size of test specimens to be made. Larger batches will produce more accurate data. Machine mixing is recommended since it more nearly represents job conditions; it is mandatory if the concrete is to contain entrained air. The mixing procedures outlined in CSA Test Method A23.2-2C (ASTM C 192) should be used.

Measurements and Calculations

Tests for slump, air content, and temperature should be made on the trial mixture, and the following measurements and calculations should also be performed.

Density, Yield and Cementing Materials Factor. The density of freshly mixed concrete is expressed in kilograms per cubic metre. The yield is the volume of fresh concrete produced in a batch, usually expressed in cubic metres, and cementing materials factor is expressed in kilograms per cubic metre. The yield is calculated by dividing the total mass of the materials batched by the density of the freshly mixed concrete. The cementing materials factor is calculated by dividing the mass of cementing materials in the batch (kg) by the yield (m³). Density, yield and cementing materials factor are determined in accordance with CSA Test Method A23.2-6C (ASTM C 138).

Absolute Volume. The absolute volume of a granular material (in the case of concrete, cementing materials and aggregates) is the volume of the solid matter in the particles; it does not include the volume of air spaces between particles. The volume (yield) of freshly mixed concrete is equal to the sum of the absolute volumes of the concrete ingredients—cementing materials, water (exclusive of that absorbed in the aggregate), aggregates, admixtures when applicable, and air. The absolute volume is computed from a material's mass and relative density as follows:

Absolute volume

$$= \frac{\text{mass of loose material}}{(\text{relative density of a material} \times \text{density of water})}$$

A value of 3.15 can be used for the relative density of portland cement. Blended cements have relative densities ranging from 2.90 to 3.15. The relative density of fly ash varies from 1.9 to 2.8, slag from 2.85 to 2.95, and silica fume from 2.20 to 2.25. The relative density of water is 1.0 and the density of water is 1000 kg/m³ at 4°C—accurate enough for mix calculations at room temperature. More accurate water density values are given in Table 9-12. Relative density of normal aggregate usually ranges between 2.4 and 2.9.

The relative density of aggregate as used in mix-design calculations is the relative density of either saturated surface-dry (SSD) material or ovendry material.

Table 9-12. Density of Water Versus Temperature

Temperature, °C	Density, kg/m³
16	998.93
18	998.58
20	998.19
22	997.75
24	997.27
26	996.75
28	996.20
30	995.61

Relative densities of admixtures, such as water reducers, can also be considered if needed. Absolute volume is usually expressed in cubic metres.

The absolute volume of air in concrete, expressed as cubic metres per cubic metre, is equal to the total air content in percent divided by 100 (for example, 7% ÷ 100) and then multiplied by the volume of the concrete batch.

The volume of concrete in a batch can be determined by either of two methods: (1) if the relative densities of the aggregates and cementing materials are known, these can be used to calculate concrete volume; or (2) if relative densities are unknown, or they vary, the volume can be computed by dividing the total mass of materials in the mixer by the density of concrete. In some cases, both determinations are made, one serving as a check on the other.

EXAMPLES OF MIXTURE PROPORTIONING

Example 1. Absolute Volume

Conditions and Specifications. Concrete is required for a pavement that will be exposed to moisture and deicing salts in a severe freeze-thaw environment (C-2 exposure class). A compressive strength, f'_c, of 35 MPa at 28 days is specified. Air entrainment is required. Slump should be between 25 mm and 75 mm. A nominal maximum size aggregate of 28 mm is required. No statistical data on previous mixes are available. The materials available are as follows:

Cement: Type 10, meeting CSA Standard A-5, with a relative density of 3.15.

Coarse aggregate: Well-graded, 28-mm nominal maximum-size rounded gravel (CSA A23.1) with an ovendry relative density of 2.68, absorption of 0.5% (moisture content at SSD condition) and ovendry rodded bulk density of 1600 kg/m³. The laboratory sample for trial batching has a moisture content of 2%.

Fine aggregate: Natural sand (CSA A23.1) with an ovendry relative density of 2.64 and absorption of 0.7%. The laboratory sample moisture content is 6%. The fineness modulus is 2.80.

Air-entraining admixture: Wood-resin type, ASTM C 260.

Water reducer: ASTM C 494. This particular admixture is known to reduce water demand by 10% when used at a dosage rate of 3 g (or 3 mL) per kg of cement. Assume that the chemical admixtures have a density close to that of water, meaning that 1 mL of admixture has a mass of 1 g.

From this information, the task is to proportion a trial mixture that will meet the above conditions and specifications.

Strength. The design strength of 35 MPa is greater than the 32 MPa required in Table 9-1 for a C-2 exposure condition. Since no statistical data is available, f'_{cr} (required compressive strength for proportioning) from Table 9-11 is equal to $f'_c + 8.5$. Therefore, $f'_{cr} = 35 + 8.5 = 43.5$ MPa.

Water to Cementing Materials Ratio. For a C-2 environment with moist freezing and thawing and deicing salts, the maximum water to cementing materials ratio should be 0.45. The recommended water to cementing materials ratio for an f'_{cr} of 43.5 MPa is 0.31 from Fig. 9-2 interpolated or from Table 9-3 [{(45 – 43.5)(0.34 – 0.30) /(45 – 40)} + 0.30 = 0.31]. Since the lower water to cementing materials ratio governs, the mix must be designed for 0.31. If a plot from trial batches or field tests had been available, the water to cementing materials ratio could have been extrapolated from that data.

Air Content. For a C-2 exposure with a corresponding air content Category **1**, Table 9-5 recommends for a 28 to 40 mm nominal maximum size of aggregate, an air content of 4 to 7%. Therefore, design the mix for 7% (the maximum allowable) for batch proportions. The trial-batch air content must be within ±0.5 percentage points of the maximum allowable air content.

Slump. The slump is specified at 25 mm to 75 mm. Use 75 mm ±20 mm for proportioning purposes.

Water Content. Table 9-5 and Fig. 9-4 recommend that a 75-mm slump, air-entrained concrete made with 28-mm nominal maximum-size aggregate should have a water content of about 175 kg/m³. However, rounded gravel should reduce the water content of the table value by about 25 kg/m³. Therefore, the water content can be estimated to be about 150 kg/m³ (175 kg/m³ minus 25 kg/m³). In addition, the water reducer will reduce water demand by 10% resulting in an estimated water demand of 135 kg/m³.

Cement Content. The cement content is based on the maximum water-cementing materials ratio and the water content. Therefore, 135 kg/m³ of water divided by a water-cementing materials ratio of 0.31 requires a cement content of 435 kg/m³; this is greater than the 335 kg/m³ required for deicer exposure (Table 9-7).

Coarse-Aggregate Content. The quantity of 28-mm nominal maximum-size coarse aggregate can be estimated from Fig. 9-3 or Table 9-4. The bulk volume of coarse aggregate recommended when using sand with a fineness modulus of 2.80 is 0.67. Since it has a bulk density of 1600 kg/m³, the ovendry mass of coarse aggregate for a cubic metre of concrete is

$$1600 \times 0.67 = 1072 \text{ kg}$$

Admixture Content. For an 8% air content, the air-entraining admixture manufacturer recommends a dosage rate of 0.5 g per kg of cement. From this information, the amount of air-entraining admixture per cubic metre of concrete is

$$0.5 \times 435 = 218 \text{ g or } 0.218 \text{ kg}$$

The water reducer dosage rate of 3 g per kg of cement results in

$$3 \times 435 = 1305 \text{ g or } 1.305 \text{ kg of water reducer}$$
$$\text{per cubic metre of concrete}$$

Fine-Aggregate Content. At this point, the amounts of all ingredients except the fine aggregate are known. In the absolute volume method, the volume of fine aggregate is determined by subtracting the absolute volumes of the known ingredients from 1 cubic metre. The absolute volume of the water, cement, admixtures and coarse aggregate is calculated by dividing the known mass of each by the product of their relative density and the density of water. Volume computations are as follows:

Water	$=$	$\dfrac{135}{1 \times 1000}$	$= 0.135 \text{ m}^3$
Cement	$=$	$\dfrac{435}{3.15 \times 1000}$	$= 0.138 \text{ m}^3$
Air	$=$	$\dfrac{7.0}{100}$	$= 0.070 \text{ m}^3$
Coarse aggregate	$=$	$\dfrac{1072}{2.68 \times 1000}$	$= 0.400 \text{ m}^3$
Total volume of known ingredients			0.743 m^3

The calculated absolute volume of fine aggregate is then

$$1 - 0.743 = 0.257 \text{ m}^3$$

The mass of dry fine aggregate is

$$0.257 \times 2.64 \times 1000 = 678 \text{ kg}$$

The mixture then has the following proportions before trial mixing for one cubic metre of concrete:

Water	135 kg
Cement	435 kg
Coarse aggregate (dry)	1072 kg
Fine aggregate (dry)	678 kg
Total mass	2320 kg
Air-entraining admixture	0.218 kg
Water reducer	1.305 kg
Slump	75 mm (±20 mm for trial batch)
Air content	7% (±0.5% for trial batch)
Estimated concrete density (using SSD aggregate)	= 135 + 435 + (1072 x 1.005*) + (678 x 1.007*) = 2330 kg/m³

---

* (0.5% absorption ÷ 100) + 1 = 1.005
(0.7% absorption ÷ 100) + 1 = 1.007

The liquid admixture volume is generally too insignificant to include in the water calculations. However, certain admixtures, such as shrinkage reducers, plasticizers, and corrosion inhibitors are exceptions due to their relatively large dosage rates; their volumes should be included.

Moisture. Corrections are needed to compensate for moisture in and on the aggregates. In practice, aggregates will contain some measurable amount of moisture. The dry-batch quantities of aggregates, therefore, have to be increased to compensate for the moisture that is absorbed in and contained on the surface of each particle and between particles. The mixing water added to the batch must be reduced by the amount of free moisture contributed by the aggregates. Tests indicate that for this example, coarse-aggregate moisture content is 2% and fine-aggregate moisture content is 6%.

With the aggregate moisture contents (MC) indicated, the trial batch aggregate proportions become

Coarse aggregate (2% MC) = 1072 x 1.02 = 1093 kg

Fine aggregate (6% MC) = 678 x 1.06 = 719 kg

Water absorbed by the aggregates does not become part of the mixing water and must be excluded from the water adjustment. Surface moisture contributed by the coarse aggregate amounts to 2% – 0.5% = 1.5%; that contributed by the fine aggregate is, 6% – 0.7% = 5.3%. The estimated requirement for added water becomes

135 – (1072 x 0.015) – (678 x 0.053) = 83 kg

The estimated batch weights for one cubic metre of concrete are revised to include aggregate moisture as follows:

Water (to be added)	83 kg
Cement	435 kg
Coarse aggregate (2% MC, wet)	1093 kg
Fine aggregate (6% MC, wet)	719 kg
Total	2330 kg
Air-entraining admixture	0.218 kg
Water reducer	1.305 kg

Trial Batch. At this stage, the estimated batch quantities should be checked by means of trial batches or by full-size field batches. Enough concrete must be mixed for appropriate air and slump tests and for casting the three cylinders required for 28-day compressive-strength tests, plus beams for flexural tests if necessary. For a laboratory trial batch it is convenient, in this case, to scale down the quantities to produce 0.1 m³ of concrete as follows:

Water	83 x 0.1 =	8.3 kg
Cement	435 x 0.1 =	43.5 kg
Coarse aggregate (wet)	1093 x 0.1 =	109.3 kg
Fine aggregate (wet)	719 x 0.1 =	71.9 kg
Total		233.0 kg
Air-entraining admixture	218 g x 0.1 = 21.8 g or 21.8 mL	
Water reducer	1305 g x 0.1 = 130 g or 130 mL	

This concrete, when mixed, had a measured slump of 100 mm, an air content of 8%, and a density of 2274 kg per cubic metre. During mixing, some of the premeasured water may remain unused or additional water may be added to approach the required slump. In this example, although 8.3 kg of water was calculated to be added, the trial batch actually used only 8.0 kg. The mixture excluding admixtures therefore becomes

Water	8.0 kg
Cement	43.5 kg
Coarse aggregate (2% MC)	109.3 kg
Fine aggregate (6% MC)	71.9 kg
Total	232.7 kg

The yield of the trial batch is

$$\frac{232.7 \text{ kg}}{2274 \text{ kg/m}^3} = 0.10233 \text{ m}^3$$

The mixing water content is determined from the added water plus the free water on the aggregates and is calculated as follows:

Water added	8.0 kg

Free water on coarse aggregate

$$= \frac{109.3}{1.02} \text{ x } 0.015^* \qquad = 1.61 \text{ kg}$$

Free water on fine aggregate

$$= \frac{71.9}{1.06} \text{ x } 0.053^* \qquad = 3.59 \text{ kg}$$

Total water	13.20 kg

The mixing water required for a cubic metre of the same slump concrete as the trial batch is

$$\frac{13.2}{0.10233} = 129 \text{ kg}$$

Batch Adjustments. The design was for a slump of 75 mm. The measured 100-mm slump of the trial batch is unacceptable (80 mm or less ±20 mm max.), the yield was slightly high, and the 8.0% air content as measured in this example is also too high (more than 0.5% above 7.5% max.). Adjust the yield and reestimate the amount of air-entraining admixture required for an 7% air content and adjust the water to obtain a 75-mm slump. Increase the mixing water content by 3 kg/m³ for each 1% by which the air content is decreased from that of the trial batch and reduce the water content by 2 kg/m³ for each 10 mm reduction in slump. The adjusted mixture water for the reduced slump and air content is

(3 kg water x 1 percentage point difference for air) – (2 kg water x 25/10 for slump change) + 129 = 127 kg of water

---

*(2% MC – 0.5% absorption)/100 = 0.015
(6% MC – 0.7% absorption)/100 = 0.053

With less mixing water needed in the trial batch, less cement also is needed to maintain the desired water-cement ratio of 0.31. The new cement content is

$$\frac{127}{0.31} = 410 \text{ kg}$$

The amount of coarse aggregate remains unchanged because workability is satisfactory. The new adjusted batch quantities based on the new cement and water contents are calculated after the following volume computations:

Water $\quad = \dfrac{127}{1 \times 1000} = 0.127 \text{ m}^3$

Cement $\quad = \dfrac{410}{3.15 \times 1000} = 0.130 \text{ m}^3$

Coarse aggregate (dry) $\quad = \dfrac{1072}{2.68 \times 1000} = 0.400 \text{ m}^3$

Air $\quad = \dfrac{7}{100} = 0.070 \text{ m}^3$

Total $\quad\quad\quad\quad\quad\quad\quad\quad\quad 0.727 \text{ m}^3$

Fine aggregate volume $\quad = 1 - 0.727 = 0.273 \text{ m}^3$

The mass of dry fine aggregate required is
0.273 x 2.64 x 1000 = 721 kg

Air-entraining admixture (the manufacturer suggests reducing the dosage by 0.1 g to reduce air 1 percentage point) $\quad = 0.4 \times 410 = 164$ g or mL

Water reducer $\quad = 3.0 \times 410 = 1230$ g or mL

Adjusted batch quantities per cubic metre of concrete are

Water	127 kg
Cement	410 kg
Coarse aggregate (dry)	1072 kg
Fine aggregate (dry)	721 kg
Total	2330 kg

Air-entraining admixture $\quad$ 164 g or mL

Water reducer $\quad$ 1230 g or mL

Estimated concrete density (aggregates at SSD) $\quad = 127 + 410 + (1072 \times 1.005) + (721 \times 1.007) = 2340 \text{ kg/m}^3$

After checking these adjusted proportions in a trial batch, it was found that the concrete had the desired slump, air content, and yield. The 28-day test cylinders had an average compressive strength of 48 MPa, which exceeds the f'_{cr} of 43.5 MPa. Due to fluctuations in moisture content, absorption rates, and relative density of the aggregate, the density determined by volume calculations may not always equal the density determined by CSA Test Method A23.2-6C (ASTM C 138). Occasionally, the proportion of fine to coarse aggregate is kept constant in adjusting the batch quantities to maintain workability or other properties obtained in the first trial batch. After adjustments to the cementing materials, water, and air content have been made, the volume remaining for aggregate is appropri-ately proportioned between the fine and coarse aggregates.

Additional trial concrete mixtures with water-cementing materials ratios above and below 0.31 should also be tested to develop a strength to water-cementing materials ratio relationship. From that data, a new more economical mixture with a compressive strength closer to f'_{cr} and a lower cement content can be proportioned and tested. The final mixture would probably look similar to the above mixture with a slump range of 25 mm to 75 mm and an air content of 4% to 7%. The amount of air-entraining admixture must be adjusted to field conditions to maintain the specified air content.

Pozzolans and Slag. Pozzolans and slag are sometimes added in addition to or as a partial replacement of cement to aid in workability and resistance to sulphate attack and alkali reactivity. If a pozzolan or slag were required for this example mixture, it would have been entered in the first volume calculation used in determining fine aggregate content. For example:

Assume that 45 kg of fly ash with a relative density of 2.5 were to be used in addition to the originally derived cement content. The ash volume would be

$$\frac{45}{2.5 \times (1000)} = 0.018 \text{ m}^3$$

The water to cementing materials ratio would be

$$\frac{W}{C + P} = \frac{135}{(435 + 45)} = 0.28$$

The water to portland cement only ratio would still be

$$\frac{W}{C} = \frac{135}{435} = 0.31 \text{ by mass}$$

The fine aggregate volume would have to be reduced by 0.018 m³ to allow for the volume of ash.

The pozzolan amount and volume computation could also have been derived in conjunction with the first cement content calculation using a water to cementing materials ratio of 0.31 (or equivalent). For example, assume 15% of the cementing material is specified to be a pozzolan and W/CM or W/(C + P) = 0.31.

Then with $\quad W = 135$ kg and $C + P = 435$ kg,

$$P = 435 \times \frac{15}{100} = 65 \text{ kg}$$

and $\quad C = 435 - 65 = 370$ kg.

Appropriate proportioning computations for these and other mix ingredients would follow.

Example 2. Laboratory Trial Mixture Using the PCA Water-Cement Ratio Method

With the following method, the mix designer develops the concrete proportions directly from the laboratory trial batch rather than the absolute volume of the constituent ingredients.

Conditions and Specifications. Concrete is required for a plain concrete pavement to be constructed in Ottawa, Ontario. The pavement specified compressive strength is 35 MPa at 28 days. The standard deviation of the concrete producer is 2.0 MPa. Type 10 cement and 20-mm nominal maximum-size coarse aggregate is locally available. Proportion a concrete mixture for these conditions and check it by trial batch. Enter all data in the blank spaces on a trial mixture data sheet (Fig. 9-5).

Durability Requirements. The pavement will have a C-2 class of exposure being exposed to deicers and freezing and thawing and therefore should have a maximum water to cementing materials ratio of 0.45 (Table 9-1) and at least 335 kg of cement per cubic metre of concrete.

Strength Requirements. Using CSA acceptance criteria, for a standard deviation of 2.0 MPA, the f'_{cr} (required compressive strength for proportioning) must be the larger of

$$f'_{cr} = f'_c + 1.4S = 35 + 1.4\,(2.0) = 37.8 \text{ MPa}$$

or

$$f'_{cr} = f'_c + 2.4S - 3.5 = 35 + 2.4\,(2.0) - 3.5 = 36.3 \text{ MPa}$$

Therefore, the required average compressive strength
= 37.8 MPa.

Aggregate Size. The 20-mm maximum-size coarse aggregate and the fine aggregate are in saturated-surface dry condition for the trial mixtures.

Air Content. The CSA A23.1 recommended air content for a C-2 class of exposure and a corresponding air content Category 1, using a 14 to 20 mm nominal maximum size of aggregate is 5% to 8% (Table 9-5).

Slump. The specified target slump for this project is 40 (±20) mm.

Batch Quantities. For convenience, a batch containing 10 kg of cement is to be made. The quantity of mixing water required is 10 x 0.45 = 4.5 kg. Representative samples of fine and coarse aggregates are measured in suitable containers. The values are entered as initial mass in Column 2 of the trial-batch data sheet (Fig. 9-5).

All of the measured quantities of cement, water, and air-entraining admixture are used and added to the mixer. Fine and coarse aggregates, previously brought to a saturated, surface-dry condition, are added until a workable concrete mixture with a slump deemed adequate for placement is produced. The relative proportions or fine and coarse aggregate for workability can readily be judged by an experienced concrete technician or engineer.

Data and Calculations for Trial Batch
(saturated surface-dry aggregates)

Batch size: 10 kg ___✓___   20 kg _____   40 kg _____ of cement

Note: Complete Columns 1 through 4, fill in items below, the complete 5 and 6.

1 Material	2 Initial mass, kg	3 Final mass, kg	4 Mass. used, (Col. 2 minus Col. 3)	5 Mass per m³ No. of batches (C) x Col. 4	6 Remarks
Cement	10.0	0	10.0	341	
Water	4.5	0	4.5	153	
Fine aggregate	37.6	17.3	20.3	691 (a)	% F.A.* = $\frac{a}{a+b}$ x 100
Coarse aggregate	44.1	11.0	33.1	1128 (b)	= 38%
Air-entraining admixture	10 mℓ	Total (T) =	67.9	2313	
		T x C = 67.9 x 34.0648 =		2313	Math check

Measured slump: _____45_____ mm   Measured air content _____7.5_____ %

Appearance: Sandy _____   Good ___✓___   Rocky _____

Workability: Good ___✓___   Fair _____   Poor _____

Mass of container + concrete = _____42.7_____ kg

Mass of container = _____8.0_____ kg

Mass of concrete (A) = _____34.7_____ kg

Volume of container (B) = _____0.015_____ m³

Density of concrete (D) = $\frac{A}{B}$ = ___34.7/0.015___ = ___2313___ kg/m³

Volume of concrete produced = $\frac{\text{Total mass of material per batch}}{\text{Density}}$ = $\frac{T}{D}$

= ___67.9/2313___ = ___0.0293558___ m³

Number of __67.9__ kg batches per m³ (C) = $\frac{1.0 \text{ m}^3}{\text{Volume}}$ = $\frac{1.0}{0.0293558}$ = __34.0648__ batches

*Percentage fine aggregate of total aggregates = $\frac{\text{Mass of fine aggregate}}{\text{Total mass of aggregates}}$ x 100

Fig. 9-5. Trial mixture data sheet.

Table 9-13. Example of Results of Laboratory Trial Mixtures*

Batch no.	Slump, mm	Air content, percent	Density, kg/m³	Cement content, kg/m³	Fine aggregate, percent of total aggregate	Workability
1	50	5.7	2341	346	28.6	Harsh
2	40	6.2	2332	337	33.3	Fair
3	45	7.5	2313	341	38.0	Good
4	36	6.8	2324	348	40.2	Good

*Water-cement ratio was 0.45.

Workability. Results of tests for slump, air content, density, and a description of the appearance and workability are noted in the data sheet and Table 9-13.

The amounts of fine and coarse aggregates not used are recorded on the data sheet in Column 3, and mass of aggregates used (Column 2 minus Column 3) are noted in Column 4. If the slump when tested had been greater than that required, additional fine or coarse aggregates (or both) would have been added to reduce slump. Had the slump been less than required, water and cement in the appropriate ratio (0.45) would have been added to increase slump. It is important that any additional quantities be measured accurately and recorded on the data sheet.

Mixture Proportions. Mixture proportions for a cubic metre of concrete are calculated in Column 5 of Fig. 9-5 by using the batch yield (volume) and density. For example, the number of kilograms of cement per cubic metre is determined by dividing one cubic metre by the volume of concrete in the batch and multiplying the result by the number of kilograms of cement in the batch. The percentage of fine aggregate by mass of total aggregate is also calculated. In this trial batch, the cement content was 341 kg/m³ and the fine aggregate made up 38% of the total aggregate by mass. The air content and slump were acceptable. The 28-day strength was 39.1 MPa, greater than f'_{cr}. The mixture in Column 5, along with slump and air content limits of 40 (±20) mm and 5% to 8%, respectively, is now ready for submission to the project engineer.

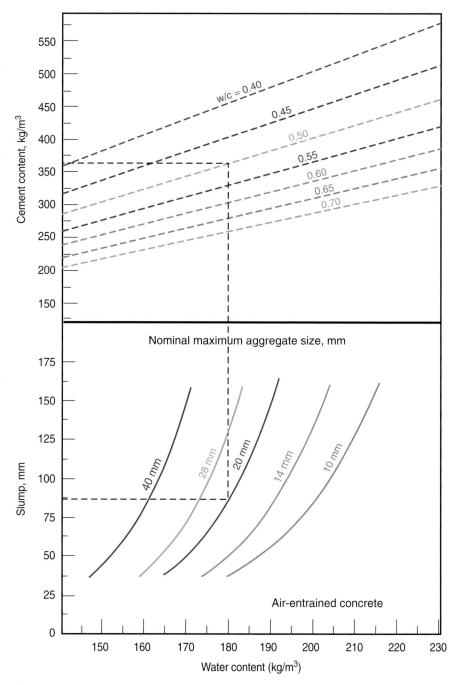

Fig. 9-6. Example graphical relationship for a particular aggregate source demonstrating the relationship between slump, aggregate size, water to cement ratio, and cement content (adapted from Hover 1995).

Mixture Adjustments. To determine the most workable and economical proportions, additional trial batches could be made varying the percentage of fine aggregate. In each batch the water-cement ratio, aggregate gradation, air content, and slump should remain about approximately the same. Results of four such trial batches are summarized in Table 9-13.

Table 9-14 illustrates the change in mix proportions for various types of concrete mixtures using a particular aggregate source. Information for concrete mixtures using particular ingredients can be plotted in several ways to illustrate the relationship between ingredients and properties. This is especially useful when optimizing concrete mixtures for best economy or to adjust to specification or material changes (Fig. 9-6).

Example 3. Absolute Volume Method Using Multiple Cementing Materials and Admixtures

The following example illustrates how to develop a mix using the absolute volume method when more than one cementing material and admixture are used.

Conditions and Specifications. Concrete with a structural design strength of 40 MPa is required for a bridge to be exposed to freezing and thawing, deicers, and moderate sulphate soils. A coulomb value not exceeding 1500 is required to minimize permeability to chlorides. Water reducers, air entrainers, and plasticizers are allowed. A shrinkage reducer is requested to keep shrinkage under

Table 9-14. Example Trial Mixtures for Air-Entrained Concrete of Medium Consistency, 80-mm to 100-mm slump

Water-cement ratio, kg per kg	Nominal maximum size of aggregate, mm	Air content, percent	Water, kg per m³ of concrete	Cement, kg per m³ of concrete	With fine sand, fineness modulus = 2.50			With coarse sand, fineness modulus = 2.90		
					Fine aggregate, percent of total aggregate	Fine aggregate, kg per m³ of concrete	Coarse aggregate, kg per m³ of concrete	Fine aggregate, percent of total aggregate	Fine aggregate, kg per m³ of concrete	Coarse aggregate, kg per m³ of concrete
0.40	10	7.5	202	505	50	744	750	54	809	684
	14	7.5	194	485	41	630	904	46	702	833
	20	6	178	446	35	577	1071	39	648	1000
	40	5	158	395	29	518	1255	33	589	1184
0.45	10	7.5	202	450	51	791	750	56	858	684
	14	7.5	194	387	43	678	904	47	750	833
	20	6	178	395	37	619	1071	41	690	1000
	40	5	158	351	31	553	1225	35	625	1184
0.50	10	7.5	202	406	53	833	750	57	898	684
	14	7.5	194	387	44	714	904	49	785	833
	20	6	178	357	38	654	1071	42	726	1000
	40	5	158	315	32	583	1225	36	654	1184
0.55	10	7.5	202	369	54	862	750	58	928	684
	14	7.5	194	351	45	744	904	49	815	833
	20	6	178	324	39	678	1071	43	750	1000
	40	5	158	286	33	613	1225	37	684	1184
0.60	10	7.5	202	336	54	886	750	58	952	684
	14	7.5	194	321	46	768	904	50	839	833
	20	6	178	298	40	702	1071	44	773	1000
	40	5	158	262	33	631	1225	37	702	1184
0.65	10	7.5	202	312	55	910	750	59	976	684
	14	7.5	194	298	47	791	904	51	863	833
	20	6	178	274	40	720	1071	44	791	1000
	40	5	158	244	34	649	1225	38	720	1184
0.70	10	7.5	202	288	55	928	750	59	994	684
	14	7.5	194	277	47	809	904	51	880	833
	20	6	178	256	41	738	1071	45	809	1000
	40	5	158	226	34	660	1225	38	732	1184

300 millionths. Some structural elements exceed a thickness of 1 metre, requiring control of heat development. The concrete producer has a standard deviation of 2 MPa for similar mixes to that required here. For difficult placement areas, a slump of 200 mm to 250 mm is required. The following materials are available:

Cement:	Type 20E-SF, portland silica fume cement (a blended cement), meeting CSA Standard A362. Relative density of 3.14. Silica fume content of 5%.
Fly ash:	Class F, CSA A23.5. Relative density of 2.60.
Slag:	Type S, CSA A23.5. Relative density of 2.90.
Coarse aggregate:	Well-graded 20-mm nominal maximum-size crushed rock with an ovendry relative density of 2.68, absorption of 0.5%, and ovendry density of 1600 kg/m³. The laboratory sample has a moisture content of 2.0%. This aggregate has a history of alkali-silica reactivity in the field.
Fine aggregate:	Natural sand with some crushed particles with an ovendry relative density of 2.64 and an absorption of 0.7%. The laboratory sample has a moisture content of 6%. The fineness modulus is 2.80.
Air entrainer:	Synthetic, ASTM C 260.
Retarding water reducer:	Type D, ASTM C 494. Dosage of 3 g per kg of cementing materials.
Plasticizer:	Type 1, ASTM C 1017. Dosage of 30 g per kg of cementing materials.
Shrinkage reducer:	Dosage of 15 g per kg of cementing materials.

Strength. For a standard deviation of 2.0 MPa, the f'_{cr} must be the greater of

$$f'_{cr} = f'_c + 1.4S = 40 + 1.4(2) = 42.8$$

or

$$f'_{cr} = f'_c + 2.4S - 3.5 = 40 + 2.4(2) - 3.5 = 41.3$$

therefore $f'_{cr} = 42.8$

Water to Cementing Materials Ratio. Past field records using these materials indicate that a water to cementing materials ratio of 0.35 is required to provide a strength level of 42.8 MPa.

For a deicer environment and to protect embedded steel from corrosion, Table 9-1 requires a maximum water to cementing materials ratio of 0.40 and a strength of at least 35 MPa. For a moderate sulphate environment, Table 9-2 requires a maximum water to cementing materials ratio of 0.50 and a strength of at least 30 MPa. Both the water to cementing materials ratio requirements and strength requirements are met and exceeded using the above determined 0.35 water to cementing materials ratio and 40 MPa design strength.

Air Content. For a C-1 class of severe exposure, and a corresponding air content Category of **1** due to the freeze-thaw environment, the total air content recommended is 5% to 8% when using a 20 mm maximum size aggregate (Table 9-5). Therefore, use 8% for batch proportions. The trial batch air content must be within ±0.5 percentage points of the maximum allowable air content.

Slump. Assume a slump of 50 mm without the plasticizer and a maximum of 200 mm to 250 mm after the plasticizer is added. Use 250 ± 20 mm for proportioning purposes.

Water Content. Table 9-5 recommends that for a 25- to 50-mm slump, air-entrained concrete with 20-mm aggregate should have a water content of about 168 kg/m³. Assume the retarding water reducer and plasticizer will jointly reduce water demand by 15% in this case, resulting in an estimated water demand of 143 kg per cubic metre, while achieving the 250-mm slump.

Cementing Materials Content. The amount of cementing materials is based on the maximum water-cementing materials ratio and water content. Therefore, 143 kg of water divided by a water-cementing materials ratio of 0.35 requires a cement content of 409 kg. Fly ash and slag will also be used as supplementary cementing materials to help control alkali-silica reactivity and control temperature rise. Local use has shown that a fly ash dosage of 15% and a slag dosage of 30% by mass of cementing materials are adequate. Therefore, the suggested cementing materials for one cubic metre of concrete are as follows:

Cement: 55% of 409 = 225 kg (5% silica fume or 11.3 kg)

Fly ash: 15% of 409 = 61 kg

Slag: 30% of 409 = 123 kg

These dosages meet the requirements of Table 9-8 (2.8% silica fume from the cement + 15% fly ash + 30% slag = 47.8% which is less than the 50% maximum allowed).

Coarse-Aggregate Content. The quantity of 20-mm nominal maximum-size coarse aggregate can be estimated from Fig. 9-3. The bulk volume of coarse aggregate recommended when using sand with a fineness modulus of 2.80 is 0.62. Since the coarse aggregate has a bulk density of 1600 kg/m³, the ovendry mass of coarse aggregate for a cubic metre of concrete is

1600 x 0.62 = 992 kg/m³

Admixture Content. For an 8% air content, the air-entraining admixture manufacturer recommends a dosage of 0.5 g per kg of cementing materials. The amount of air entrainer is then

0.5 x 409 = 205 g = 0.205 kg

The retarding water reducer dosage rate is 3 g per kg of cementing materials. This results in

3 x 409 = 1227 g or 1.227 kg of water reducer per cubic metre of concrete.

The plasticizer dosage rate is 30 g per kg of cementing materials. This results in

30 x 409 = 12,270 g or 12.270 kg of plasticizer per cubic metre of concrete.

The shrinkage reducer dosage rate is 15 g per kg of cementing materials. This results in

15 x 409 = 6135 g or 6.135 kg of shrinkage reducer per cubic metre of concrete.

Fine-Aggregate Content. At this point, the amounts of all ingredients except the fine aggregate are known. The volume of fine aggregate is determined by subtracting the absolute volumes of all known ingredients from 1 cubic metre. The absolute volumes of the ingredients is calculated by dividing the known mass of each by the product of their relative density and the density of water. Assume a relative density of 1.0 for the chemical admixtures. Assume a density of water of 997.75 kg/m³ as all materials in the laboratory are maintained at a room temperature of 22°C (Table 9-12). Volumetric computations are as follows:

Water (including chemical admixtures) $= \dfrac{143}{1.0 \times 997.75} = 0.143 \text{ m}^3$

Cement $= \dfrac{225}{3.14 \times 997.75} = 0.072 \text{ m}^3$

Fly ash $= \dfrac{61}{2.60 \times 997.75} = 0.024 \text{ m}^3$

Slag $= \dfrac{123}{2.90 \times 997.75} = 0.043 \text{ m}^3$

Air $= \dfrac{8.0}{100} = 0.080 \text{ m}^3$

Coarse aggregate $= \dfrac{992}{2.68 \times 997.75} = 0.371 \text{ m}^3$

Total $= 0.733 \text{ m}^3$

The calculated absolute volume of fine aggregate is then
1 – 0.733 = 0.267 m³

The mass of dry fine aggregate is
0.267 x 2.64 x 997.75 = 703 kg

The admixture volumes are

Air entrainer $= \dfrac{0.205}{(1.0 \times 997.75)} = 0.0002 \text{ m}^3$

Water reducer $= \dfrac{1.227}{(1.0 \times 997.75)} = 0.0012 \text{ m}^3$

Plasticizer $= \dfrac{12.270}{(1.0 \times 997.75)} = 0.0123 \text{ m}^3$

Shrinkage reducer $= \dfrac{6.135}{(1.0 \times 997.75)} = 0.0061 \text{ m}^3$

Total = 19.84 kg of admixture with a volume of 0.0198 m³

Consider the admixtures part of the mixing water

Mixing water minus admixtures = 143 – 19.84 = 123 kg

The mixture then has the following proportions before trial mixing for 1 cubic metre of concrete:

Water	123 kg
Cement	225 kg
Fly ash	61 kg
Slag	123 kg
Coarse aggregate (dry)	992 kg
Fine aggregate (dry)	703 kg
Air entrainer	0.205 kg
Water reducer	1.227 kg
Plasticizer	12.27 kg
Shrinkage reducer	6.135 kg
Total	= 2247 kg
Slump	= 250 mm (± 20 mm for trial batch)
Air content	= 8% (± 0.5% for trial batch)

Estimated concrete density using SSD aggregate (adding absorbed water)

= 123 + 225 + 61 + 123 + (992 x 1.005) + (703 x 1.007) + 20 (admixtures) = 2257 kg/m³

Moisture. The dry batch quantities of aggregates have to be increased to compensate for the moisture on and in the aggregates and the mixing water reduced accordingly. The coarse aggregate and fine aggregate have moisture contents of 2% and 6%, respectively. With the moisture contents indicated, the trial batch aggregate proportions become

Coarse aggregate (2% MC) = 992 x 1.02 = 1012 kg
Fine aggregate (6% MC) = 703 x 1.06 = 745 kg

Absorbed water does not become part of the mixing water and must be excluded from the water adjustment. Surface moisture contributed by the coarse aggregate amounts to 2% – 0.5% = 1.5% and that contributed by the fine aggregate, 6% – 0.7% = 5.3%. The estimated added water becomes

123 – (992 x 0.015) – (703 x 0.053) = 71 kg

The batch quantities for one cubic metre of concrete are revised to include aggregate moisture as follows:

Water (to be added)	71 kg
Cement	225 kg
Fly ash	61 kg
Slag	123 kg
Coarse aggregate (2% MC)	1012 kg
Fine aggregate (6% MC)	745 kg
Air entrainer	0.205 kg
Water reducer	1.227 kg
Plasticizer	12.27 kg
Shrinkage reducer	6.14 kg

Trial Batch. The above mixture is tested in a 0.1 m³ batch in the laboratory (multiply above quantities by 0.1 to obtain batch quantities). The mixture had an air content of 7.8%, a slump of 240 mm, a density of 2257 kg/m³, a yield of 0.1 m³, and a compressive strength of 44 MPa. Rapid chloride testing resulted in a coulomb value of 990 (ASTM C 1202). A modified version of CSA A23.2-25A (ASTM C 1260) was used to evaluate the potential of the mix for alkali-silica reactivity, resulting in an acceptable expansion of 0.02% (expansion less than 0.1% at 14 days, therefore use of SCMs was effective). Temperature rise was acceptable and shrinkage was within specifications. The water-soluble chloride content was 0.06%, meeting the requirements of Table 9-9. The following mix proportions meet all applicable requirements and are ready for submission to the project engineer for approval:

Water added	123 kg (143 kg total including admixtures)
Cement, Type 20E-SF	225 kg
Fly ash, Class F	61 kg
Slag, Type S	123 kg
Coarse aggregate	992 kg (ovendry) or 997 kg (SSD)
Fine aggregate	703 kg (ovendry) or 708 kg (SSD)
Air entrainer*	0.205 kg
Water reducer*	1.227 kg
Plasticizer*	12.27 kg
Shrinkage reducer*	6.14 kg
Slump	200 mm to 250 mm
Air content	5% to 8%
Density (SSD agg.)	2257 kg/m³
Yield	1 m³
Water-cementing materials ratio	0.35

*Liquid admixture dosages can also be given in L and mL in mix proportion documents.

CONCRETE FOR SMALL JOBS

Although well-established ready mixed concrete mixtures are used for most construction, ready mix is not always practical for small jobs, especially those requiring one cubic metre or less. Small batches of concrete mixed at the site are required for such jobs.

If mixture proportions or mixture specifications are not available, Tables 9-15 and 9-16 can be used to select proportions for concrete for small jobs. Recommendations with respect to exposure conditions discussed earlier should be followed.

The proportions in Tables 9-15 and 9-16 are only a guide and may need adjustments to obtain a workable mix with locally available aggregates (PCA 1988). Packaged, combined, dry concrete ingredients (ASTM C 387) are also available.

DESIGN REVIEW

In practice, concrete mixture proportions will be governed by the limits of data available on the properties of materials, the degree of control exercised over the production of concrete at the plant, and the amount of supervision at the jobsite. It should not be expected that field results will be an exact duplicate of laboratory trial batches. An adjustment of the selected trial mixture is usually necessary on the job.

Table 9-15. Proportions by Mass to Make One Tenth Cubic Metre of Concrete for Small Jobs

Nominal maximum size coarse aggregate, mm	Air-entrained concrete				Non-air-entrained concrete			
	Cement, kg	Wet fine aggregate, kg	Wet coarse aggregate, kg*	Water, kg	Cement, kg	Wet fine aggregate, kg	Wet coarse aggregate, kg	Water, kg
10	46	85	74	16	46	94	74	18
14	43	74	88	16	43	85	88	18
20	40	67	104	16	40	75	104	16
28	38	62	112	15	38	72	112	15
40	37	61	120	14	37	69	120	14

*If crushed stone is used, decrease coarse aggregate by 5 kg and increase fine aggregate by 5 kg.

Table 9-16. Proportions by Bulk Volume* of Concrete for Small Jobs

Nominal maximum size coarse aggregate, mm	Air-entrained concrete				Non-air-entrained concrete			
	Cement	Wet fine aggregate	Wet coarse aggregate	Water	Cement	Wet fine aggregate	Wet coarse aggregate	Water
10	1	2¼	1½	½	1	2½	1½	½
14	1	2¼	2	½	1	2½	2	½
20	1	2¼	2½	½	1	2½	2½	½
28	1	2¼	2¾	½	1	2½	2¾	½
40	1	2¼	3	½	1	2½	3	½

*The combined volume is approximately ⅔ of the sum of the original bulk volumes.

The mixture design and proportioning procedures presented here and summarized in Fig. 9-7 are applicable to normal-density concrete. For concrete requiring some special property, using special admixtures or materials—low-density aggregates, for example—different proportioning principles may be involved.

Internet web sites also provide assistance with designing and proportioning concrete mixtures (Bentz 2001). Many of these web sites are internationally oriented and assume principles not used in North America. Therefore, appropriate cautions should be taken when using the internet to design concrete mixtures.

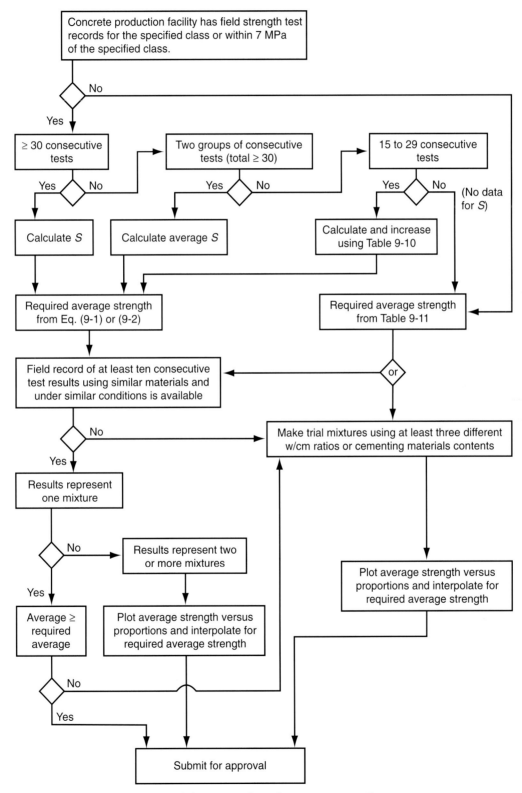

Fig. 9-7. Flowchart for selection and documentation of concrete proportions.

REFERENCES

Abrams, D. A., *Design of Concrete Mixtures,* Lewis Institute, Structural Materials Research Laboratory, Bulletin No. 1, PCA LS001, Chicago, http://www.portcement.org/pdf_files/LS001.pdf, 1918, 20 pages.

ACI Committee 211, *Standard Practice for Selecting Proportions for Normal, Heavyweight and Mass Concrete,* ACI 211.1-91, American Concrete Institute, Farmington Hills, Michigan, 1991.

ACI Committee 211, *Guide for Selecting Proportions for High-Strength Concrete with Portland Cement and Fly Ash,* ACI 211.4R-93, American Concrete Institute, Farmington Hills, Michigan, 1993.

ACI Committee 211, *Guide for Submittal of Concrete Proportions,* ACI 211.5R-96, American Concrete Institute, Farmington Hills, Michigan, 1996.

ACI Committee 211, *Guide for Selecting Proportions for No-Slump Concrete,* ACI 211.3R-97, American Concrete Institute, Farmington Hills, Michigan, 1997.

ACI Committee 211, *Standard Practice for Selecting Proportions for Structural Lightweight Concrete,* ACI 211.2-98, American Concrete Institute, Farmington Hills, Michigan, 1998.

ACI Committee 214, *Recommended Practice for Evaluation of Strength Test Results of Concrete,* ACI 214-77, reapproved 1997, American Concrete Institute, Farmington Hills, Michigan, 1977.

ACI Committee 301, *Specifications for Structural Concrete,* ACI 301-99, American Concrete Institute, Farmington Hills, Michigan, 1999.

ACI Committee 302, *Guide for Concrete Floor and Slab Construction,* ACI 302.1R-96, American Concrete Institute, Farmington Hills, Michigan, 1996.

ACI Committee 318, *Building Code Requirements for Structural Concrete,* ACI 318-99, and *Commentary,* ACI 318R-99, American Concrete Institute, Farmington Hills, Michigan, 1999.

ACI Committee 357, *Guide for the Design and Construction of Fixed Offshore Concrete Structures,* ACI 357R-84, American Concrete Institute, Farmington Hills, Michigan, 1984.

Bentz, Dale, *Concrete Optimization Software Tool,* http://ciks.cbt.nist.gov/bentz/fhwa, National Institute of Standards and Technology, 2001.

CSA Standard A23.1-00/A23.2-00, *Concrete Materials and Methods of Concrete Construction/Methods of Test for Concrete,* Canadian Standards Association, Toronto, 2000.

Hover, Ken, "Graphical Approach to Mixture Proportioning by ACI 211.1-91," *Concrete International,* American Concrete Institute, Farmington Hills, Michigan, September, 1995, pages 49 to 53.

Hover, Kenneth C., "Concrete Design: Part 1, Finding Your Perfect Mix," http://www.cenews.com/edconc0998.html, *CE News,* September 1998.

Hover, Kenneth C., "Concrete Design: Part 2, Proportioning Water, Cement, and Air," http://www.cenews.com/edconc1098.html, *CE News,* October 1998.

Hover, Kenneth C., "Concrete Design: Part 3, Proportioning Aggregate to Finish the Process," http://www.enews.com/edconc1198.html, *CE News,* November 1998.

PCA, *Concrete for Small Jobs,* IS174, Portland Cement Association, http://www.portcement.org/pdf_files/IS174.pdf, 1988.

Shilstone, James M., Sr., "Concrete Mixture Optimization," *Concrete International,* American Concrete Institute, Farmington Hills, Michigan, June 1990, pages 33 to 39.

CHAPTER 10
Batching, Mixing, Transporting, and Handling Concrete

The specification, production, and delivery of concrete are achieved in different ways. The basic processes and common techniques are explained here. CSA Standard A23.1 (ASTM C 94) provides standard specifications for the manufacture and delivery of freshly mixed concrete. Standards of the Concrete Plant Manufacturers Bureau, Truck Mixer Manufacturers Bureau, and Volumetric Mixer Manufacturers Bureau can be found on the National Ready Mixed Concrete Association's website at http://www.nrmca.org.

Three alternative options for ordering or specifying concrete are described in CSA Standard A23.1 (ASTM C 94):

Alternative (1) Common: When the owner requires the concrete supplier to assume responsibility for the concrete mix proportions. The owner shall specify the type of cementing materials, the class of exposure the concrete must withstand, compressive strength needed for structural requirements, maximum size of coarse aggregate, air content, the admixtures and any other properties that may be required, such as slump or minimum flexural strength. The supplier must certify that the plant, equipment, and materials used meet the Standard, that the mix proportions will produce concrete of the specified quality and yield, and that the strengths when evaluated, will meet the requirements of the Standard.

Alternative (2) Prescription: When the owner assumes responsibility for the mix proportions and properties of the concrete. The owner shall specify the types and amounts of cementing materials, the nominal maximum size of coarse aggregate and proportions by mass of fine and coarse aggregate, the maximum total water content by mass—all on a per cubic metre of concrete basis. In addition, the owner must specify the air content, type of admixtures, and the slump at the point of discharge. The supplier specifies that the plant, equipment, and all materials to be used meet the requirements of the standard.

Alternative (3) Performance: When the owner requires the concrete supplier to assume responsibility for the concrete "as delivered". The owner must specify the class of concrete, based on the requirements for concrete given in CSA A23.1, to meet the appropriate exposure condition(s), and also must specify the architectural, structural, and durability criteria. The supplier must certify that the quality plan ensures the performance criteria will be measured and recorded and that the concrete complies with the specified performance criteria prior to the supply of the concrete. The quality plan must also designate the specified slump or some other measure of concrete workability.

BATCHING

Batching is the process of measuring concrete mix ingredients by either mass or volume and introducing them into the mixer. To produce concrete of uniform quality, the ingredients must be measured accurately for each batch. Most specifications require that batching be done by mass rather than by volume (CSA A23.1 or ASTM C 94). Water

Fig. 10-1. Control room for batching equipment in a typical ready mixed concrete plant. (69894)

and liquid admixtures can be measured accurately by either volume or mass. Volumetric batching (ASTM C 685) is used for concrete mixed in continuous mixers.

CSA Standard A23.1 provides guidance on the measurement of materials. Percentages of accuracy for measurement of concrete materials are as follows.

Cement: When the quantity to be batched exceeds 30% of the scale capacity, the scale reading shall be within 1% of the required mass. For smaller batches, the scale reading for the amount used shall not be less than the required quantity and not more than 4% in excess.

Supplementary Cementing Materials: When the quantity to be batched exceeds 30% of the scale capacity, the scale reading shall be within 1% of the required mass. For smaller batches, the scale reading shall be within 4% of the required quantity.

Aggregate: When individual aggregate weigh batchers are used, the scale reading for each material shall be within 2% of the specified mass. In a cumulative aggregate weigh batcher, the cumulative mass after each measurerment shall be within 1% of the required cumulative amount when the scale is used in excess of 30% of its capacity. For cumulative measurement less than 30% of scale capacity, the allowable variation shall be ±0.3% of scale capacity, or ±3% of the required cumulative mass, whichever is less.

Mixing Water: The measuring equipment shall be accurate to ±1% of the required mass. As mixing water is considered to consist of the water added to the batch, surface moisture on the aggregate, water contained in admixture solutions, and ice used as a concrete coolant, the total amount of mixing water obtained from all sources shall be within ±3% of the specified quantity.

Admixtures: Powdered admixtures shall be measured by mass, and liquid admixtures by mass or volume. Volumetric measurement shall be within an accuracy of ±3% of the required amount, or 30 mL, whichever is greater. Mass measurement accuracy shall be within ±3% of the required amount.

Equipment should be capable of measuring quantities within these tolerances for the smallest batch regularly used, as well as for larger batches (Fig. 10-1). The accuracy of scales and batching equipment should be checked periodically and adjusted when necessary.

MIXING CONCRETE

All concrete should be mixed thoroughly until it is uniform in appearance, with all ingredients evenly distributed. Mixers should not be loaded above their rated capacities and should be operated at the mixing speed recommended by the manufacturer. Increased output should be obtained

by using a larger mixer or additional mixers, rather than by speeding up or overloading the equipment on hand. If the blades of a mixer become worn or coated with hardened concrete, mixing action will be less efficient. These conditions should be corrected.

If concrete has been adequately mixed, samples taken from different portions of a batch will have essentially the same density, air content, slump, and coarse-aggregate content. Maximum allowable differences to evaluate mixing uniformity within a batch of ready mixed concrete are given in CSA Standard A23.1 (ASTM C 94).

Structural low-density concrete can be mixed the same way as normal-density concrete when the aggregates have less than 10% total absorption by mass or when the absorption is less than 2% by mass during the first hour after immersion in water. For aggregates not meeting these limits, mixing procedures are described in PCA (1986).

Stationary Mixing

Concrete is sometimes mixed at the jobsite in a stationary mixer or a paving mixer (Fig. 10-2). Stationary mixers include both onsite mixers and central mixers in ready mix plants. They are available in sizes up to 9.0 m³ and can be of the tilting or nontilting type or the open-top revolving blade or paddle type. All types may be equipped with loading skips and some are equipped with a swinging discharge chute. Many stationary mixers have timing devices, some of which can be set for a given mixing time and locked so that the batch

Fig. 10-2. Concrete can be mixed at the jobsite in a stationary mixer. (58642)

cannot be discharged until the designated mixing time has elapsed.

Careful attention should be paid to the required mixing time. Many specifications require a minimum mixing time of one minute plus 15 seconds for every cubic metre, unless mixer performance tests demonstrate that shorter periods are acceptable and will provide a uniform concrete mixture. Short mixing times can result in nonhomogenous mixtures, poor distribution of air voids (resulting in poor frost resistance), poor strength gain, and early stiffening problems. The mixing period should be measured from the time all cementing materials and aggregates are in the mixer drum, provided all the water is added before one-fourth of the mixing time has elapsed (ACI 304R-00).

Under usual conditions, up to about 10% of the mixing water should be placed in the drum before the solid materials are added. Water then should be added uniformly with the solid materials, leaving about 10% to be added after all other materials are in the drum. When heated water is used in cold weather, this order of charging may require some modification to prevent possible rapid stiffening when hot water contacts the cement. In this case, addition of the cementing materials should be delayed until most of the aggregate and water have intermingled in the drum. Where the mixer is charged directly from a batch plant, the materials should be added simultaneously at such rates that the charging time is about the same for all materials. If supplementary cementing materials are used, they should be added after the cement.

If retarding or water-reducing admixtures are used, they should be added in the same sequence in the charging cycle each time. If not, significant variations in the time of initial setting and percentage of entrained air may result. Addition of the admixture should be completed not later than one minute after addition of water to the cement has been completed or prior to the start of the last three-fourths of the mixing cycle, whichever occurs first. If two or more admixtures are used in the same batch of concrete, they should be added separately; this is intended to avoid any interaction that might interfere with the efficiency of any of the admixtures and adversely affect the concrete properties. In addition, the sequence in which they are added to the mix can be important too.

Ready Mixed Concrete

Ready mixed concrete is proportioned and mixed off the project site and is delivered to the construction area in a freshly mixed and unhardened state. It can be manufactured by any of the following methods:

1. Central-mixed concrete is mixed completely in a stationary mixer (Fig. 10-3). Fig. 10-4 illustrates a central mix ready mix plant. The concrete is delivered either in a truck agitator (Fig. 10-5 bottom), a truck mixer operating at agitating speed (Fig. 10-6), or a nonagitating truck (Fig. 10-5 top).

2. Shrink-mixed concrete is mixed partially in a stationary mixer and completed in a truck mixer.

3. Truck-mixed concrete is mixed completely in a truck mixer (Fig. 10-6).

CSA Standard A23.1 (ASTM C 94) notes that when a truck mixer is used for complete mixing, 70 to 100 revolutions of the drum or blades at the rate of rotation designated by the manufacturer as *mixing speed* are usually required to produce the specified uniformity of concrete. All revolutions after 100 should be at a rate of rotation designated by the manufacturer as *agitating speed*. Agitating speed is usually about 2 to 6 rpm, and mixing speed is generally about 6 to 18 rpm. Mixing at high speeds for long periods of time, about 1 or more hours, can result in concrete strength loss, temperature rise, excessive loss of entrained air, and accelerated slump loss.

When truck mixers are used, CSA Standard A23.1 (ASTM C 94) limits the time between batching and complete discharge of the concrete at the job site; this time is 2 hours after introduction of water to the cement and aggregates or the cement to the aggregates. It also requires that the concrete subjected to exposure classifications C-1, C-2, and F-1 be retested for conformance to air-content requirements when more than 90 minutes have elapsed since batching. Mixers and agitators should always be operated within the limits for volume and speed of rotation designated by the equipment manufacturer.

Fig. 10-3. Central mixing in a stationary mixer of the tilting drum type with delivery by a truck mixer operating at agitating speed. (69926)

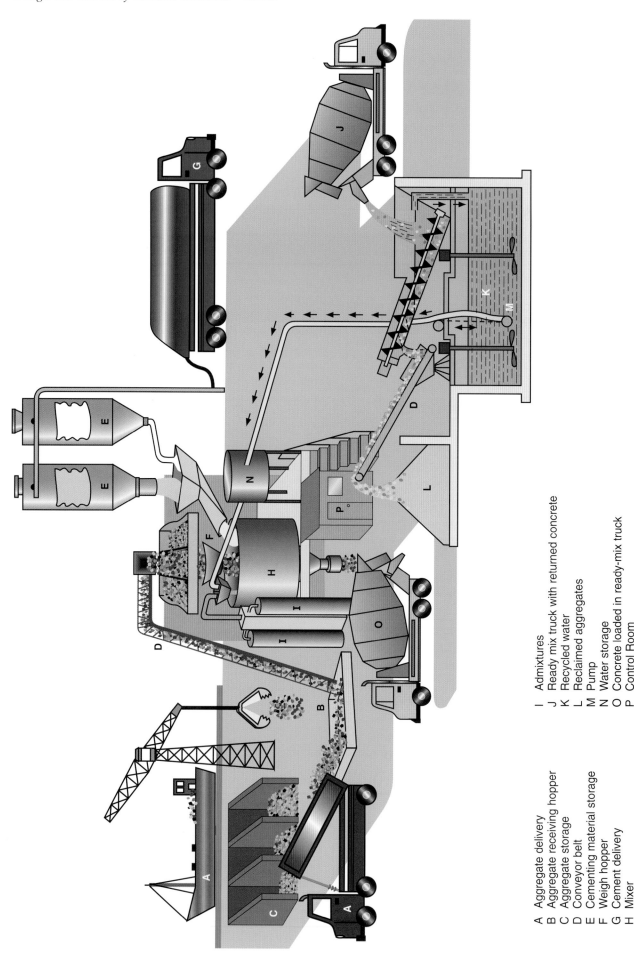

A   Aggregate delivery
B   Aggregate receiving hopper
C   Aggregate storage
D   Conveyor belt
E   Cementing material storage
F   Weigh hopper
G   Cement delivery
H   Mixer
I   Admixtures
J   Ready mix truck with returned concrete
K   Recycled water
L   Reclaimed aggregates
M   Pump
N   Water storage
O   Concrete loaded in ready-mix truck
P   Control Room

Fig. 10-4. Schematic of a central mix ready mix plant.

Fig. 10-6. Truck-mixed concrete is mixed completely in a truck mixer. (1153)

Mobile Batcher Mixed Concrete (Continuous Mixer)

Mobile volumetric mixers are special trucks (Fig. 10-7) that batch by volume and continuously mix concrete as the dry concrete ingredients, water, and admixtures are continuously fed into a mixing trough, typically an auger system. The concrete must conform to ASTM C 685 specifications and is proportioned and mixed at the jobsite in the quantities needed. The concrete mixture is also easily adjusted for project placement and weather conditions.

Fig 10-5. (top) Nonagitating trucks are used with central-mix batch plants where short hauls and quick concrete discharge allows the rapid placement of large volumes of concrete. (bottom) Truck agitators are also used with central-mix batch plants. Agitation mixing capabilities allow truck agitators to supply concrete to projects with slow rates of concrete placement and at distances greater than nonagitating trucks. (69897, 69898)

Fig. 10-7. Mobile batcher measures materials by volume and continuously mixes concrete as the dry ingredients, water, and admixtures are fed into a mixing trough at the rear of the vehicle. (54087)

177

Remixing Concrete

Fresh concrete that is left to agitate in the mixer drum tends to stiffen before initial set develops. Such concrete may be used if upon remixing it becomes sufficiently plastic to be compacted in the forms. CSA Standard A23.1 (ASTM C 94) allows the addition of extra water at the jobsite only when the measured slump at the start of discharge is less than that specified, and no more than 60 minutes have elapsed from the time of batching to the start of discharge. The extra water added can only be an amount to bring the concrete up to the specified slump, and in no case be more than the lesser of either 16 L/m^3 or 10% of the mixing water. In doing so, the specified water-to-cementing materials ratio also cannot be exceeded. The water may be added on the instruction of the producer when the concrete is supplied on the basis of Alternative 1 or 3 and on instruction of the owner when supplied on the basis of Alternative 2. The mixer drum shall be turned at mixing speed for at least 30 revolutions after the addition of the water, or until the uniformity of the concrete is within the limits as described in CSA Standard A23.1 (ASTM C 94). If early setting becomes a persistent problem, a retarder may be used to control early hydration, especially in high-cement-content mixes. Mixture adjustments at the jobsite for air entrainment, and the addition of other admixtures, is permitted, followed by sufficient mixing.

Indiscriminate addition of water to make concrete more fluid should not be allowed because this lowers the quality of concrete. The later addition of water and remixing to retemper the mixture can result in marked strength reduction. When superplasticized concrete falls below the designated slump due to delay, it should be retempererd with superplasticizing admixtures only, not water.

TRANSPORTING AND HANDLING CONCRETE

Good advanced planning can help choose the appropriate handling method for an application. Consider the following three occurrences that, should they occur during handling and placing, could seriously affect the quality of the finished work:

Delays. The objective in planning any work schedule is to produce the fastest work with the best labour force and the proper equipment for the work at hand. Machines for transporting and handling concrete are being improved all the time. The greatest productivity will be achieved if the work is planned to get the most out of personnel and equipment and if the equipment is selected to reduce the delay time during concrete placement.

Early Stiffening and Drying Out. Concrete begins to stiffen as soon as the cementing materials and water are mixed, but the degree of stiffening that occurs in the first 30 minutes is not usually a problem; concrete that is kept agi-

tated generally can be placed and compacted within 2 hours after mixing unless hot concrete temperatures or high cement contents speed up hydration excessively. Planning should eliminate or minimize any variables that would allow the concrete to stiffen to the extent that full consolidation is not achieved and finishing becomes difficult. Less time is available during conditions that hasten the stiffening process, such as hot and dry weather, use of accelerators, and use of heated concrete.

Segregation. Segregation is the tendency for coarse aggregate to separate from the sand-cement mortar. This results in part of the batch having too little coarse aggregate and the remainder having too much. The former is likely to shrink more and crack and have poor resistance to abrasion. The latter may be too harsh for full consolidation and finishing and is a frequent cause of honeycombing. The method and equipment used to transport and handle the concrete must not result in segregation of the concrete materials.

Methods and Equipment for Transporting and Handling Concrete

Table 10-1 summarizes the most common methods and equipment for moving concrete to the point where it is needed.

There have been few, if any, major changes in the principles of conveying concrete during the last 75 years. What has changed is the technology that led to development of better machinery to do the work more efficiently. The wheelbarrow and buggy, although still used, have advanced to become the power buggy (Fig. 10-8); the horse-drawn wagon is now the ready mixed concrete truck (Figs. 10-9 and 10-10); and the bucket hauled over a pulley wheel has become the bucket and crane (Fig. 10-11).

Years ago concrete was placed in reinforced concrete buildings by means of a tower and long chutes. This was a guyed tower centrally placed on the site with a hopper at

Fig. 10-8. Versatile power buggy can move all types of concrete over short distances. (54088)

178

Table 10-1. Methods and Equipment for Transporting and Handling Concrete

Equipment	Type and range of work for which equipment is best suited	Advantages	Points to watch for
Belt conveyors	For conveying concrete horizontally or to a higher or lower level. Usually positioned between main discharge point and secondary discharge point.	Belt conveyors have adjustable reach, traveling diverter, and variable speed both forward and reverse. Can place large volumes of concrete quickly when access is limited.	End-discharge arrangements needed to prevent segregation and leave no mortar on return belt. In adverse weather (hot, windy) long reaches of belt need cover.
Belt conveyors mounted on truck mixers	For conveying concrete to a lower, horizontal, or higher level.	Conveying equipment arrives with the concrete. Adjustable reach and variable speed.	End-discharge arrangements needed to prevent segregation and leave no mortar on return belt.
Buckets	Used with cranes, cableways, and helicopters for construction of buildings and dams. Convey concrete directly from central discharge point to formwork or to secondary discharge point.	Enables full versatility of cranes, cableways, and helicopters to be exploited. Clean discharge. Wide range of capacities.	Select bucket capacity to conform to size of the concrete batch and capacity of placing equipment. Discharge should be controllable.
Chutes on truck mixers	For conveying concrete to a lower level, usually below ground level, on all types of concrete construction.	Low cost and easy to maneuver. No power required; gravity does most of the work.	Slopes should range between 1 to 2 and 1 to 3 and chutes must be adequately supported in all positions. End-discharge arrangements (downpipe) needed to prevent segregation.
Cranes and buckets	The right equipment for work above ground level.	Can handle concrete, reinforcing steel, formwork, and sundry items in bridges and concrete-framed buildings.	Has only one hook. Careful scheduling between trades and operations is needed to keep crane busy.
Dropchutes	Used for placing concrete in vertical forms of all kinds. Some chutes are one piece tubes made of flexible rubberized canvas or plastic, others are assembled from articulated metal cylinders (elephant trunks).	Dropchutes direct concrete into formwork and carry it to bottom of forms without segregation. Their use avoids spillage of grout and concrete on reinforcing steel and form sides, which is harmful when off-the-form surfaces are specified. They also will prevent segregation of coarse particles.	Dropchutes should have sufficiently large, splayed-top openings into which concrete can be discharged without spillage. The cross section of dropchute should be chosen to permit inserting into the formwork without interfering with reinforcing steel.
Mobile batcher mixers	Used for intermittent production of concrete at jobsite, or where only small quantities are required.	A combined materials transporter and mobile batching and mixing system for quick, precise proportioning of specified concrete. One-man operation.	Trouble-free operation requires good preventive maintenance program on equipment. Materials must be identical to those in original mix design.
Nonagitating trucks	Used to transport concrete on short hauls over smooth roadways.	Capital cost of nonagitating equipment is lower than that of truck agitators or mixers.	Concrete slump should be limited. Possibility of segregation. Height is needed for high lift of truck body upon discharge.
Pneumatic guns (shotcrete)	Used where concrete is to be placed in difficult locations and where thin sections and large areas are needed.	Ideal for placing concrete in freeform shapes, for repairing structures, for protective coatings, thin linings, and building walls with one-sided forms.	Quality of work depends on skill of those using equipment. Only experienced nozzlemen should be employed.
Pumps	Used to convey concrete directly from central discharge point at jobsite to formwork or to secondary discharge point.	Pipelines take up little space and can be readily extended. Delivers concrete in continuous stream. Pump can move concrete both vertically and horizontally. Truck-mounted pumps can be delivered when necessary to small or large projects. Tower-crane mounted pump booms provide continuous concrete for tall building construction.	Constant supply of freshly-mixed concrete is needed with average consistency and without any tendency to segregate. Care must be taken in operating pipeline to ensure an even flow and to clean out at conclusion of each operation. Pumping vertically, around bends, and through flexible hose will considerably reduce the maximum pumping distance.

Table 10-1. Methods and Equipment for Transporting and Handling Concrete (Continued)

Equipment	Type and range of work for which equipment is best suited	Advantages	Points to watch for
Screw spreaders	Used for spreading concrete over large flat areas, such as in pavements and bridge decks.	With a screw spreader a batch of of concrete discharged from a bucket or truck can be quickly spread over a wide area to a uniform depth. The spread concrete has good uniformity of compaction before vibration is used for final compaction.	Screw spreaders are normally used as part of a paving train. They should be used for spreading before vibration is applied.
Tremies	For placing concrete under-water.	Can be used to funnel concrete down through the water into the foundation or other part of the structure being cast.	Precautions are needed to ensure that the tremie discharge end is always buried in fresh concrete, so that a seal is preserved between water and concrete mass. Diameter should be 250 to 300 mm unless pressure is available. Concrete mixture needs more cement, 390 kg/m³, and greater slump, 150 to 225 mm, because concrete must flow and consolidate without any vibration.
Truck agitators	Used to transport concrete for all uses in pavements, structures, and buildings. Haul distances must allow discharge of concrete within 2 hours, but limit may be waived under certain circumstances.	Truck agitators usually operate from central mixing plants where quality concrete is produced under controlled conditions. Discharge from agitators is well controlled. There is uniformity and homogeneity of concrete on discharge.	Timing of deliveries should suit job organization. Concrete crew and equipment must be ready onsite to handle concrete.
Truck mixers	Used to transport concrete for uses in pavements, structures, and buildings. Haul distances must allow discharge of concrete within 2 hours, but limit may be waived under certain circumstances.	No central mixing plant needed, only a batching plant, since concrete is completely mixed in truck mixer. Discharge is same as for truck agitator.	Timing of deliveries should suit job organization. Concrete crew and equipment must be ready onsite to handle concrete. Control of concrete quality is not as good as with central mixing.
Wheelbarrows and buggies	For short flat hauls on all types of onsite concrete construction, especially where accessibility to work area is restricted.	Very versatile and therefore ideal inside and on jobsites where placing conditions are constantly changing.	Slow and labour intensive.

Fig. 10-9. Ready mixed concrete can often be placed in its final location by direct chute discharge from a truck mixer. (54955)

Fig. 10-10. In comparison to conventional rear-discharge trucks, front-discharge truck mixers provide the driver with more mobility and control for direct discharge into place. (70006)

Fig. 10-11. Concrete is easily lifted to its final location by bucket and crane. (69687)

Fig. 10-12. The tower crane and bucket can easily handle concrete for tall-building construction. (69969)

Fig. 10-13. The conveyor belt is an efficient, portable method of handling concrete. A dropchute prevents concrete from segregating as it leaves the belt; a scraper prevents loss of mortar. Conveyor belts can be operated in series and on extendable booms of hydraulic cranes. (69896)

the top to which concrete was hauled by winch. A series of chutes suspended from the tower allowed the concrete to flow by gravity directly to the point required. As concrete-framed buildings became taller, the need to hoist reinforcement and formwork as well as concrete to higher levels led to the development of the tower crane—a familiar sight on the building skyline today (Fig. 10-12). It is fast and versatile, but the fact that it has only one hook must be considered when planning a job.

The conveyer belt is old in concept and much changed over the years (Fig. 10-13). Recently, truck-mixer-mounted conveyor belts have come into use (Fig. 10-14). The pneumatic process for shotcreting was patented in 1911 and is literally unchanged (see Chapter 18). The first mechanical concrete pump was developed and used in the 1930s and the hydraulic pump was developed in the 1950s. The advanced mobile pump with hydraulic placing boom (Fig. 10-15) is probably the single most important innovation in concrete handling equipment. It is economical to use in placing both large and small quantities of concrete, depending on jobsite conditions. For small to medium size projects, a combination of truck mixer and boom pump can be used to transport and place concrete. The screw spreader (Fig. 10-16) has been very effective in placing and distributing concrete for pavements. Screw spreaders can place a uniform depth of concrete quickly and efficiently. See Panarese (1987) for extensive information on methods to transport and handle concrete.

Fig. 10-14. A conveyor belt mounted on a truck mixer places concrete up to about 12 metres without the need for additional handling equipment. (53852)

Fig. 10-15. (top) A truck-mounted pump and boom can conveniently move concrete vertically or horizontally to the desired location. (bottom) View of concrete discharging from flexible hose connected to rigid pipeline leading from the pump. Rigid pipe is used in pump booms and in pipelines to move concrete over relatively long distances. Up to 8 m of flexible hose may be attached to the end of a rigid line to increase placement mobility. (69968, 69966)

Fig. 10-16. The screw spreader quickly spreads concrete over a wide area to a uniform depth. Screw spreaders are used primarily in pavement construction. (69895)

Choosing the Best Method

The first thing to look at is the type of job, its physical size, the total amount of concrete to be placed, and the time schedule. Studying the job details further will tell how much of the work is below, at, or above ground level. This aids in choosing the concrete handling equipment necessary for placing concrete at the required levels.

Concrete must be moved from the mixer to the point of placement as rapidly as possible without segregation or loss of ingredients. The transporting and handling equipment must have the capacity to move sufficient concrete so that cold joints are eliminated.

Work At and Below Ground Level

The largest volumes of concrete in a typical job usually are either below or at ground level and therefore can be placed by methods different from those employed on the superstructure. Concrete work below ground can vary enormously—from filling large-diameter bored piles or massive mat foundations to the intricate work involved in basement and subbasement walls. A crane can be used to handle formwork, reinforcing steel, and concrete. However, the crane may be fully employed erecting formwork and reinforcing steel in advance of the concrete, and other methods of handling the concrete may have to be used to place the largest volume in the least time.

Possibly the concrete can be chuted directly from the truck mixer to the point needed. Chutes should be metal or metal lined. They must not slope greater than 1 vertical to 2 horizontal or less than 1 vertical to 3 horizontal. Long chutes, over 6 metres, or those not meeting slope standards must discharge into a hopper before distribution to point of need.

Alternatively, a concrete pump can move the concrete to its final position (Fig. 10-15). Pumps must be of adequate capacity and capable of moving concrete without segregation. The loss of slump caused by pressure that forces mix water into the aggregates as the mix travels from pump hopper to discharge at the end of the pipeline must be minimal—not greater than 50 mm. The air content generally should not be reduced by more than 2 percentage points during pumping. Air loss greater than this may be caused by a boom configuration that allows the concrete to fall excessively. In view of this, specifications for both slump and air content should be met at the discharge end of the pump. Pipelines must not be made of aluminum or aluminum alloys to avoid excessive entrainment of air; aluminum reacts with cement alkali hydroxides to form hydrogen gas which can result in serious reduction in concrete strength.

Belt conveyors are very useful for work near ground level. Since placing concrete below ground is frequently a matter of horizontal movement assisted by gravity, lightweight portable conveyors can be used for high output at relatively low cost.

Work Above Ground Level

Conveyor belt, crane and bucket, hoist, pump, or the ultimate sky-hook, the helicopter, can be used for lifting concrete to locations above ground level (Fig. 10-17). The tower crane (Fig. 10-12) and pumping boom (Fig. 10-18) are the right tools for tall buildings. The volume of concrete needed per floor as well as boom placement and length affect the use of a pump; large volumes minimize pipeline movement in relation to output.

The specifications and performance of transporting and handling equipment are being continuously improved. The best results and lowest costs will be realized if the work is planned to get the most out of the equipment and if the equipment is flexibly employed to reduce total job cost. Any method is expensive if it does not get the job done. Panarese (1987) is very helpful in deciding which method to use based on capacity and range information for various methods and equipment.

Fig. 10-18. A pump boom mounted on a mast and located near the center of a structure can frequently reach all points of placement. It is especially applicable to tall buildings where tower cranes cannot be tied up with placing concrete. Concrete is supplied to the boom through a pipeline from a ground-level pump. Concrete can be pumped hundreds of metres vertically with these pumping methods. (49935)

Fig. 10-17. For work aboveground or at inaccessible sites, a concrete bucket can be lifted by helicopter. (Source: Paschel)

REFERENCES

ACI Committee 304, *Guide for Measuring, Mixing, Transporting, and Placing Concrete*, ACI 304R-00, ACI Committee 304 Report, American Concrete Institute, Farmington Hills, Michigan, 2000.

ACI Committee 304, *Placing Concrete by Pumping Methods*, ACI 304.2R-96, ACI Committee 304 Report, American Concrete Institute, Farmington Hills, Michigan, 1996.

ACI Committee 304, *Placing Concrete with Belt Conveyors*, ACI 304.4R-95, ACI Committee 304 Report, American Concrete Institute, Farmington Hills, Michigan, 1995.

ACI Committee 301, *Specifications for Structural Concrete*, ACI 301-99, ACI Committee 301 Report, American Concrete Institute, Farmington Hills, Michigan, 1999.

CSA Standard A23.1-00, *Concrete Materials and Methods of Concrete Construction*, Canadian Standards Association, Toronto, 2000.

Haney, James T., and Meyers, Rodney A., *Ready Mixed Concrete—Plant and Truck Mixer Operations and Quality Control,* NRMCA Publication No. 172, National Ready Mixed Concrete Association, Silver Spring, Maryland, May 1985.

Panarese, William C., *Transporting and Handling Concrete,* IS178, Portland Cement Association, 1987.

PCA, *Concrete for Small Jobs,* IS174, Portland Cement Association, http://www.portcement.org/pdf_files/IS174.pdf, 1988.

PCA, *Structural Lightweight Concrete,* IS032, Portland Cement Association, http://www.portcement.org/pdf_files/IS032.pdf, 1986.

CHAPTER 11
Placing and Finishing Concrete

PREPARATION BEFORE PLACING

Preparation prior to placing concrete for pavements or slabs on grade includes compacting, trimming, and moistening the subgrade (Figs. 11-1, 11-2, and 11-3); erecting the forms; and setting the reinforcing steel and other embedded items securely in place. Moistening the subgrade is important, especially in hot, dry weather to keep the dry subgrade from drawing too much water from the concrete; it also increases the immediate air-moisture level thereby decreasing the amount of evaporation from the concrete surface. The strength or bearing capacity of the subgrade should be adequate to support anticipated structural loads.

In cold weather, concrete must not be placed on a frozen subgrade. Snow, ice, and other debris must be removed from within the forms before concrete is placed. Where concrete is to be deposited on rock or hardened concrete, all loose material must be removed, and cut faces should be nearly vertical or horizontal rather than sloping.

Recently placed concrete requiring an overlay is usually roughened shortly after hardening to produce a better

Fig. 11-2. Water trucks with spray-bars are used to moisten subgrades and base course layers to achieve adequate compaction and to reduce the amount of water drawn out of concrete as it's placed. (69931)

Fig. 11-1. A base course foundation for concrete pavement is shaped by an auto-trimmer to design grades, cross section and alignment by automatic sensors that follow string lines. (69939)

Fig. 11-3. (top) Adequate compaction of a base course foundation for concrete pavement can be achieved by using a vibratory roller. (bottom) Vibratory plate compactors are also used to prepare subgrades under slabs. (69934, 69930)

bond with the next placement. As long as no laitance (a weak layer of concrete), dirt, or loose particles are present, newly hardened concrete requires little preparation prior to placing freshly mixed concrete on it. When in service for a period of time, old hardened concrete usually requires mechanical cleaning and roughening prior to placement of new concrete. The subject of placing freshly mixed concrete on hardened concrete is discussed in more detail under the sections entitled "Placing on Hardened Concrete" and "Construction Joints."

Forms should be accurately set, clean, tight, adequately braced, and constructed of or lined with materials that will impart the desired off-the-form finish to the hardened concrete. Wood forms, unless oiled or otherwise treated with a form-release agent, should be moistened before placing concrete, otherwise they will absorb water from the concrete and swell. Forms should be made for removal with minimum damage to the concrete. With wood forms, use of too large or too many nails should be avoided to facilitate removal and reduce damage. For architectural concrete, the form-release agent should be a nonstaining material. See Hurd (1979) and ACI Committee 347 (1997) for more information on formwork.

Reinforcing steel should be clean and free of loose rust or mill scale when concrete is placed. Unlike subgrades, reinforcing steel can be colder than 0°C with special considerations. See "Concreting Aboveground" in Chapter 14 for more details. Mortar splattered on reinforcing bars from previous placements need not be removed from steel and other embedded items if the next lift is to be completed within a few hours; loose, dried mortar, however, must be removed from items that will be encased by later lifts of concrete.

All equipment used to place concrete must be clean and in good working condition. Standby equipment should be available in the event of a breakdown.

DEPOSITING THE CONCRETE

Concrete should be deposited continuously as near as possible to its final position without objectionable segregation (Figs. 11-4, 11-5, 11-6, 11-7, and 11-8). In slab construction, placing should be started along the perimeter at one end of the work with each batch discharged against previously placed concrete. The concrete should not be dumped in separate piles and then leveled and worked together; nor should the concrete be deposited in large piles and moved horizontally into final position. Such practices result in segregation because mortar tends to flow ahead of the coarser material.

In general, concrete should be placed in walls, thick slabs, or foundations in horizontal layers of uniform thickness; each layer should be thoroughly consolidated before the next is placed. The rate of placement should be rapid enough so that previously placed concrete has not yet set when the next layer of concrete is placed upon it. Timely placement and adequate consolidation will prevent flow lines, seams, and planes of weakness (cold joints) that result

Fig. 11-4. Wheelbarrows are used to place concrete in areas that are not easily accessed by other placement methods. (69929)

Fig. 11-5. The swing arm on a conveyor belt allows fresh concrete to be placed fairly evenly across a deck. (70002)

Fig. 11-6. Dump trucks deposit concrete ahead of a slip-form paver that places the entire width of a street in one pass. Epoxy coated dowels on metal chairs are positioned at a joint and spiked down to the base course just ahead of the paver. (69936)

Fig. 11-7. Curb machines continuously extrude low-slump concrete into a shape that immediately stands without support of formwork. (69937)

Fig. 11-8. Concrete should be placed as near as possible to its final position. (70009)

from placing freshly mixed concrete on concrete past initial set. Layers should be about 150 to 500 mm thick for reinforced members and 375 to 500 mm thick for mass work; the thickness will depend on the width between forms and the amount of reinforcement.

To avoid segregation, concrete should not be moved horizontally over too long a distance as it is being placed in forms or slabs. In some work, such as placing concrete in sloping wingwalls or beneath window openings in walls, it is necessary to move the concrete horizontally within the forms, but this should be kept to a minimum.

Where standing water is present, concrete should be placed in a manner that displaces the water ahead of the concrete but does not allow water to be mixed in with the concrete; to do so will reduce the quality of the concrete. In all cases, water should be prevented from collecting at the ends, in corners, and along faces of forms. Care should be taken to avoid disturbing saturated subgrade soils so they maintain sufficient bearing capacity to support structural loads.

Chutes and dropchutes are used to move concrete to lower elevations without segregation and spattering of mortar on reinforcement and forms. Slopes of chutes (vertical to horizontal) should range between 1 to 2 and 1 to 3. Properly designed concrete has been allowed to drop by free fall into caissons. Results of a field test to determine if concrete could be dropped vertically 15 metres into a caisson without segregation proved that there was no significant difference in aggregate gradation between control samples as delivered and free-fall samples taken from the bottom of the caisson (Turner 1970). More recent field studies indicate that free fall of concrete from heights of up to 46 m—directly over reinforcing steel or at a high slump—does not result in segregation of the concrete ingredients nor reduce compressive strength (Suprenant 2001). However, if a baffle is not used to control the flow of concrete onto sloped surfaces at the end of an inclined chute, segregation can occur.

Concrete is sometimes placed through openings, called windows, in the sides of tall, narrow forms. When a chute discharges directly through the opening without controlling concrete flow at the end of the chute there is danger of segregation. A collecting hopper should be used outside the opening to permit the concrete to flow more smoothly through the opening; this will decrease the tendency to segregate.

When concrete is placed in tall forms at a fairly rapid rate, some bleed water may collect on the top surface, especially with non-air-entrained concrete. Bleeding can be reduced by placing more slowly and by using concrete of a stiffer consistency, particularly in the lower portion of the form. When practical, concrete should be placed to a level 300 mm to 400 mm below the top of tall forms and an hour or so allowed for the concrete to partially set. Placing should resume before the surface hardens to avoid formation of a cold joint. If practical to work around vertical reinforcing steel, it is good practice to overfill the form by 25 mm or so and cut off the excess concrete after it has stiffened and bleeding has ceased.

In monolithic placement of deep beams, walls, or columns, to avoid cracks between structural elements, a delay in concrete placement of up to 2 hours should be scheduled to allow settlement of the deep element before concreting is continued in any slabs, beams, or girders framing into them. The delay should be short enough however, to allow the next layer of concrete to knit with the previous layer by vibration, thus preventing cold joints and honeycombing (ACI Committee 304 1997). Haunches and column capitals are considered part of the floor or roof slab and should be placed integrally with them.

PLACING CONCRETE UNDERWATER

Concrete should be placed in the air rather than underwater whenever possible. When it must be placed underwater, the work should be done under experienced supervision. The basic principles for normal concrete work in the dry apply, with common sense, to underwater concreting. The following special points, however, should be observed:

Concrete should not be placed in water having a temperature below 5°C except when the strength gain of the concrete is sufficient when determined by special test specimens cured under identical conditions as the structure.

The slump of the concrete should be specified at 150 to 225 mm and the mixture should have a maximum water-cementing materials ratio of 0.45. Generally, the cementitious materials content will be 390 kg/m³ or more.

It is important that the concrete flow without segregation; therefore, the aim in proportioning should be to obtain a cohesive mixture with high workability. Antiwashout admixtures can be used to make concrete cohesive enough to be placed in limited depths of water, even without tremies. Using rounded aggregates, a higher percentage of fines, and entrained air should help to obtain the desired consistency.

The velocity of the current in the water through which the concrete is deposited should not exceed 3 m per minute.

Methods for placing concrete underwater include the following: tremie, concrete pump, bottom-dump buckets, grouting preplaced aggregate, toggle bags, bagwork, and the diving bell.

A tremie is a smooth, straight pipe long enough to reach the lowest point to be concreted from a working platform above the water. The diameter of the tremie pipe should be at least 8 times the diameter of the maximum size of aggregate. A hopper to receive the concrete is attached to the top of the pipe. CSA Standard A23.1 recommends that the specified slump for the tremie method of placing be 170 mm. Another means of measuring tremie concrete consistency and other applications where high flowability is desired, is by the slump flow method. Slump flow is tested in accordance with the Standards of the Japanese Society of Civil Engineers. (For a commentary on the Slump Flow Test refer to Chapter 18 and Appendix K– CSA Standard A23.1). Tremie pipes should be capable of being raised vertically and positioned at a maximum of 6 metres apart. To maintain a watertight tremie it is recommended to keep the discharge end buried at least 0.3 m in the previously placed concrete, and while raising the tremie, maintain a seal below the rising top surface, forcing the concrete to flow in beneath it by pressure. Placing should be continuous with as little disturbance to the previously placed concrete as possible. The top surface should be kept as level as possible. See ACI Committee 304 (2000) for additional information.

Mobile concrete pumps with a variable radius boom makes easy work of placing concrete underwater. Because the flexible hose on a concrete pump is similar to a tremie, the same placement techniques apply.

With the grouting preplaced aggregate method, the forms are first filled with clean coarse aggregate, then the voids in the coarse aggregate are filled with a grout to produce concrete. Grouting preplaced aggregate has advantages when placing concrete in flowing water.

Concrete can be placed more quickly and economically than by conventional placement methods. However, the method is very specialized and should only be performed by qualified experienced personnel.

Sand bags half full of plastic concrete can be used for small jobs, filling gaps, or temporary work. The tied end should face away from the outside.

SPECIAL PLACING TECHNIQUES

Concrete may be placed by methods other than the usual cast-in-place method. These methods, such as shotcreting, are described in Chapter 18. No matter what method is used, the basics of mixing, placing, consolidating, and curing apply to all portland cement concretes.

CONSOLIDATING CONCRETE

Consolidation is the process of compacting fresh concrete; to mold it within the forms and around embedded items and reinforcement; and to eliminate stone pockets, honeycomb, and entrapped air (Fig. 11-9). It should not remove significant amounts of intentionally entrained air in air-entrained concrete.

Consolidation is accomplished by hand or by mechanical methods. The method chosen depends on the consistency of the mixture and the placing conditions, such as complexity of the formwork and amount and spacing of reinforcement. Generally, mechanical methods using either internal or external vibration are the preferred methods of consolidation.

Workable, flowing mixtures can be consolidated by hand rodding, that is, thrusting a tamping rod or other suitable tool repeatedly into the concrete. The rod should be long enough to reach the bottom of the form or lift and thin enough to easily pass between the reinforcing steel and the forms. Low-slump concrete can be transformed

Fig. 11-9. Honeycomb and rock pockets are the results of inadequate consolidation. (50207)

into flowing concrete for easier consolidation through the use of superplasticizers without the addition of water to the concrete mixture.

Spading can be used to improve the appearance of formed surfaces. A flat, spadelike tool should be repeatedly inserted and withdrawn adjacent to the form. This forces the larger coarse aggregates away from the forms and assists entrapped air voids in their upward movement toward the top surface where they can escape. A mixture designed to be readily consolidated by hand methods should not be consolidated by mechanical methods; otherwise, the concrete is likely to segregate under intense mechanical action.

Even in highly reinforced elements, proper mechanical consolidation makes possible the placement of stiff mixtures with the low water-cementing materials ratios and high coarse-aggregate contents associated with high-quality concrete (Fig. 11-10). Among the mechanical methods are centrifugation, used to consolidate moderate-to-high-slump concrete in making pipes, poles, and piles; shock or drop tables, used to compact very stiff low-slump concrete in the manufacture of architectural precast units; and vibration—internal and external.

Fig. 11-10. Proper vibration makes possible the placement of stiff concrete mixtures, even in heavily-reinforced concrete members. (55806)

Vibration

Vibration, either internal or external, is the most widely used method for consolidating concrete. When concrete is vibrated, the internal friction between the aggregate particles is temporarily destroyed and the concrete behaves like a liquid; it settles in the forms under the action of gravity and the large entrapped air voids rise more easily to the surface. Internal friction is reestablished as soon as vibration stops.

Vibrators, whether internal or external, are usually characterized by their frequency of vibration, expressed as

the number of vibrations per second, hertz (Hz), or vibrations per minute (vpm); they are also designated by the amplitude of vibration, which is the deviation in millimetres from the point of rest. The frequency of vibration can be measured using a vibrating reed tachometer.

When vibration is used to consolidate concrete, a standby vibrator should be on hand at all times in the event of a mechanical breakdown.

Internal Vibration. Internal or immersion-type vibrators, often called spud or poker vibrators (Figs. 11-10 and 11-11), are commonly used to consolidate concrete in walls, columns, beams, and slabs. Flexible-shaft vibrators consist of a vibrating head connected to a driving motor by a flexible shaft. Inside the head, an unbalanced weight connected to the shaft rotates at high speed, causing the head to revolve in a circular orbit. The motor can be powered by electricity, gasoline, or air. The vibrating head is usually cylindrical with a diameter ranging from 20 to 180 mm. Some vibrators have an electric motor built right into the head, which is generally at least 50 mm in diameter. The dimensions of the vibrator head as well as its frequency and amplitude in conjunction with the workability of the mixture affect the performance of a vibrator.

Small-diameter vibrators have high frequencies ranging from 170 to 250 Hz (9,000 to 15,000 vpm) and low amplitudes ranging between 0.4 and 0.8 mm. As the diameter of the head increases, the frequency decreases and the amplitude increases. The effective radius of action of a vibrator increases with increasing diameter. Vibrators with a diameter of 20 to 40 mm have a radius of action in freshly mixed concrete ranging between 75 and 150 mm whereas the radius of action for vibrators of 50- to 75-mm diameter ranges between 175 and 350 mm. Table 11-1 shows the range of characteristics and applications for internal vibrators for various applications.

Proper use of internal vibrators is important for best results. Vibrators should not be used to move concrete hor-

Fig. 11-11. Internal vibrators are commonly used to consolidate concrete in walls, columns, beams, and slabs. (69970)

Table 11-1. Range of Characteristics, Performance, and Applications of Internal* Vibrators

Group	Diameter of head, mm	Recommended frequency, vibrations per minute**	Suggested values of			Approximate values of		Application
			Eccentric moment, mm-kg**	Average amplitude, mm	Centrifugal force, kg	Radius of action,† mm	Rate of concrete placement, m³/h‡	
1	20-40	9000-15,000	3.5-12	0.4-0.8	45-180	75-150	1-4	Plastic and flowing concrete in very thin members and confined places. May be used to supplement larger vibrators, especially in prestressed work where cables and ducts cause congestion in forms. Also used for fabricating laboratory test specimens.
2	30-60	8500-12,500	9-29	0.5-1.0	140-400	125-250	2-8	Plastic concrete in thin walls, columns, beams, precast piles, thin slabs, and along construction joints. May be used to supplement larger vibrators in confined areas.
3	50-90	8000-12,000	23-81	0.6-1.3	320-900	175-350	5-15	Stiff plastic concrete (less than 80-mm slump) in general construction such as walls, columns, beams, pre-stressed piles, and heavy slabs. Auxiliary vibration adjacent to forms of mass concrete and pavements. May be gang mounted to provide full-width internal vibration of pavement slabs.
4	80-150	7000-10,500	8-290	0.8-1.5	680-1800	300-500	10-30	Mass and structural concrete up to 50-mm slump deposited in quantities up to 3 m³ in relatively open forms of heavy construction (powerhouses, heavy bridge piers, and foundations). Also used for auxiliary vibration in dam construction near forms and around embedded items and reinforcing steel.
5	130-150	5500-8500	260-400	1.0-2.0	1100-2700	400-600	20-40	Mass concrete in gravity dams, large piers, massive walls, etc. Two or more vibrators will be required to operate simultaneously to mix and consolidate quantities of concrete of 3 m³ or more deposited at one time into the form.

* Generally, extremely dry or very stiff concrete does not respond well to internal vibrators.
** While vibrator is operating in concrete.
† Distance over which concrete is fully consolidated.
‡ Assumes the insertion spacing is 1½ times the radius of action, and that vibrator operates two-thirds of time concrete is being placed. These ranges reflect not only the capability of the vibrator but also differences in workability of the mix, degree of deaeration desired, and other conditions experienced in construction.

Adapted from ACI 309.

izontally since this causes segregation. Whenever possible, the vibrator should be lowered vertically into the concrete at regularly spaced intervals and allowed to descend by gravity. It should penetrate to the bottom of the layer being placed and at least 150 mm into any previously placed layer. The height of each layer or lift should be about the length of the vibrator head or generally a maximum of 500 mm in regular formwork.

In thin slabs, the vibrator should be inserted at an angle or horizontally in order to keep the vibrator head completely immersed. However, the vibrator should not be dragged around randomly in the slab. For slabs on grade, the vibrator should not make contact with the subgrade. The distance between insertions should be about 1½ times the radius of action so that the area visibly affected by the vibrator overlaps the adjacent previously vibrated area by a few millimetres.

The vibrator should be held stationary until adequate consolidation is attained and then slowly withdrawn. An insertion time of 5 to 15 seconds will usually provide adequate consolidation. The concrete should move to fill the hole left by the vibrator on withdrawal. If the hole does not refill, reinsertion of the vibrator at a nearby point should solve the problem.

Adequacy of internal vibration is judged by experience and by changes in the surface appearance of the concrete. Changes to watch for are the embedment of large aggregate particles, general batch leveling, the appearance of a thin film of mortar on the top surface, and the cessation of large bubbles of entrapped air escaping at the surface. Internal vibration may significantly affect the entrained-air-void system in concrete (Stark 1986, and Hover 2001). Detailed guidance for proper vibration should be followed (ACI Committee 309).

Allowing a vibrator to remain immersed in concrete after paste accumulates over the head is bad practice and can result in nonuniformity. The length of time that a vibrator should be left in the concrete will depend on the workability of the concrete, the power of the vibrator, and the nature of the section being consolidated.

In heavily-reinforced sections where an internal vibrator cannot be inserted, it is sometimes helpful to vibrate the reinforcing bars by attaching a form vibrator to their exposed portions. This practice eliminates air and water trapped under the reinforcing bars and increases bond between the bars and surrounding concrete; use this method only if the concrete is still workable under the action of vibration. Internal vibrators should not be attached to reinforcing bars for this purpose because the vibrators may be damaged.

Revibration of previously compacted concrete can be done to both fresh concrete as well as any underlying layer that has partially hardened. Revibration is used to improve bond between concrete and reinforcing steel, release water trapped under horizontal reinforcing bars, and remove additional entrapped air voids. In general, if concrete becomes workable under revibration, the practice is not harmful and may be beneficial.

External Vibration. External vibrators can be form vibrators, vibrating tables, or surface vibrators such as vibratory screeds, plate vibrators, vibratory roller screeds, or vibratory hand floats or trowels. Form vibrators, designed to be securely attached to the outside of the forms, are especially useful (1) for consolidating concrete in members that are very thin or congested with reinforcement, (2) to supplement internal vibration, and (3) for stiff mixes where internal vibrators cannot be used.

Attaching a form vibrator directly to the form generally is unsatisfactory. Rather, the vibrator should be attached to a steel plate that in turn is attached to steel I-beams or channels passing through the form stiffeners themselves in a continuous run. Loose attachments can result in significant vibration energy losses and inadequate consolidation.

Form vibrators can be either electrically or pneumatically operated. They should be spaced to distribute the intensity of vibration uniformly over the form; optimum spacing is best found by experimentation. Sometimes it may be necessary to operate some of the form vibrators at a different frequency for better results; therefore, it is recommended that form vibrators be equipped with controls to regulate their frequency and amplitude. Duration of external vibration is considerably longer than for internal vibration—generally between 1 and 2 minutes. A reed tachometer can not only determine frequency of vibration, but also give a rough estimate of amplitude of vibration by noting the oscillation of the reed at various points along the forms. This will assist in identifying dead spots or weak areas of vibration. A vibrograph

could be used if more reliable measurements of frequency and amplitude are needed.

Form vibrators should not be applied within the top metre of vertical forms. Vibration of the top of the form, particularly if the form is thin or inadequately stiffened, causes an in-and-out movement that can create a gap between the concrete and the form. Internal vibrators are recommended for use in this area of vertical forms.

Vibrating tables are used in precasting plants. They should be equipped with controls so that the frequency and amplitude can be varied according to the size of the element to be cast and the consistency of the concrete. Stiffer mixtures generally require lower frequencies (below 6000 vpm) and higher amplitudes (over 0.13 mm) than more workable mixtures. Increasing the frequency and decreasing the amplitude as vibration progresses will improve consolidation.

Surface vibrators, such as vibratory screeds (Figs. 11-12, 11-13, and 11-14), are used to consolidate concrete in floors and other flatwork. Vibratory screeds give positive control of the strikeoff operation and save a great deal of labour. When using this equipment, concrete need not have slumps in excess of 75 mm. For greater than 75 mm slumps, care should be taken because surface vibration of such concrete will result in an excessive accumulation of mortar and fine material on the surface; this may reduce wear resistance. For the same reason, surface vibrators should not be operated after the concrete has been adequately consolidated.

Because surface vibration of concrete slabs is least effective along the edges, a spud or poker-type vibrator should be used along the edge forms immediately before the vibratory screed is applied.

Vibratory screeds are used for consolidating slabs up to 250 mm thick, provided such slabs are nonreinforced or

Fig. 11-12. Vibratory screeds such as this truss-type unit reduce the work of strikeoff while consolidating the concrete. (55801)

Fig. 11-13. Where floor tolerances are not critical, an experienced operator using this vibratory screed does not need screed poles supported by chairs to guide the screed. Instead, he visually matches elevations to forms or previous passes. The process is called wet screeding. (69938)

Fig. 11-14. A laser level stimulating the sensors on this screed guides the operator as he strikes off the concrete. Screed poles and chairs are not needed and fewer workers are required to place concrete. Laser screeds interfaced with total station surveying equipment can also strike off sloped concrete surfaces. (69939)

only lightly reinforced (welded-wire fabric). Internal vibration or a combination of internal and surface vibration is recommended for reinforced slabs. More detailed information regarding internal and external vibration of concrete can be obtained from ACI Committee 309.

Consequences of Improper Vibration. Following are some of the worst defects caused by undervibration: (1) honeycomb; (2) excessive amount of entrapped air voids, often called bugholes; (3) sand streaks; (4) cold joints; (5) placement lines; and (6) subsidence cracking.

Honeycomb results when the spaces between coarse aggregate particles do not become filled with mortar.

Faulty equipment, improper placement procedures, a concrete mix containing too much coarse aggregate, or congested reinforcement can cause honeycomb.

Excessive entrapped air voids are similar to, but not as severe as honeycomb. Vibratory equipment and operating procedures are the primary causes of excessive entrapped air voids, but the other causes of honeycomb apply too.

Sand streaks results when heavy bleeding washes mortar out from along the form. A wet, harsh mixture that lacks workability because of an insufficient amount of mortar or fine aggregate may cause sand streaking. Segregation from striking reinforcing steel without adequate vibration may also contribute to streaking.

Cold joints are a discontinuity resulting from a delay in placement that allowed one layer to harden before the adjacent concrete was placed. The discontinuity can reduce the structural integrity of a concrete member if the successive lifts did not properly bond together. The concrete can be kept alive by revibrating it every 15 minutes or less depending on job conditions. However, once the time of initial setting approaches, vibration should be discontinued and the surface should be suitably prepared for the additional concrete.

Placement lines or "pour" lines are dark lines between adjacent placements of concrete batches. They may occur if, while vibrating the overlying layer, the vibrator did not penetrate the underlying layer enough to knit the layers together.

Subsidence cracking may occur at or near the initial setting time as concrete settles over reinforcing steel in relatively deep elements that have not been adequately vibrated. Revibration at the latest time that the vibrator will sink into the concrete under its own weight may eliminate these cracks.

Defects from overvibration include: (1) segregation as vibration and gravity causes heavier aggregates to settle while lighter aggregates rise; (2) sand streaks; (3) loss of entrained air in air-entrained concrete; (4) excessive form deflections or form damage; and (5) form failure caused by excessive pressure from vibrating the same location too long and/or placing concrete more quickly than the designed rate of pour.

Undervibration is more often a problem than overvibration.

CONCRETE SLABS

Concrete slabs can be finished in many ways, depending on the intended service use. Various colours and textures may be called for, such as an exposed-aggregate or a pattern-stamped surface. Some surfaces may require only strikeoff and screeding to proper contour and elevation, while for other surfaces a broomed, floated, or troweled finish may be specified. Details are given in ACI Committee 302, Kosmatka (1991), Panarese (1995), PCA (1980a), and Farny (2001).

The mixing, transporting, and handling of concrete for slabs should be carefully coordinated with the finishing operations. Concrete should not be placed on the subgrade or into forms more rapidly than it can be spread, struck off, consolidated, and bullfloated or darbied. In fact, concrete should not be spread over too large an area before strikeoff, nor should a large area be struck off and bleed water allowed to accumulate before bullfloating or darbying.

Finishing crews should be large enough to correctly place, finish, and cure concrete slabs with due regard for the effects of concrete temperature and atmospheric conditions on the setting time of the concrete and the size of the placement to be completed.

Subgrade Preparation

Cracks, slab settlement, and structural failure can often be traced to an inadequately prepared and poorly compacted subgrade. The subgrade on which a slab on ground is to be placed should be well drained, of uniform bearing capacity, level or properly sloped, and free of sod, organic matter, and frost. The three major causes of nonuniform support are: (1) the presence of soft unstable saturated soils or hard rocky soils, (2) backfilling without adequate compaction, and (3) expansive soils. Uniform support cannot be achieved by merely dumping granular material on a soft area. To prevent bridging and settlement cracking, soft or saturated soil areas and hard spots (rocks) should be dug out and filled with soil similar to the surrounding subgrade or when not available, with granular material such as sand, gravel, or crushed stone. All fill materials must be compacted to provide the same uniform support as the rest of the subgrade. Proof rolling the subgrade using a fully-loaded dump truck or similar heavy equipment is commonly used to identify areas of unstable soils that need additional attention.

During subgrade preparation, it should be remembered that undisturbed soil is generally superior to compacted material for supporting concrete slabs. Expansive, compressible, and potentially troublesome soils should be evaluated by a geotechnical engineer; a special slab design may be required.

The subgrade should be moistened with water in advance of placing concrete, but should not contain puddles or wet, soft, muddy spots when the concrete is placed.

Subbase

A satisfactory slab on ground can be built without a subbase. However, a subbase is frequently placed on the subgrade as a leveling course to equalize minor surface irregularities, enhance uniformity of support, bring the site to the desired grade, and serve as a capillary break between the slab and the subgrade.

Where a subbase is used, the contractor should place and compact to near maximum density a 100-mm thick layer of granular material such as sand, gravel, crushed stone, or slag. If a thicker subbase is needed for achieving the desired grade, the material should be compacted in thin layers about 100 mm deep unless tests determine compaction of thicker a lift is possible (Fig. 11-15). Subgrades and subbases can be compacted with small plate vibrators, vibratory rollers, or hand tampers. Unless the subbase is well compacted, it is better not to use a subbase; simply leave the subgrade uncovered and undisturbed.

Fig. 11-15. Nuclear gauges containing radioactive sources used to measure soil density and moisture can determine if a subbase has been adequately compacted. (69932)

Vapor Retarders and Moisture-Problem Prevention

Many of the moisture problems associated with enclosed slabs on ground (floors) can be minimized or eliminated by: (1) sloping the landscape away from buildings; (2) using a 100-mm thick granular subbase to form a capillary break between the soil and the slab; (3) providing drainage for the granular subbase to prevent water from collecting under the slab; (4) installing foundation drain tile; and (5) by installing a vapour retarder, often polyethylene sheeting.

For years vapour retarders have been mistakenly called vapour barriers. A vapour retarder slows the movement of water vapour by use of a 0.15 to 0.25 mm polyethylene film that is overlapped approximately 150 mm at the edges. A vapour retarder does not stop 100% of vapour migration; a vapour barrier does. Vapour barriers are thick, rugged multiple-ply-reinforced membranes that are sealed at the edges. Vapour retarders are discussed in this text because they are more commonly used; but many of the following principles apply to vapour barriers as well.

A vapour retarder should be placed under all concrete floors on ground that are likely to receive an impermeable floor covering such as sheet vinyl tile or be used for any purpose where the passage of water vapour through the floor might damage moisture-sensitive equipment or materials in contact with the floor. However, a few project sites with deep groundwater tables and sandy soils containing very low silt or clay contents may not require the use of a vapour retarder under concrete slabs.

Vapor retarders placed directly under concrete slabs may increase the time delay before final finishing due to longer bleeding times, particularly in cold weather. To minimize this effect, a minimum 75-mm thick layer of approved granular, self-draining compactible subbase material should be placed over the vapour retarder (or insulation if present) (ACI Committee 302). Some contractors find only 75 mm of granular sand over polyethylene sheeting to be slippery, somewhat dangerous, and difficult to keep in place while concreting. A 150 to 200 mm-thick subbase will alleviate this problem. The subbase over a vapour retarder must be kept from getting saturated by rain or construction activities to prevent excessive vapour migration after the concrete slab is placed.

If concrete is placed directly on a vapour retarder, the water-cementing materials ratio should be kept low (0.45 or less) because excess mix water can only escape to the surface as bleed water. Because of a longer bleeding period, settlement cracking over reinforcement and shrinkage cracking is more likely. For more information see ACI (2001) and ACI Committee 302.

Good quality, well-consolidated concrete at least 100-mm thick is practically impermeable to the passage of liquid water unless the water is under considerable pressure; however, such concrete—even concrete several times as thick—is not impermeable to the passage of water vapour.

Water vapour that passes through a concrete slab evaporates at the surface if it is not sealed. Floor coverings such as linoleum, vinyl tile, carpeting, wood, and synthetic surfacing effectively seal the moisture within the slab; eventually this moisture may deteriorate latex adhesives causing the floor covering to loosen, buckle, or blister.

To prevent problems with floor covering materials caused by moisture within the concrete, the following steps should be taken: (1) use a low water-cementing materials ratio concrete; (2) moist-cure the slab for 7 days; (3) allow the slab a 2-or-more-month drying period (Hedenblad 1997 and 1998); and (4) test the slab moisture condition before installing the floor covering.

In one commonly used test (ASTM F 1869), the moisture vapour emission rate from a concrete slab is determined by taping a domed plastic vapour barrier with a desiccant (drying substance) under it to the floor. After about 72 hours the mass of the desiccant is determined and the moisture vapour emission rate is calculated. The slab is considered dry enough for placing a flooring material if the moisture vapour emission rate is below either 1.4 or 2.3 $kg/m^2/1000\ m^2$ depending on the type of floor covering to be installed. Flooring-material manufacturers often have their own recommended test and specified moisture limits for installing their product. For more information and additional tests for water vapour transmission, see "Moisture Testing" in Chapter 16, Kosmatka (1985), and PCA (2000).

Insulation is sometimes installed over the vapour retarder to assist in keeping the temperature of a concrete floor above the dew point; this helps prevent moisture in the air from condensing on the slab surface. This practice also creates a warm floor for thermal comfort. Codes and specifications often require insulation at the perimeter of a floor slab. Placing insulation under the entire slab on ground for energy conservation alone usually cannot be justified economically. For more details, see PCA (1985).

Forms

Edge forms and intermediate screeds should be set accurately and firmly to the specified elevation and contour for the finished surface. Slab edge forms are usually metal or wood braced firmly with wood or steel stakes to keep them in horizontal and vertical alignment. The forms should be straight and free from warping and have sufficient strength to resist concrete pressure without bulging. They should also be strong enough to support any mechanical placing and finishing equipment used.

Rain Protection

Prior to commencing placing of concrete, the owner and contractor should be aware of procedures to be followed in the event of rain during the placing operation. Protective coverings such as polyethylene sheets or tarpaulins should be available and onsite at all times. When rain occurs, all batching and placing operations should stop and the fresh concrete should be covered to the extent that the rain does not indent the surface of the concrete or wash away the cement paste. When rain ceases, the covering should be removed and remedial measures taken such as surface retexturing or reworking in-place plastic concrete, before concrete placing resumes.

Placing and Spreading

Placement should start at the far point of a slab and proceed toward the concrete supply source. The concrete, which should be placed as close as possible to its final position, should slightly overfill the forms and be roughly leveled with square ended shovels or concrete rakes. Large voids trapped in the concrete during placing should be removed by consolidation.

Screeding (Strikeoff)

Screeding or strikeoff is the process of cutting off excess concrete to bring the top surface of a slab to proper grade. The template used in the manual method is called a straightedge although the lower edge may be straight or slightly curved, depending on the surface specified. It should be moved across the concrete with a sawing motion while advancing forward a short distance with each movement. There should be a surplus (surcharge) of concrete against the front face of the straightedge to fill in low areas as the straightedge passes over the slab. A 150-mm slab needs a surcharge of about 25 mm. Straightedges are sometimes equipped with vibrators that consolidate the concrete and assist in reducing the strikeoff work. This combination of straightedge and vibrator is called a vibratory screed (Fig. 11-12). Vibratory screeds are discussed earlier in this chapter under "Consolidating Concrete." Screeding, consolidation, and bullfloating must be completed before excess bleed water collects on the surface.

Bullfloating or Darbying

To eliminate high and low spots and to embed large aggregate particles, a bullfloat or darby (Fig. 11-16 top) should be used immediately after strikeoff. The long-handle bullfloat (Fig. 11-16 bottom) is used on areas too large to reach with a short-handle darby. Highway straightedges are often used to obtain very flat surfaces (Fig. 11-17). For non-air-entrained concrete, these tools can be made of wood; for air-entrained concrete they should be of aluminum or magnesium alloy.

Bullfloating or darbying must be completed before bleed water accumulates on the surface. Care must be taken not to overwork the concrete as this could result in a less durable surface.

The preceeding operations should level, shape, and smooth the surface and work up a slight amount of cement paste. Although sometimes no further finishing is required, on most slabs bullfloating or darbying is followed by one or more of the following finishing operations: edging, jointing, floating, troweling, and brooming. A slight hardening of the concrete is necessary before the start of any of these finishing operations. When the bleed-water sheen has evaporated and the concrete will sustain foot pressure with only about 6-mm indentation, the surface is ready for continued finishing operations (Fig. 11-18).

Warning: One of the principal causes of surface defects in concrete slabs is finishing while bleed water is on the surface. If bleed water is worked into the surface, the water-cementing materials ratio is significantly increased which reduces strength, entrained-air content, and watertightness of the surface. Any finishing operation performed on the surface of a concrete slab while bleed water is present can cause crazing, dusting, or scaling (PCA 2001). Floating and troweling the concrete (discussed later) before the bleeding process is completed may also trap bleed water under the finished surface producing a weakened zone or void under the finished surface; this occasionally results in delaminations. The use of low-slump concrete with an adequate cementing materials content and properly graded fine aggregate will minimize

Fig. 11-16. (top) Darbying brings the surface to the specified level and is done in tight places where a bullfloat cannot reach. (bottom) Bullfloating must be completed before any bleed water accumulates on the surface. (70010, 69940, 70011)

Fig. 11-17. Highway straightedges are used on highway pavement and floor construction where very flat surfaces are desired. (69941)

Fig. 11-18. Power floating using walk-behind and ride-on equipment. Footprints indicate proper timing. When the bleedwater sheen has evaporated and the concrete will sustain foot pressure with only slight indentation, the surface is ready for floating and final finishing operations. (69942)

bleeding and help ensure maintenance-free slabs. For exterior slabs, air entrainment also reduces bleeding. ACI Committee 302 and Farny (2001) present the placing and finishing techniques in more detail and PCA (2001) discusses defects.

Edging and Jointing

Edging is required along all edge forms and isolation and construction joints in floors and outdoor slabs such as walks, driveways, and patios. Edging densifies and compacts concrete next to the form where floating is less effective, making it more durable and less vulnerable to scaling, chipping, and popouts.

In the edging operation, the concrete should be cut away from the forms to a depth of 25 mm using a pointed mason trowel or a margin trowel. Then an edger should be held almost flat on the surface and run with the front of the tool slightly raised to prevent the edger from leaving too deep an impression. Edging may be required after each subsequent finishing operation for interior slabs.

Proper jointing practices can eliminate unsightly random cracks. Contraction joints, also called control joints, are made with a hand groover or by inserting strips of plastic, wood, metal, or preformed joint material into the unhardened concrete. When hand methods are used to form control joints in exterior concrete slabs, mark the forms to accurately locate the joints. Prior to bullfloating, the edge of a thin strip of wood or metal may be used to knock down the coarse aggregate where the joint will be hand tooled. The slab should then be jointed immediately after bullfloating or in conjunction with the edging operation. Control joints also can be made in hardened concrete by sawing. Jointing is discussed further under the heading "Making Joints in Floors and Walls" later in this chapter.

Floating

After the concrete has been hand-edged and hand-jointed, it should be floated with a metal or wood hand float or with a finishing machine using float blades (Fig. 11-18).

The purpose of floating is threefold: (1) to embed aggregate particles just beneath the surface; (2) to remove slight imperfections, humps, and voids; and (3) to compact the mortar at the surface in preparation for additional finishing operations. The concrete should not be overworked as this may bring an excess of water and fine material to the surface and result in subsequent surface defects.

Hand floats usually are made of fibreglass, magnesium, or wood. The metal float reduces the amount of work required because drag is reduced as the float slides more readily over the concrete surface. A magnesium float is essential for hand-floating air-entrained concrete because a wood float tends to stick to and tear the concrete surface. The light metal float also forms a smoother surface than the wood float.

The hand float should be held flat on the concrete surface and moved with a slight sawing motion in a sweeping arc to fill in holes, cut off lumps, and smooth ridges. When finishing large slabs, power floats can be used to reduce finishing time.

Floating produces a relatively even (but not smooth) texture that has good slip resistance and is often used as a final finish, especially for exterior slabs. Where a float finish is the desired final finish, it may be necessary to float the surface a second time after it has hardened a little more.

Marks left by hand edgers and groovers are normally removed during floating unless the marks are desired for decorative purposes; in such cases the edger and groover should be used again after final floating.

Troweling

Where a smooth, hard, dense surface is desired, floating should be followed by steel troweling (Fig. 11-19). Troweling should not be done on a surface that has not been floated; troweling after only bullfloating or darbying is not an adequate finishing procedure.

It is customary when hand-finishing large slabs to float and immediately trowel an area before moving the kneeboards. These operations should be delayed until the concrete has hardened sufficiently so that water and fine material are not brought to the surface. Too long a delay, of course, will result in a surface that is too hard to float and trowel. The tendency, however, is to float and trowel the surface too soon. Premature floating and troweling can cause scaling, crazing, or dusting and a surface with reduced wear resistance.

Spreading dry cement on a wet surface to take up excess water is a bad practice and can cause crazing. Such wet spots should be avoided, if possible, by adjustments in aggregate gradation, mix proportions, and consistency. When wet spots do occur, finishing operations should be delayed until the water either evaporates or is

Fig. 11-19. Hand floating (right hand) the surface with a hand float held flat on the concrete surface and moved in a sweeping arc with a slight sawing motion. Troweling (left hand) with blade tilted is performed before moving the kneeboards. (69933)

removed with a rubber floor squeegee or by dragging a soft rubber garden hose across the slab surface. If a squeegee or hose is used, care must be taken so that excess cement paste is not removed with the water.

The first troweling may produce the desired surface free of defects. However, surface smoothness, density, and wear resistance can all be improved by additional trowelings. There should be a lapse of time between successive trowelings to permit the concrete to become harder. As the surface stiffens, each successive troweling should be made with smaller trowels, using progressively more tilt and pressure on the trowel blade. The final pass should make a ringing sound as the trowel moves over the hardening surface.

A power trowel is similar to a power float, except that the machine is fitted with smaller, individual steel trowel blades that are adjustable for tilt and pressure on the concrete surface. When the first troweling is done by machine, at least one additional troweling by hand should be done to remove small irregularities. If necessary, tooled edges and joints should be rerun after troweling to maintain uniformity and true lines.

Exterior concrete should not be troweled for several reasons: (1) because it can lead to a loss of entrained air caused by overworking the surface, and (2) troweled surfaces can be slippery when wet. Floating and brooming can provide the non-slip surface texture needed for exterior concrete slabs.

Brooming

Brooming should be performed before the concrete has thoroughly hardened, but it should be sufficiently hard to retain the scoring impression to produce a slip-resistant surface (Fig. 11-20). Rough scoring can be achieved with a rake, a steel-wire broom, or a stiff, coarse, fibre broom; such coarse-textured brooming usually follows floating. If a finer texture is desired, the concrete should be floated to a smooth surface and then brushed with a soft-bristled broom. Interior concrete could also be troweled before

brooming. Best results are obtained with brooms that are specially made for texturing concrete. Slabs are usually broomed transversely to the main direction of traffic.

Highway pavements are textured by "tining" the surface with stiff wires; this improves traction and provides vehicles with a surface that significantly reduces the chance of hydroplaning (Fig. 11-21).

Fig. 11-20. Brooming provides a slip-resistant surface mainly used on exterior concrete. (69943)

Fig. 11-21. (top) This machine is tining the surface of fresh concrete. (bottom) Tining of pavements improves tire traction and reduces the risk of hydroplaning. (69944, 69945)

Curing and Protection

All newly placed and finished concrete slabs should be cured and protected from drying, from extreme changes in temperature, and from damage by subsequent construction and traffic.

Curing should begin immediately after finishing (Fig. 11-22). Curing is needed to ensure continued hydration of the cement, strength gain of the concrete, and a minimum of early drying shrinkage.

Special precautions are necessary when concrete work continues during periods of adverse weather. In cold weather, arrangements should be made in advance for heating, covering, insulating, or enclosing the concrete. Hot-weather work may require special precautions against rapid evaporation and drying and high temperatures.

Fig. 11-22. An excellent method of wet curing is to completely cover the surface with wet burlap and keep it continuously wet during the curing period. (69946)

PLACING ON HARDENED CONCRETE

Bonded Construction Joints in Structural Concrete

A bonded construction joint is needed between two structural concrete placements. When freshly mixed concrete is placed in contact with existing hardened concrete, a high-quality bond and watertight joint are required. Poorly bonded construction joints are usually the result of (1) lack of bond between old and new concrete, or (2) a weak porous layer in the hardened concrete at the joint. The quality of a bonded joint therefore depends on the quality of the hardened concrete and preparation of its surface.

In columns and walls, the concrete near the top surface of a lift is often of inferior quality to the concrete below. This may be due to poor consolidation or use of badly pro-

portioned or high-slump mixtures that cause excessive laitance, bleeding, and segregation. Even in well-proportioned and carefully consolidated mixtures, some aggregate particle settlement and water gain (bleeding) at the top surface is unavoidable; this is particularly true with high rates of placement. Also, the encasing formwork prevents the escape of moisture from the fresh concrete. While formwork provides adequate curing as long as it remains in place, the top surface where there is no encasing formwork may dry out too rapidly; this may result in a weak porous layer unless protection and curing are provided.

Preparing Hardened Concrete

When freshly mixed concrete is placed on recently hardened concrete, certain precautions must be taken to secure a well-bonded, watertight joint. The hardened concrete must be clean, sound, and reasonably rough with some coarse aggregate particles exposed. Any laitance, soft mortar, dirt, wood chips, form oil, or other foreign materials must be removed since they could interfere with proper bonding of the subsequent placement.

The surface of old concrete upon which fresh concrete is to be placed must be thoroughly roughened and cleaned of all dust, surface films, deposits, loose particles, grease, oil, and other foreign material. In most cases it will be necessary to remove the entire surface down to sound concrete. Roughening and cleaning with lightweight chipping hammers, waterblasting, scarifiers, sandblasting (Fig. 11-23), shotblasting and hydrojetting are some of the satisfactory methods for exposing sound concrete. Care must

Fig. 11-23. Sandblasting can clean any size or shape surface – horizontal, vertical or overhead. Consult local environmental regulations regarding sandblasting. (55805)

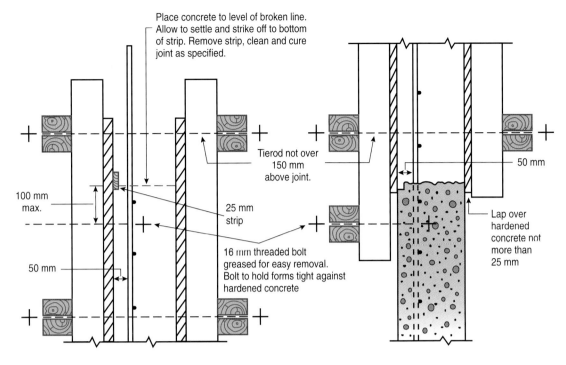

Fig. 11-24. A straight, horizontal construction joint can be built using this detail.

be taken to avoid contamination of the clean surface before a bonding grout and overlay concrete are placed.

Partially set or recently hardened concrete may only require stiff-wire brushing. In some types of construction such as dams, the surface of each concrete lift is cut with a high-velocity air-water jet to expose clean, sound concrete just before final set. This is usually done 4 to 12 hours after placing. The surface must then be protected and continuously cured until concreting is resumed for the next lift.

For two-course floors, the top surface of the base slab can be roughened with a steel or stiff fibre broom. The surface should be level, heavily scored, and free of laitance; then it should be protected until it is thoroughly cleaned just before the grout coat and topping mix are placed. When placing a bonded topping on a floor slab, the base slab should be cleaned of all laitance, dust, debris, grease or other foreign substances by using one of the following methods:

a. Wet- or dry-grit sandblasting
b. High-pressure water blasting
c. Mechanical removal by scabblers, or grinding wheels
d. Power brooming and vacuuming

Hardened concrete may be left dry or be moistened before new concrete is placed on it; however, the surface should not be wet or have any free-standing water. Laboratory studies indicate a slightly better bond is obtained on a dry surface than on a damp surface; however, the increased moisture level in the hardened concrete and in the environment around the concrete reduces water loss

from the concrete mixture. This can be very beneficial, especially on hot, dry days.

For making a horizontal construction joint in reinforced concrete wall construction, good results have been obtained by constructing the forms to the level of the joint, overfilling the forms a few millimetres, and then removing the excess concrete just before hardening occurs; the top surface then can be manually roughened with stiff brushes. The procedure is illustrated in Fig. 11-24.

In the case of vertical construction joints cast against a bulkhead, the concrete surface generally is too smooth to permit proper bonding. So, particular care should be given to removal of the smooth surface finish before reerecting the forms for placing freshly mixed concrete against the joint. Stiff-wire brushing may be sufficient if the concrete is less than three days old; otherwise, bushhammering or sandblasting may be needed, followed by washing with clean water to remove all dust and loose particles.

Bonding New to Previously Hardened Concrete

Care must be used when making horizontal construction joints in wall sections where freshly-mixed concrete is to be placed on hardened concrete. A good bond can be obtained by placing a rich concrete (higher cement and sand content than normal) in the bottom 150 mm of the new lift and thoroughly vibrating the joint interface. Alternatively, a cement-sand grout can be scrubbed into a clean surface immediately ahead of concreting.

A topping concrete mix for slabs can be bonded to the previously prepared base slab by one of the following procedures:

1. **Portland cement-sand grouting:** A 1 to 1 cement-sand grout having a water-cement ratio of not greater than 0.45, mixed to a creamlike consistency, is scrubbed into the prepared dry or damp (no free water) base slab surface.

2. **Latex.** A latex-bonding agent is added to the cement-sand grout and is spread in accordance with the latex manufacturer's direction.

3. **Epoxy.** An approved epoxy-bonding agent placed on the base concrete, prepared in accordance with the epoxy manufacturer's direction.

The bonding procedure should produce tensile bond strength with the base concrete in excess of 1.0 MPa.

Grout is placed just a short distance ahead of the overlay or top-course concrete (Fig. 11-25). This method may also be applicable to horizontal joints in walls. The grout should not be allowed to dry out prior to the overlay placement; otherwise, the dry grout may act as a poor surface for bonding. The surface of the base slab should have been prepared by one of the methods discussed previously. Overlays are discussed further under "Patching, Cleaning, and Finishing" later in this chapter.

Fig. 11-25. Application of a bonding grout just ahead of the overlay concrete. The grout must not dry out before the concrete is placed. (51995)

MAKING JOINTS IN FLOORS AND WALLS

The following three types of joints are common in concrete construction: isolation joints, contraction joints, and construction joints.

Isolation Joints

Isolation joints (Fig. 11-26) permit both horizontal and vertical differential movements at adjoining parts of a structure. They are used, for example, around the perimeter of a floor on ground, around columns, and around machine foundations to separate the slab from the more rigid parts of the structure.

Isolation-joint material (often called expansion-joint material) can be as thin as 6 mm or less, but 13-mm material is commonly used. Care should be taken to ensure that all the edges for the full depth of the slab are isolated from adjoining construction; otherwise cracking can occur.

Columns on separate footings are isolated from the floor slab either with a circular or square-shaped isolation joint. The square shape should be rotated to align its corners with control and construction joints.

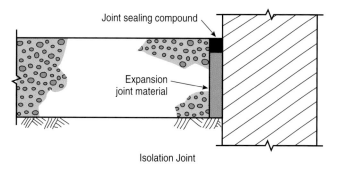

Isolation Joint

Fig. 11-26. Isolation joints permit horizontal and vertical movements between abutting faces of a slab and fixed parts of a structure.

Contraction Joints

Contraction joints (Fig. 11-27) provide for movement in the plane of a slab or wall and induce controlled cracking caused by drying and thermal shrinkage at preselected locations. Contraction joints (also sometimes called control joints) should be constructed to permit transfer of loads perpendicular to the plane of a slab or wall. If no contraction joints are used, or if they are too widely spaced in slabs on ground or in lightly reinforced walls, random cracks may occur; cracks are most likely when drying and thermal shrinkage produce tensile stresses in excess of the concrete's tensile strength.

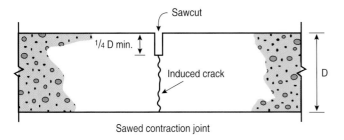

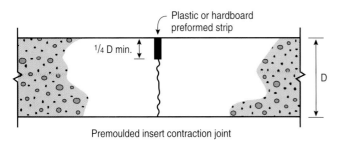

Fig. 11-27. Contraction joints provide for horizontal movement in the plane of a slab or wall and induce controlled cracking caused by drying and thermal shrinkage.

Contraction joints in slabs on ground can be made in several ways. One of the most common methods is to saw a continuous straight slot in the top of the slab (Fig. 11-28). This creates a plane of weakness in which a crack will form. Vertical loads are transmitted across a contraction joint by aggregate interlock between the opposite faces of the crack providing the crack is not too wide and the spacing between joints is not too great. Crack widths at saw-cut contraction joints that exceed 0.9 mm do not reliably transfer loads. The effectiveness of load transfer by aggregate interlock depends on more than crack width. Other factors include: slab thickness, subgrade support,

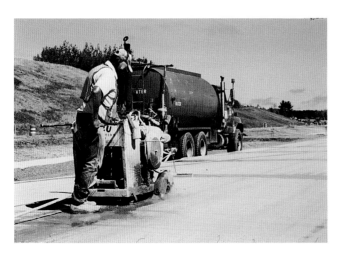

Fig. 11-28. Sawing a continuous cut in the top of a slab is one of the most economical methods for making a contraction joint. (69947)

load magnitude, repetitions of load, and aggregate angularity. Steel dowels (Figs. 11-6 and 11-29b) may be used to increase load transfer at contraction joints when heavy wheel loads are anticipated. They are placed at the centre of the slab and aligned perpendicular to the joint to permit horizontal movement. Dowel bars should be smooth and ½ of each bar coated with a bond breaker or sleeve so slippage can occur at one end. Sizes and spacing of dowels are shown in Farny (2001). See ACI Committee 302 and PCA (1982) for further discussions on doweled joints.

Sawing must be coordinated with the setting time of the concrete. It should be started as soon as the concrete has hardened sufficiently to prevent aggregates from being dislodged by the saw (usually within 4 to 12 hours after the concrete hardens); sawing should be completed before drying shrinkage stresses become large enough to produce cracking. The timing depends on factors such as mix proportions, ambient conditions, and type and hardness of aggregates. New dry-cut sawing techniques allow saw cutting to take place shortly after final finishing is completed. Generally, the slab should be cut before the concrete cools, when the concrete sets enough to prevent raveling or tearing while saw cutting, and before drying-shrinkage cracks start to develop.

Contraction joints also can be formed in the fresh concrete with hand groovers or by placing strips of wood, metal, or preformed joint material at the joint locations. The top of the strips should be flush with the concrete surface. Contraction joints, whether sawed, grooved, or preformed, should extend into the slab to a depth of at least *one-fourth* the slab thickness or a minimum of 25 mm deep. It is recommended that the joint depth not exceed one-third the slab thickness if load transfer from aggregate interlock is important.

Contraction joints in walls are also planes of weakness that permit movements in the plane of the wall. The thickness of the wall at a contraction joint should be reduced by 25%, preferably 30%. Under the guidance of the design engineer, in lightly reinforced walls, half of the horizontal steel rebars should be cut at the joint. Care must be taken to cut alternate bars precisely at the joint. At the corners of openings in walls where contraction joints are located, extra diagonal or vertical and horizontal reinforcement should be provided to control cracking. Contraction joints in walls should be spaced not more than about 6 metres apart. In addition, contraction joints should be placed where abrupt changes in wall thickness or height occur, and near corners—if possible, within 3 to 4 metres. Depending on the structure, these joints may need to be caulked to prevent the passage of water through the wall. Instead of caulking, a waterstop (or both) can be used to prevent water from leaking through the crack that occurs in the joint.

The spacing of contraction joints in floors on ground depends on (1) slab thickness, (2) shrinkage potential of the concrete, (3) subgrade friction, (4) environment, and (5) the

Table 11-2. Spacing of Contraction Joints in Metres*

Slab thickness, mm	Maximum-size aggregate less than 20 mm	Maximum-size aggregate 20 mm and larger
100	2.4	3.0
125	3.0	3.75
150	3.75	4.5
175	4.25	5.25**
200	5.0**	6.0**
225	5.5**	6.75**
250	6.0**	7.5**

* Spacings are appropriate for slumps between 100 mm and 150 mm. If concrete cools at an early age, shorter spacings may be needed to control random cracking. (A temperature difference of only 6°C may be critical.) For slumps less than 100 mm, joint spacing can be increased by 20%.

** When spacings exceed 4.5 m, load transfer by aggregate interlock decreases markedly.

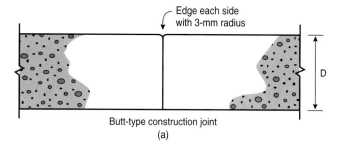

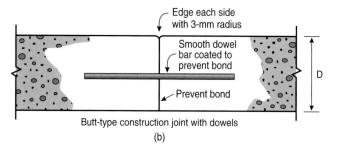

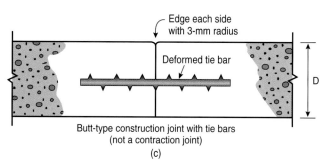

Fig. 11-29. Construction joints are stopping places in the process of construction. Construction-joint types (a) and (b) are also used as contraction joints.

absence or presence of steel reinforcement. Unless reliable data indicate that more widely spaced joints are feasible, the suggested intervals given in Table 11-2 should be used for well-proportioned concrete with aggregates having normal shrinkage characteristics. Joint spacing should be decreased for concrete suspected of having high shrinkage characteristics. The panels created by contraction joints should be approximately square. Panels with excessive length-to-width ratio (more than 1½ to 1) are likely to crack at an intermediate location. In joint layout design it is also important to remember that contraction (control) joints should only terminate at a free edge or at an isolation joint. Contraction joints should never terminate at another contraction joint as cracking will be induced from the end of the terminated joint into the adjacent panel. This is sometimes referred to as sympathetic cracking. Refer to Fig. 11-31, which illustrates one possible joint layout solution to eliminate the potential for induced sympathetic cracking.

Construction Joints

Construction joints (Fig. 11-29) are stopping places in the process of construction. A true construction joint should bond new concrete to existing concrete and permit no movement. Deformed tiebars are often used in construction joints to restrict movement. Because extra care is needed to make a true construction joint, they are usually designed and built to function as contraction or isolation joints. For example, in a floor on ground the construction joints align with columns and function as contraction joints and therefore are purposely made unbonded. The structural designer of suspended slabs should decide the location of construction joints. Oils, form-release agents, and paints are used as debonding materials. In thick, heavily-

loaded floors, unbonded doweled construction joints are commonly used. For thin slabs, the flat-faced butt-type joint will suffice.

On most structures it is desirable to have wall joints that will not detract from appearance. When properly made, wall joints can become inconspicuous or hidden by rustication strips. They thus can become an architectural as well as a functional feature of the structure. However, if rustication strips are used in structures that may be exposed to deicing salts, such as bridge columns and abutments, care should be taken to ensure that the reinforcing steel has the required depth of concrete cover to prevent corrosion.

Horizontal joints in walls should be made straight, exactly horizontal, and should be placed at suitable locations. A straight horizontal construction joint can be made by nailing a 25 mm wood strip to the inside face of the form near the top (see Fig. 11-24). Concrete should then be placed to a level slightly above the bottom of the strip. After the concrete has settled and before it becomes too hard, any laitance that has formed on the top surface should be removed. The strip can then be removed and any irregularities in the joint leveled off. The forms are removed and then reerected above the construction joint for the next lift of concrete. To prevent concrete leakage from staining the wall below, gaskets should be used where forms contact previously placed hardened concrete.

A variation of this procedure makes use of a rustication strip instead of the 25 mm wood strip to form a groove in the concrete for architectural effect (Fig. 11-30). Rustication strips can be V-shaped, rectangular, or slightly beveled. If V-shaped, the joint should be made at the point of the V. If rectangular or beveled, the joint should be made at the top edge of the inner face of the strip.

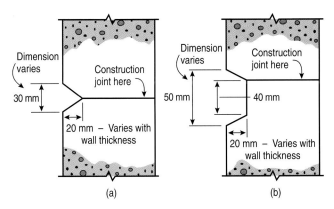

Fig. 11-30. Horizontal construction joints in walls with V-shaped (a) and beveled (b) rustication strips.

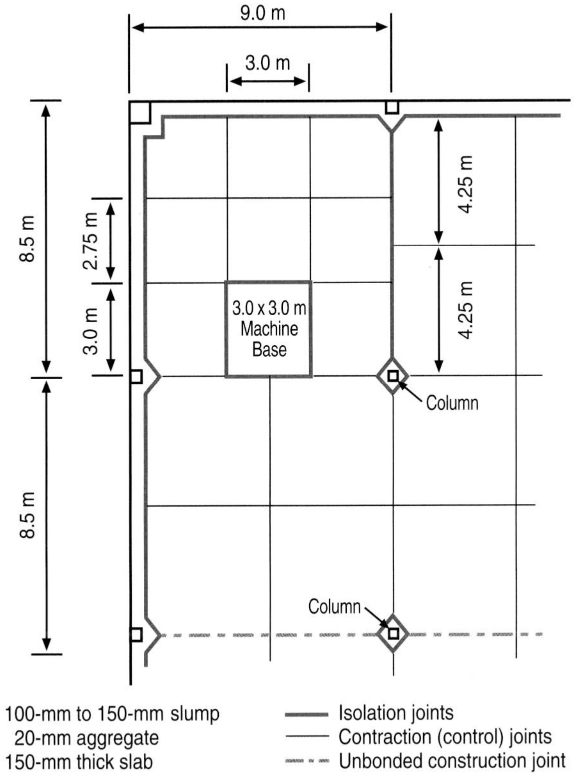

100-mm to 150-mm slump
20-mm aggregate
150-mm thick slab

—— Isolation joints
—— Contraction (control) joints
—-- Unbonded construction joint

Fig. 11-31. Typical joint layout for a 150-mm thick concrete floor on ground.

JOINT LAYOUT FOR FLOORS

A typical joint layout for all three joint types—isolation, contraction, and construction—is illustrated in Fig. 11-31. Isolation joints are provided around the perimeter of the floor where it abuts the walls and around all fixed elements that may restrain movement of the slab. This includes columns and machinery bases that penetrate the floor slab. With the slab isolated from other building elements, the remaining task is to locate and correctly space contraction joints to eliminate random cracking. Construction joint locations are coordinated with the floor contractor to accommodate work schedules and crew size. Unbonded construction joints should coincide with the contraction joint pattern and act as contraction joints. Construction joints should be planned to provide long-strips for each placement rather than a checker-board pattern. Contraction joints are then placed to divide the long-strips into relatively square panels, with panel length not exceeding 1.5 times the width. Contraction joints only stop at free edges or isolation joints. For more information on joints, see ACI Committee 302 (1996), PCA (1982), and Farny (2001). For joints in walls, see PCA (1982), PCA (1982a), PCA (1984), PCA (1984a), and PCA (1984b).

FILLING FLOOR JOINTS

There are three options for treating joints: they can be filled, sealed, or left open. The movement at contraction joints in a floor is generally very small. For some industrial and commercial uses, these joints can be left unfilled or unsealed. Where there are wet conditions, hygienic and dust-control requirements, or considerable traffic by small, hard-wheel vehicles, joint filling is necessary.

The difference between a filler and a sealer is the hardness of the material; fillers are more rigid than sealers and provide support to joint edges. In many places where traffic loading is light, a resilient material such as a polyurethane elastomeric sealant is satisfactory. However, heavy-traffic areas require support for joint edges to prevent spalling at saw-cuts; in such cases a good quality, semi-rigid epoxy or polyurea filler with a Shore Hardness of A-80 or D-50 (ASTM D 2240) should be used. The material should be installed full depth in the saw cut, without a backer rod, and flush with the floor surface.

Isolation joints are intended to accommodate movement; thus a flexible, elastomeric sealant should be used to keep foreign materials out of the joint.

UNJOINTED FLOORS

An unjointed floor, or one with a limited number of joints, can be constructed when joints are unacceptable. Three unjointed floor methods are suggested:

1. A prestressed floor can be built through the use of post-tensioning. With this method, steel strands in ducts are tensioned after the concrete hardens to produce compressive stress in the concrete. This compressive stress will counteract the development of tensile stresses in the concrete and provide a crack-free floor. Large areas, 1000 m² and more, can be constructed in this manner without intermediate joints.

2. Large areas—a single day of slab placement, usually 800 to 1000 m²—can be cast without contraction joints when the amount of distributed steel in the floor is about one-half of one percent of the cross-sectional area of the slab. Special effort should be made to reduce subgrade friction in floors without contraction joints. Farny (2001) discusses use of distributed steel in floors.

3. Where available, concrete made with expansive cement meeting the requirements of ASTM C 845 can be used to offset the amount of drying shrinkage to be anticipated after curing. Contraction joints are not needed when construction joints are used at intervals of 10 to 35 metres. Large areas, up to 2000 m², have been cast in this manner without joints. Steel reinforcement is needed in order to produce compressive stresses during and after the expansion period—this is a form of prestressing.

REMOVING FORMS

It is advantageous to leave forms in place as long as possible to continue the curing period. However, there are times when it is necessary to remove forms as soon as possible. For example, where a rubbed finish is specified, forms must be removed early to permit the first rubbing before the concrete becomes too hard. Furthermore, it is often necessary to remove forms quickly to permit their immediate reuse.

In any case, shoring should not be removed until the concrete is strong enough to satisfactorily carry the stresses from both the dead load of the structure and any imposed construction loads. The concrete should be hard enough so that the surfaces will not be injured in any way when reasonable care is used in removing forms. In general, for concrete temperatures above 10°C, the side forms of reasonably thick, supported sections can be removed 24 hours after concreting. Beam and floor slab forms and supports (shoring) may be removed between 3 and 21 days, depending on the size of the member and the strength gain of the concrete. For most conditions, it is better to rely on the strength of the concrete as determined by in situ or

field-cured specimen testing rather than arbitrarily selecting an age at which forms may be removed. Advice on reshoring is provided in CSA Standard S269.1, *Falsework for Construction Purposes*, and by ACI Committee 347.

For form removal, the designer should specify the minimum strength requirements for various members. The age-strength relationship should be determined from representative samples of concrete used in the structure and field-cured under job conditions. It should be remembered, however, that strengths are affected by the materials used, temperature, and other conditions. The time required for form removal, therefore, will vary from job to job.

A pinch bar or other metal tool should not be placed against the concrete to wedge forms loose. If it is necessary to wedge between the concrete and the form, only wooden wedges should be used. Stripping should be started some distance away from and move toward a projection. This relieves pressure against projecting corners and reduces the chance of edges breaking off.

Recessed forms require special attention. Wooden wedges should be gradually driven behind the form and the form should be tapped lightly to break it away from the concrete. Forms should not be pulled off rapidly after wedging has been started at one end; this is almost certain to break the edges of the concrete.

PATCHING, CLEANING, AND FINISHING

After forms are removed, all bulges, fins, and small projections can be removed by chipping or tooling. Undesired bolts, nails, ties, or other embedded metal can be removed or cut back to a depth of 13 mm from the concrete surface. When required, the surface can be rubbed or ground to provide a uniform appearance. Any cavities such as tierod holes should be filled unless they are intended for decorative purposes. Honeycombed areas must be repaired and stains removed to present a concrete surface that is uniform in colour. All of these operations can be minimized by exercising care in constructing the formwork and placing the concrete. In general, repairs are easier to make and more successful if they are made as soon as practical, preferably as soon as the forms are removed. However, the procedures discussed below apply to both new and old hardened concrete.

Holes, Defects, and Overlays

Patches usually appear darker than the surrounding concrete; therefore, some white cement should be used in mortar or concrete for patching where appearance is important. Samples should be applied and cured in an inconspicuous location, perhaps a basement wall, several days in advance of patching operations to determine the most suitable proportions of white and gray cements. Steel troweling should be avoided since this darkens the patch.

Bolt holes, tierod holes, and other cavities that are small in area but relatively deep should be filled with a dry-pack mortar. The mortar should be mixed as stiff as is practical: use 1 part cement, 2½ parts sand passing a 1.25 mm sieve, and just enough water to form a ball when the mortar is squeezed gently in the hand. The cavity should be clean with no oil or loose material and kept damp with water for several hours. A neat-cement paste should be scrubbed onto the void surfaces, but not allowed to dry before the mortar is placed. The mortar should be tamped into place in layers about 13 mm thick. Vigorous tamping and adequate curing will ensure good bond and minimum shrinkage of the patch.

Concrete used to fill large patches and thin-bonded overlays should have a low water-cementing materials ratio, often with a cementing materials content equal to or greater than the concrete to be repaired. Cementing materials contents range from 360 to 500 kg per cubic metre and the water-cementing materials ratio is usually 0.45 or less. The aggregate size should be no more than ⅓ the patch or overlay thickness. A 10-mm nominal maximum size coarse aggregate is commonly used. The sand proportion can be higher than usual, often equal to the amount of coarse aggregate, depending on the desired properties and application.

Before the patching concrete is applied, the surrounding concrete should be clean and sound (Fig. 11-32). Abrasive methods of cleaning (sandblasting, hydrojetting, waterblasting, scarification, or shotblasting) are usually required. For overlays, a cement-sand grout, a cement-sand-latex grout, or an epoxy bonding agent should be applied to the prepared surface with a brush or broom (see the earlier section "Bonding New to Previously Hardened Concrete"). Typical grout mix proportions are 1 part cement and 1 part fine sand and latex or epoxy admixtures. The grout should be applied immediately before the new concrete is placed. The grout should not be allowed to dry before the freshly mixed concrete is placed; otherwise bond may be impaired. The concrete may be dry or damp when the grout is applied *but not wet* with free-standing water. The minimum thickness for most patches and overlays is 20 mm. Some structures, like bridge decks, should have a minimum repair thickness of 40 mm. A superplasticizer is one of many admixtures often added to overlay or repair concrete to reduce the water-cementing materials ratio and to improve workability and ease of consolidation (Kosmatka 1985).

Honeycombed and other defective concrete should be cut out to expose sound material. If defective concrete is left adjacent to a patch, moisture may get into the voids; in time, weathering action will cause the patch to spall. The edges of the defective area should be cut or chipped straight and at right angles to the surface, or slightly undercut to provide a key at the edge of the patch. No featheredges should be permitted (Fig. 11-33). Based on the size of the patch, either a mortar or a concrete patching mixture should be used.

Shallow patches can be filled with a dry-pack mortar as described earlier. This should be placed in layers not more than 13 mm thick, with each layer given a scratch finish to improve bond with the subsequent layer. The final layer can be finished to match the surrounding concrete by floating, rubbing, or tooling, or on formed surfaces by pressing a section of form material against the patch while still plastic.

Deep patches can be filled with concrete held in place by forms. Such patches should be reinforced and doweled to the hardened concrete (*Concrete Manual* 1981). Large, shallow vertical or overhead repairs may best be accomplished by shotcreting. Several proprietary low-shrinkage cementitious repair products are also available.

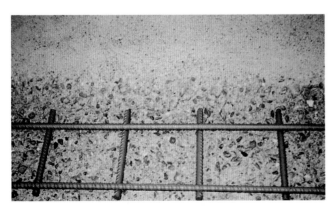

Fig. 11-32. Concrete prepared for patch installation. (69972)

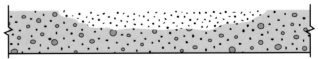

(a) Incorrectly installed patch. The feathered edges can break down under traffic or weather away.

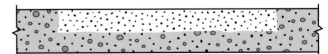

(b) Correctly installed patch. The chipped area should be at least 20 mm deep with the edges at right angles or undercut to the surface.

Fig. 11-33. Patch installation.

Fig. 11-34. Good curing is essential to successful patching. This patch is covered with polyethylene sheeting plus rigid insulation to retain moisture and heat for rapid hydration and strength gain. (40434)

Curing Patches

Following patching, good curing is essential (Fig. 11-34). Curing should be started immediately to avoid early drying. Wet burlap, wet sand, plastic sheets, curing paper, tarpaulins, or a combination of these can be used. In locations where it is difficult to hold these materials in place, an application of two coats of membrane-curing compound is often the most convenient method.

Cleaning Concrete Surfaces

Concrete surfaces are not always uniform in colour when forms are removed; they may have a somewhat blotchy appearance and there may be a slight film of form-release agent in certain areas. There may be mortar stains from leaky forms or there may be rust stains. Flatwork can also become discoloured during construction. Where appearance is important, all surfaces should be cleaned after construction has progressed to the stage where there will be no discolouration from subsequent construction activities.

There are three techniques for cleaning concrete surfaces: water, chemical, and mechanical (abrasion). Water dissolves dirt and rinses it from the surface. Chemical cleaners, usually mixed with water, react with dirt to separate it from the surface, and then the dirt and chemicals are rinsed off with clean water. Mechanical methods—sandblasting is most common—remove dirt by abrasion.

Before selecting a cleaning method, it should be tested on an inconspicuous area to be certain it will be helpful and not harmful. If possible, identify the characteristics of the discolouration because some treatments are more effective than others in removing certain materials.

Water cleaning methods include low-pressure washes, moderate-to-high-pressure waterblasting, and steam. Low-pressure washing is the simplest, requiring only that water run gently down the concrete surface for a day or two. The softened dirt then is flushed off with a slightly higher pressure rinse. Stubborn areas can be scrubbed with a nonmetallic-bristle brush and rinsed again. High-pressure waterblasting is used effectively by experienced operators. Steam cleaning must be performed by skilled operators using special equipment. Water methods are the least harmful to concrete, but they are not without potential problems. Serious damage may occur if the concrete surface is subjected to freezing temperatures while it is still wet; and water can bring soluble salts to the surface, forming a chalky, white deposit called efflorescence.

Chemical cleaning is usually done with water-based mixtures formulated for specific materials such as brick, stone, and concrete. An organic compound called a surfactant (surface-active agent), which acts as a detergent to wet the surface more readily, is included in most chemical cleaners. A small amount of acid or alkali is included to separate the dirt from the surface. For example, hydrochloric (muriatic) acid is commonly used to clean masonry walls and remove efflorescence. There can be problems related to the use of chemical cleaners. Their acid or alkaline properties can lead to reaction between cleaner and concrete as well as mortar, painted surfaces, glass, metals, and other building materials. Since chemical cleaners are used in the form of water-diluted solutions, they too can liberate soluble salts from within the concrete to form efflorescence. Some chemicals can also expose the aggregate in concrete. Chemicals commonly used to clean concrete surfaces and remove discolouration include weak solutions (1% to 10% concentration) of hydrochloric, acetic, or phosphoric acid. Diammonium citrate (20% to 30% water solution) is especially useful in removing discolouration stains and efflorescence on formed and flatwork surfaces. Chemical cleaners should be used by skilled operators taking suitable safety precautions. See Greening (1966) and PCA (1988) for more information.

Mechanical cleaning includes sandblasting, shotblasting, scarification, power chipping, and grinding. These methods wear the dirt off the surface rather than separate it from the surface. They, in fact, wear away both the dirt and some of the concrete surface; it is inevitable that there will be some loss of decorative detail, increased surface roughness, and rounding of sharp corners. Abrasive methods may also reveal defects (voids) hidden just beneath the formed surface.

Chemical and mechanical cleaning can each have an abrading effect on the concrete surface that may change the appearance of a surface compared to that of an adjacent uncleaned surface.

Finishing Formed Surfaces

Many off-the-form concrete surfaces require little or no additional treatment when they are carefully constructed with the proper forming materials. These surfaces are divided into two general classes: smooth and textured or patterned. Smooth surfaces are produced with plastic-coated forms, steel forms, fibreglass-reinforced plastic forms, formica forms, or tempered-hardboard forms. Textured or patterned surfaces are achieved with form liners, rough-sawn lumber, special grades and textures of plywood, or by fracturing the projections of a striated surface.

Rough-form finishes require patching of all tieholes and defects, unless tieholes are left open for architectural effect. Otherwise, these surfaces need no further work since texture and finish are imparted by the forms.

For a *smooth off-the-form finish*, it is important to arrange the smooth-facing forming material and tierods in a symmetrical pattern. Studs and wales that are capable of preventing excessive deflections must support smooth-finish forms that are somewhat lightweight.

A *smooth, rubbed finish* is produced on a newly hardened concrete surface no later than the day following form removal. The forms are removed and necessary patching completed as soon as possible. The surface is then wetted and rubbed with a carborundum brick or other abrasive until a satisfactory uniform colour and texture are produced.

A *sand-floated finish* can also be produced on newly hardened concrete surfaces. No later than 5 to 6 hours following form removal, the surface should be thoroughly wetted and rubbed with a wood float in a circular motion, working fine sand into the surface until the resulting finish is even and uniform in texture and colour.

A *grout cleandown (sack-rubbed finish)* can be used to impart a uniform colour and appearance to a smooth surface. After defects have been repaired, the surface should be saturated thoroughly with water and kept wet at least one hour before finishing operations begin. Next a grout of 1 part cement, 1½ to 2 parts of fine sand passing a 630 μm sieve, and sufficient water for a thick, creamy consistency should be prepared. It should be preshrunk by mixing at least one hour before use and then remixed without the addition of water and applied uniformly by brush, plasterer's trowel, or rubber float to completely fill all air bubbles and holes.

The surface should be vigorously floated with a wood, sponge rubber, or cork float immediately after applying the grout to fill any small air holes (bugholes) that are left; any remaining excess grout should be scraped off with a sponge-rubber float. If the float pulls grout from the holes, a sawing motion of the tool should correct the difficulty; any grout remaining on the surface should be allowed to stand undisturbed until it loses some of its plasticity but not its damp appearance. Then the surface should be rubbed with clean, dry burlap to remove all excess grout. All air holes should remain filled, but no visible film of grout should remain after the rubbing. Any section being cleaned with grout must be completed in one day, since grout remaining on the surface overnight is difficult to remove.

If possible, work should be done in the shade and preferably during cool, damp weather. During hot or dry weather, the concrete can be kept moist with a fine fog spray.

The completed surface should be moist-cured by keeping the area wet for 36 hours following the clean down. When completely dry, the surface should have a uniform colour and texture.

SPECIAL SURFACE FINISHES

Patterns and Textures

A variety of patterns and textures can be used to produce decorative finishes. Patterns can be formed with divider strips or by scoring or stamping the surface just before the concrete hardens. Textures can be produced with little effort and expense with floats, trowels, and brooms; more elaborate textures can be achieved with special techniques (Fig. 11-35). See Kosmatka (1991).

Exposed-Aggregate Concrete

An exposed-aggregate finish provides a rugged, attractive surface in a wide range of textures and colours. Select aggregates are carefully chosen to avoid deleterious substances; they are usually of uniform size such as 10 to 13 mm or larger. They should be washed thoroughly before use to assure satisfactory bond. Flat or elongated aggregate particles should not be used since they are easily dislodged when the aggregate is exposed. Caution should be exercised when using crushed stone; it not only has a greater tendency to stack during the seeding operation (requiring more labour), but it also may be undesirable in some applications (pool decks, for example).

The aggregate should be evenly distributed or seeded in one layer onto the concrete surface immediately after the slab has been bullfloated or darbied. The particles must be completely embedded in the concrete. This can be done by lightly tapping with a wooden hand float, a darby, or the broad side of a piece of lumber; then, when the concrete can support a finisher on kneeboards, the surface should be hand-floated with a magnesium float or darby until the mortar completely surrounds and slightly covers all the aggregate particles.

Methods of exposing the aggregate usually include washing and brushing, using retarders, and scrubbing. When the concrete has hardened sufficiently, simultaneously brushing and flushing with water should expose the aggregate. In washing and brushing, the surface layer of mortar should be carefully washed away with a light spray of water and brushed until the desired exposure is achieved.

Fig. 11-35. Patterned, textured, and coloured concretes are very attractive. (59031, 53598, 59003, 47835, 44898)

Since timing is important, test panels should be made to determine the correct time for exposing the aggregate without dislodging the particles. On large jobs, a water-insoluble retarder can be sprayed or brushed on the surface immediately after floating, but on small jobs this may not be necessary. When the concrete becomes too hard to produce the required finish with normal washing and brushing, a dilute hydrochloric acid can be used. Surface preparation should be minimized and applicable local environmental laws should be followed.

Two other methods for obtaining an exposed aggregate surface are: (1) the monolithic technique where a select aggregate, usually gap-graded, is mixed throughout the batch of concrete, and (2) the topping technique in which the select exposed-aggregate is mixed into a topping that is placed over a base slab of conventional concrete.

The aggregate in exposed-aggregate concrete can also be exposed by methods other than those already discussed. The following techniques expose the aggregate after the concrete has hardened to a compressive strength of around 30 MPa.

Abrasive blasting is best applied to a gap-graded aggregate concrete. The nozzle should be held perpendicular to the surface and the concrete removed to a maximum depth of about one-third the diameter of the coarse aggregate.

Waterblasting can also be used to texture the surface of hardened concrete, especially where local ordinances prohibit the use of sandblasting for environmental reasons. High-pressure water jets are used on surfaces that have or have not been treated with retarders.

In tooling or bushhammering, a layer of hardened concrete is removed and the aggregate is fractured at the surface. The surfaces attained can vary from a light scaling to a deep, bold texture obtained by jackhammering with a single-pointed chisel. Combs and multiple points can be used to produce finishes similar to some finishes used on cut stone.

Grinding and polishing will produce an exposed-aggregate concrete such as terrazzo, which is primarily used indoors. This technique is done in several successive steps using either a stone grinder or diamond-disk grinder. Each successive step uses finer grit than the preceding one. A polishing compound and buffer can then be used for a honed finish.

For special architectural concrete finishes, CSA Standard A23.4 requires that the owner provide a reference sample indicating the surface texture, quality, and type of finish required for bidding purposes. It also requires that the contractor make a preconstruction mockup field sample for approval by the owner. For more information see Kosmatka (1991), PCA (1972), and PCA (1995).

Coloured Finishes

Coloured concrete finishes for decorative effects in both interior and exterior applications can be achieved by four different methods: (1) the one-course or integral method, (2) the two-course method, (3) the dry-shake method, and (4) stains and paints (discussed below).

Colour pigments added to the concrete in the mixer to produce a uniform colour is the basis for the one-course method. Both natural and synthetic pigments are satisfactory if they are: (1) insoluble in water, (2) free from soluble salts and acids, (3) fast to sunlight, (4) fast to alkali and weak acids, (5) limited to small amounts of calcium sulphate, and (6) ground fine enough so that 90% passes a 45 micron screen. Use only the minimum amount necessary to produce the desired colour and not more than 10% by mass of the cement.

In the two-course method, a base slab is placed and left with a rough texture to bond better to a coloured topping layer. As soon as the base slab can support the weight of a cement mason, the topping course can be placed. If the base slab has hardened, prepare a bonding grout for the base slab prior to placing the topping mix. The topping mix is normally 13 mm to 25 mm thick, with a ratio of cement to sand of 1:3 or 1:4. The mix is floated and troweled in the prescribed manner. The two-course method is more commonly used because it is more economical than the one-course method.

A prepackaged dry-colour material is cast onto the surface of a concrete slab in the dry-shake method. After the slab has been floated once, two-thirds of the dry colouring material should be broadcast evenly by hand over the surface. After the material has absorbed water from the fresh concrete, it should be floated into the surface. Immediately after, the rest of the material is applied at right angles to the initial application, so that a uniform colour is obtained. The slab again should be floated to work the remaining material into the surface. Other finishing operations may follow depending on the exposure and type of finish desired. *Exterior surfaces that will be exposed to freezing and thawing should not be troweled because it can lead to a loss of entrained air at the surface of the slab. Proper floating and brooming can provide a non-slip surface texture needed for exterior slab surfaces.*

Stains, Paints, and Clear Coatings

Many types of stains, paints and clear coatings can be applied to concrete surfaces. Among the principal paints used are portland cement base, latex-modified portland cement, and latex (acrylic and polyvinyl acetate) paints (PCA 1992). However, stains and paints are used only when it is necessary to colour existing concrete. It is difficult to obtain a uniform colour with dyes or stains; therefore, the manufacturer's directions should be closely followed.

Portland cement based paints can be used on either interior or exterior exposures. The surface of the concrete should be damp at the time of application and each coat should be dampened as soon as possible without disturbing the paint. Damp curing of conventional portland cement paint is essential. On open-textured surfaces, such as concrete masonry, the paint should be applied with stiff-bristle brushes (scrub brushes). Paint should be worked well into the surface. For concrete with a smooth or sandy finish, whitewash or Dutch-type calcimine brushes are best.

The latex materials used in latex-modified portland cement paints retard evaporation, thereby retaining the necessary water for hydration of the portland cement. When using latex-modified paints, moist curing is not required.

Most latex paints are resistant to alkali and can be applied to new concrete after 10 days of good drying weather. The preferred method of application is by long-fibre, tapered nylon brushes 100 to 150 mm wide; however, roller or spray methods can also be used. The paints may be applied to damp, but not wet surfaces. If the surface is moderately porous, or if extremely dry conditions prevail, prewetting the surface is advisable.

Clear coatings are frequently used on concrete surfaces to (1) prevent soiling or discolouration of the concrete by air pollution, (2) to facilitate cleaning the surface if it does become dirty, (3) to brighten the colour of the aggregates, and (4) to render the surface water-repellent and thus prevent colour change due to rain and water absorption. The better coatings often consist of methyl methacrylate forms of acrylic resin, as indicated by a laboratory evaluation of commercial clear coatings (Latvin 1968). The methyl methacrylate coatings should have a higher viscosity and solids content when used on smooth concrete, since the original appearance of smooth concrete is more difficult to maintain than the original appearance of exposed-aggregate concrete.

Other materials, such as silane and siloxane penetrating sealers, are commonly used as water repellents for many exterior concrete applications.

PRECAUTIONS

Protect Your Head and Eyes. Construction equipment and tools represent constant potential hazards to busy construction personnel. That's why hard hats are required on construction projects. It is therefore recommended that some sort of head protection, such as a hard hat or safety hat, be worn when working any construction job, large or small.

Proper eye protection is essential when working with cement or concrete. Eyes are particularly vulnerable to blowing dust, splattering concrete, and other foreign objects. On some jobs it may be advisable to wear full-cover goggles or safety glasses with side shields. Actions that cause dust to become airborne should be avoided. Local or general ventilation can control exposures below applicable exposure limits; respirators may be used in poorly ventilated areas, where exposure limits are exceeded, or when dust causes discomfort or irritation.

Protect Your Back. All materials used to make concrete—portland cement, sand, coarse aggregate, and water—can be quite heavy, even in small quantities. When lifting heavy materials, the back should be straight, legs bent, keeping the lifting load between the legs and as close to the body as possible. Mechanical equipment should be used to place concrete as close as possible to its final position. After the concrete is deposited in the desired area by chute, pump, or wheelbarrow, it should be pushed—not lifted—into final position with a shovel; a short-handled, square-end shovel is an effective tool for spreading concrete, but special concrete rakes or come-alongs also can be used. Excessive horizontal movement of the concrete should be avoided; it not only requires extra effort, but may also lead to segregation of the concrete ingredients.

Protect Your Skin. When working with fresh concrete, care should be taken to avoid skin irritation or chemical burns (see warning statement in the box). Prolonged contact between fresh concrete and skin surfaces, eyes, and clothing may result in burns that are quite severe, including third-degree burns. Eyes and skin that come in contact with fresh concrete should be flushed thoroughly with clean water. If irritation persists, consult a physician. For deep burns or large affected skin areas, seek medical attention immediately.

The A-B-Cs of fresh concrete's effect on skin are:

Abrasive   Sand contained in fresh concrete is abrasive to bare skin.

Basic & Caustic   Portland cement is alkaline in nature, so wet concrete and other cement mixtures are strongly basic (pH of 12 to 13). Strong bases—like strong acids—are harmful, or caustic to skin.

Drying   Portland cement is hygroscopic—it absorbs water. In fact, portland cement needs water to harden. It will draw water away from any material it contacts—including skin.

Clothing worn as protection from fresh concrete should not be allowed to become saturated with moisture from fresh concrete because saturated clothing can transmit alkaline or hygroscopic effects to the skin. Clothing that becomes saturated from contact with fresh concrete should be rinsed out promptly with clear water to prevent continued contact with skin surfaces. Waterproof gloves, a long-sleeved shirt, and long pants should be worn. If you must stand in fresh concrete while it is being placed, screeded, or floated, wear rubber boots high enough to prevent concrete from getting into them (PCA 1998).

WARNING: Contact with wet (unhardened) concrete, mortar, cement, or cement mixtures can cause SKIN IRRITATION, SEVERE CHEMICAL BURNS (THIRD-DEGREE), or SERIOUS EYE DAMAGE. Frequent exposure may be associated with irritant and/or allergic contact dermatitis. Wear waterproof gloves, a long-sleeved shirt, full-length trousers, and proper eye protection when working with these materials. If you have to stand in wet concrete, use waterproof boots that are high enough to keep concrete from flowing into them. Wash wet concrete, mortar, cement, or cement mixtures from your skin immediately. Flush eyes with clean water immediately after contact. Indirect contact through clothing can be as serious as direct contact, so promptly rinse out wet concrete, mortar, cement, or cement mixtures from clothing. Seek immediate medical attention if you have persistent or severe discomfort.

REFERENCES

ACI Committee 207, *Roller Compacted Mass Concrete*, ACI 207.5R-99, American Concrete Institute, Farmington Hills, Michigan, 1999, 47 pages.

ACI Committee 301, *Specifications for Structural Concrete*, ACI 301-99, American Concrete Institute, Farmington Hills, Michigan, 1999, 49 pages.

ACI Committee 302, *Guide for Concrete Floor and Slab Construction*, ACI 302.1R-96, American Concrete Institute, Farmington Hills, Michigan, 1996, 65 pages.

ACI Committee 304, *Guide for Measuring, Mixing, Transporting and Placing Concrete*, ACI 304R-00, American Concrete Institute, Farmington Hills, Michigan, 2000, 41 pages.

ACI Committee 309, *Guide for Consolidation of Concrete*, ACI 309R-96, American Concrete Institute, Farmington Hills, Michigan, 1996, 39 pages.

ACI Committee 347, *Guide to Formwork for Concrete*, ACI 347R-94, reapproved 1999, American Concrete Institute, Farmington Hills, Michigan, 1999, 34 pages.

ACI, "Vapor Retarder Location," *Concrete International*, American Concrete Institute, Farmington Hills, Michigan, April 2001, pages 72 and 73.

Colley, B. E., and Humphrey, H. A., *Aggregate Interlock at Joints in Concrete Pavements*, Development Department Bulletin DX124, Portland Cement Association, http://www.portcement.org/pdf_files/DX124.pdf, 1967.

Concrete Manual, 8th ed., U.S. Bureau of Reclamation, Denver, revised 1981.

Farny, James A., *Concrete Floors on Ground*, EB075, Portland Cement Association, 2001, 136 pages.

Greening, N. R., and Landgren, R., *Surface Discoloration of Concrete Flatwork*, Research Department Bulletin RX203, Portland Cement Association, http://www.portcement. org/pdf_files/RX203.pdf, 1966.

Hedenblad, Göran, *Drying of Construction Water in Concrete-Drying Times and Moisture Measurement*, LT229, Swedish Council for Building Research, Stockholm, 1997, 54 pages.

Hedenblad, Göran, "Concrete Drying Time," *Concrete Technology Today*, PL982, Portland Cement Association, http://www.portcement.org/pdf_files/PL982.pdf, 1998, pages 4 and 5.

Hover, K. C., "Vibration Tune-up," *Concrete International*, American Concrete Institute, Farmington Hills, Michigan, September 2001, pages 31 to 35.

Hurd, M. K., *Formwork for Concrete*, SP-4, 6th edition, American Concrete Institute, Farmington Hills, Michigan, 1995, 500 pages.

Kosmatka, Steven H., *Finishing Concrete Slabs with Color and Texture*, PA124, Portland Cement Association, 1991, 34 pages.

Kosmatka, Steven H., "Floor-Covering Materials and Moisture in Concrete," *Concrete Technology Today*, PL853, Portland Cement Association, http://www.portcement. org/pdf_files/PL853.pdf, September 1985.

Kosmatka, Steven H., "Repair with Thin-Bonded Overlay," *Concrete Technology Today*, PL851, Portland Cement Association, http://www.portcement.org/pdf_files/ PL851.pdf, March 1985.

Litvin, Albert, *Clear Coatings for Exposed Architectural Concrete*, Development Department Bulletin DX137, Portland Cement Association, http://www.portcement.org/pdf_ files/DX137.pdf, 1968.

Panarese, William C., *Cement Mason's Guide*, PA122, Portland Cement Association, 1995, 24 pages.

PCA, *Bonding Concrete or Plaster to Concrete*, IS139, Portland Cement Association, 1976.

PCA, *Bridge Deck Renewal with Thin-Bonded Concrete Resurfacing*, IS207, Portland Cement Association, 1980.

PCA, *Building Movements and Joints*, EB086, Portland Cement Association, http://www.portcement.org/pdf_ files/EB086.pdf, 1982, 68 pages.

PCA, *Bushhammering of Concrete Surfaces*, IS051, Portland Cement Association, 1972.

PCA, *Color and Texture in Architectural Concrete*, SP021, Portland Cement Association, 1995, 36 pages.

PCA, *Concrete Basements for Residential and Light Building Construction*, IS208, Portland Cement Association, 1980a.

PCA, *Concrete Slab Surface Defects: Causes, Prevention, Repair*, IS177, Portland Cement Association, 2001, 12 pages.

PCA, "Foam Insulation Under Basement Floors," *Concrete Technology Today*, PL853, Portland Cement Association, http://www.portcement.org/pdf_files/PL853.pdf, September 1985.

PCA, *Joint Design for Concrete Highway and Street Pavements*, IS059, Portland Cement Association, 1980.

PCA, *Joints in Walls Below Ground*, CR059, Portland Cement Association, 1982a.

PCA, "Joints to Control Cracking in Walls," *Concrete Technology Today*, PL843, Portland Cement Association, http://www.portcement.org/pdf_files/PL843.pdf, September 1984.

PCA, *Painting Concrete*, IS134, Portland Cement Association, 1992, 8 pages.

PCA, *Removing Stains and Cleaning Concrete Surfaces*, IS214, Portland Cement Association, http://www.portcement. org/pdf_files/IS214.pdf, 1988, 16 pages.

PCA, *Resurfacing Concrete Floors*, IS144, Portland Cement Association, 1996, 8 pages.

PCA, "Sealants for Joints in Walls," *Concrete Technology Today*, PL844, Portland Cement Association, http://www. portcement.org/pdf_files/PL844.pdf, December 1984a.

PCA, *Subgrades and Subbases for Concrete Pavements*, IS029, Portland Cement Association, 1975.

PCA, *Understanding Concrete Floors and Moisture Issues*, CD-ROM Version 1, CD014, Portland Cement Association, 2000.

PCA, "Why Concrete Walls Crack," *Concrete Technology Today*, PL842, Portland Cement Association, http:// www.portcement.org/pdf_files/PL842.pdf, June 1984b.

PCA, *Working Safely with Concrete*, MS271, Portland Cement Association, 1998, 6 pages.

Stark, David C., *Effect of Vibration on the Air-Void System and Freeze-Thaw Durability of Concrete*, Research and Development Bulletin RD092, Portland Cement Association, http://www.portcement.org/pdf_files/RD092.pdf, 1986.

Suprenant, Bruce A., "Free Fall of Concrete," *Concrete International*, American Concrete Institute, Farmington Hills, Michigan, June 2001, pages 44 and 45.

Turner, C. D., "Unconfined Free-Fall of Concrete," *Journal of the American Concrete Institute*, American Concrete Institute, Detroit, December 1970, pages 975 to 976.

CHAPTER 12
Curing Concrete

Curing is the maintenance of a satisfactory moisture content and temperature in concrete for a period of time immediately following placing and finishing so that the desired properties may develop (Fig. 12-1). The need for adequate curing of concrete cannot be overemphasized. Curing has a strong influence on the properties of hardened concrete; proper curing will increase durability, strength, watertightness, abrasion resistance, volume stability, and resistance to freezing and thawing and deicers. Exposed slab surfaces are especially sensitive to curing as strength development and freeze-thaw resistance of the top surface of a slab can be reduced significantly when curing is defective.

When portland cement is mixed with water, a chemical reaction called hydration takes place. The extent to which this reaction is completed influences the strength and durability of the concrete. Freshly mixed concrete normally contains more water than is required for hydration of the cement; however, excessive loss of water by evaporation can delay or prevent adequate hydration. The surface is particularly susceptible to insufficient hydration because it dries first. If temperatures are favorable, hydration is relatively rapid the first few days after concrete is placed; however, it is important for water to be retained in the concrete during this period, that is, for evaporation to be prevented or substantially reduced.

With proper curing, concrete becomes stronger, more impermeable, and more resistant to stress, abrasion, and freezing and thawing. The improvement is rapid at early ages but continues more slowly thereafter for an indefinite period. Fig. 12-2 shows the strength gain of concrete with age for different moist curing periods and Fig. 12-3 shows the relative strength gain of concrete cured at different temperatures.

Fig. 12-1. Curing should begin as soon as the concrete hardens sufficiently to prevent marring or erosion of the surface. Burlap sprayed with water is an effective method for moist curing. (69973)

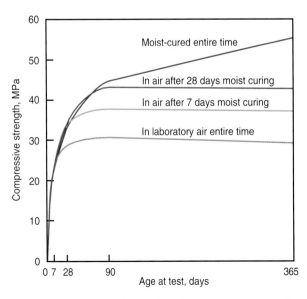

Fig. 12-2. Effect of moist curing time on strength gain of concrete (Gonnerman and Shuman 1928).

213

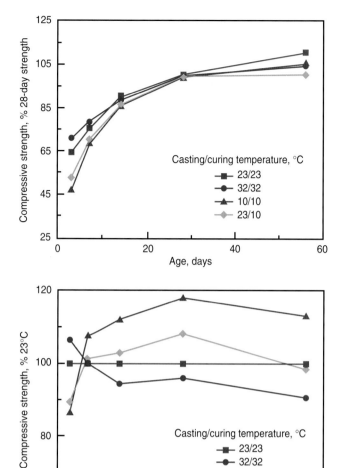

Fig. 12-3. Effect of curing temperature on strength gain (top) relative to 28-day strength and (bottom) relative to the strength of concrete at 23°C (Burg 1996).

The most effective method for curing concrete depends on the materials used, method of construction, and the intended use of the hardened concrete. For most jobs, curing generally involves applying curing compounds, or covering the freshly placed and finished concrete with impermeable sheets or wet burlap. In some cases, such as in hot and cold weather, special care using other precautions is needed.

Concrete mixtures with high cementing material contents and low water-cementing materials ratios (less than 0.40) may require special curing needs. As cement hydrates (chemically combining with water) the internal relative humidity decreases causing the paste to self-desiccate (dry out) if no external water is provided. The paste can self-desiccate to a level where hydration stops. This may influence desired concrete properties, especially if the internal relative humidity drops below 80% within the first seven days. In view of this, membrane-forming curing com-

pounds may not retain enough water in the concrete. Therefore, fogging and wet curing become necessary to maximize hydration (Copeland and Bragg 1955). Fogging during and after placing and finishing also helps minimize plastic cracking in concretes with very low water-cementing materials ratios (especially around 0.30 or less).

When moist curing is interrupted, the development of strength continues for a short period and then stops after the concrete's internal relative humidity drops to about 80%. However, if moist curing is resumed, strength development will be reactivated, but the original potential strength may not be achieved. Although it can be done in a laboratory, it is difficult to resaturate concrete in the field. Thus, it is best to moist-cure the concrete continuously from the time it is placed and finished until it has gained sufficient strength, impermeability, and durability.

Loss of water will also cause the concrete to shrink, thus creating tensile stresses within the concrete. If these stresses develop before the concrete has attained adequate tensile strength, surface cracking can result. All exposed surfaces, including exposed edges and joints, must be protected against moisture evaporation.

Hydration proceeds at a much slower rate when the concrete temperature is low. Temperatures below 10°C are unfavorable for the development of early strength; below 4°C the development of early strength is greatly retarded; and at or below freezing temperatures, down to -10°C, little or no strength develops.

In recent years, a maturity concept has been introduced to evaluate the development of strength when there is variation in the curing temperature of the concrete. Maturity is the product of the age of the concrete and its average curing temperature above a certain base temperature. Refer to Chapter 14 for more information on the maturity concept. It follows that concrete should be protected so that its temperature remains favorable for hydration and moisture is not lost during the early hardening period.

CURING METHODS AND MATERIALS

Concrete can be kept moist (and in some cases at a favorable temperature) by three curing methods:

1. Methods that maintain the presence of mixing water in the concrete during the early hardening period. These include ponding or immersion, spraying or fogging, and saturated wet coverings. These methods afford some cooling through evaporation, which is beneficial in hot weather.

2. Methods that reduce the loss of mixing water from the surface of the concrete. This can be done by covering the concrete with impervious paper or plastic sheets, or by applying membrane-forming curing compounds.

3. Methods that accelerate strength gain by supplying heat and additional moisture to the concrete. This is usually accomplished with live steam, heating coils, or electrically heated forms or pads.

The method or combination of methods chosen depends on factors such as availability of curing materials, size, shape, and age of concrete, production facilities (in place or in a plant), esthetic appearance, and economics. As a result, curing often involves a series of procedures used at a particular time as the concrete ages. For example, fog spraying or plastic covered wet burlap can precede application of a curing compound. The timing of each procedure depends on the degree of hardening of the concrete needed to prevent the particular procedure from damaging the concrete surface (ACI 308 1997).

Ponding and Immersion

On flat surfaces, such as pavements and floors, concrete can be cured by ponding. Earth or sand dikes around the perimeter of the concrete surface can retain a pond of water. Ponding is an ideal method for preventing loss of moisture from the concrete; it is also effective for maintaining a uniform temperature in the concrete. The curing water should not be more than about 11°C cooler than the concrete to prevent thermal stresses that could result in cracking. Since ponding requires considerable labour and supervision, the method is generally used only for small jobs.

The most thorough method of curing with water consists of total immersion of the finished concrete element. This method is commonly used in the laboratory for curing concrete test specimens. Where appearance of the concrete is important, the water used for curing by ponding or immersion must be free of substances that will stain or discolour the concrete. The material used for dikes may also discolour the concrete.

Fogging and Sprinkling

Fogging (Fig. 12-4) and sprinkling with water are excellent methods of curing when the ambient temperature is well above freezing and the humidity is low. A fine fog mist is frequently applied through a system of nozzles or sprayers to raise the relative humidity of the air over flatwork, thus slowing evaporation from the surface. Fogging is applied to minimize plastic shrinkage cracking until finishing operations are complete. Once the concrete has set sufficiently to prevent water erosion, ordinary lawn sprinklers are effective if good coverage is provided and water runoff is of no concern. Soaker hoses are useful on surfaces that are vertical or nearly so.

The cost of sprinkling may be a disadvantage. The method requires an ample water supply and careful supervision. If sprinkling is done at intervals, the concrete must be prevented from drying between applications of water by using burlap or similar materials; otherwise alternate cycles of wetting and drying can cause surface crazing or cracking.

Fig. 12-4. Fogging minimizes moisture loss during and after placing and finishing of concrete. (69974)

Wet Coverings

Fabric coverings saturated with water, such as burlap, cotton mats, rugs, or other moisture-retaining fabrics, are commonly used for curing (Fig. 12-5). Treated burlaps that reflect light and are resistant to rot and fire are available. The requirements for burlap are described in the *Specification for Burlap Cloths Made from Jute or Kenaf* (AASHTO M 182), and those for white burlap-polyethylene sheeting are described in ASTM C 171.

Burlap must be free of any substance that is harmful to concrete or causes discolouration. New burlap should be thoroughly rinsed in water to remove soluble substances and to make the burlap more absorbent.

Fig. 12-5. Lawn sprinklers saturating burlap with water keep the concrete continuously moist. Intermittent sprinkling is acceptable if no drying of the concrete surface occurs. (50177)

Wet, moisture-retaining fabric coverings should be placed as soon as the concrete has hardened sufficiently to prevent surface damage. During the waiting period other curing methods are used, such as fogging or the use of membrane forming finishing aids. Care should be taken to cover the entire surface with wet fabric, including the edges of slabs. The coverings should be kept continuously moist so that a film of water remains on the concrete surface throughout the curing period. Use of polyethylene film over wet burlap is a good practice; it will eliminate the need for continuous watering of the covering. Periodically rewetting the fabric under the plastic before it dries out should be sufficient. Alternate cycles of wetting and drying during the early curing period may cause crazing of the surface.

Wet coverings of earth, sand, or sawdust are effective for curing and are often useful on small jobs. Sawdust from most woods is acceptable, but oak and other woods that contain tannic acid should not be used since deterioration of the concrete may occur. A layer about 50 mm thick should be evenly distributed over the previously moistened surface of the concrete and kept continuously wet.

Wet hay or straw can be used to cure flat surfaces. If used, it should be placed in a layer at least 150 mm thick and held down with wire screen, burlap, or tarpaulins to prevent its being blown off by wind.

A major disadvantage of moist earth, sand, sawdust, hay, or straw coverings is the possibility of discolouring the concrete.

Impervious Paper

Impervious paper for curing concrete consists of two sheets of kraft paper cemented together by a bituminous adhesive with fibre reinforcement. Such paper, conforming to ASTM C 171, is an efficient means of curing horizontal surfaces and structural concrete of relatively simple shapes. An important advantage of this method is that periodic additions of water are not required. Curing with impervious paper enhances the hydration of cement by preventing loss of moisture from the concrete (Fig. 12-6).

As soon as the concrete has hardened sufficiently to prevent surface damage, it should be thoroughly wetted and the widest paper available applied. Edges of adjacent sheets should be overlapped about 150 mm and tightly sealed with sand, wood planks, pressure-sensitive tape, mastic, or glue. The sheets must be weighted to maintain close contact with the concrete surface during the entire curing period.

Impervious paper can be reused if it effectively retains moisture. Tears and holes can easily be repaired with curing-paper patches. When the condition of the paper is questionable, additional use can be obtained by using it in double thickness.

In addition to curing, impervious paper provides some protection to the concrete against damage from sub-

Fig.12-6. Impervious curing paper is an efficient means of curing horizontal surfaces. (69994)

sequent construction activity as well as protection from the direct sun. It should be light in colour and nonstaining to the concrete. Paper with a white upper surface is preferable for curing exterior concrete during hot weather.

Plastic Sheets

Plastic sheet materials, such as polyethylene film, can be used to cure concrete (Fig. 12-7). Polyethylene film is an effective moisture retarder and is easily applied to complex as well as simple shapes. Its application is the same as described for impervious paper.

Curing with polyethylene film (or impervious paper) can cause patchy discolouration, especially if the concrete contains calcium chloride and has been finished by hard-steel troweling. This discolouration is more pronounced when the film becomes wrinkled, but it is difficult and time consuming on a large project to place sheet materials without wrinkles. Flooding the surface under the covering may prevent discolouration, but other means of curing should be used when uniform colour is important.

Polyethylene film should conform to ASTM C 171, which specifies a 0.10-mm thickness for curing concrete, but lists only clear and white opaque film. However, black film is available and is satisfactory under some conditions. White film should be used for curing exterior concrete during hot weather to reflect the sun's rays. Black film can be used during cool weather or for interior locations. Clear film has little effect on heat absorption.

ASTM C 171 also includes a sheet material consisting of burlap impregnated on one side with white opaque polyethylene film. Combinations of polyethylene film bonded to an absorbent fabric such as burlap help retain moisture on the concrete surface.

216

Fig. 12-7. Polyethylene film is an effective moisture barrier for curing concrete and easily applied to complex as well as simple shapes. To minimize discolouration, the film should be kept as flat as possible on the concrete surface. (70014)

Polyethylene film may also be placed over wet burlap or other wet covering materials to retain the water in the wet covering material. This procedure eliminates the labour-intensive need for continuous watering of wet covering materials.

Membrane-Forming Curing Compounds

Liquid membrane-forming compounds consisting of waxes, resins, chlorinated rubber, and other materials can be used to retard or reduce evaporation of moisture from concrete. They are the most practical and most widely used method for curing not only freshly placed concrete but also for extending curing of concrete after removal of forms or after initial moist curing. However, the most effective methods of curing concrete are wet coverings or water spraying that keeps the concrete continually damp. Curing compounds should be able to maintain the relative humidity of the concrete surface above 80% for seven days to sustain cement hydration.

Membrane-forming curing compounds are of two general types: clear, or translucent; and white pigmented. Clear or translucent compounds may contain a fugitive dye that makes it easier to check visually for complete coverage of the concrete surface when the compound is applied. The dye fades away soon after application. On hot, sunny days, use of white-pigmented compounds are recommended; they reduce solar-heat gain, thus reducing the concrete temperature. Pigmented compounds should be kept agitated in the container to prevent pigment from settling out.

Curing compounds should be applied by hand-operated or power-driven spray equipment immediately after final finishing of the concrete (Fig. 12-8). The concrete surface should be damp when the coating is applied. On dry, windy days, or during periods when adverse weather conditions could result in plastic shrinkage cracking, application of a curing compound immediately after final finishing and before all free water on the surface has evaporated will help prevent the formation of cracks. Power-driven spray equipment is recommended for uniform application of curing compounds on large paving projects. Spray nozzles and windshields on such equipment should be arranged to prevent wind-blown loss of curing compound.

Normally only one smooth, even coat is applied at a typical rate of 3 to 4 m² per litre; but products may vary, so manufacturer's recommended application rates should be followed. If two coats are necessary to ensure complete coverage, for effective protection the second coat should be applied at right angles to the first. Complete coverage of the surface must be attained because even small pinholes in the membrane will increase the evaporation of moisture from the concrete.

Curing compounds might prevent bonding between hardened concrete and a freshly placed concrete overlay. And, most curing compounds are not compatible with adhesives used with floor covering materials. Consequently, they should either be tested for compatibility, or not used when bonding of overlying materials is necessary. For example, a curing compound should not be applied to the base slab of a two-course floor. Similarly, some curing compounds may affect the adhesion of paint to concrete floors. Curing compound manufacturers should be consulted to determine if their product is suitable for the intended application.

Curing compounds should be uniform and easy to maintain in a thoroughly mixed solution. They should not sag, run off peaks, or collect in grooves. They should form

Fig. 12-8. Liquid membrane-forming curing compounds should be applied immediately following final finishing, with uniform and adequate coverage over the entire surface and edges for effective, extended curing of concrete. (69975)

a tough film to withstand early construction traffic without damage, be nonyellowing, and have good moisture-retention properties.

Caution is necessary when using curing compounds containing solvents of high volatility in confined spaces or near sensitive occupied spaces such as hospitals because evaporating volatiles may cause respiratory problems. Applicable local environmental laws concerning volatile organic compound (VOC) emissions should be followed.

Curing compounds should conform to ASTM C 309. A method for determining the efficiency of curing compounds, waterproof paper, and plastic sheets is described in ASTM C 156. Curing compounds with sealing properties are specified under ASTM C 1315.

Internal Moist Curing

Internal moist curing refers to methods of providing moisture from within the concrete as opposed to outside the concrete. This water should not effect the initial water to cementing materials ratio of the fresh concrete. Low-density fine aggregate or absorbent polymer particles with an ability to retain a significant amount of water may provide additional moisture for concretes prone to self desiccation. When more complete hydration is needed for concretes with low water to cementing materials ratios (around 0.30 or less), 60 kg/m³ to 180 kg/m³ of saturated low density aggregate can provide additional moisture to extend hydration, resulting in increased strength and durability. All of the fine aggregate in a mixture can be replaced with saturated low-density fine aggregate to maximize internal moist curing. Internal moist curing must be accompanied by external curing methods.

Forms Left in Place

Forms provide satisfactory protection against loss of moisture if the top exposed concrete surfaces are kept wet. A soaker hose is excellent for this. The forms should be left on the concrete as long as practical.

Wood forms left in place should be kept moist by sprinkling, especially during hot, dry weather. If this cannot be done, they should be removed as soon as practical and another curing method started without delay. Colour variations may occur from formwork and uneven water curing of walls.

Steam Curing

Steam curing, a method for the accelerated curing of concrete, is advantageous where early strength gain in concrete is important or where additional heat is required to accomplish hydration, as in cold weather. Requirements for steam curing are contained in CSA Standard A23.4, *Precast Concrete—Materials and Construction*. The Standard covers the requirements for curing at elevated tempera-

tures and the application and control of heat for accelerated curing of concrete.

Two methods of steam curing are used: live steam at atmospheric pressure (for enclosed cast-in-place structures and large precast concrete units) and high-pressure steam in autoclaves (for small manufactured units). Only live steam at atmospheric pressure will be discussed here.

A typical steam-curing cycle consists of (1) an initial delay prior to steaming, (2) a period for increasing the temperature, (3) a period for holding the maximum temperature constant, and (4) a period for decreasing the temperature. A typical atmospheric steam-curing cycle is shown in Fig. 12-9.

Steam curing at atmospheric pressure is generally done in an enclosure to minimize moisture and heat losses. Tarpaulins are frequently used to form the enclosure. Application of steam to the enclosure should be delayed until initial set occurs or delayed at least 3 hours after final placement of concrete to allow for some hardening of the concrete. However, a 3- to 5-hour delay period prior to steaming will achieve maximum early strength, as shown in Fig. 12-10.

Steam temperature in the enclosure should be kept at about 60°C until the desired concrete strength has developed. Strength will not increase significantly if the maximum steam temperature is raised from 60°C to 70°C. Steam-curing temperatures above 70°C should be avoided; they are uneconomical and may result in damage.

Besides early strength gain, there are other advantages of curing concrete at temperatures of around 60°C; for example, there is reduced drying shrinkage and creep as

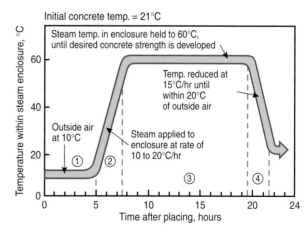

① Initial delay prior to steaming   3 to 5 hours
② Temperature increase period   2½ hours
③ Constant temperature period   6 to 12 hours*
④ Temperature decrease period   2 hours

*Type 30 (III) or high-early-strength cement, longer for other types

Fig. 12-9. A typical atmospheric steam-curing cycle.

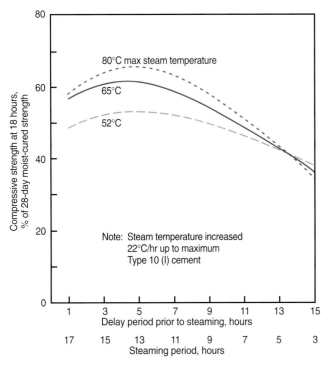

Fig. 12-10. Relationship between strength at 18 hours and delay period prior to steaming. In each case, the delay period plus the steaming period totaled 18 hours (Hanson 1963).

compared to concrete cured at 23°C for 28 days (Klieger 1960 and Tepponen and Eriksson 1987).

CSA A23.4 categorizes concrete elements to receive accelerated curing on the basis of ambient conditions to which they will be exposed. The categories are (a) "moisture category dry" or (b) "moisture category damp."

"Moisture category dry" applies to concrete elements that, after curing, will not be subjected to a moist environment during service. Unless such units are air-entrained, they are also not to be exposed to a moist environment for a period of more than three months prior to service.

"Moisture category damp" applies to concrete elements that will be subjected to a moist environment during service or non-air-entrained elements that will be exposed to a moist environment for a period of more than three months prior to service. Examples include: (a) external elements of buildings or structures that are exposed to precipitation, surface water, or ground water, such as bridge girders, unprotected roof slabs, building facades, basement walls, railway ties; (b) internal building elements subject to moist conditions such as indoor swimming pools and parking garages; and (c) elements, or parts of which may frequently drop below their dew point, such as chimneys and internal voids in bridges.

Excessive rates of heating and cooling should be avoided to prevent damaging volume changes. CSA A23.4 states that for both "moisture category dry" and "moisture category damp" the maximum heating rate in the enclo-

sure surrounding the concrete should not exceed 20°C/ hour. Similarly, the maximum cooling rate for both moisture categories is 15°C/hour and is to continue until the concrete temperature is no more than 20°C above the ambient temperature outside the enclosure.

The maximum concrete temperature allowed by CSA A23.4 is 70°C for "moisture category dry" and 60°C for "moisture category damp". It is recommended that an internal temperature of concrete of 70°C, the maximum allowable stated for "moisture category dry", should not be exceeded, to avoid heat induced delayed expansion and undue reduction in ultimate strength. Concrete temperatures are commonly monitored at the exposed ends of the element. Monitoring air temperatures alone is not sufficient because the heat of hydration may cause the concrete to exceed 70°C.

The temperature of the concrete prior to casting can be increased by injecting live steam into the mixture along with the addition of mixing water. This practice of preheating the concrete materials as a means of accelerated curing has different requirements to be met than those for steam curing. These are covered in CSA A23.4 under *Heated Concrete* including those requirements of Table 3 – *Heated Concrete Cycle.*

The curing temperature in the enclosure should be held until the concrete has reached the desired strength. The time required will depend on the concrete mixture and steam temperature in the enclosure (ACI Committee 517 1992).

Insulating Blankets or Covers

Layers of dry, porous material such as straw or hay can be used to provide insulation against freezing of concrete when temperatures fall below 0°C.

Formwork can be economically insulated with commercial blanket or batt insulation that has a tough moistureproof covering. Suitable insulating blankets are manufactured of fibreglass, sponge rubber, cellulose fibres, mineral wool, vinyl foam, and open-cell polyurethane foam. When insulated formwork is used, care should be taken to ensure that concrete temperatures do not become excessive.

Framed enclosures of canvas tarpaulins, reinforced polyethylene film, or other materials can be placed around the structure and heated by space heaters or steam. Portable hydronic heaters are used to thaw subgrades as well as heat concrete without the use of an enclosure

Curing concrete in cold weather should follow the recommendations in Chapter 14, CSA Standard A23.1, and ACI 306 (1997), *Cold-Weather Concreting.* Recommendations for curing concrete in hot weather can be found in Chapter 13, CSA Standard A23.1, and ACI 305 (1999), *Hot-Weather Concreting.*

Electrical, Oil, Microwave, and Infrared Curing

Electrical, hot oil, microwave and infrared curing methods have been available for accelerated and normal curing of concrete for many years. Electrical curing methods include a variety of techniques: (1) use of the concrete itself as the electrical conductor, (2) use of reinforcing steel as the heating element, (3) use of a special wire as the heating element, (4) electric blankets, and (5) the use of electrically heated steel forms (presently the most popular method). Electrical heating is especially useful in cold-weather concreting. Hot oil may be circulated through steel forms to heat the concrete. Infrared rays and microwave have had limited use in accelerated curing of concrete. Concrete that is cured by infrared methods is usually under a covering or enclosed in steel forms. Electrical, oil, and infrared curing methods are used primarily in the precast concrete industry.

CURING PERIOD AND TEMPERATURE

The period of time that concrete should be protected from freezing, abnormally high temperatures, and against loss of moisture depends upon a number of factors: the type of cementing materials used; mixture proportions; required strength, size and shape of the concrete member; ambient weather; and future exposure conditions. The curing period may be 3 weeks or longer for lean concrete mixtures used in massive structures such as dams; conversely, it may be only a few days for rich mixes, especially if Type 30 (III or HE) cement is used. Steam-curing periods are normally much shorter, ranging from a few hours to 3 days; but generally 24-hour cycles are used.

Since all the desirable properties of concrete are improved by curing, the curing period should be as long as necessary. CSA Standard A23.1 defines the basic curing period for all concrete surfaces as either 3-days at a minimum temperature of 10°C, or for the time necessary to attain 40% of the specified compressive strength of the concrete. Additional curing, beyond that specified as basic curing, is required by CSA Standard A23.1 for durability, structural safety, and mass concrete.

For structural safety, CSA Standard A23.1 requires that the basic curing period be extended until the concrete has attained a compressive strength level considered adequate for structural safety as specified by the owner.

For concrete subjected to exposure classifications F-1, C-1, C-2, S-1, and S-2 (see Tables 2-2, 8-2, and 8-3), and concrete exposed to abrasion and air pollution in heavy industrial areas (contact local environment agencies for guidance), additional curing is required for durability. Immediately following the basic curing period, an additional 4 consecutive days at a minimum temperature of 10°C, or for the time necessary to attain 70% of the specified compressive strength, is required. If high-performance is being used, the concrete is to be moist cured.

A higher curing temperature provides earlier strength gain in concrete than a lower temperature, but it may decrease 28-day strength as shown in Fig. 12-11. If strength tests are made to establish the time when curing can cease or forms can be removed, representative concrete test cylinders or beams should be fabricated in the field, kept adjacent to the structure or pavement they represent, and cured by the same methods. Cores, cast-in-place removable cylinders, and nondestructive testing methods may also be used to determine the strength of a concrete member.

Since the rate of hydration is influenced by cement type and the presence of supplementary cementing materials, the curing period should be prolonged for concretes made with cementing materials possessing slow-strength-gain characteristics. For mass concrete (large piers, locks, abutments, dams, heavy footings, massive columns and transfer girders), CSA Standard A23.1 requires additional curing beyond the basic curing period. For reinforced massive sections, the basic curing period must be extended an additional 4 consecutive days: and for unreinforced massive sections an additional 7 consecutive days.

During cold weather, additional heat is often required to maintain favorable curing temperatures of 10°C to 20°C. Vented gas or oil-fired heaters, heating coils, portable

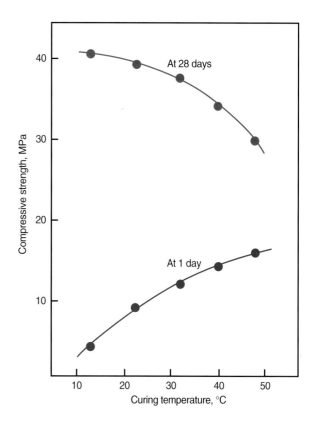

Fig. 12-11. One-day strength increases with increasing curing temperature, but 28-day strength decreases with increasing curing temperature (Verbeck and Helmuth 1968).

hydronic heaters, or live steam can be used to supply the required heat. In all cases, care must be taken to avoid loss of moisture from the concrete. Exposure of fresh concrete to heater or engine exhaust gases must be avoided as this can result in surface deterioration and dusting (rapid carbonation). If water curing is being used, CSA A23.1 requires it to be terminated 12 hours before the end of the protection period.

High-early-strength concrete can be used in cold weather to speed-up setting time and strength development. This can reduce the curing period, but the requirements for the basic curing period as defined in CSA Standard A23.1 must be followed.

For adequate deicer scale resistance of concrete, the minimum curing period generally corresponds to the time required to develop the design strength of the concrete at the surface (CSA A23.1 requirement is 70% of design strength). *A period of air-drying after curing will enhance resistance to scaling. This drying period should be at least 1 month of relatively dry weather before the application of deicing salts.*

CSA Standard A23.1 also has requirements for curing during hot weather conditions. The standard requires that when the air temperature is at or above 27°C, curing during the basic curing period is to be accomplished by water spray or by using saturated absorptive fabric, in order to achieve cooling by evaporation. Curing of mass concrete during the basic curing period shall also be by water when the air temperature is at or above 20°C, in order to minimize the temperature rise of the concrete. It is common, however, to see white–pigmented curing compounds used in some hot-weather applications, such as for pavements, either where water is not readily available or where water curing is not considered a practical solution. In such cases the use of a curing compound is either specified or approved by the owner.

SEALING COMPOUNDS

Sealing compounds (sealers) are liquids applied to the surface of hardened concrete to reduce the penetration of liquids or gases such as water, deicing solutions, and carbon dioxide that cause freeze-thaw damage, corrosion of reinforcing steel, and acid attack. In addition, sealing compounds used on interior floor slabs reduce dusting and the absorption of spills while making the surface easier to clean.

Sealing compounds differ in purpose from curing compounds; they should not be confused as being the same. The primary purpose of a curing compound is to retard the loss of water from newly placed concrete and it is applied immediately after finishing. Surface sealing compounds on the other hand retard the penetration of harmful substances into hardened concrete and are typically not applied until the concrete is 28 days old. Surface sealers are generally classified as either film-forming or penetrating.

Sealing exterior concrete is an optional procedure generally performed to help protect concrete from freeze-thaw damage and chloride penetration from deicers. Curing is not optional when using a sealer; curing is necessary to produce properties needed for concrete to perform adequately for its intended purpose. Satisfactory performance of exterior concrete still primarily depends on an adequate air-void system, sufficient strength, and the use of proper placing, finishing and curing techniques. However, not all concrete placed meets those criteria; surface sealers can help improve the durability of these concretes.

Film-forming sealing compounds remain mostly on the surface with only a slight amount of the material penetrating the concrete. The relatively large molecular structure of these compounds limits their ability to penetrate the surface. Thinning them with solvents will not improve their penetrating capability. These materials not only reduce the penetration of water, they also protect against mild chemicals; furthermore, they prevent the absorption of grease and oil as well as reduce dusting under pedestrian traffic.

Surface sealers consist of acrylic resins, chlorinated rubber, urethanes, epoxies, and alpha methyl styrene. The effectiveness of film-forming sealers depends on the continuity of the layer formed. Abrasive grit and heavy traffic can damage the layer requiring the reapplication of the material. Consult manufacturers' application recommendations because some of these materials are intended for interior use only and may yellow and deteriorate under exposure to ultraviolet light.

The penetrating sealer that has been used most extensively for many years is a mixture of 50 percent boiled linseed oil and 50 percent mineral spirits (AASHTO M 233). Although this mixture is an effective sealer, it has two main disadvantages: it darkens the concrete, and periodic reapplication is necessary for long-term protection.

A new generation of water-repellent penetrating sealers have a very small molecular size that allows penetration and saturation of the concrete as deep as 3 mm. The two most common are silane and siloxane, compounds which are derived from the silicone family. These sealers allow the concrete to breath, thus preventing a buildup of vapour pressure between the concrete and sealer that can occur with some film-forming materials. Because the sealer is embedded within the concrete, making it more durable to abrasive forces or ultraviolet deterioration, it can provide longer lasting protection than film-forming sealers. However, periodic retreatment is recommended. They are popular for protecting bridge decks and other exterior structures from corrosion of reinforcing steel caused by chloride infiltration from deicing chemicals or sea spray (Fig 12-12).

Application of any sealer should only be done on concrete that is clean and allowed to dry for at least 24 hours at temperatures above 16°C. At least 28 days should be allowed to elapse before applying sealers to new concrete. Penetrating sealers cannot fill surface voids if they are filled with water. Some surface preparation may be necessary if the concrete is old and dirty. Concrete placed in the late fall should not be sealed until spring because the sealer may cause the concrete to retain water that may exacerbate freeze-thaw damage.

The precautions outlined earlier regarding volatile solvents in curing compounds also apply to sealing compounds. The effectiveness of water-based surface sealers is still being determined. The scale resistance provided by concrete sealers should be evaluated based on criteria established in ASTM C 672. For more information on surface sealing compounds, see AASHTO M 224, ACI Committee 330 and ACI Committee 362.

Fig 12-12. Penetrating sealers help protect reinforcing steel in bridge decks from corrosion due to chloride infiltration without reducing surface friction. (69976)

REFERENCES

ACI Committee 305, *Hot-Weather Concreting*, ACI 305R-99, American Concrete Institute, Farmington Hills, Michigan, 1999, 17 pages.

ACI Committee 306, *Cold-Weather Concreting*, ACI 306R-88, Reapproved 1997, American Concrete Institute, Farmington Hills, Michigan, 1997, 23 pages.

ACI Committee 308, *Standard Practice for Curing Concrete*, ACI 308-92, Reapproved 1997, American Concrete Institute, Farmington Hills, Michigan, 1997, 11 pages.

ACI Committee 330, *Guide for Design and Construction of Concrete Parking Lots*, ACI 330R-92, Reapproved 1997, American Concrete Institute, Farmington Hills, Michigan, 1997, 27 pages.

ACI Committee 362, *Guide for the Design of Durable Parking Structures*, ACI 362.1R-97, American Concrete Institute, Farmington Hills, Michigan, 1997, 40 pages.

ACI Committee 516, "High Pressure Steam Curing: Modern Practice and Properties of Autoclaved Products," *Proceedings of the American Concrete Institute*, American Concrete Institute, Farmington Hills, Michigan, August 1965, pages 869 to 908.

ACI Committee 517, *Accelerated Curing of Concrete at Atmospheric Pressure*, ACI 517.2R-87, revised 1992, American Concrete Institute, Farmington Hills, Michigan, 1992, 17 pages.

AASHTO M 182-91 (1996), *Standard Specification for Burlap Cloth Made from Jute or Kenaf*, American Association of State Highway and Transportation Officials, Washington, D.C., 1996.

AASHTO M 224-91 (1996), *Standard Specification for Use of Protective Sealers for Portland Cement Concrete*, American Association of State Highway and Transportation Officials, Washington, D.C., 1996.

Burg, Ronald G., *The Influence of Casting and Curing Temperature on the Properties of Fresh and Hardened Concrete*, Research and Development Bulletin RD113, Portland Cement Association, 1996, 20 pages.

Copeland, L. E., and Bragg, R. H., *Self Desiccation in Portland Cement Pastes*, Research Department Bulletin RX052, Portland Cement Association, http://www.portcement. org/pdf_files/RX052.pdf, 1955, 13 pages.

CSA Standard A23.1-00, *Concrete Materials and Methods of Concrete Construction*, Canadian Standards Association, Toronto, 2000.

CSA Standard A23.4-00, *Precast Concrete — Materials and Construction*, Canadian Standards Association, Toronto, 2000.

German Committee for Reinforced Concrete, *Recommendation on the Heat Treatment of Concrete*, Deutscher Ausschuss fuer Stahlbeton, Deutsches Institut fuer Normung (DIN), Berlin, September 1989, 13 pages.

Gonnerman, H. F. and Shuman, E. C., "Flexure and Tension Tests of Plain Concrete," Major Series 171, 209, and 210, *Report of the Director of Research*, Portland Cement Association, November 1928, pages 149 and 163.

Greening, N. R., and Landgren, R., *Surface Discoloration of Concrete Flatwork*, Research Department Bulletin RX203, Portland Cement Association, http://www.portcement.org/pdf_files/RX203.pdf, 1966, 19 pages.

Hanson, J. A., *Optimum Steam Curing Procedure in Precasting Plants*, with discussion, Development Department Bulletins DX062 and DX062A, Portland Cement Association, http://www.portcement.org/pdf_files/DX062.pdf and http://www.portcement.org/pdf_files/DX062A.pdf, 1963, 28 pages and 19 pages, respectively.

Hanson, J. A., *Optimum Steam Curing Procedures for Structural Lightweight Concrete*, Development Department Bulletin DX092, Portland Cement Association, http://www.portcement.org/pdf_files/DX092.pdf, 1965.

Highway Research Board, *Curing of Concrete Pavements*, Current Road Problems No. 1-2R, Highway Research Board, Washington, D.C., May 1963.

Klieger, Paul, *Curing Requirements for Scale Resistance of Concrete*, Research Department Bulletin RX082, Portland Cement Association, http://www.portcement.org/pdf_files/RX082.pdf, 1957, 17 pages.

Klieger, Paul, *Some Aspects of Durability and Volume Change of Concrete for Prestressing*, Research Department Bulletin RX118, Portland Cement Association, http://www.portcement.org/pdf_files/RX118.pdf, 1960, 15 pages.

Klieger, Paul, and Perenchio, William, *Silicone Influence on Concrete to Freeze-Thaw and De-icer Damage*, Research Department Bulletin RX169, Portland Cement Association, http://www.portcement.org/pdf_files/RX169.pdf, 1963, 15 pages.

Lerch, William, *Plastic Shrinkage*, Research Department Bulletin RX081, Portland Cement Association, http://www.portcement.org/pdf_files/RX081.pdf, 1957, 7 pages.

Pierce, James S., "Mixing and Curing Water for Concrete," *Significance of Tests and Properties of Concrete and Concrete-Making Materials*, STP 169C, American Society for Testing and Materials, West Conshohocken, Pennsylvania, 1994, pages 473 to 477.

Powers, T. C., *A Discussion of Cement Hydration in Relation to the Curing of Concrete*, Research Department Bulletin RX025, Portland Cement Association, http://www.portcement.org/pdf_files/RX025.pdf, 1948, 15 pages.

Senbetta, Ephraim, "Curing and Curing Materials," *Significance of Tests and Properties of Concrete and Concrete-Making Materials*, STP 169C, American Society for Testing and Materials, West Conshohocken, Pennsylvania, 1994, pages 478 to 483.

Tepponen, Pirjo, and Eriksson, Bo-Erik, "Damages in Concrete Railway Sleepers in Finland," *Nordic Concrete Research*, Publication No. 6, The Nordic Concrete Federation, Oslo, 1987.

Verbeck, George J., and Helmuth, R. A., "Structures and Physical Properties of Cement Pastes," *Proceedings, Fifth International Symposium on the Chemistry of Cement*, vol. III, The Cement Association of Japan, Tokyo, 1968, page 9.

CHAPTER 13
Hot-Weather Concreting

Weather conditions at a jobsite—hot or cold, windy or calm, dry or humid—may be vastly different from the optimum conditions assumed at the time a concrete mix is specified, designed, or selected, or from laboratory conditions in which concrete specimens are stored and tested. Hot weather conditions adversely influence concrete quality primarily by accelerating the rate of moisture loss and rate of cement hydration that occur at higher temperatures. Detrimental hot weather conditions include:

- high ambient temperature
- high concrete temperature
- low relative humidity
- high wind speed
- solar radiation

Hot weather conditions can create difficulties in fresh concrete, such as:

- increased water demand
- accelerated slump loss leading to the addition of water on the jobsite
- increased rate of setting resulting in placing and finishing difficulties
- increased tendency for plastic cracking
- critical need for prompt early curing
- difficulties in controlling entrained air
- increased concrete temperature resulting in long-term strength loss
- increased potential for thermal cracking

Adding water to the concrete at the jobsite can adversely affect properties and serviceability of the hardened concrete, resulting in:

- decreased strength from higher water to cementing materials ratio
- decreased durability
- increased permeability
- nonuniform surface appearance
- increased tendency for drying shrinkage and uncontrolled cracking
- reduced wear resistance

Only by taking precautions to alleviate these difficulties in anticipation of hot-weather conditions can concrete work proceed smoothly. For more information on the above topics, see ACI Committee 305 (1999).

WHEN TO TAKE PRECAUTIONS

During hot weather the most favorable temperature for achieving high quality freshly mixed concrete is usually lower than can be obtained without artificial cooling. A concrete temperature of 10°C to 15°C is desirable to maximize beneficial mix properties, but such temperature are not always practical. Many specifications require only that concrete when placed should have a temperature of less than 29°C to 32°C.

Fig. 13-1. Liquid nitrogen added directly into a truck mixer is an effective method of reducing concrete temperature for delivery to mass concrete placements or during hot-weather concreting. (69954)

Precautions should be planned in advance to counter the effects of a high concrete temperature when the concrete placed is somewhere between 25°C and 35°C. See Table 13-1 for the permissible concrete temperatures at placing permitted in CSA Standard A23.1. Last-minute attempts to prevent hot-weather damage are rarely performed soon enough. If acceptable field data is not available, the maximum temperature limit should be established for conditions at the jobsite; this should be based on trial-batch tests at the temperature and for the typical concrete section thickness anticipated, rather than on ideal temperatures cited in many specifications and reference publications. If possible, large batches should be made to measure mix properties at various batch temperatures and time intervals to establish the relationship for particular properties as a function of time and batch temperature. This process will establish the maximum allowable time to discharge concrete after batching for various concrete temperatures.

More than controlling the maximum temperature is required to determine when to employ precautions to produce concrete with the required strength and durability. For most work it is too complex to simply limit only the maximum temperature of concrete as placed; circumstances and concrete requirements vary too widely. For example, a temperature limit that would serve successfully at one jobsite could be highly restrictive at another. Atmospheric conditions, including air temperature, relative humidity and wind speed, in conjunction with site conditions influence the precautions needed. For example, flatwork done under a roof that blocks solar radiation with exterior walls in place that screen the wind could be completed using a high temperature concrete; this concrete would cause difficulty if placed outdoors on the same day where it would be exposed to direct sun and wind.

Which precautions to use and when to use them will depend on: the type of construction; characteristics of the materials being used; and the experience of the placing and finishing crew in dealing with the atmospheric conditions on the site. The following list of precautions will reduce or avoid the potential problems of hot-weather concreting:

- use materials and mix proportions that have a good record in hot-weather conditions
- cool the concrete or one or more of its ingredients (Fig. 13-1)

Table 13-1. Permissible Concrete Temperatures at Placing

Thickness of section, m	Temperatures, °C	
	Minimum	Maximum
Less than 0.3	10	35
0.3-1	10	30
1-2	5	25
More than 2	5	20

Source: CSA Standard A23.1.

- use a concrete consistency that allows rapid placement and consolidation
- reduce the time of transport, placing and finishing as much as possible
- schedule concrete placements to limit exposure to atmospheric conditions, such as at night or during favorable weather conditions
- consider methods to limit moisture loss during placing and finishing, such as sunshades, windscreens, fogging, or spraying
- apply temporary moisture-retaining films after screeding
- organize a preconstruction conference to discuss the precautions required for the project

The above precautions are discussed in further detail throughout this chapter.

EFFECTS OF HIGH CONCRETE TEMPERATURES

As concrete temperature increases there is a loss in slump that is often unadvisedly compensated for by adding water to the concrete at the jobsite. At higher temperatures a greater amount of water is required to hold slump constant than is needed at lower temperatures. Adding water without adding cement results in a higher water-cementing materials ratio, thereby lowering the strength at all ages and adversely affecting other desirable properties of the hardened concrete. This is in addition to the adverse effect on strength at later ages due to the higher temperature, even without the addition of water. Adding cement to compensate for the use of additional mix water may not be enough to achieve the desired concrete properties because additional cement will further increase the concrete temperature and water demand.

As shown in Fig. 13-2, if the temperature of freshly mixed concrete is increased from 10°C to 38°C, about 20 kg of additional water is needed per cubic metre to maintain the same 75-mm slump. This additional water could reduce strength by 12% to 15% and produce a compressive strength cylinder test result that may not comply with specifications.

High temperatures of freshly mixed concrete increase the rate of setting and shorten the length of time within which the concrete can be transported, placed, and finished. Setting time can be reduced by 2 or more hours with a 10°C increase in concrete temperature (Fig. 13-3). Concrete should remain plastic long enough so that each layer can be placed without development of cold joints or discontinuities in the concrete. Retarding admixtures, ASTM C 494 Type B, and hydration control admixtures can be beneficial in offsetting the accelerating effects of high temperature.

In hot weather, there is an increased tendency for cracks to form both before and after hardening. Rapid evaporation of water from freshly placed concrete can

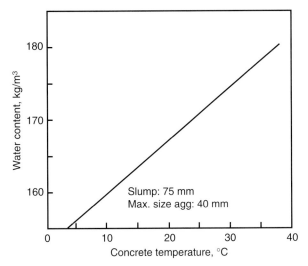

Fig. 13-2. The water requirement of a concrete mixture increases with an increase in concrete temperature (Bureau of Reclamation 1981).

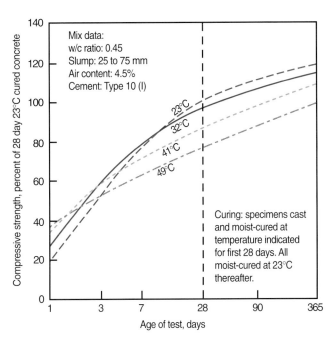

Fig. 13-4. Effect of high concrete temperatures on compressive strength at various ages (Klieger 1958).

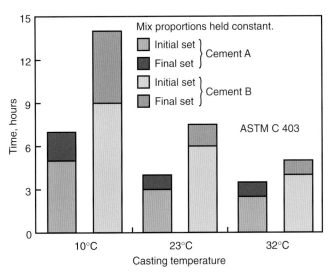

Fig. 13-3. Effect of concrete temperature on setting time (Burg 1996).

cause plastic-shrinkage cracks before the surface has hardened (discussed in more detail later in this chapter). Cracks may also develop in the hardened concrete because of increased drying shrinkage due to higher water contents or thermal volume changes as the concrete cools.

Air entrainment is also affected in hot weather. At elevated temperatures, an increase in the amount of air-entraining admixture is required to produce a given air content.

Fig. 13-4 shows the effect of high initial concrete temperatures on compressive strength. The concrete temperatures at the time of mixing, casting, and curing were 23°C, 32°C, 41°C, and 49°C. After 28 days, the specimens were all moist-cured at 23°C until the 90-day and one-year test

ages. The tests, using identical concretes of the same water-cementing materials ratio, show that while higher concrete temperatures give higher early strength than concrete at 23°C, at later ages concrete strengths are lower. If the water content had been increased to maintain the same slump (without increasing cement content), the reduction in strength would have been even greater than shown.

The proper fabrication, curing, and testing of compression test specimens during hot weather is critical. Steps should be taken to make sure CSA Standard A23.2-3C (ASTM C 31) procedures are followed regarding initial curing of strength specimens for acceptance or quality control testing at 15°C to 25°C. If the initial 24 hour curing is at 38°C, the 28-day compressive strength of the test specimens may be 10% to 15% lower than if cured at the required CSA A23.2-3C (ASTM C 31) curing temperatures (Gaynor 1985).

Because of the detrimental effects of high concrete temperatures, all operations in hot weather should be directed toward keeping the concrete as cool as possible.

COOLING CONCRETE MATERIALS

The usual method of cooling concrete is to lower the temperature of the concrete materials before mixing. One or more of the ingredients should be cooled. In hot weather the aggregates and mixing water should be kept as cool as practicable; these materials have a greater influence on concrete temperature after mixing than other ingredients.

The contribution of each ingredient in a concrete mixture to the temperature of the freshly mixed concrete is related to the temperature, specific heat, and quantity of

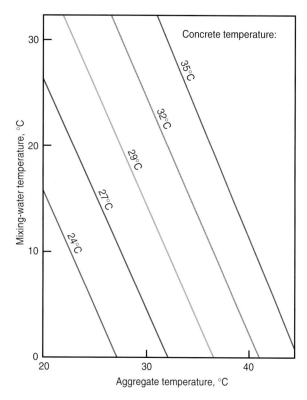

Fig. 13-5. Temperature of freshly mixed concrete as affected by temperature of its ingredients. Although the chart is based on the following mixture, it is reasonably accurate for other typical mixtures:

Aggregate	**1360 kg**
Moisture in aggregate	**27 kg**
Added mixing water	**109 kg**
Cement at 66°C	**256 kg**

each material. Fig. 13-5 shows graphically the effect of temperature of materials on the temperature of fresh concrete. It is evident that although concrete temperature is primarily dependent upon the aggregate temperature, cooling the mixing water can be effective.

The approximate temperature of concrete can be calculated from the temperatures of its ingredients by using the following equation (NRMCA 1962):

$$T = \frac{0.22\,(T_a M_a + T_c M_c) + T_w M_w + T_{wa} M_{wa}}{0.22\,(M_a + M_c) + M_w + M_{wa}} \quad \text{where}$$

T = temperature of the freshly mixed concrete, °Celsius

T_a, T_c, T_w, and T_{wa} = temperature in °Celsius of aggregates, cement, added mixing water, and free water on aggregates, respectively

M_a, M_c, M_w, and M_{wa} = mass, kilograms, of aggregates, cementing materials, added mixing water, and free water on aggregates, respectively

Example calculations for initial concrete temperature are shown in Table 13-2A.

Of all the materials in concrete, water is the easiest to cool. Even though it is used in smaller quantities than the other ingredients, cold water will produce a moderate reduction in the concrete temperature. Mixing water from a cool source should be used. It should be stored in tanks that are not exposed to the direct rays of the sun. Tanks and pipelines carrying mixing water should be buried, insulated, shaded, or painted white to keep water as cool as practical. Water can be cooled by refrigeration, liquid nitrogen, or ice. Cooling the mix water temperature 2.0°C to 2.2°C will usually lower the concrete temperature about 0.5°C. However, because mix water is such a small percentage of the total mixture, it is difficult to lower concrete temperatures more than about 4.5°C by cooling the water alone.

Ice can be used as part of the mixing water provided it is completely melted by the time mixing is completed. When using crushed ice, care must be taken to store it at a temperature that will prevent the formation of lumps.

When ice is added as part of the mixing water, the effect of the heat of fusion of the ice must be considered; so the equation for temperature of fresh concrete is modified as follows:

Table 13-2A. Effect of Temperature of Materials on Initial Concrete Temperatures

Material	Mass, *M*, kg	Specific heat kJ/kg • °C	Joules to vary temperature, 1°C	Initial temperature of material, *T*, °C	Total joules in material
	(1)	(2)	(3) Col.1 x Col. 2	(4)	(5) Col. 3 x Col. 4
Cement	335 (M_c)	0.92	308	66 (T_c)	20,328
Water	123 (M_w)	4.184	515	27 (T_w)	13,905
Total aggregate	1839 (M_a)	0.92	1692	27 (T_a)	45,684
			2515		79,917

Initial concrete temperature = $\dfrac{79,917}{2515}$ = 31.8°C

To achieve 1°C reduction in initial concrete temperature:

Cement temperature must be lowered = $\dfrac{2515}{308}$ = 8.2°C

Or water temperature dropped = $\dfrac{2515}{515}$ = 4.9°C

Or aggregate temperature cooled = $\dfrac{2515}{1692}$ = 1.5°C

$$T\ (C°) = \frac{0.22\,(T_a M_a + T_c M_c) + T_w M_w + T_{wa} M_{wa} - 80 M_i}{0.22\,(M_a + M_c) + M_w + M_{wa} + M_i}$$

where M_i is the mass in kilograms of ice (NRMCA 1962 and Mindess, Sidney, and Young 1981).

The heat of fusion of ice is 335 kJ per kg. Calculations in Table 13-2B show the effect of 44 kg of ice in reducing the temperature of concrete. Crushed or flaked ice is more effective than chilled water in reducing concrete temperature. The amount of water and ice must not exceed the total mixing-water requirements.

Fig. 13-6 shows crushed ice being charged into a truck mixer prior to the addition of other materials. Mixing time should be long enough to completely melt the ice. The volume of ice should not replace more than approximately 75% of the total batch water. The maximum temperature reduction from the use of ice is limited to about 11°C.

If a greater temperature reduction is required, the injection of liquid nitrogen into the mixer may be the best alternative method.

The liquid nitrogen can be added directly into a central mixer drum or the drum of a truck mixer to lower concrete temperature. Fig. 13-1 shows liquid nitrogen added directly into a truck mixer near a ready mix plant. Care should be taken to prevent the liquid nitrogen from contacting the metal drum; the super cold liquid nitrogen may crack the drum. The addition of liquid nitrogen does not in itself influence the amount of mix water required except that lowering the concrete temperature can reduce water demand.

Aggregates have a pronounced effect on the fresh concrete temperature because they represent 70% to 85% of the total mass of concrete. To lower the temperature of concrete 0.5°C requires only a 0.8°C to 1.1°C reduction in the temperature of the coarse aggregate.

There are several simple methods of keeping aggregates cool. Stockpiles should be shaded from the sun and kept moist by sprinkling. Do not spray salt water on aggregate stockpiles. Since evaporation is a cooling process, sprinkling provides effective cooling, especially when the relative humidity is low.

Fig. 13-6. Substituting ice for part of the mixing water will substantially lower concrete temperature. A crusher delivers finely crushed ice to a truck mixer reliably and quickly. (44236)

Sprinkling of coarse aggregates should be adjusted to avoid producing excessive variations in the surface moisture content and thereby causing a loss of slump uniformity. Refrigeration is another method of cooling materials. Aggregates can be immersed in cold-water tanks, or cooled air can be circulated through storage bins. Vacuum cooling can reduce aggregate temperatures to as low as 1°C.

Cement temperature has only a minor effect on the temperature of the freshly mixed concrete because of cement's low specific heat and the relatively small amount of cement in a concrete mixture. A cement temperature change of 5°C generally will change the concrete temperature by only 0.5°C. Because cement loses heat slowly during storage, it may still be warm when delivered. (This heat is produced in grinding the cement clinker during manufacture.) Since the temperature of cement does affect the temperature of the fresh concrete to some extent, some specifications place a limit on its

Table 13-2B. Effect of Ice (44 kg) on Temperature of Concrete

Material	Mass, *M*, kg	Specific heat kJ/kg • °C	Joules to vary temperature, 1°C	Initial temperature of material, *T*, °C	Total joules in material
	(1)	(2)	(3) Col.1 x Col. 2	(4)	(5) Col. 3 x Col. 4
Cement	335 (M_c)	0.92	308	66 (T_c)	20,328
Water	123 (M_w)	4.184	515	27 (T_w)	13,905
Total aggregate	1839 (M_a)	0.92	1692	27 (T_a)	45,684
Ice	44 (M_i)	4.184	184	0 (T_i)	0
			2699		
minus	44 (M_i) x heat of fusion, (335 kJ/kg) =				–14,740
					65,177

Concrete temperature = $\dfrac{65{,}177}{2699}$ = 24.1°C

temperature at the time of use. This limit varies from 66°C to 82°C (ACI Committee 305). However, it is preferable to specify a maximum temperature for freshly mixed concrete rather than place a temperature limit on individual ingredients (Lerch 1955).

SUPPLEMENTARY CEMENTING MATERIALS

Many concrete producers consider the use of supplementary cementing materials to be essential in hot weather conditions. The materials of choice are fly ash, other pozzolans, and ground granulated blast-furnace slag meeting the requirements of CSA Standard A23.5 (ASTM C 618 and ASTM C 989). These materials generally slow both the rate of setting as well as the rate of slump loss. However, some caution regarding finishing is needed; because the rate of bleeding can be slower than the rate of evaporation, plastic shrinkage cracking or crazing may result. This is discussed in greater detail under "Plastic Shrinkage Cracking."

PREPARATION BEFORE CONCRETING

Before concrete is placed, certain precautions should be taken during hot weather to maintain or reduce concrete temperature. CSA Standard A23.1 states, that when the air temperature is at or above 27°C, or when there is a probability of it rising to 27°C during the placing period, facilities should be provided for protection of the concrete in place from the effects of hot and/or drying weather conditions. Under severe drying conditions, mixers, chutes, conveyor belts, hoppers, pump lines, and other equipment for handling concrete should also be shaded, painted white, or covered with wet burlap to reduce solar heat.

Forms, reinforcing steel, and subgrade should be fogged or sprinkled with cool water just before the concrete is placed. Fogging the area during placing and finishing operations not only cools the contact surfaces and surrounding air but also increases its relative humidity. This reduces the temperature rise of the concrete and minimizes the rate of evaporation of water from the concrete after placement. For slabs on ground, it is a good practice to moisten the subgrade the evening before concreting. There should be no standing water or puddles on forms or subgrade at the time concrete is placed.

During extremely hot periods, improved results can be obtained by restricting concrete placement to early morning, evening, or nighttime hours. This practice has resulted in less thermal shrinkage and cracking of thick slabs and pavements.

TRANSPORTING, PLACING, FINISHING

Transporting and placing concrete should be done as quickly as practical during hot weather. Delays contribute to loss of slump and an increase in concrete temperature. Sufficient labour and equipment must be available at the jobsite to handle and place concrete immediately upon delivery.

Prolonged mixing, even at agitating speed, should be avoided. If delays occur, stopping the mixer and then agitating intermittently can minimize the heat generated by mixing. CSA Standard A23.1 requires that discharge of concrete be completed within 2 hours. During hot weather the time limit can be reasonably reduced to 1 hour or even 45 minutes. If specific time limitations on the completion of discharge of the concrete are desired, they should be included in the project specifications. It is also reasonable to obtain test data from a trial batch simulating the time, mixing, and anticipated concrete temperature to document, if necessary, a reduction in the time limit.

Since the setting of concrete is more rapid in hot weather, extra care must be taken with placement techniques to avoid cold joints. For placement of walls, shallower layers can be specified to assure enough time for consolidation with the previous lift. Temporary sunshades and windbreaks help to minimize cold joints.

Floating of slabs should be done promptly after the water sheen disappears from the surface or when the concrete can support a finisher with his foot making no more than a 5-mm indentation in the slab surface. Finishing on dry and windy days requires extra care. Rapid drying of the concrete at the surface may cause plastic shrinkage cracking.

PLASTIC SHRINKAGE CRACKING

Plastic shrinkage cracks sometimes occur in the surface of freshly mixed concrete soon after it has been placed, while it is being finished or shortly thereafter (Fig. 13-7). These

Fig. 13-7. Typical plastic shrinkage cracks. (1311)

cracks which appear mostly on horizontal surfaces can be substantially eliminated if preventive measures are taken.

Plastic shrinkage cracking is usually associated with hot-weather concreting; however, it can occur any time ambient conditions produce rapid evaporation of moisture from the concrete surface. These cracks occur when water evaporates from the surface faster than it can rise to the surface during the bleeding process. This creates rapid drying shrinkage and tensile stresses in the surface that often result in short, irregular cracks. The following conditions, singly or collectively, increase evaporation of surface moisture and increase the possibility of plastic shrinkage cracking:

1. Low air temperature
2. High concrete temperature
3. Low humidity
4. High wind speed

The crack length is generally 50 to 1000 mm in length and they are usually spaced in an irregular pattern from 50 to 600 mm apart. Fig. 13-8 is useful for determining when precautionary measures should be taken. There is no way to predict with certainty when plastic shrinkage cracking will occur.

CSA Standard A23.1 requires that when the rate of evaporation exceeds 1 kg/m² per hour, precautionary measures such as windscreens are required around all sides of concrete elements. With some concrete mixtures, such as those containing pozzolans, cracking is possible if the rate of evaporation exceeds 0.5 kg/m² per hour. Concrete containing silica fume is particularly prone to plastic shrinkage because bleeding rates are commonly only 0.25 kg/m² per hour. Therefore, protection from premature drying is essential at lower bleeding rates. At some point in the process of setting, bleeding goes to zero and the surface begins to dry at evaporation rates much lower than the typically specified 1.0 kg/m² per hour; in such cases, further protection becomes necessary regardless of the type of concrete mixture.

One or more of the following precautions listed can minimize the occurrence of plastic shrinkage cracking.

To use this chart:

1. Enter with air temperature, move up to relative humidity.

2. Move right to concrete temperature.

3. Move down to wind velocity.

4. Move left: read approximate rate of evaporation.

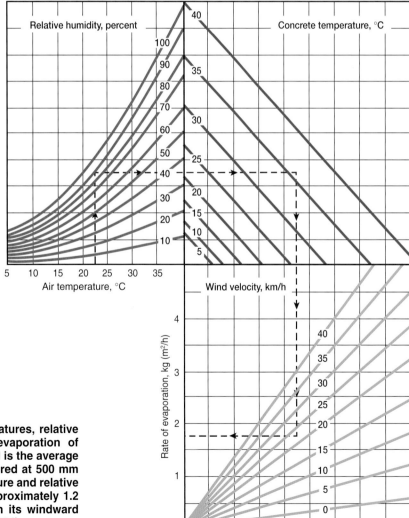

Fig. 13-8. Effect of concrete and air temperatures, relative humidity, and wind velocity on rate of evaporation of surface moisture from concrete. Wind speed is the average horizontal air or wind speed in km/h measured at 500 mm above the evaporating surface. Air temperature and relative humidity should be measured at a level approximately 1.2 to 1.8 m above the evaporating surface on its windward side shielded from the sun's rays (Menzel 1954).

They should be considered while planning for hot-weather concrete construction or while dealing with the problem after construction has started. They are listed in the order in which they should be done during construction.

1. Moisten concrete aggregates that are dry and absorptive.
2. Keep the concrete temperature low by cooling aggregates and mixing water.
3. Dampen the subgrade (Fig. 13-9) and fog forms prior to placing concrete.
4. Erect temporary windbreaks to reduce wind velocity over the concrete surface.
5. Erect temporary sunshades to reduce concrete surface temperatures.
6. Protect the concrete with temporary coverings, such as polyethylene sheeting, during any appreciable delay between placing and finishing.
7. Fog the slab immediately after placing and before finishing, taking care to prevent the accumulation of water that may reduce the quality of the cement paste in the slab surface.
8. Add plastic fibres to the concrete mixture to help reduce plastic shrinkage crack formation.

Fogging the concrete before and after final finishing is the most effective way to minimize evaporation and reduce plastic shrinkage cracking. Use of a fog spray will raise the relative humidity of the ambient air over the slab, thus reducing evaporation from the concrete. Fog nozzles atomize water using air pressure (Figs. 13-10 and 13-11) to create a fog blanket. A fog spray can be produced with a 15 to 20 MPa pressure washer in combination with an atomizing type nozzle. They should not be confused with garden-hose nozzles, which leave an excess amount of

Fig. 13-10. Fog nozzle. (9853)

Fig. 13-11. Fogging cools the air and raises the relative humidity above flatwork to lessen rapid evaporation from the concrete surface, thus reducing cracking and improving surface durability. (69956)

water on the slab. Fogging should be continued until a suitable curing material such as a curing compound, wet burlap, or curing paper can be applied.

Other methods to prevent the rapid loss of moisture from the concrete surface include:

• Spray application of temporary moisture-retaining films (usually polymers); these compounds can be applied immediately after screeding to reduce water evaporation before final finishing operations and curing commences. These materials are floated and troweled into the surface during finishing and should

Fig. 13-9. Dampening the subgrade, yet keeping it free of standing water will lessen drying of the concrete and reduce problems from hot weather conditions. (69955)

have no adverse effect on the concrete or inhibit the adhesion of membrane-curing compounds.

- Reduction of time between placing and the start of curing by eliminating delays during construction.

If plastic shrinkage cracks should appear during finishing, striking each side of the crack with a float and refinishing can close the cracks. However, the cracking may reoccur unless the causes are corrected.

CURING AND PROTECTION

Curing and protection are more critical in hot weather than in temperate periods. Retaining forms in place cannot be considered a satisfactory substitute for curing in hot weather; they should be loosened as soon as practical without damage to the concrete. The need for moist curing of concrete slabs is greatest during the first few hours after finishing. To prevent the drying of exposed concrete surfaces, moist curing should commence as soon as the surfaces are finished.

CSA A23.1 requires that when the air temperature is at or above 27°C, curing during the basic curing period should be accomplished by water spray or by using saturated absorptive fabric, in order to achieve cooling by evaporation. For mass concrete, curing should be by water for the basic curing period when the air temperature is at or above 20°C, in order to minimize the temperature rise of the concrete. The basic curing period is defined by CSA Standard A23.1 as curing for a period of 3-days at a minimum temperature of 10°C or for the time necessary to attain 40% of the specified 28-day compressive strength. The additional curing requirements for durability, structural safety, and mass concrete are covered in detail in Chapter 12. On hardened concrete and on flat concrete surfaces in particular, curing water should not be more than about 11°C cooler than the concrete. This will minimize cracking caused by thermal stresses due to temperature differentials between the concrete and curing water.

White-pigmented curing compounds can be used on horizontal surfaces in hot weather if approved by the owner. If approved, the application of the curing compound should be preceded by 24 hours of moist curing. If this is not practical, the compound should be applied immediately after final finishing. The concrete surfaces should be moist.

Moist-cured surfaces should dry out slowly after the curing period to reduce the possibility of surface crazing and cracking. Crazing, a network pattern of fine cracks that do not penetrate much below the surface, is caused by minor surface shrinkage. Crazing cracks are very fine and barely visible except when the concrete is drying after the surface has been wet. The cracks encompass small concrete areas less than 50 mm in dimension, forming a chicken-wire like pattern.

ADMIXTURES

For unusual cases in hot weather and where careful inspection is maintained, a retarding admixture may be beneficial in delaying the setting time, despite the somewhat increased rate of slump loss resulting from their use. A hydration control admixture can be used to stop cement hydration and setting. Hydration is resumed, when desired, with the addition of a special accelerator (reactivator).

Retarding admixtures should conform to the requirements of ASTM C 494 Type B. Admixtures should be tested with job materials under job conditions before construction begins; this will determine their compatibility with the basic concrete ingredients and their ability under the particular conditions to produce the desired results.

HEAT OF HYDRATION

Heat generated during cement hydration raises the temperature of concrete to a greater or lesser extent depending on the size of the concrete placement, its surrounding environment, and the amount of cement in the concrete. As a general rule a 5°C to 9°C temperature rise per 45 kg of portland cement can be expected from the heat of hydration (ACI Committee 211 1997). There may be instances in hot-weather-concrete work and massive concrete placements when measures must be taken to cope with the generation of heat from cement hydration and attendant thermal volume changes to control cracking (see Chapters 15 and 18).

REFERENCES

ACI Committee 211, *Standard Practice for Selecting Proportions for Normal, Heavyweight, and Mass Concrete,* ACI 211.1-91, reapproved 1997, American Concrete Institute, Farmington Hills, Michigan, 1997, 38 pages.

ACI Committee 305, *Hot-Weather Concreting,* ACI 305R-99, American Concrete Institute, Farmington Hills, Michigan, 1999, 17 pages.

ACI Committee 308, *Standard Specification for Curing Concrete,* ACI 308.1-98, American Concrete Institute, Farmington Hills, Michigan, 1998, 9 pages.

Burg, Ronald G., *The Influence of Casting and Curing Temperature on the Properties of Fresh and Hardened Concrete,* Research and Development Bulletin RD113, Portland Cement Association, 1996, 13 pages.

Bureau of Reclamation, *Concrete Manual,* 8th ed., Denver, revised 1981.

CSA Standard A23.1-00, *Concrete Materials and Methods of Concrete Construction,* Canadian Standards Association, Toronto, 2000.

Gaynor, Richard D.; Meininger, Richard C.; and Khan, Tarek S., *Effect of Temperature and Delivery Time on Concrete Proportions,* NRMCA Publication No. 171, National Ready Mixed Concrete Association, Silver Spring, Maryland, June 1985.

Klieger, Paul, *Effect of Mixing and Curing Temperature on Concrete Strength,* Research Department Bulletin RX103, Portland Cement Association, http://www.portcement. org/pdf_files/RX103.pdf, 1958.

Lerch, William, *Hot Cement and Hot Weather Concrete Tests,* IS015, Portland Cement Association, http://www.port cement.org/pdf_files/IS015.pdf, 1955.

Menzel, Carl A., "Causes and Prevention of Crack Development in Plastic Concrete," *Proceedings of the Portland Cement Association,* 1954, pages 130 to 136.

Mindess, Sidney, and Young, J. Francis, *Concrete,* Prentice Hall, Englewood Cliffs, New Jersey, 1981.

NRMCA, *Cooling Ready Mixed Concrete,* NRMCA Publication No. 106, National Ready Mixed Concrete Association, Silver Spring, Maryland, 1962.

CHAPTER 14
Cold-Weather Concreting

Concrete can be placed safely without damage from freezing throughout the winter months in cold climates if certain precautions are taken. CSA Standard A23.1 requires that when the air temperature is at or below 5°C, or when there is a probability of it falling below 5°C within 24 hours of placing, all materials and equipment needed for adequate protection and curing shall be on hand and ready for use before concrete placement is started. Normal concreting practices can be resumed once there is no possibility of encountering these conditions.

During cold weather, the concrete mixture and its temperature should be adapted to the construction procedure and ambient weather conditions. Preparations should be made to protect the concrete; enclosures, windbreaks, portable heaters, insulated forms, and blankets should be ready to maintain the concrete temperature (Fig. 14-1). Forms, reinforcing steel, and embedded fixtures must be clear of snow and ice at the time concrete is placed. Thermometers and proper storage facilities for test cylinders should be available to verify that precautions are adequate.

EFFECT OF FREEZING FRESH CONCRETE

Concrete gains very little strength at low temperatures. Freshly mixed concrete must be protected against the disruptive effects of freezing (Fig.14-2) until the degree of saturation of the concrete has been sufficiently reduced by the process of hydration. The time at which this reduction is accomplished corresponds roughly to the time required for the concrete to attain a compressive strength of 3.5 MPa (Powers 1962). At normal temperatures and water-cementing materials ratios less than 0.60, this occurs within the first 24 hours after placement. Significant ultimate strength reductions, up to about 50%, can occur if concrete is frozen within a few hours after placement or before it attains a compressive strength of 3.5 MPa (McNeese 1952). CSA A23.1 requires that plain concrete to be exposed to deicers should attain a strength of 32 MPa prior to repeated cycles of freezing and thawing.

Concrete that has been frozen just once at an early age can be restored to nearly normal strength by providing favorable subsequent curing conditions. Such concrete, however, will not be as resistant to weathering nor as watertight as concrete that had not been frozen. The critical period after which concrete is not seriously damaged by one or two freezing cycles is dependent upon the concrete ingredients and conditions of mixing, placing, curing, and subsequent drying. For example, air-entrained concrete is less susceptible to damage by early freezing than non-air-

Fig. 14-1. When suitable preparations to build enclosures and insulate equipment have been made, cold weather is no obstacle to concrete construction. (69876, 43464)

Fig. 14-2. Closeup view of ice impressions in paste of frozen fresh concrete. The ice crystal formations occur as unhardened concrete freezes. They do not occur in adequately hardened concrete. The disruption of the paste matrix by freezing can cause reduced strength gain and increased porosity. (44047)

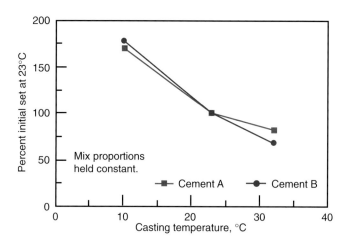

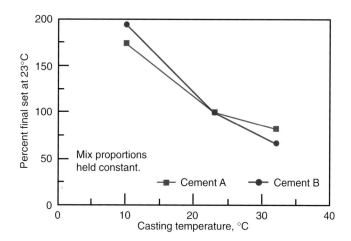

Fig. 14-3. Initial set characteristics as a function of casting temperature (top), and final set characteristics as a function of casting temperature (bottom) (Burg 1996).

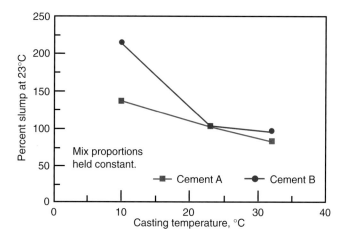

Fig. 14-4 Slump characteristics as a function of casting temperature (Burg 1996).

entrained concrete. See Chapter 8, "Air-Entrained Concrete," for more information.

STRENGTH GAIN OF CONCRETE AT LOW TEMPERATURES

Temperature affects the rate at which hydration of cement occurs—low temperatures retard hydration and consequently retard the hardening and strength gain of concrete.

If concrete is frozen and kept frozen above about -10°C, it will gain strength slowly. Below that temperature, cement hydration and concrete strength gain cease. Fig. 14-3 illustrates the effect of cool temperatures on setting time. Fig. 14-4 illustrates the effects of casting temperature on slump. Figs. 14-5 and 14-6 show the age-compressive strength relationship for concrete that has been cast and cured at various temperatures. Note in Fig. 14-6 that concrete cast and cured at 4°C and 13°C had relatively low strengths for the first week; but after 28 days—when all specimens were moist-cured at 23°C—strengths for the 4°C and 13°C concretes grew faster than the 23°C concrete and at one year they were slightly higher.

Higher-early strengths can be achieved through use of Type 30 (III) High-early-strength cement as illustrated in Fig. 14-7. Principal advantages occur during the first 7 days. At a 4°C curing temperature, the advantages of Type 30 (III) cement are more pronounced and persist longer than at the higher temperature.

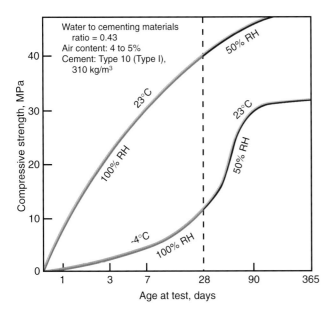

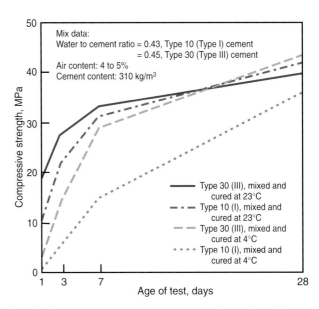

Fig. 14-7. Early-age compressive-strength relationships for Type 10 (I) and Type 30 (III) portland cement concretes mixed and cured at 4°C compared to 23°C (Klieger 1958).

Fig. 14-5. Effect of temperature conditions on the strength development of concrete. Concrete for the lower curve was cast at 4°C and placed immediately in a curing room at -4°C. Both concretes received 100% relative-humidity curing for first 28 days followed by 50% relative-humidity curing (Klieger 1958).

HEAT OF HYDRATION

Concrete generates heat during hardening as a result of the chemical process by which cement reacts with water to form a hard, stable paste. The heat generated is called heat of hydration; it varies in amount and rate for different cements. Dimensions of the concrete placement, ambient air temperature, initial concrete temperature, water-cementing materials ratio, admixtures, and the composition, fineness, and amount of cementing materials all affect heat generation and buildup.

Heat of hydration is useful in winter concreting as it contributes to the heat needed to provide a satisfactory curing temperature; often without other temporary heat sources, particularly in more massive elements.

Concrete must be delivered at the proper temperature and account must be taken of the temperature of forms, reinforcing steel, the ground, or other concrete on which the fresh concrete is cast. Concrete should not be cast on frozen concrete or on frozen ground.

Fig. 14-8 shows a concrete pedestal being covered with a tarpaulin just after the concrete was placed. Tarpaulins and insulated blankets are often necessary to retain the heat of hydration more efficiently and keep the concrete as warm as possible. Thermometer readings of the concrete's temperature will tell whether the covering is adequate. The heat liberated during hydration will offset to a considerable degree the loss of heat during placing, finishing, and early curing operations. As the heat of hydration slows down, the need to cover the concrete becomes more important.

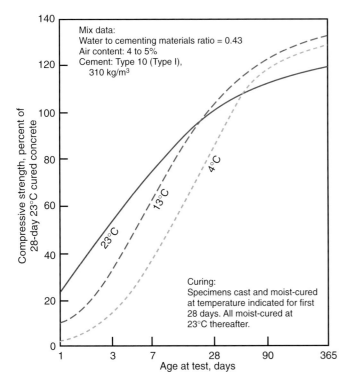

Fig. 14-6. Effect of low temperatures on concrete compressive strength at various ages. Note that for this particular mixture made with Type 10 (I) cement, the best temperature for long-term strength (1 year) was 13°C (Klieger 1958).

Fig. 14-8. Concrete footing pedestal being covered with a tarpaulin to retain the heat of hydration. (69870)

SPECIAL CONCRETE MIXTURES

High strength at an early age is desirable in winter construction to reduce the length of time temporary protection is required. The additional cost of high-early-strength concrete is often offset by earlier reuse of forms and shores, savings in the shorter duration of temporary heating, earlier setting times that allows the finishing of flatwork to begin sooner, and earlier use of the structure. High-early-strength concrete can be obtained by using one or a combination of the following:

1. Type 30 (III) High-early-strength cement
2. Additional portland cement (60 to 120 kg/m³)
3. Chemical accelerators

Small amounts of an accelerator such as calcium chloride (at a maximum dosage of 2% by mass of portland cement) can be used to accelerate the setting and early-age strength development of concrete in cold weather. Accelerators containing chlorides should not be used where there is an in-service potential for corrosion, such as in concrete members containing steel reinforcement or where aluminum or galvanized inserts will be used. Chlorides are not recommended for concretes exposed to soil or water containing sulphates or for concretes susceptible to alkali-aggregate reaction.

Accelerators must not be used as a substitute for proper curing and frost protection. Specially designed accelerating admixtures allow concrete to be placed at temperatures down to -7°C. The purpose of these admixtures is to reduce the time of initial setting, but not necessarily to speed up strength gain. Covering concrete to keep out moisture and to retain heat of hydration is still necessary. Furthermore, traditional antifreeze solutions, as used in automobiles, should never be used; the quantity of these materials needed to appreciably lower the freezing point of concrete is so great that strength and other properties can be seriously affected.

Since the goal of using special mixtures during cold weather concreting is to reduce the time of setting, a low water-cementing materials ratio, low-slump concrete is particularly desirable, especially for cold-weather flatwork; concrete mixtures with higher slumps usually take longer to set. In addition, evaporation is minimized so that finishing can be accomplished quicker (Fig. 14-9).

AIR-ENTRAINED CONCRETE

Entrained air is particularly desirable in any concrete placed during freezing weather. Concrete that is not air entrained can suffer strength loss and internal as well as surface damage as a result of freezing and thawing (Fig. 14-10). Air entrainment provides the capacity to absorb stresses due to ice formation within the concrete. See Chapter 8, "Air-Entrained Concrete."

Air entrainment should always be used for construction during the freezing months. The exception is concrete work done under roof where there is no chance that rain, snow, or water from other sources can saturate the concrete and where there is no chance of freezing.

The likelihood of water saturating a concrete floor during construction is very real. Fig. 14-11 shows conditions in the upper story of an apartment building during

Fig. 14-9. Finishing this concrete flatwork can proceed because a windbreak has been provided, there is adequate heat under the slab, and the concrete has low slump. (43465)

winter construction. Snow fell on the top deck. When heaters were used below to warm the deck, the snow melted. Water ran through floor openings down to a level that was not being heated. The water-saturated concrete froze, which caused a strength loss, particularly at the floor surface. This could also result in greater deflection of the floor and a surface that is less wear-resistant than it might have been.

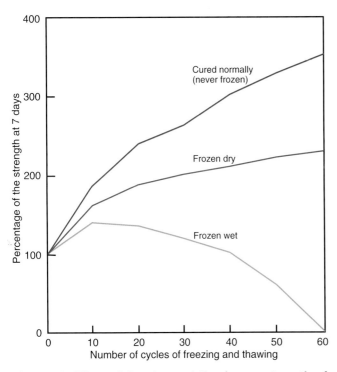

Fig. 14-10. Effect of freezing and thawing on strength of concrete that does not contain entrained air (cured 7 days before first freeze) (Powers 1956).

Fig. 14-11. Example of a concrete floor that was saturated with rain, snow, or water and then frozen, showing the need for air entrainment. (69869)

TEMPERATURE OF CONCRETE

Temperature of Concrete as Mixed

The temperature of fresh concrete as *placed* should not be less than the minimums given in Table 14-1 for the respective thickness of section. Note that lower concrete temperatures are recommended for more massive concrete sections because heat generated during hydration is dissipated less rapidly in heavier sections. Also, at lower ambient air temperatures more heat is lost from concrete during transporting and placing; hence, the concrete temperature as mixed must be higher for colder weather.

There is little advantage in using fresh concrete at a temperature much above 20°C. Higher concrete temperatures do not afford proportionately longer protection from freezing because the rate of heat loss is greater. Also, high concrete temperatures are undesirable since they increase thermal shrinkage after hardening, require more mixing water for the same slump, and contribute to the possibility of plastic-shrinkage cracking (caused by rapid moisture loss through evaporation). Therefore, the temperature of the concrete as *mixed* should not be more than 5°C above the minimums recommended in Table 14-1.

Aggregate Temperature. The temperature of aggregates varies with weather and type of storage. Aggregates usually contain frozen lumps and ice when the temperature is below freezing. Frozen aggregates must be thawed to avoid pockets of aggregate in the concrete after batching, mixing, and placing. If thawing takes place in the mixer, excessively high water contents in conjunction with the cooling effect due to the ice melting must be avoided.

At temperatures above freezing it is seldom necessary to heat aggregates, the desired concrete temperature can usually be obtained by heating only the mixing water. At temperatures below freezing, in addition to heating the mixing water, often only the fine aggregate needs to be heated to produce concrete of the required temperature, provided the coarse aggregate is free of frozen lumps.

Table 14-1. Permissible Concrete Temperatures at Placing

Thickness of section, m	Temperatures, °C	
	Minimum	Maximum
Less than 0.3	10	35
0.3-1	10	30
1-2	5	25
More than 2	5	20

Source: CSA Standard A23.1.

Three of the most common methods for heating aggregates are: (1) storing in bins or weigh hoppers heated by steam coils or live steam; (2) storing in silos heated by hot air or steam coils; and (3) stockpiling over heated slabs, stem vents or pipes. Although heating aggregates stored in bins or weigh hoppers is most commonly used, the volume of aggregate that can be heated is often limited and quickly consumed during production. Circulating steam through pipes over which aggregates are stockpiled is a recommended method for heating aggregates. Stockpiles can be covered with tarpaulins to retain and distribute heat and to prevent formation of ice. Live steam, preferably at pressures of 500 to 900 kPa, can be injected directly into the aggregate pile to heat it, but the resultant variable moisture content in aggregates might result in erratic mixing-water control.

On small jobs aggregates can be heated by stockpiling over metal culvert pipes in which fires are maintained. Care should be taken to prevent scorching the aggregates.

Mixing-Water Temperature. Of the ingredients used to make concrete, mixing water is the easiest and most practical to heat. The mass of aggregates and cement in concrete is much greater than the mass of water; however for the same mass, water can store about five times as much heat as can cement and aggregate. For cement and aggregates, the average specific heat (that is, heat units required to raise the temperature 1°C per kg of material can be assumed as 0.925 kJ compared to 4.187 kJ for water.

Fig. 14-12 shows the effect of temperature of materials on temperature of fresh concrete. The chart is based on the equation

$$T = \frac{0.22\left(T_a\,M_a + T_c\,M_c\right) + T_w\,M_w + T_{wa}\,M_{wa}}{0.22\left(M_a + M_c\right) + M_w + M_{wa}}$$

where

T = temperature in degrees Celsius of the fresh concrete

T_a, T_c, T_w, and T_{wa} = temperature in degrees Celsius of the aggregates, cement, added mixing water, and free moisture on aggregates, respectively; generally $T_a = T_{wa}$

M_a, M_c, M_w, and M_{wa} = mass in kilograms of the aggregates, cement, free moisture on aggregates, and mixing water, respectively

If the weighted average temperature of aggregates and cement is above 0°C, the proper mixing-water temperature for the required concrete temperature can be selected from Fig. 14-12.

To avoid the possibility of a quick or flash set of the concrete when either water or aggregates are heated to above 38°C, they should be combined in the mixer first before the cement is added. If this mixer-loading sequence is followed, water temperatures up to the boiling point can be used, provided the aggregates are cold enough to reduce the final temperature of the aggregates and water mixture to appreciably less than 38°C.

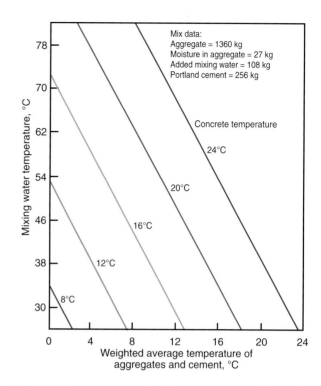

Fig. 14-12. Temperature of mixing water needed to produce heated concrete of required temperature. Temperatures are based on the mixture shown but are reasonably accurate for other typical mixtures.

Fluctuations in mixing-water temperature from batch to batch should be avoided. The temperature of the mixing water can be adjusted by blending hot and cold water.

Temperature of Concrete as Placed and Maintained

There will be some temperature loss after mixing while the truck mixer is traveling to the construction site and waiting to discharge its load. The concrete should be placed in the forms before its temperature drops below that given in Table 14-1; that concrete temperature should be maintained for the duration of the protection period given in Chapter 12 under "Curing Period and Temperature."

Cooling After Protection

CSA Standard A23.1 requires that, to avoid cracking of the concrete due to sudden temperature change near the end of the curing period, the protection should not be completely removed until the concrete has cooled to the temperature differential given in Table 14-2. Appendix D of the Standard provides guidance for the permissible stripping time for insulated formwork. The determination and recording of air and concrete temperatures is the responsibility of the owner.

CONTROL TESTS

Thermometers are needed to check the concrete temperatures as delivered, as placed, and as maintained. An inexpensive pocket thermometer is shown in Fig. 14-13.

After the concrete has hardened, temperatures can be checked with special surface thermometers or with an ordinary thermometer that is kept covered with insulating blankets. A simple way to check temperature below the concrete surface is shown in Fig. 14-14. Instead of filling the hole shown in Fig. 14-14 with a fluid, it can be fitted with insulation except at the bulb.

Concrete test cylinders must be maintained at a temperature between 15°C and 25°C at the jobsite for at least 20 hours until they are taken to a laboratory for curing. During this 20-hour period cylinders should be kept in a curing box and covered with a nonabsorptive, nonreactive plate or impervious plastic bag, with the temperature accurately controlled by a thermostat (Fig. 14-15). When stored in an insulated curing box outdoors, cylinders are less likely to be jostled by vibrations than if they were left on the floor of a trailer. If kept in a trailer where the heat may be turned off at night or over a weekend or holiday, the cylinders would not be at the prescribed curing temperatures during this critical period.

In addition to laboratory-cured cylinders, it is useful to field-cure some test cylinders in order to monitor actual curing conditions on the job in cold weather. It is sometimes difficult to find the right locations for field curing. Differences in the surface to volume ratios between cylinders and the structure, in conjunction with differences in mass, make correlating field-cured cylinder strengths to in-place strengths difficult. A preferred location is in a boxout in a floor slab or wall with thermal insulation for cover. When placed on a formwork ledge just below a heated, suspended floor, possible high temperatures there will not duplicate the average temperature in the slab, nor the lowest temperature on top of the slab. Still, field-cured cylinders are more indicative of actual concrete strength

Fig. 14-13. A bimetallic pocket thermometer with a metal sensor suitable for checking fresh concrete temperatures. (69881, 69882)

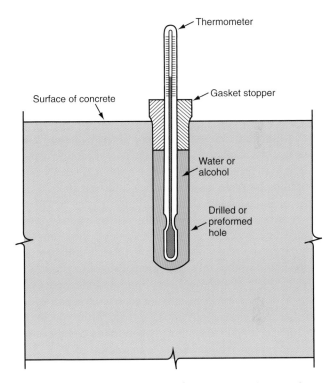

Fig. 14-14. Scheme for measuring concrete temperatures below the surface with a glass thermometer.

Table 14-2. Maximum Permissible Temperature Differential Between Concrete Surface and Ambient Air with Wind up to 25 km/h

| Thickness of concrete, m | Maximum permissible termperature differential, °C | | | | |
| | Length to height ratio of structure* | | | | |
	0**	3	5	7	20 or more
0.3	29	22	19	17	12
0.6	22	18	16	15	12
0.9	18	16	15	14	12
1.2	17	15	14	13	12
1.5	16	14	13	13	12

  * Length shall be the longer restrained dimension and the height shall be considered the unrestrained dimension.
 ** Very high, narrow structure such as columns.
Source: CSA Standard A23.1.

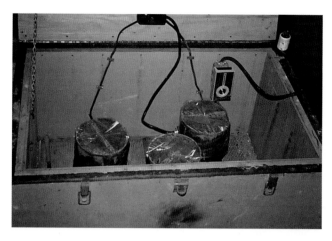

Fig. 14-15. Insulated curing box with thermostat for curing test cylinders. Heat is supplied by electric rubber heating mats on the bottom. A wide variety of designs are possible for curing boxes. (43463)

than laboratory-cured cylinders. Particular care should be taken to protect compressive strength test cylinders from freezing; their small mass may not generate enough heat of hydration to protect them.

Molds stripped from cylinders after the first 20 ± 4 hours must be wrapped tightly in plastic bags or laboratory curing started immediately. When cylinders are picked up for delivery to the laboratory, they must be maintained at a temperature of 15°C to 25°C until they are placed in the laboratory curing room. Demolding time and transportation to the laboratory may be extended to 76 hours for cylinders representing concrete having a specified strength of less than 35 MPa, provided the specimens are stored in an environmentally controlled facility at the project site that maintains the temperature at 20 ± 5 °C immediately adjacent to the specimens, and prevents loss of moisture from the specimens.

Cast-in-place cylinders (ASTM C 873) and non-destructive testing methods discussed in Chapter 16, as well as maturity techniques discussed later in this chapter, are helpful in monitoring in-place concrete strength.

CONCRETING ON GROUND

Concreting on ground during cold weather involves some extra effort and expense, but many contractors find that it more than pays for itself. In winter, the site around the structure may be frozen rather than a morass of mud. The concrete will furnish some if not all of the heat needed for proper curing. Internal concrete temperatures should be monitored. Insulated blankets or simple enclosures are easily provided. Embankments are frozen and require less bracing. With a good start during the winter months, construction gets above the ground before warmer weather arrives.

Placing concrete on the ground involves different procedures from those used at an upper level: (1) the ground must be thawed before placing concrete; (2) cement hydration will furnish some of the curing heat; (3) construction of enclosures is much simpler and use of insulating blankets may be sufficient; (4) in the case of a floor slab, a *vented* heater is required if the area is enclosed; and (5) hydronic heaters can be used to thaw subgrades using insulated blankets or to heat enclosures without concern for carbonation. For more on hydronic heaters, see "Heaters" later in this chapter.

Once cast, footings should be backfilled as soon as possible with unfrozen fill. Concrete should never be placed on a frozen subgrade or backfilled with frozen fill; otherwise once they thaw, uneven settlements may occur causing cracking.

CSA A23.1 (ACI Committee 306) requires that concrete not be placed on any surface that would lower the temperature of the concrete in place below the minimum values shown in Table 14-1, except when nonchloride, noncorrosive accelerators are being used in accordance with the Standard. Concrete placement temperatures should be kept as close as possible to the suggested minimum values as higher temperatures result in an increase in mixing water, increased slump loss, and an increase in thermal shrinkage.

When the subgrade is frozen to a depth of approximately 80 mm, the surface region can be thawed by (1) steaming; (2) spreading a layer of hot sand, gravel, or other granular material where the grade elevations allow it; (3) removing and replacing with unfrozen fill; (4) covering the subgrade with insulation for a few days; or (5) by the use of hydronic heaters under insulated blankets which can thaw frozen ground at a rate of 0.3 m per 24 hours to a depth up to 3 m (Grochoski 2000). Placing concrete for floor slabs and exposed footings should be delayed until the ground thaws and warms sufficiently to ensure that it will not freeze again during the protection and curing period.

Slabs can be cast on ground at ambient temperatures as low as 2°C as long as the minimum concrete temperature as placed is not less than shown in Table 14-1. Although surface temperatures need not be higher than a few degrees above freezing, they also should preferably not be more than 5°C higher than the minimum placement temperature either. The duration of curing should not be less than that described in Chapter 12 for the appropriate exposure classification. Because of the risk of surface imperfections that might occur on exterior concrete placed in late fall and winter, many concrete contractors choose to delay concrete placement until spring. By waiting until spring, temperatures will be more favorable for cement hydration; this will help generate adequate strengths along with sufficient drying so the concrete can resist freeze-thaw damage.

CONCRETING ABOVEGROUND

Working aboveground in cold weather usually involves several different approaches compared to working at ground level:

1. The concrete mixture need not be changed to generate more heat because portable heaters can be used to heat the undersides of floor and roof slabs. Nevertheless, there are advantages to having a mix that will produce a high strength at an early age; for example, artificial heat can be cut off sooner, and forms can be recycled faster. (See Table 14-3 which has been adapted from ACI 306, *Cold Weather Concreting.*)
2. Enclosures must be constructed to retain the heat under floor and roof slabs.
3. Portable heaters used to warm the underside of formed concrete can be direct-fired heating units (without venting).

Before placing concrete, the heaters under a formed deck should be turned on to preheat the forms and melt any snow or ice remaining on top. Temperature requirements for surfaces in contact with fresh concrete are the same as those outlined in the previous section "Concreting on Ground." Metallic embedments at temperatures below the freezing point may result in local freezing that decreases the bond between concrete and steel reinforce-ment. ACI Committee 306 suggests that a reinforcing bar having a cross-sectional area of about 650 mm² should have a temperature of at least -12°C immediately before being surrounded by fresh concrete at a temperature of at least 13°C. Caution and additional study are required before definitive recommendations can be formulated. See ACI 306 for additional information.

When slab finishing is completed, insulating blankets or other insulation must be placed on top of the slab to ensure that proper curing temperatures are maintained. The insulation value (R) necessary to maintain the concrete surface temperature of walls and slabs aboveground at 10°C or above for 7 days may be estimated from Fig. 14-16. To maintain a temperature for longer periods, more insulation is required. ACI 306 has additional graphs and tables for slabs placed on ground at a temperature of 2°C. Insulation can be selected based on R values provided by insulation manufacturers or by using the information in Table 14-4.

When concrete strength development is not determined, a conservative estimate can be made if adequate protection at the recommended temperature is provided for the duration of time found in Table 14-3. However, the actual amount of insulation and length of the protection period should be determined from the monitored in-place concrete temperature and the desired strength. A correla-

Table 14-3.

A. Recommended Duration of Concrete Temperature in Cold Weather–Air-Entrained Concrete*

Service category	Protection from early-age freezing		For safe stripping strength	
	Conventional concrete,** days	High-early strength concrete,† days	Conventional concrete,** days	High-early-strength concrete,† days
No load, not exposed‡ favourable moist-curing	2	1	2	1
No load, exposed, but later has favourable moist-curing	3	2	3	2
Partial load, exposed			6	4
Fully stressed, exposed			See Table B below	

B. Recommended Duration of Concrete Temperature for Fully Stressed, Exposed, Air-Entrained Concrete

Required percentage of standard-cured 28-day strength	Days at 10°C			Days at 21°C		
	Type of portland cement			Type of portland cement		
	10 (I or GU)	20 (II or MS)	30 (III or HE)	10 (I or GU)	20 (II or MS)	30 (III or HE)
50	6	9	3	4	6	3
65	11	14	5	8	10	4
85	21	28	16	16	18	12
95	29	35	26	23	24	20

* Adapted from Tables 5.1 and 5.3 of ACI 306. Cold weather is defined as that in which average daily temperature is less than 4°C for 3 successive days except that if temperatures above 10°C occur during at least 12 hours in any day, the concrete should no longer be regarded as winter concrete and normal curing practice should apply. For recommended concrete temperatures, see Table 14-1. For concrete that is *not* air entrained, ACI Committee 306 states that protection for durability should be at least twice the number of days listed in Table A.

    Part B was adapted from Table 6.8 of ACI 306R-88. The values shown are approximations and will vary according to the thickness of concrete, mix proportions, etc. They are intended to represent the ages at which supporting forms can be removed. For recommended concrete temperatures, see Table 14-1.

** Made with Type 10, Type 20 (ASTM Type I, II, GU, or MS) portland cement.

† Made with Type 10, Type 30 (ASTM Type III or HE) cement, or an accelerator, or an extra 60 kg/m³ of cement.

‡ "Exposed" means subject to freezing and thawing.

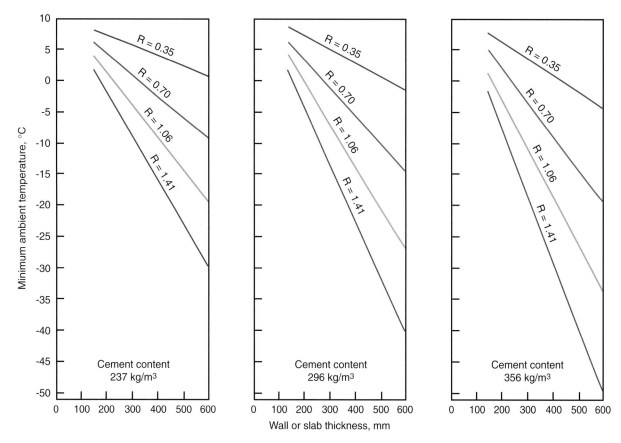

Fig. 14-16. Thermal resistance (R) of insulation required to maintain the concrete surface temperature of walls and slabs aboveground at 10°C or above for 7 days. Concrete temperature as placed: 10°C. Maximum wind velocity: 24 km/h. Note that in order to maintain a certain minimum temperature for a longer period of time, more insulation or a higher R value is required (adapted from ACI 306).

tion between curing temperature, curing time, and compressive strength can be determined from laboratory testing of the particular concrete mix used in the field (see "Maturity Concept" discussed later in this chapter). CSA Standard A23.1 suggests that when the concrete reaches a compressive strength of 7.0 MPa (ACI 306 states 3.5 MPa), concrete will normally have sufficient strength to resist early frost damage. If the concrete will be in a saturated condition when frozen (Exposure Class F-1), the concrete should be properly air entrained and must have developed a compressive strength of 30 MPa.

Corners and edges are particularly vulnerable during cold weather. As a result, the thickness of insulation for these areas, especially on columns, should be about three times the thickness that is required to maintain the same for walls or slabs. On the other hand, if the ambient temperature rises much above the temperature assumed in selecting insulation values, the temperature of the concrete may become excessive. This increases the probability of thermal shock and cracking when forms are removed.

Temperature readings of insulated concrete should therefore be taken at regular intervals and should not vary from ambient air temperatures by more than the values given in Table 14-2. In addition, insulated concrete temperatures should not be allowed to rise much above 27°C. In case of a sudden increase in concrete temperature, up to say 35°C, it may be necessary to remove some of the insulation or loosen the formwork. The maximum temperature differential between the concrete interior and the concrete surface should be about 20°C to minimize cracking. The weather forecast should be checked and appropriate action taken for expected temperature changes.

Columns and walls should not be cast on frozen foundations having a temperature at or below 0°C because chilling of concrete in the bottom of the column or wall will retard strength development. Concrete should not be placed on any surface that would lower the temperature of the as-placed concrete below the minimum values shown in Table 14-1.

Table 14-4. Insulation Values of Various Materials

Material	Density, kg/m³	Thermal resistance, *R*, for 10-mm thickness of material,* (m² · °C)/W
Board and Slabs		
Expanded polyurethane	24	0.438
Expanded polystyrene, extruded smooth-skin surface	29 to 56	0.347
Expanded polystyrene, extruded cut-cell surface	29	0.277
Glass fiber, organic bonded	64 to 144	0.277
Expanded polystyrene, molded beads	16	0.247
Mineral fiber with resin binder	240	0.239
Mineral fiberboard, wet felted	256 to 272	0.204
Vegetable fiberboard sheathing	288	0.182
Cellular glass	136	0.201
Laminated paperboard	480	0.139
Particle board (low density)	590	0.128
Plywood	545	0.087
Loose fill		
Wood fiber, soft woods	32 to 56	0.231
Perlite (expanded)	80 to 128	0.187
Vermiculite (exfoliated)	64 to 96	0.157
Vermiculite (exfoliated)	112 to 131	0.148
Sawdust or shavings	128 to 240	0.154

Material	Material thickness, mm	Thermal resistance, *R*, for thickness of material,* (m² · °C)/W
Mineral fiber blanket, fibrous form (rock, slag, or glass) 5 to 32 kg/m³	50 to 70	1.23
	75 to 85	1.90
	90 to 165	3.34
Mineral fiber loose fill (rock, slag, or glass) 10 to 32 kg/m³	95 to 125	1.90
	165 to 220	3.34
	190 to 250	3.87
	260 to 350	5.28

* Values are from *ASHRAE Handbook of Fundamentals,* American Society of Heating, Refrigerating, and Air-conditioning Engineers, Inc., New York, 1977 and 1981.
  R values are the reciprocal of U values (conductivity).

ENCLOSURES

Heated enclosures are very effective for protecting concrete in cold weather, but are probably the most expensive too (Fig. 14-17). Enclosures can be of wood, canvas tarpaulins, or polyethylene film (Fig. 14-18). Prefabricated, rigid-plastic enclosures are also available. Plastic enclosures that admit daylight are the most popular but temporary heat in these enclosures can be an expensive option.

When enclosures are being constructed below a deck, the framework can be extended above the deck to serve as a windbreak. Typically, a height of 2 m will protect concrete and construction personnel against biting winds that cause temperature drops and excessive evaporation. Wind breaks could be taller or shorter depending on anticipated wind velocities, ambient temperatures, relative humidity, and concrete placement temperatures.

Fig. 14-17. Even in the winter, an outdoor swimming pool can be constructed if a heated enclosure is used. (43453)

Fig. 14-18. (top) Tarpaulin heated enclosure maintains an adequate temperature for proper curing and protection during severe and prolonged winter weather. (bottom) Polyethylene plastic sheets admitting daylight are used to fully enclose a building frame. The temperature inside is maintained at 10°C with space heaters. (69877, 69878)

Enclosures can be made to be moved with flying forms; more often, though, they must be removed so that the wind will not interfere with maneuvering the forms into position. Similarly, enclosures can be built in large panels like gang forms with the windbreak included (Fig. 14-1).

INSULATING MATERIALS

Heat and moisture can be retained in the concrete by covering it with commercial insulating blankets or batt insulation (Fig. 14-19). The effectiveness of insulation can be determined by placing a thermometer under it and in contact with the concrete. If the temperature falls below the minimum required in Table 14-1, additional insulating material, or material with a higher R value, should be applied. Corners and edges of concrete are most vulnerable to freezing. In view of this, temperatures at these locations should be checked often.

The thermal resistance (R) values for common insulating materials are given in Table 14-4. For maximum efficiency, insulating materials should be kept dry and in close contact with concrete or formwork.

Fig. 14-19. Stack of insulating blankets. These blankets trap heat and moisture in the concrete, providing beneficial curing. (43460)

Concrete pavements can be protected from cold weather by insulating blankets or by spreading 300 mm or more of dry straw or hay on the surface for insulation. Tarpaulins, polyethylene film, or waterproof paper should be used as a protective cover over the straw or hay to make the insulation more effective and prevent it from blowing away. The straw or hay should be kept dry or its insulation value will drop considerably.

Stay-in-place insulating concrete forms (ICF) became popular in North America in the 1990s. They have the inherent benefit of being able to provide an insulation value for cold-weather construction (Fig. 14-20). Forms built for repeated use often can be economically insulated with commercial blanket or batt insulation. The insulation should have a tough moisture-proof covering to withstand handling abuse and exposure to the weather. Rigid insulation can also be used (Fig. 14-21).

Insulating blankets for construction are made of fiberglass, sponge rubber, open-cell polyurethane foam, vinyl

Fig. 14-20. Insulating concrete forms (ICF) permit concreting in cold weather. (69699)

Fig. 14-21. With air temperatures down to -23°C, concrete was cast in this insulated column form made of 19-mm high-density plywood inside, 25-mm rigid polystyrene in the middle, and 13-mm rough plywood outside. RSI value: 1.0 m² · °C/W. (43461)

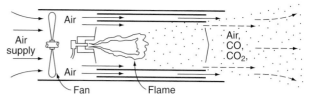

2) Direct-fired heater

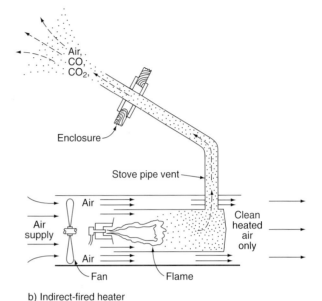

b) Indirect-fired heater

Fig. 14-22. Two types of air heaters.

foam, mineral wool, or cellulose fibers. The outer covers are made of canvas, woven polyethylene, or other tough fabrics that will withstand rough handling. The R value for a typical insulating blanket is about 1.2 m² · °C/W for 50 to 70 mm thickness, but since R values are not marked on the blankets, their effectiveness should be checked with a thermometer. If necessary, they can be used in two or three layers to attain the desired insulation.

HEATERS

Three types of heaters are used in cold-weather concrete construction: direct fired, indirect fired, and hydronic systems (Figs. 14-22 to 14-25). Indirect-fired heaters are vented to remove the products of combustion. Where heat is to be supplied to the top of fresh concrete—for example, a floor slab—vented heaters are required. Carbon dioxide (CO_2) in the exhaust must be vented to the outside and prevented from reacting with the fresh concrete (Fig. 14-23). Direct-fired units can be used to heat the enclosed space beneath concrete placed for a floor or a roof deck (Fig. 14-24).

 Hydronic systems transfer heat by circulating a glycol/water solution in a closed system of pipes or hoses

Fig. 14-23. An indirect-fired heater. Notice vent pipe that carries combustion gases outside the enclosure. (43459)

Fig. 14-24. A direct-fired heater installed through the enclosure, thus using a fresh air supply. (69875)

Fig. 14-25. Hydronic system showing hoses (top) laying on soil to defrost subgrade and (bottom) warming the forms while fresh concrete is pumped in. (68345, 68344)

(see Fig. 14-25). These systems transfer heat more efficiently than forced air systems without the negative effects of exhaust gases and drying of the concrete from air movement. The specific heat of water/glycol solutions is more than six times greater than air. As a result, hydronic heaters can deliver very large quantities of heat at low temperature differentials of 5°C or less between the heat transfer hose and the concrete. Cracking and curling induced by temperature gradients within the concrete are almost eliminated along with the danger of accidentally overheating the concrete and damaging long-term strength gain.

Typical applications for hydronic systems include thawing and preheating subgrades. They are also used to cure elevated and on-grade slabs, walls, foundations, and columns. To heat a concrete element, hydronic heating hoses are usually laid on or hung adjacent to the structure and covered with insulated blankets and sometimes plastic sheets. Usually, construction of temporary enclosures is not necessary. Hydronic systems can be used over areas much larger than would be practical to enclose. If a heated enclosure is necessary for other work, hydronic hoses can be sacrificed (left under a slab on grade) to make the slab a radiant heater for the structure built above (Grochoski 2000).

Any heater burning a fossil fuel produces carbon dioxide (CO_2); this gas will combine with calcium hydroxide on the surface of fresh concrete to form a weak layer of calcium carbonate that interferes with cement hydration (Kauer and Freeman 1955). The result is a soft, chalky surface that will dust under traffic. Depth and degree of carbonation depend on concentration of CO_2, curing temperature, humidity, porosity of the concrete, length of exposure, and method of curing. Direct-fired heaters, therefore, should not be permitted to heat the air over concreting operations—at least until 24 hours have elapsed. In addi-

tion, the use of gasoline-powered construction equipment should be restricted in enclosures during that time. If unvented heaters are used, immediate wet curing or the use of a curing compound will minimize carbonation.

Carbon monoxide (CO), another product of combustion, is not usually a problem unless the heater is using recirculated air. Four hours of exposure to 200 parts per million of CO will produce headaches and nausea. Three hours of exposure to 600 ppm can be fatal. The American National Standard Safety Requirements for Temporary and Portable Space Heating Devices and Equipment Used in the Construction Industry (ANSI A10.10) limits concentrations of CO to 50 ppm at worker breathing levels. The standard also establishes safety rules for ventilation and the stability, operation, fueling, and maintenance of heaters.

A salamander is an inexpensive combustion heater without a fan that discharges its combustion products into the surrounding air; heating is accomplished by radiation from its metal casing. Salamanders are fueled by coke, oil, wood, or liquid propane. They are but one form of a direct-fired heater. A primary disadvantage of salamanders is the high temperature of their metal casing, a definite fire hazard. Salamanders should be placed so that they will not overheat formwork or enclosure materials. When placed on floor slabs, they should be elevated to avoid scorching the concrete.

Some heaters burn more than one type of fuel. The approximate heat values of fuels are as follows:

No. 1 fuel oil	37,700 kJ/L
Kerosene	37,400 kJ/L
Gasoline	35,725 kJ/L
Liquid-propane gas	25,500 kJ/L
Natural gas	37,200 kJ/m³

The output rating of a portable heater is usually the heat content of the fuel consumed per hour. A rule of thumb is that about 134,000 kJ are required for each 100 m³ of air to develop a 10°C temperature rise.

Electricity can also be used to cure concrete in winter. The use of large electric blankets equipped with thermostats is one method. The blankets can also be used to thaw subgrades or concrete foundations.

Use of electrical resistance wires that are cast into the concrete is another method. The power supplied is under 50 volts, and from 7.0 to 23.5 MJ of electricity per cubic metre of concrete is required, depending on the circumstances. The method has been used in the Montreal, Quebec, area for many years. Where electrical resistance wires are used, insulation should be included during the initial setting period. If insulation is removed before the recommended time, the concrete should be covered with an impervious sheet and the power continued for the required time.

Steam is another source of heat for winter concreting. Live steam can be piped into an enclosure or supplied through radiant heating units. In choosing a heat source, it must be remembered that the concrete itself supplies heat through hydration of cement; this is often enough for curing needs if the heat can be retained within the concrete with insulation.

DURATION OF HEATING

After concrete is in place, it should be protected and kept at the recommended curing temperatures and moisture conditions discussed in Chapter 12, "Curing Concrete." These curing temperatures should be maintained until sufficient strength is gained to withstand exposure to low temperatures, anticipated environment, and construction and service loads. The length of protection required to accomplish this will depend on the cement type and amount, whether accelerating admixtures were used, and the loads that must be carried. Recommended minimum periods of protection are given in Table 14-3. The duration of heating structural concrete that requires the attainment of full service loading before forms and shores are removed should be based on the adequacy of in-place compressive strengths rather than an arbitrary time period. If no data are available, a conservative estimate of the

length of time for heating and protection can be made using Table 14-3.

Moist Curing

Strength gain stops when moisture required for curing is no longer available. Concrete retained in forms or covered with insulation seldom loses enough moisture at 5°C to 15°C to impair curing. However, a positive means of providing moist curing is needed to offset drying from low wintertime humidities and heaters used in enclosures during cold weather.

Live steam exhausted into an enclosure around the concrete is an excellent method of curing because it provides both heat and moisture. Steam is especially practical in extremely cold weather because the moisture provided offsets the rapid drying that occurs when very cold air is heated.

Liquid membrane-forming compounds can be used for early curing of concrete surfaces within heated enclosures.

Terminating the Heating Period

Rapid cooling of concrete at the end of the heating period should be avoided. Sudden cooling of the concrete surface while the interior is still warm may cause thermal cracking, especially in massive sections such as bridge piers, abutments, dams, and large structural members; thus cooling should be gradual. A safe temperature differential between a concrete surface and the ambient air temperature with wind up to 25 km/h can be obtained in Table 14-2 Protection should not be removed until the concrete has reached these temperature differentials. In the case of insulated formwork, the permissible stripping times can be determined by using Fig. D2 – *Appendix D, CSA Standard A23.1.* Gradual cooling can be accomplished by lowering the heat or by simply shutting off the heat and allowing the enclosure to cool to outside ambient air temperature.

FORM REMOVAL AND RESHORING

It is good practice in cold weather to leave forms in place as long as possible. Even within heated enclosures, forms serve to distribute heat more evenly and help prevent drying and local overheating.

The basic curing period as defined in CSA A23.1 is that concrete surfaces shall be cured for either 3 days at a minimum temperature of 10°C or for the time necessary to attain 40% of the specified 28-day compressive strength of the concrete. For structural safety the basic curing period is to be extended until the compressive strength level required for safety, as determined by the owner, is achieved. Table 14-3A can be used to determine the minimum time in days that vertical support for forms should be left in place. Before shores and forms are removed, fully stressed structural concrete should be tested to determine if in-place strengths are adequate, rather than waiting an arbitrary time period. In-place strengths can be monitored using one of the following: (1) field-cured cylinders, CSA A23.2-3C (ASTM C 31); (2) probe penetration tests (ASTM C 803); (3) cast-in-place cylinders (ASTM C 873); (4) pullout testing (ASTM C 900); or (5) maturity testing (ASTM C 1074). Many of these tests are indirect methods of measuring compressive strength; they require correlation in advance with standard cylinders before estimates of in-place strengths can be made.

If in-place compressive strengths are not documented, Table 14-3B lists conservative time periods in days to achieve various percentages of the standard laboratory-cured 28-day strength. The engineer issuing project drawings and specifications in cooperation with the formwork contractor must determine what percentage of the design strength is required (see ACI Committee 306). Side forms can be removed sooner than shoring and temporary false-work (ACI Committee 347).

MATURITY CONCEPT

The maturity concept is based on the principle that strength gain in concrete is a function of curing time and temperature. The maturity concept, as described in ACI 306R-88 and ASTM C 1074 can be used to evaluate strength development when the prescribed curing temperatures have not been maintained for the required time or when curing temperatures have fluctuated. The concept is expressed by the equation:

$$M = \Sigma \, (C + 10) \, \Delta t$$

where

    M = maturity factor
    Σ = summation
    C = concrete temperature, degrees Celsius
    Δt = duration of curing at temperature C, usually in hours

The equation is based on the premise that concrete gains strength (that is, cement continues to hydrate) at temperatures as low as -10°C.

Before construction begins, a calibration curve is drawn plotting the relationship between compressive strength and the maturity factor for a series of test cylinders (of the particular concrete mixture proportions) cured in a laboratory and tested for strength at successive ages.

The maturity concept is not precise and has some limitations. But, the concept is useful in checking the curing of concrete and estimating strength in relation to time and temperature. It presumes that all other factors affecting concrete strength have been properly controlled. With these limitations in mind, the maturity concept has gained greater acceptance for representing the compressive strength of the concrete for removal of shoring or opening a pavement to traffic; but it is no substitute for quality con-

trol and proper concreting practices (Malhotra 1974 and ACI Committee 347).

To monitor the strength development of concrete in place using the maturity concept, the following information must be available:

1. The strength-maturity relationship of the concrete used in the structure. The results of compressive strength tests at various ages on a series of cylinders made of a concrete similar to that used in the structure; this must be done to develop a strength-maturity curve. These cylinders are cured in a laboratory at 23°C ± 2°C.
2. A time-temperature record of the concrete in place. Temperature readings are obtained by placing expendable thermistors or thermocouples at varying depths in the concrete. The location giving the lowest values provides the series of temperature readings to be used in the computation (Fig. 14-26).

See ACI 306R-88 for sample calculations using the maturity concept.

Fig. 14-26. (top) Automatic temperature recorder. (bottom) Thermocouples and wiring at various depths in a caisson. (57368, 57366)

REFERENCES

ACI Committee 306, *Cold-Weather Concreting,* ACI 306R-88, reapproved 1997, American Concrete Institute, Farmington Hills, Michigan, 1997, 23 pages.

ACI Committee 347, *Guide to Formwork for Concrete,* ACI 347R-94, reapproved 1999, American Concrete Institute, Farmington Hills, Michigan, 1999, 34 pages.

Brewer, Harold W., *General Relation of Heat Flow Factors to the Unit Weight of Concrete,* Development Department Bulletin DX114, Portland Cement Association, http://www.portcement.org/pdf_files/DX114.pdf, 1967.

Burg, Ronald G., *The Influence of Casting and Curing Temperature on the Properties of Fresh and Hardened Concrete,* Research and Development Bulletin RD113, Portland Cement Association, 1996, 20 pages.

Copeland, L. E.; Kantro, D. L.; and Verbeck, George, *Chemistry of Hydration of Portland Cement,* Research Department Bulletin RX153, Portland Cement Association, http://www.portcement.org/pdf_files/RX153.pdf, 1960.

CSA Standard A23.1/A23.2-00, *Concrete Materials and Methods of Concrete Construction/Methods of Test for Concrete,* Canadian Standards Association, Toronto, 2000.

Grochoski, Chet, "Cold-Weather Concreting with Hydronic Heaters," *Concrete International,* American Concrete Institute, Farmington Hills, Michigan, April 2000, pages 51 to 55.

Kauer, J. A., and Freeman, R. L., "Effect of Carbon Dioxide on Fresh Concrete," *Journal of the American Concrete Institute Proceedings,* vol. 52, December 1955, pages 447 to 454. Discussion: December 1955, Part II, pages 1299 to 1304, American Concrete Institute, Farmington Hills, Michigan.

Klieger, Paul, *Curing Requirements for Scale Resistance of Concrete,* Research Department Bulletin RX082, Portland Cement Association, http://www.portcement.org/pdf_files/RX082.pdf, 1957.

Klieger, Paul, *Effect of Mixing and Curing Temperature on Concrete Strength,* Research Department Bulletin RX103, Portland Cement Association, http://www.portcement.org/pdf_files/RX103.pdf, 1958.

Malhotra, V. M., "Maturity Concept and the Estimation of Concrete Strength: A Review," Parts I and II, *Indian Concrete Journal,* vol. 48, Associated Cement Companies, Ltd., Bombay, April and May 1974.

McNeese, D. C., "Early Freezing of Non-Air-Entrained Concrete," *Journal of the American Concrete Institute Proceedings,* vol. 49, American Concrete Institute, Farmington Hills, Michigan, December 1952, pages 293 to 300.

NRMCA, *Cold Weather Ready Mixed Concrete,* Publication No. 130, National Ready Mixed Concrete Association, Silver Spring, Maryland, 1968.

Powers, T. C., *Resistance of Concrete to Frost at Early Ages,* Research Department Bulletin RX071, Portland Cement Association, http://www.portcement.org/pdf_files/RX071.pdf, 1956.

Powers, T. C., *Prevention of Frost Damage to Green Concrete,* Research Department Bulletin RX148, Portland Cement Association, http://www.portcement.org/pdf_files/RX148.pdf, 1962, 18 pages.

U.S. Bureau of Reclamation, *Concrete Manual,* 8th ed., U.S. Bureau of Reclamation, Denver, 1981.

CHAPTER 15
Volume Changes of Concrete

Concrete changes slightly in volume for various reasons, and understanding the nature of these changes is useful in planning or analyzing concrete work. If concrete were free of any restraints to deform, normal volume changes would be of little consequence; but since concrete in service is usually restrained by foundations, subgrades, reinforcement, or connecting members, significant stresses can develop. This is particularly true of tensile stresses.

Cracks develop because concrete is relatively weak in tension but quite strong in compression. Controlling the variables that affect volume changes can minimize high stresses and cracking. Tolerable crack widths should be considered in the structural design.

Volume change is defined merely as an increase or decrease in volume. Most commonly, the subject of concrete volume changes deals with linear expansion and contraction due to temperature and moisture cycles. But chemical effects such as carbonation shrinkage, sulphate attack, and the disruptive expansion of alkali-aggregate reactions also cause volume changes. Also, creep is a volume change or deformation caused by sustained stress or load. Equally important is the elastic or inelastic change in dimensions or shape that occurs instantaneously under applied load.

For convenience, the magnitude of volume changes is generally stated in linear rather than volumetric units. Changes in length are often expressed as a coefficient of length in parts per million, or simply as millionths. It is applicable to any length unit (for example, m/m); one millionth is 0.000001 m/m and 600 millionths is 0.000600 m/m. Change of length can also be expressed as a percentage; thus 0.06% is the same as 0.000600, which incidentally is approximately the same as 6 mm per 10 m. The volume changes that ordinarily occur in concrete are small, ranging in length change from perhaps 10 millionths up to about 1000 millionths.

EARLY AGE VOLUME CHANGES

The volume of concrete begins to change shortly after it is cast. Early volume changes, within 24 hours, can influence the volume changes (such as drying shrinkage) and crack formation in hardened concrete, especially for low water to cementing materials ratio concrete. Following are discussions on various forms of early volume change:

Chemical Shrinkage

Chemical shrinkage refers to the reduction in absolute volume of solids and liquids in paste resulting from cement hydration. The absolute volume of hydrated cement products is less than the absolute volume of cement and water before hydration. This change in volume of cement paste during the plastic state is illustrated by the first two bars in Fig. 15-1. This does not include air bubbles from mixing. Chemical shrinkage continues to occur at a microscopic scale as long as cement hydrates. After initial set, the paste cannot deform as much as when it was in a plastic state.

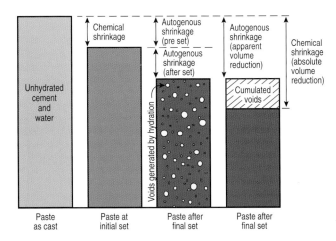

Fig. 15-1. Chemical shrinkage and autogenous shrinkage volume changes of fresh and hardened paste. Not to scale.

Therefore, further hydration and chemical shrinkage is compensated by the formation of voids in the microstructure (Fig. 15-1). Most of this volume change is internal and does not significantly change the visible external dimensions of a concrete element.

The amount of volume change due to chemical shrinkage can be estimated from the hydrated cement phases and their crystal densities or it can be determined by physical test as illustrated in Fig. 15-2. The Japan Concrete Institute has a test method for chemical shrinkage of cement paste (Tazawa 1999). An example of long-term chemical shrinkage for portland cement paste is illustrated in Fig. 15-3. Early researchers sometimes referred to chemical shrinkage as the absorption of water during hydration

(Powers 1935). Le Chatelier (1900) was the first to study chemical shrinkage of cement pastes.

Autogenous Shrinkage

Autogenous shrinkage is the macroscopic volume reduction (visible dimensional change) of cement paste, mortar, or concrete caused by cement hydration. The macroscopic volume reduction of autogenous shrinkage is much less than the absolute volume reduction of chemical shrinkage because of the rigidity of the hardened paste structure. Chemical shrinkage is the driving force behind autogenous shrinkage. The relationship between autogenous shrinkage and chemical shrinkage is illustrated in Figs. 15-1, 15-4, and 15-5. Some researchers and organizations consider that

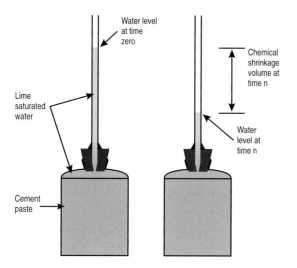

Fig. 15-2. Test for chemical shrinkage of cement paste showing flask for cement paste and pipet for absorbed water measurement.

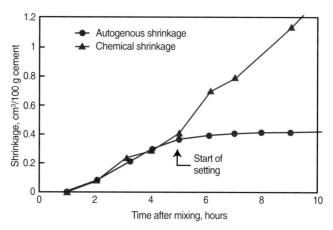

Fig. 15-4. Relationship between autogenous shrinkage and chemical shrinkage of cement paste at early ages (Hammer 1999).

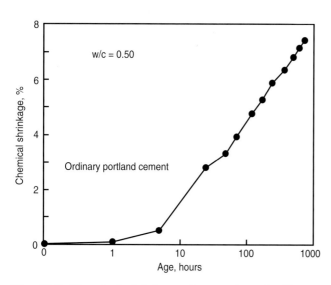

Fig. 15-3. Chemical shrinkage of cement paste (Tazawa 1999).

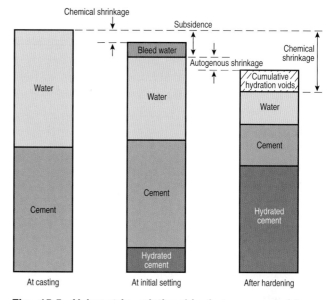

Fig. 15-5. Volumetric relationship between subsidence, bleed water, chemical shrinkage, and autogenous shrinkage. Only autogenous shrinkage after initial set is shown. Not to scale.

autogenous shrinkage starts at initial set while others evaluate autogenous shrinkage from time of placement.

When external water is available, autogenous shrinkage cannot occur. When external water is not available, cement hydration consumes pore water resulting in self desiccation of the paste and a uniform reduction of volume (Copeland and Bragg 1955). Autogenous shrinkage increases with a decrease in water to cementing materials ratio and with an increase in the amount of cement paste. Normal concrete has negligible autogenous shrinkage; however, autogenous shrinkage is most prominent in concrete with a water to cementing materials ratio under 0.42 (Holt 2001). High-strength, low water to cementing materials ratio (0.30) concrete can experience 200 to 400 millionths of autogenous shrinkage. Autogenous shrinkage can be half that of drying shrinkage for concretes with a water to cementing materials ratio of 0.30.

Recent use of high performance, low water to cementing materials ratio concrete in bridges and other structures has renewed interest in autogenous shrinkage to control crack development. Concretes susceptible to large amounts of autogenous shrinkage should be cured with external water for at least 7 days to help control crack development. Fogging should be provided as soon as the concrete is cast. The hydration of supplementary cementing materials also contributes to autogenous shrinkage, although at different levels than portland cement. In addition to adjusting paste content and water to cementing materials ratios, autogenous shrinkage can be reduced by using shrinkage reducing admixtures or internal curing techniques. Some cementitious systems may experience autogenous expansion. Tazawa (1999) and Holt (2001) review techniques to control autogenous shrinkage.

Test methods for autogenous shrinkage and expansion of cement paste, mortar, and concrete and tests for autogenous shrinkage stress of concrete are presented by Tazawa (1999).

Subsidence

Subsidence refers to the vertical shrinkage of fresh cementitious materials before initial set. It is caused by bleeding (settlement of solids relative to liquids), air voids rising to the surface, and chemical shrinkage. Subsidence is also called settlement shrinkage. Subsidence of well-consolidated concrete with minimal bleed water is insignificant. The relationship between subsidence and other shrinkage mechanisms is illustrated in Fig. 15-5. Excessive subsidence is often caused by a lack of consolidation of fresh concrete. Excessive subsidence over embedded items, such as supported steel reinforcement, can result in cracking over embedded items. Concretes made with air entrainment, sufficient fine materials, and low water contents will minimize subsidence cracking. Also, plastic fibres have been reported to reduce subsidence cracking (Suprenant and Malisch 1999).

Plastic Shrinkage

Plastic shrinkage refers to volume change occurring while the concrete is still fresh, before hardening. It is usually observed in the form of plastic shrinkage cracks occurring before or during finishing (Fig. 15-6). The cracks often resemble tears in the surface. Plastic shrinkage results from a combination of chemical and autogenous shrinkage and rapid evaporation of moisture from the surface that exceeds the bleeding rate. Plastic shrinkage cracking can be controlled by minimizing surface evaporation through use of fogging, wind breaks, shading, plastic sheet covers, wet burlap, spray-on finishing aids (evaporation retarders), and plastic fibres.

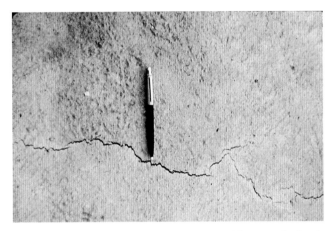

Fig. 15-6. Plastic shrinkage cracks resemble tears in fresh concrete. (1312)

Swelling

Concrete, mortar, and cement paste swell in the presence of external water. When water drained from capillaries by chemical shrinkage is replaced by external water, the volume of the concrete mass increases. As there is no self desiccation, there is no autogenous shrinkage. External water can come from wet curing or submersion. Swelling occurs due to a combination of crystal growth, absorption of water, and osmotic pressure. The swelling is not large, only about 50 millionths at early ages (Fig. 15-7). When the

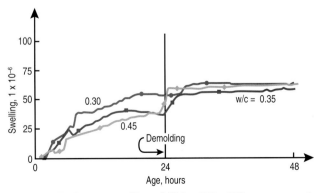

Fig. 15-7. Early age swelling of 100 x 100 x 375-mm concrete specimens cured under water (Aitcin 1999).

external water source is removed, autogenous shrinkage and drying shrinkage reverse the volume change.

Early Thermal Expansion

As cement hydrates, the exothermic reaction provides a significant amount of heat. In large elements the heat is retained, rather than dissipated as happens with thin elements. This temperature rise, occurring over the first few hours and days, can induce a small amount of expansion that counteracts autogenous and chemical shrinkage (Holt 2001).

MOISTURE CHANGES (DRYING SHRINKAGE) OF HARDENED CONCRETE

Hardened concrete expands slightly with a gain in moisture and contracts with a loss in moisture. The effects of these moisture cycles are illustrated schematically in Fig. 15-8. Specimen A represents concrete stored continuously in water from time of casting. Specimen B represents the same concrete exposed first to drying in air and then to alternate cycles of wetting and drying. For comparative purposes, it should be noted that the swelling that occurs during continuous wet storage over a period of several years is usually less than 150 millionths; this is about one-fourth of the shrinkage of air-dried concrete for the same period. Fig. 15-9 illustrates swelling of concretes wet cured for 7 days following by shrinkage when sealed or exposed

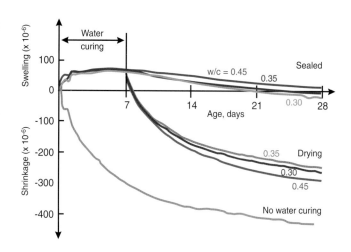

Fig. 15-9. Length change of concrete samples exposed to different curing regimes (Aitcin 1999).

to air drying. Autogenous shrinkage reduces the volume of the sealed concretes to a level about equal to the amount of swelling at 7 days. Note that the concretes wet cured for 7 days had less shrinkage due to drying and autogenous effects than the concrete that had no water curing. This illustrates the importance of early, wet curing to minimize shrinkage (Aitcin 1999).

Tests indicate that the drying shrinkage of small, plain concrete specimens (without reinforcement) ranges from about 400 to 800 millionths when exposed to air at 50% humidity. Concrete with a unit drying shrinkage of 550 millionths shortens about the same amount as the thermal contraction caused by a decrease in temperature of 55°C. Preplaced aggregate concrete has a drying shrinkage of 200 to 400 millionths; this is considerably less than normal concrete due to point-to-point contact of aggregate particles in preplaced aggregate concrete. The drying shrinkage of structural low-density concrete ranges from slightly less than to 30 percent more than that of normal-density concrete, depending on the type of aggregate used.

The drying shrinkage of reinforced concrete is less than that for plain concrete, the difference depending on the amount of reinforcement. Steel reinforcement restricts but does not prevent drying shrinkage. In reinforced concrete structures with normal amounts of reinforcement, drying shrinkage is assumed to be 200 to 300 millionths. Similar values are found for slabs on ground restrained by subgrade.

For many outdoor applications, concrete reaches its maximum moisture content in winter; so in winter the volume changes due to increase in moisture content and the decrease in average temperature tend to offset each other.

The amount of moisture in concrete is affected by the relative humidity of the ambient air. The free moisture content of concrete elements after drying in air at relative humidities of 50% to 90% for several months is about 1% to

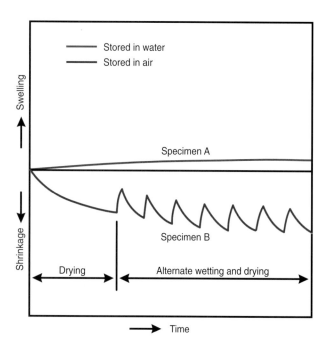

Fig. 15-8. Schematic illustration of moisture movements in concrete. If concrete is kept continuously wet, a slight expansion occurs. However, drying usually takes place, causing shrinkage. Further wetting and drying causes alternate cycles of swelling and shrinkage (Roper 1960).

2% by mass of the concrete; the actual amount depends on the concrete's constituents, original water content, drying conditions, and the size and shape of the concrete element.

After concrete has dried to a constant moisture content at one relative humidity condition, a decrease in humidity causes it to lose moisture while an increase causes it to gain moisture. The concrete shrinks or swells with each such change in moisture content due primarily to responses of the cement paste to moisture changes. Most aggregates show little response to changes in moisture content, although there are a few aggregates that swell or shrink in response to such changes.

As drying takes place, concrete shrinks. Where there is no restraint, movement occurs freely and no stresses or cracks develop (Fig. 15-10a top). If the tensile stress that results from restrained drying shrinkage exceeds the tensile strength of the concrete, cracks can develop (Fig. 15-10a bottom). Random cracks may develop if joints are not properly provided and the concrete element is restrained from shortening (Fig. 15-10b). Contraction joints for slabs on ground should be spaced at distances of 24 to 36 times the slab thickness to control random cracks (Fig. 15-10c). Joints in walls are equally important for crack control (Fig. 15-10d). Fig. 15-11 illustrates the relationship between drying rate at different depths, drying shrinkage, and mass loss for normal-density concrete (Hanson 1968).

Shrinkage may continue for a number of years, depending on the size and shape of the concrete mass. The rate and ultimate amount of shrinkage are usually smaller for large masses of concrete than for small masses; on the other hand, shrinkage continues longer for large masses. Higher volume-to-surface ratios (larger masses) experience lower shrinkage as shown in Fig. 15-12.

The rate and amount of drying shrinkage for small concrete specimens made with various cements are shown in Fig. 15-13. Specimens were initially moist-cured for 14 days at 21°C, then stored for 38 months in air at the same temperature and 50% relative humidity. Shrinkage recorded at the age of 38 months ranged from 600 to 790 millionths. An average of 34% of this shrinkage occurred within the first month. At the end of 11 months an average of 90% of the 38-month shrinkage had taken place.

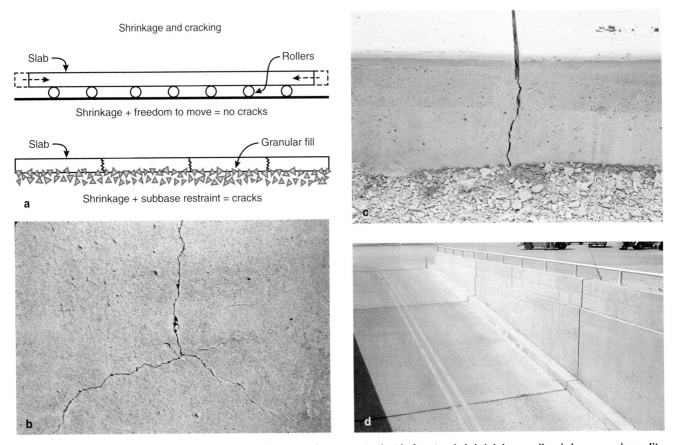

Fig. 15-10. (a) Illustration showing no crack development in concrete that is free to shrink (slab on rollers); however, in reality a slab on ground is restrained by the subbase (or other elements) creating tensile stresses and cracks. (b) Typical shrinkage cracks in a slab on ground. (c) A properly functioning contraction joint controls the location of shrinkage cracking. (d) Contraction joints in the slabs and walls shown will minimize the formation of cracks. (A-5271, 4434, 1144)

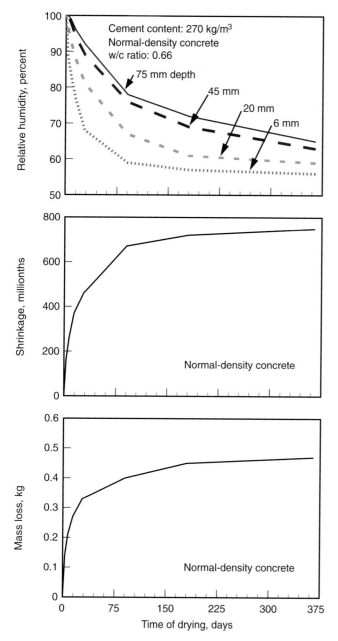

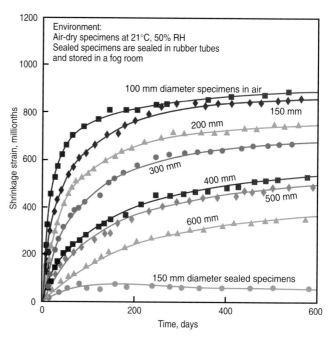

Fig. 15-12. Drying shrinkage of various sizes of cylindrical specimens made of Elgin, Illinois gravel concrete (Hansen and Mattock 1966).

Fig. 15-11. Relative humidity distribution at various depths, drying shrinkage, and mass loss of 150 × 300-mm cylinders moist-cured for 7 days followed by drying in laboratory air at 23°C and 50% RH (Hanson 1968).

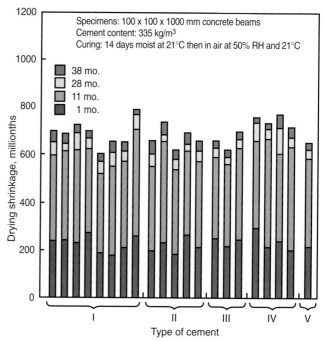

Fig. 15-13. Results of long-term drying shrinkage tests by the U.S. Bureau of Reclamation. Shrinkage ranged from 600 to 790 millionths after 38 months of drying. The shrinkage of concretes made with air-entraining cements was similar to that for non-air-entrained concretes in this study (Bureau of Reclamation 1947 and Jackson 1955).

Effect of Concrete Ingredients on Drying Shrinkage

The most important controllable factor affecting drying shrinkage is the amount of water per unit volume of concrete. The results of tests illustrating the water content to shrinkage relationship are shown in Fig. 15-14. Shrinkage can be minimized by keeping the water content of concrete as low as possible. This is achieved by keeping the total coarse aggregate content of the concrete as high as possible (minimizing paste content). Use of low slumps and placing methods that minimize water requirements are thus major factors in controlling concrete shrinkage. Any practice that increases the water requirement of the cement paste, such as the use of high slumps (without superplasticizers), excessively high freshly mixed concrete temperatures, high fine-aggregate contents, or use of small-size coarse aggregate, will increase shrinkage. A small amount of water can be added to ready mixed concrete at the jobsite without affecting drying shrinkage properties as long as the additions are within mix specifications (Suprenant and Malisch 2000).

The general uniformity of shrinkage of concretes with different types of cement at different ages is illustrated in Fig. 15-13. However, this does not mean that all cements or cementing materials have similar shrinkage. Supplemen-

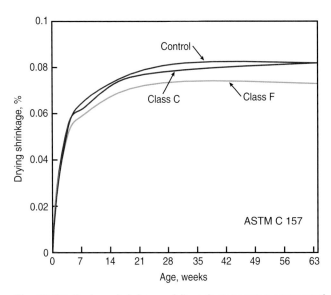

Fig. 15-15. Drying shrinkage of fly ash concretes compared to a control mixture. The graphs represent the average of four Class C ashes and six Class F ashes, with the range in drying shrinkage rarely exceeding 0.01 percentage points. Fly ash dosage was 25% of the cementing material (Gebler and Klieger 1986).

tary cementing materials usually have little effect on shrinkage at normal dosages. Fig. 15-15 shows that concretes with normal dosages of selected fly ashes performed similar to the control concrete made with only portland cement as the cementing material.

Aggregates in concrete, especially coarse aggregate, physically restrain the shrinkage of hydrating cement paste. Paste content affects the drying shrinkage of mortar more than that of concrete. Drying shrinkage is also dependent on the type of aggregate. Hard, rigid aggregates are difficult to compress and provide more restraint to shrinkage than softer, less rigid aggregates. As an extreme example, if steel balls were substituted for ordinary coarse aggregate, shrinkage would be reduced 30% or more. Drying shrinkage can also be reduced by avoiding aggregates that have high drying shrinkage properties and aggregates containing excessive amounts of clay. Quartz, granite, feldspar, limestone, and dolomite aggregates generally produce concretes with low drying shrinkages (ACI Committee 224). Steam curing will also reduce drying shrinkage.

Most chemical admixtures have little effect on shrinkage. The use of accelerators such as calcium chloride will increase drying shrinkage of concrete. Despite reductions in water content, some water-reducing admixtures can increase drying shrinkage, particularly those that contain an accelerator to counteract the retarding effect of the admixture. Air entrainment has little or no effect on drying shrinkage. High-range water reducers usually have little effect on drying shrinkage (Fig. 15-16). Drying shrinkage can be evaluated in accordance with ASTM C 157.

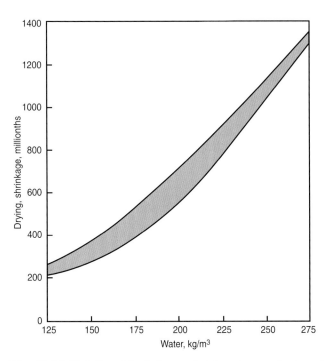

Fig. 15-14. Relationship between total water content and drying shrinkage. A large number of mixtures with various proportions is represented within the shaded area of the curves. Drying shrinkage increases with increasing water contents.

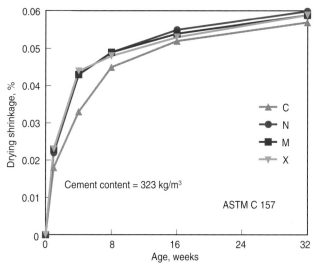

Fig. 15-16. Drying shrinkage of concretes made with selected high-range water reducers (N,M, and X) compared to a control mixture (C) (Whiting and Dziedzic 1992).

Effect of Curing on Drying Shrinkage

The amount and type of curing can effect the rate and ultimate amount of drying shrinkage. Curing compounds, sealers, and coatings can trap free moisture in the concrete for long periods of time, resulting in delayed shrinkage. Wet curing methods, such as fogging or wet burlap, hold off shrinkage until curing is terminated, after which the concrete dries and shrinks at a normal rate. Cooler initial

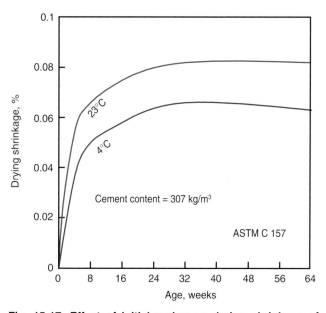

Fig. 15-17. Effect of initial curing on drying shrinkage of portland cement concrete prisms. Concrete with an initial 7-day moist cure at 4°C had less shrinkage than concrete with an initial 7-day moist cure at 23°C. Similar results were found with concretes containing 25% fly ash as part of the cementing material (Gebler and Klieger 1986).

curing temperatures can reduce shrinkage (Fig. 15-17). Steam curing will also reduce drying shrinkage. Computer software is available to predict the effect of curing and environmental conditions on shrinkage and cracking (FHWA and Transtec 2001). Hedenblad (1997) provides tools to predict the drying of concrete as effected by different curing methods and type of construction.

TEMPERATURE CHANGES OF HARDENED CONCRETE

Concrete expands slightly as temperature rises and contracts as temperature falls, although it can expand slightly as free water in the concrete freezes. Temperature changes may be caused by environmental conditions or by cement hydration. An average value for the coefficient of thermal expansion of concrete is about 10 millionths per degree Celsius, although values ranging from 6 to 13 millionths per degree Celsius have been observed. This amounts to a length change of 5 mm for 10 m of concrete subjected to a rise or fall of 50°C. The coefficient of thermal expansion for structural low-density concrete varies from 7 to 11 millionths per degree Celsius. The coefficient of thermal expansion of concrete can be determined by AASHTO TP 60.

Thermal expansion and contraction of concrete varies with factors such as aggregate type, cement content, water-cementing materials ratio, temperature range, concrete age, and relative humidity. Of these, aggregate type has the greatest influence.

Table 15-1 shows some experimental values of the thermal coefficient of expansion of concretes made with aggregates of various types. These data were obtained from tests on small concrete specimens in which all factors were the same except aggregate type. In each case, the fine aggregate was of the same material as the coarse aggregate.

The thermal coefficient of expansion for steel is about 12 millionths per degree Celsius, which is comparable to

Table 15-1. Effect of Aggregate Type on Thermal Coefficient of Expansion of Concrete

Aggregate type (from one source)	Coefficient of expansion, millionths per °C
Quartz	11.9
Sandstone	11.7
Gravel	10.8
Granite	9.5
Basalt	8.6
Limestone	6.8

Coefficients of concretes made with aggregates from different sources may vary widely from these values, especially those for gravels, granites, and limestones (Davis 1930).

that for concrete. The coefficient for reinforced concrete can be assumed as 11 millionths per degree Celsius, the average for concrete and steel.

Temperature changes that result in shortening can crack concrete members that are highly restrained by another part of the structure or by ground friction. Consider a long restrained concrete member cast without joints that, after moist curing, is allowed to drop in temperature. As the temperature drops, the concrete wants to shorten, but cannot because it is restrained longitudinally. The resulting tensile stresses cause the concrete to crack. Tensile strength and modulus of elasticity of concrete both may be assumed proportional to the square root of concrete compressive strength. And calculations show that a large enough temperature drop will crack concrete regardless of its age or strength, provided the coefficient of expansion does not vary with temperature and the concrete is fully restrained (FHWA and Transtec 2001 and PCA 1982).

Precast wall panels and slabs and pavements on ground are susceptible to bending and curling caused by temperature gradients that develop when concrete is cool on one side and warm on the other. The calculated amount of curling in a wall panel is illustrated in Fig. 15-18.

For the effect of temperature changes in mass concrete due to heat of hydration, see Chapter 18.

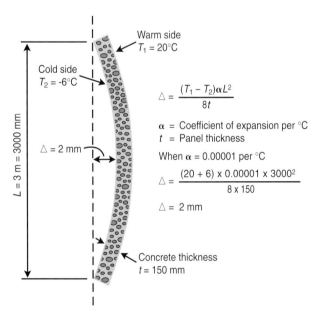

Fig. 15-18. Curling of a plain concrete wall panel due to temperature that varies uniformly from inside to outside.

Warm side
$T_1 = 20°C$

Cold side
$T_2 = -6°C$

$L = 3 \text{ m} = 3000 \text{ mm}$

$\triangle = 2 \text{ mm}$

$$\triangle = \frac{(T_1 - T_2)\alpha L^2}{8t}$$

α = Coefficient of expansion per °C
t = Panel thickness

When $\alpha = 0.00001$ per °C

$$\triangle = \frac{(20 + 6) \times 0.00001 \times 3000^2}{8 \times 150}$$

$\triangle = 2 \text{ mm}$

Concrete thickness
$t = 150 \text{ mm}$

Low Temperatures

Concrete continues to contract as the temperature is reduced below freezing. The amount of volume change at subfreezing temperatures is greatly influenced by the moisture content, behavior of the water (physical state—ice or liquid), and type of aggregate in the concrete. In one study, the coefficient of thermal expansion for a temperature range of 24°C to -155°C varied from 6×10^{-6} per °C for a low density aggregate concrete to 8.2×10^{-6} per °C for a sand and gravel mixture. Subfreezing temperatures can significantly increase the compressive and tensile strength and modulus of elasticity of moist concrete. Dry concrete properties are not as affected by low temperatures. In the same study, moist concrete with an original compressive strength of 35 MPa at 24°C achieved over 117 MPa at –100°C. The same concrete tested ovendry or at a 50% internal relative humidity had strength increases of only about 20%. The modulus of elasticity for sand and gravel concrete with 50% relative humidity was only 8% higher at –150°C than at 24°C, whereas the moist concrete had a 50% increase in modulus of elasticity. Going from 24°C to –170°C, the thermal conductivity of normal-density concrete also increased, especially for moist concrete. The thermal conductivity of low-density aggregate concrete is little affected (Monfore and Lentz 1962 and Lentz and Monfore 1966).

High Temperatures

Temperatures greater than 95°C that are sustained for several months or even several hours can have significant effects on concrete. The total amount of volume change of concrete is the sum of volume changes of the cement paste and aggregate. At high temperatures, the paste shrinks due to dehydration while the aggregate expands. For normal-aggregate concrete, the expansion of the aggregate exceeds the paste shrinkage resulting in an overall expansion of the concrete. Some aggregates such as expanded shale, andesite, or pumice with low coefficients of expansion can produce a very volume-stable concrete in high-temperature environments (Fig. 15-19). On the other hand, some aggregates undergo extensive and abrupt volume changes at a particular temperature, causing disruption in the concrete. For example, in one study a dolomitic limestone aggregate contained an iron sulphide impurity caused severe expansion, cracking, and disintegration in concrete exposed to a temperature of 150°C for four months; at temperatures above and below 150°C there was no detrimental expansion (Carette, Painter, and Malhotra 1982). The coefficient of thermal expansion tends to increase with temperature rise.

Besides volume change, sustained high temperatures can also have other, usually irreversible, effects such as a reduction in strength, modulus of elasticity, and thermal conductivity. Creep increases with temperature. Above

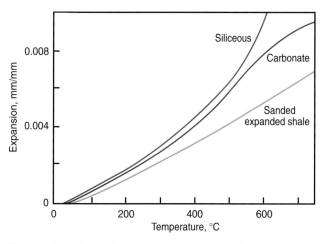

Fig. 15-19. Thermal expansion of concretes containing various types of aggregate (Abrams 1977).

100°C, the paste begins to dehydrate (lose chemically combined water of hydration) resulting in significant strength losses. Strength decreases with increases in temperature until the concrete loses essentially all its strength. The effect of high-temperature exposure on compressive strength of concretes made with various types of aggregate is illustrated in Fig. 15-20. Several factors including concrete moisture content, aggregate type and stability, cementing materials content, exposure time, rate of temperature rise, age of concrete, restraint, and existing stress all influence the behavior of concrete at high temperatures.

If stable aggregates are used and strength reduction and the effects on other properties are accounted for in the

mix design, high quality concrete can be exposed to temperatures of 90°C to 200°C for long periods. Some concrete elements have been exposed to temperatures up to 250°C for long periods of time; however, special steps should be taken or special materials (such as heat-resistant calcium aluminate cement) should be considered for exposure temperatures greater than 200°C. Before any structural concrete is exposed to high temperatures (greater than 90°C), laboratory testing should be performed to determine the particular concrete's thermal properties. This will avoid any unexpected distress.

CURLING (WARPING)

In addition to horizontal movement caused by changes in moisture and temperature, curling of slabs on ground can be a problem; this is caused by differences in moisture content and temperature between the top and bottom of slabs (Fig. 15-21).

The edges of slabs at the joints tend to curl upward when the surface of a slab is drier or cooler than the bottom. A slab will assume a reverse curl when the surface is wetter or warmer than the bottom. However, enclosed slabs, such as floors on ground, curl only upward. When the edges of an industrial floor slab are curled upward they lose support from the subbase and become a cantilever. Lift-truck traffic passing over joints causes a repetitive vertical deflection that creates a great potential for fatigue cracking in the slab. The amount of vertical upward curl (curling) is small for a short, thick slab.

Curling can be reduced or eliminated by using design and construction techniques that minimize shrinkage differentials and by using techniques described earlier to

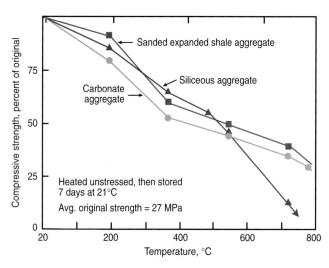

Fig. 15-20. Effect of high temperatures on the residual compressive strength of concretes containing various types of aggregate (Abrams 1973).

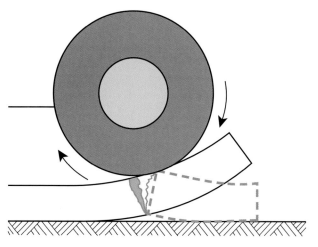

Fig. 15-21. Illustration of curling of a concrete slab on ground. The edge of the slab at a joint or free end lifts off the subbase creating a cantilevered section of concrete that can break off under heavy wheel loading.

reduce temperature and moisture-related volume changes. Thickened edges, shorter joint spacings, permanent vapour-impermeable sealers, and large amounts of reinforcing steel placed 50 mm below the surface all help reduce curling (Ytterberg 1987).

ELASTIC AND INELASTIC DEFORMATION

Compression Strain

The series of curves in Fig. 15-22 illustrate the amount of compressive stress and strain that results instantaneously due to loading of unreinforced concrete. With water-cementing materials ratios of 0.50 or less and strains up to 1500 millionths, the upper three curves show that strain is closely proportional to stress; in other words, the concrete is almost elastic. The upper portions of the curves and beyond show that the concrete is inelastic. The curves for high-strength concrete have sharp peaks, whereas those for lower-strength concretes have long and relatively flat peaks. Fig. 15-22 also shows the sudden failure characteristics of higher strength, low water to cementing materials ratio, concrete cylinders.

When load is removed from concrete in the inelastic zone, the recovery line usually is not parallel to the original line for the first load application. Therefore, the amount of permanent set may differ from the amount of inelastic deformation (Fig. 15-23).

The term "elastic" is not favored for general discussion of concrete behavior because frequently the strain may be

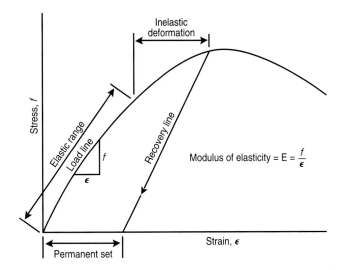

Fig. 15-23. Generalized stress-strain curve for concrete.

in the inelastic range. For this reason, the term "instantaneous strain" is often used.

Modulus of Elasticity

The ratio of stress to strain in the elastic range of a stress-strain curve for concrete defines the modulus of elasticity (*E*) of that concrete (Fig. 15-23). Normal-density concrete has a modulus of elasticity of 14,000 to 41,000 MPa, depending on factors such as compressive strength and aggregate type. For normal-density concrete with com-

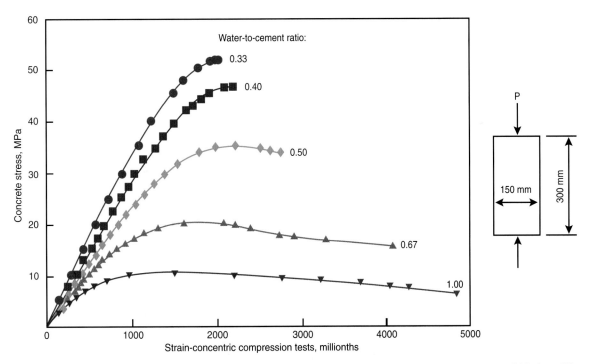

Fig. 15-22. Stress-strain curves for compression tests on 150 × 300-mm concrete cylinders at an age of 28 days (Hognestad, Hanson, and McHenry 1955).

pressive strengths (f'_c) between 20 MPa and 35 MPa, the modulus of elasticity can be estimated as 5000 times the square root of f'_c. The modulus of elasticity for structural low-density concrete is between 7000 MPa and 17,000 MPa. E for any particular concrete can be determined in accordance with ASTM C 469.

Deflection

Deflection of concrete beams and slabs is one of the more common and obvious building movements. The deflections are the result of flexural strains that develop under dead and live loads and that may result in cracking in the tensile zone of concrete members. Reinforced concrete structural design anticipates these tension cracks. Concrete members are often cambered, that is, built with an upward bow, to compensate for the expected later deflection.

Poisson's Ratio

When a block of concrete is loaded in uniaxial compression, as in Fig. 15-24, it will shorten and at the same time develop a lateral strain or bulging. The ratio of lateral to axial strain is called Poisson's ratio, μ. A common value used is 0.20 to 0.21, but the value may vary from 0.15 to 0.25 depending upon the aggregate, moisture content, concrete age, and compressive strength. Poisson's ratio (ASTM C 469) is generally of no concern to the structural designer; it is used in advanced structural analysis of flat-plate floors, shell roofs, arch dams, and mat foundations.

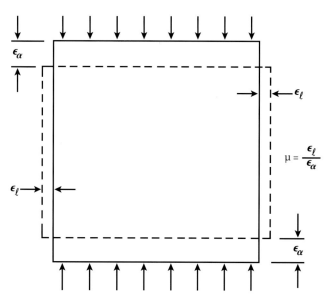

Fig. 15-24. Ratio of lateral to axial strain is Poisson's ratio, μ.

Shear Strain

Concrete, like other materials, deforms under shear forces. The shear strain produced is important in determining the load paths or distribution of forces in indeterminate structures—for example where shear-walls and columns both participate in resisting horizontal forces in a concrete building frame. The amount of movement, while not large, is significant in short, stubby members; in larger members it is overshadowed by flexural strains. Calculation of the shear modulus (modulus of rigidity), G, is shown in Fig. 15-25; G varies with the strength and temperature of the concrete.

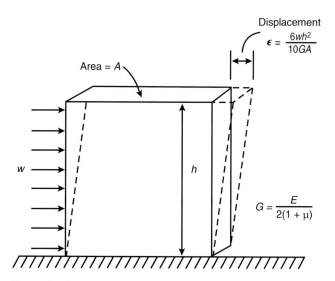

Fig. 15-25. Strain that results from shear forces on a body. G = shear modulus. μ = Poisson's ratio. Strain resulting from flexure is not shown.

Torsional Strain

Plain rectangular concrete members can also fail in torsion, that is, a twisting action caused by bending about an axis parallel to the wider face and inclined at an angle of about 45 degrees to the longitudinal axis of a member. Microcracks develop at low torque; however, concrete behaves reasonably elastic up to the maximum limit of the elastic torque (Hsu 1968).

CREEP

When concrete is loaded, the deformation caused by the load can be divided into two parts: a deformation that occurs immediately (elastic strain) and a time-dependent deformation that begins immediately but continues at a decreasing rate for as long as the concrete is loaded. This latter deformation is called creep.

The amount of creep is dependent upon (1) the magnitude of stress, (2) the age and strength of the concrete when stress is applied, and (3) the length of time the concrete is stressed. It is also affected by other factors related to the quality of the concrete and conditions of exposure, such as: (1) type, amount, and maximum size of aggregate; (2) type of cementing materials; (3) amount of cement paste; (4) size and shape of the concrete element; (5) volume to surface ratio of the concrete element; (6) amount of steel reinforcement; (7) prior curing conditions; and (8) the ambient temperature and humidity.

Within normal stress ranges, creep is proportional to stress. In relatively young concrete, the change in volume or length due to creep is largely unrecoverable; in older or drier concrete it is largely recoverable.

The creep curves shown in Fig. 15-26 are based on tests conducted under laboratory conditions in accordance with ASTM C 512. Cylinders were loaded to almost 40% of their compressive strength. Companion cylinders not subject to load were used to measure drying shrinkage; this was then deducted from the total deformation of the loaded specimens to determine creep. Cylinders were allowed to dry while under load except for those marked "sealed." The two 28-day curves for each concrete strength in Fig. 15-26 show that creep of concrete loaded under drying conditions is greater than creep of concrete sealed against drying. Concrete specimens loaded at a late age will creep less than those loaded at an early age. It can be seen that as concrete strength decreases, creep increases. Fig. 15-27

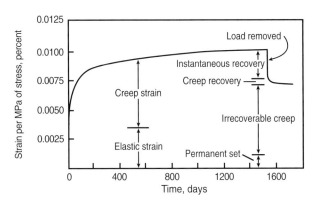

Fig. 15-27. Combined curve of elastic and creep strains showing amount of recovery. Specimens (cylinders) were loaded at 8 days immediately after removal from fog curing room and then stored at 21°C and 50% RH. The applied stress was 25% of the compressive strength at 8 days (Hansen and Mattock 1966).

illustrates recovery from the elastic and creep strains after load removal.

A combination of strains occurring in a reinforced column is illustrated in Fig. 15-28. The curves represent deformations and volume changes in a 14th-story column of a 76-story reinforced concrete building while under construction. The 400 x 1200-mm column contained 2.08% vertical reinforcement and was designed for 60 MPa concrete.

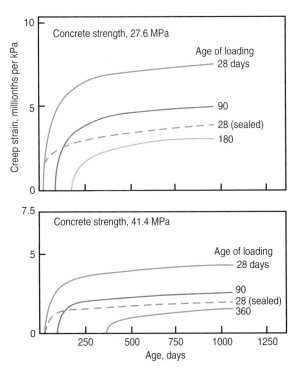

Fig. 15-26. Relationship of time and age of loading to creep of two different strength concretes. Specimens were allowed to dry during loading, except for those labeled as sealed (Russell and Corley 1977).

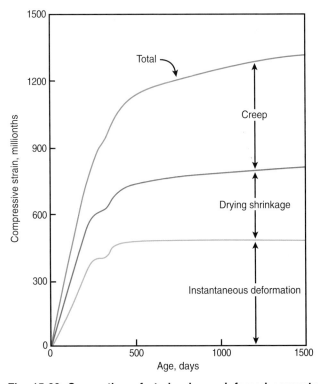

Fig. 15-28. Summation of strains in a reinforced concrete column during construction of a tall building (Russell and Corley 1977).

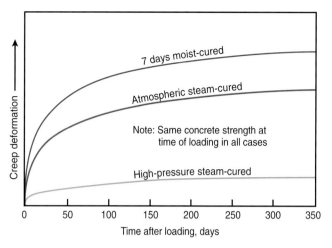

Fig. 15-29. Effect of curing method on magnitude of creep for typical normal-density concrete (Hanson 1964).

The method of curing prior to loading has a marked effect on the amount of creep in concrete. The effects on creep of three different methods of curing are shown in Fig. 15-29. Note that very little creep occurs in concrete that is cured by high-pressure steam (autoclaving). Note also that atmospheric steam-cured concrete has considerably less creep than 7-day moist-cured concrete. The two methods of steam curing shown in Fig. 15-29 reduce drying shrinkage of concrete about half as much as they reduce creep.

CHEMICAL CHANGES AND EFFECTS

Some volume changes of concrete result from chemical reactions; these may take place shortly after placing and finishing or later due to reactions within the hardened concrete in the presence of water or moisture.

Carbonation

Hardened concrete containing some moisture reacts with carbon dioxide present in air, a reaction that results in a slight shrinkage of the surface paste of the concrete. The effect, known as carbonation, is not destructive but actually increases the chemical stability and strength of the concrete. However, carbonation also reduces the pH of concrete. If steel is present in the carbonated area, steel corrosion can occur due to the absence of the protective oxide film provided by concrete's high pH. Rust is an expansive reaction and results in cracking and spalling of the concrete. The depth of carbonation is very shallow in dense, high-quality concrete, but can penetrate deeply in porous, poor-quality concrete. Because so little of a concrete element carbonates, carbonation shrinkage of cast-in-place concrete is insignificant and does not have to be considered in engineering practice.

Carbonation of paste proceeds slowly and produces little direct shrinkage at relative humidities of 100% and

25%. Maximum carbonation and carbonation shrinkage occurs at about 50% relative humidity. Irreversible shrinkage and increase in mass occurs during carbonation. And the carbonated product may show improved volume stability to subsequent moisture change and reduced permeability (Verbeck 1958).

During manufacture some concrete masonry units are deliberately exposed to carbon dioxide after reaching 80% of their rated strength; this introduction to carbonation shrinkage makes the units more dimensionally stable. Future drying shrinkage is reduced 30% or more (Toennies and Shideler 1963).

One of the causes of surface crazing of concrete is the shrinkage that accompanies natural air carbonation of young concrete. More research is needed on the effect of early carbonation on deicer scaling resistance.

Carbonation of another kind also can occur in freshly placed, unhardened concrete. This carbonation causes a soft, chalky surface called dusting; it usually takes place during cold-weather concreting when there is an unusual amount of carbon dioxide in the air due to unvented heaters or gasoline-powered equipment operating in an enclosure.

Sulphate Attack

Sulphate attack of concrete can occur where soil and groundwater have a high sulphate content and measures to reduce sulphate attack, such as use of a low water to cementing materials ratio, have not been taken. The attack is greater in concrete that is exposed to wetting and drying, such as foundation walls and posts. Sulphate attack usually results in an expansion of the concrete because of the formation of solids from the chemical action or salt crystallization. The amount of expansion in severe circumstances has been significantly higher than 0.1%, and the disruptive effect within the concrete can result in extensive cracking and disintegration. The amount of expansion cannot be accurately predicted.

Alkali-Aggregate Reactions

Certain aggregates can react with alkali hydroxides in concrete, causing expansion and cracking over a period of years. The reaction is greater in those parts of a structure exposed to moisture. A knowledge of the characteristics of local aggregates is essential. There are two types of alkali-reactive aggregates, siliceous and carbonate. Alkali-aggregate reaction expansion may exceed 0.5% in concrete and can cause the concrete to fracture and break apart.

Structural design techniques cannot counter the effects of alkali-aggregate expansion, nor can the expansion be controlled by jointing. In areas where deleteriously reactive aggregates are known to exist, special measures must be taken to prevent the occurrence of alkali-aggregate reaction.

REFERENCES

Abrams, M. S., *Compressive Strength of Concrete at Temperatures to 1600°F*, Research and Development Bulletin RD016, Portland Cement Association, http://www.portcement.org/pdf_files/RD016.pdf, 1973.

Abrams, M. S., *Performance of Concrete Structures Exposed to Fire*, Research and Development Bulletin RD060, Portland Cement Association, http://www.portcement.org/pdf_files/RD060.pdf, 1977.

Abrams, M. S., *Behavior of Inorganic Materials in Fire*, Research and Development Bulletin RD067, Portland Cement Association, http://www.portcement.org/pdf_files/RD067.pdf, 1979.

Abrams, M. S., and Orals, D. L., *Concrete Drying Methods and Their Effects on Fire Resistance*, Research Department Bulletin RX181, Portland Cement Association, http://www.portcement.org/pdf_files/RX181.pdf, 1965.

ACI Committee 209, *Prediction of Creep, Shrinkage, and Temperature Effects in Concrete Structures*, ACI 209R-92, reapproved 1997, American Concrete Institute, Farmington Hills, Michigan, 1997, 47 pages.

ACI Committee 224, *Control of Cracking in Concrete Structures*, ACI 224R-01, American Concrete Institute, Farmington Hills, Michigan, 2001, 43 pages.

ACI Committee 224, *Causes, Evaluation, and Repair of Cracks in Concrete Structures*, ACI 224.1R-93, reapproved 1998, American Concrete Institute, Farmington Hills, Michigan, 1998, 22 pages.

Aitcin, Pierre-Claude, "Does Concrete Shrink or Does it Swell?," *Concrete International*, American Concrete Institute, Farmington Hills, Michigan, December 1999, pages 77 to 80.

Brewer, H. W., *Moisture Migration—Concrete Slab-on-Ground Construction*, Development Department Bulletin DX089, Portland Cement Association, http://www.portcement.org/pdf_files/DX089.pdf, 1965.

Brewer, Harold W., *General Relation of Heat Flow Factors to the Unit Weight of Concrete*, Development Department Bulletin DX114, Portland Cement Association, http://www.portcement.org/pdf_files/DX114.pdf, 1967.

Bureau of Reclamation, "Long-Time Study of Cement Performance in Concrete—Tests of 28 Cements Used in the Parapet Wall of Green Mountain Dam," *Materials Laboratories Report No. C-345*, U.S. Department of the Interior, Bureau of Reclamation, Denver, 1947.

Carette, G. G.; Painter, K. E.; and Malhotra, V. M., "Sustained High Temperature Effect on Concretes Made with Normal Portland Cement, Normal Portland Cement and Slag, or Normal Portland Cement and Fly Ash," *Concrete International*, American Concrete Institute, Farmington Hills, Michigan, July 1982.

Carlson, Roy W., "Drying Shrinkage of Concrete as Affected by Many Factors," *Proceedings of the Forty-First Annual Meeting of the American Society for Testing and Materials*, vol. 38, part II, Technical Papers, American Society for Testing and Materials, West Conshohocken, Pennsylvania, 1938, pages 419 to 440.

Copeland, L. E., and Bragg, R. H., *Self Desiccation in Portland Cement Pastes*, Research Department Bulletin RX052, Portland Cement Association, http://www.portcement.org/pdf_files/RX052.pdf, 1955.

Cruz, Carlos R., *Elastic Properties of Concrete at High Temperatures*, Research Department Bulletin RX191, Portland Cement Association, http://www.portcement.org/pdf_files/RX191.pdf, 1966.

Cruz, C. R., and Gillen, M., *Thermal Expansion of Portland Cement Paste, Mortar, and Concrete at High Temperatures*, Research and Development Bulletin RD074, Portland Cement Association, http://www.portcement.org/pdf_files/RD074.pdf, 1980.

Davis, R. E., "A Summary of the Results of Investigations Having to Do with Volumetric Changes in Cements, Mortars, and Concretes Due to Causes Other Than Stress," *Proceedings of the American Concrete Institute*, American Concrete Institute, Farmington Hills, Michigan, vol. 26, 1930, pages 407 to 443.

FHWA and Transtec, HIPERPAV, http://www.hiperpav.com, 2001.

Gebler, Steven H., and Klieger, Paul, *Effect of Fly Ash on Some of the Physical Properties of Concrete*, Research and Development Bulletin RD089, Portland Cement Association, http://www.portcement.org/pdf_files/RD089.pdf, 1986.

Hammer, T. A., "Test Methods for Linear Measurement of Autogenous Shrinkage Before Setting," *Autogenous Shrinkage of Concrete*, edited by E. Tazawa, E&FN Spon and Routledge, New York, 1999, pages 143 to 154. Also available through PCA as LT245.

Hanson, J. A., *Prestress Loss As Affected by Type of Curing*, Development Department Bulletin DX075, Portland Cement Association, http://www.portcement.org/pdf_files/DX075.pdf, 1964.

Hanson, J. A., *Effects of Curing and Drying Environments on Splitting Tensile Strength of Concrete*, Development Department Bulletin DX141, Portland Cement Association, http://www.portcement.org/pdf_files/DX141.pdf, 1968.

Hansen, Torben C., and Mattock, Alan H., *Influence of Size and Shape of Member on the Shrinkage and Creep of Concrete*, Development Department Bulletin DX103, Portland Cement Association, http://www.portcement.org/pdf_files/DX103.pdf, 1966.

Hedenblad, Göran, "Concrete Drying Time," PL982, *Concrete Technology Today,* Portland Cement Association, http://www.portcement.org/pdf_files/PL982.pdf, July 1998.

Hedenblad, Göran, *Drying of Construction Water in Concrete,* Byggforskningsradet, The Swedish Council for Building Research, Stockholm, 1997 [PCA LT229].

Helmuth, Richard A., and Turk, Danica H., *The Reversible and Irreversible Drying Shrinkage of Hardened Portland Cement and Tricalcium Silicate Pastes,* Research Department Bulletin RX215, Portland Cement Association, http://www.portcement.org/pdf_files/RX215.pdf, 1967.

Holt, Erika E., *Early Age Autogenous Shrinkage of Concrete,* VTT Publication 446, Technical Research Center of Finland, Espoo, 2001, 194 pages. Also available through PCA as LT257.

Holt, Erika E., and Janssen, Donald J., "Influence of Early Age Volume Changes on Long-Term Concrete Shrinkage," *Transportation Research Record 1610,* Transportation Research Board, National Research Council, Washington, D.C., 1998, pages 28 to 32.

Hognestad, E.; Hanson, N. W.; and McHenry, D., *Concrete Stress Distribution in Ultimate Strength Design,* Development Department Bulletin DX006, Portland Cement Association, http://www.portcement.org/pdf_files/DX006.pdf, 1955.

Hsu, Thomas T. C., *Torsion of Structural Concrete—Plain Concrete Rectangular Sections,* Development Department Bulletin DX134, Portland Cement Association, http://www.portcement.org/pdf_files/DX134.pdf, 1968.

Jackson, F. H., *Long-Time Study of Cement Performance in Concrete—Chapter 9. Correlation of the Results of Laboratory Tests with Field Performance Under Natural Freezing and Thawing Conditions,* Research Department Bulletin RX060, Portland Cement Association, http://www.portcement.org/pdf_files/RX060.pdf, 1955.

Kaar, P. H.; Hanson, N. W.; and Capell, H. T., *Stress-Strain Characteristics of High-Strength Concrete,* Research and Development Bulletin RD051, Portland Cement Association, http://www.portcement.org/pdf_files/RD051.pdf, 1977.

Kosmatka, Steven H., "Floor-Covering Materials and Moisture in Concrete," *Concrete Technology Today,* PL853, Portland Cement Association, http://www.portcement.org/pdf_files/PL853.pdf, 1985.

Le Chatelier, H., "Sur les Changements de Volume qui Accompagent le durcissement des Ciments," Bulletin Societe de l'Encouragement pour l'Industrie Nationale, 5eme serie, tome 5, Paris, 1900.

Lentz, A. E., and Monfore, G. E., *Thermal Conductivities of Portland Cement Paste, Aggregate, and Concrete Down to Very Low Temperatures,* Research Department Bulletin RX207, Portland Cement Association, http://www.portcement.org/pdf_files/RX207.pdf, 1966.

Lerch, William, *Studies of Some Methods of Avoiding the Expansion and Pattern Cracking Associated with the Alkali-Aggregate Reaction,* Research Department Bulletin RX031, Portland Cement Association, http://www.portcement.org/pdf_files/RX031.pdf, 1950.

Malhotra, M. L., "The Effect of Temperature on the Compressive Strength of Concrete," *Magazine of Concrete Research,* vol. 8., no. 23, Cement and Concrete Association, Wexham Springs, Slough, England, August 1956, pages 85 to 94.

Monfore, G. E., *A Small Probe-Type Gage for Measuring Relative Humidity,* Research Department Bulletin RX160, Portland Cement Association, http://www.portcement.org/pdf_files/RX160.pdf, 1963.

Monfore, G. E., and Lentz, A. E., *Physical Properties of Concrete at Very Low Temperatures,* Research Department Bulletin RX145, Portland Cement Association, http://www.portcement.org/pdf_files/RX145.pdf, 1962.

Nawy, Edward G.; Barth, Florian G.; and Frosch, Robert J., *Design and Construction Practices to Mitigate Cracking,* SP-204, American Concrete Institute, Farmington Hills, Michigan, 2001, 284 pages.

PCA, *Effect of Long Exposure of Concrete to High Temperature,* ST32, Portland Cement Association, 1969.

PCA, *Building Movements and Joints,* EB086, Portland Cement Association, http://www.portcement.org/pdf_files/EB086.pdf, 1982.

PCA, *Concrete Slab Surface Defects: Causes, Prevention, Repair,* IS177, Portland Cement Association, 2001.

Philleo, Robert, *Some Physical Properties of Concrete at High Temperatures,* Research Department Bulletin RX097, Portland Cement Association, http://www.portcement.org/pdf_files/RX097.pdf, 1958.

Pickett, Gerald, *The Effect of Change in Moisture Content on the Creep of Concrete Under a Sustained Load,* Research Department Bulletin RX020, Portland Cement Association, http://www.portcement.org/pdf_files/RX020.pdf, 1947.

Powers, T. C., "Absorption of Water by Portland Cement Paste during the Hardening Process," *Industrial and Engineering Chemistry,* Vol. 27, No. 7, July 1935, pages 790 to 794.

Powers, Treval C., "Causes and Control of Volume Change," Volume 1, Number 1, *Journal of the PCA Research and Development Laboratories,* Portland Cement Association, January 1959.

Roper, Harold, *Volume Changes of Concrete Affected by Aggregate Type,* Research Department Bulletin RX123, Portland Cement Association, http://www.portcement.org/pdf_files/RX123.pdf, 1960.

Russell, H. G., and Corley, W. G., *Time-Dependent Behavior of Columns in Water Tower Place*, Research and Development Bulletin RD052, Portland Cement Association, http://www.portcement.org/pdf_files/RD052.pdf, 1977.

Steinour, Harold H., "Some Effects of Carbon Dioxide on Mortars and Concrete—Discussion," *Proceedings of the American Concrete Institute,* vol. 55, American Concrete Institute, Farmington Hills, Michigan, 1959, pages 905 to 907.

Suprenant, Bruce A., and Malisch, Ward R., "The Fiber Factor," *Concrete Construction,* Addison, Illinois, October 1999, pages 43 to 46.

Suprenant, Bruce A., and Malisch, Ward R., "A New Look at Water, Slump, and Shrinkage," *Concrete Construction,* Addison, Illinois, April 2000, pages 48 to 53.

Tazawa, Ei-ichi, *Autogenous Shrinkage of Concrete*, E & FN Spon and Routledge, New York, 1999, 428 pages [PCA LT245].

Toennies, H. T., and Shideler, J. J., *Plant Drying and Carbonation of Concrete Block—NCMA-PCA Cooperative Program,* Development Department Bulletin DX064, Portland Cement Association, http://www.portcement.org/pdf_files/DX064.pdf, 1963.

Tremper, Bailey, and Spellman, D. L., "Shrinkage of Concrete—Comparison of Laboratory and Field Performance," *Highway Research Record Number 3, Properties of Concrete,* Transportation Research Board, National Research Council, Washington, D.C., 1963.

Verbeck, G. J., *Carbonation of Hydrated Portland Cement,* Research Department Bulletin RX087, Portland Cement Association, http://www.portcement.org/pdf_files/RX087.pdf, 1958.

Whiting, David A.; Detwiler, Rachel J.; and Lagergren, Eric S., "Cracking Tendency and Drying Shrinkage of Silica Fume Concrete for Bridge Deck Applications," *ACI Materials Journal,* American Concrete Institute, Farmington Hills, Michigan, January–February, 2000, pages 71 to 77.

Whiting, D., and Dziedzic, W., *Effects of Conventional and High-Range Water Reducers on Concrete Properties,* Research and Development Bulletin RD107, Portland Cement Association, 1992.

Ytterberg, Robert F., "Shrinkage and Curling of Slabs on Grade, Part I—Drying Shrinkage, Part II—Warping and Curling, and Part III— Additional Suggestions," *Concrete International,* American Concrete Institute, Farmington Hills, Michigan, April 1987, pages 22 to 31; May 1987, pages 54 to 61; and June 1987, pages 72 to 81.

CHAPTER 16
Control Tests for Concrete

Satisfactory concrete construction and performance requires concrete with specific properties. To assure that these properties are obtained, quality control and acceptance testing are indispensable parts of the construction process. Test results provide important feedback used to base decisions regarding mix adjustments. However, past experience and sound judgment must be relied on in evaluating tests and assessing their significance in controlling the design, batching and placement processes that influence the ultimate performance of the concrete.

Specifiers are moving toward performance-based specifications (also called end-result or end-property specifications) that require the final performance of concrete be achieved independent of the process used to achieve the performance. Physical tests and concrete properties are used to measure acceptance. Such specifications may not have acceptance limits for process control tests such as slump or limits on the quantities of concrete ingredients as do prescriptive specifications. The end result of compressive strength, low permeability, documented durability, and a minimal number of cracks, for example, would be the primary measure of acceptance. Of course, even though process control tests may not be specified, the wise concrete producer would use them to guide the product to a successful end result. However, most specifications today are still a combination of prescriptive and performance requirements (Parry 2000).

CLASSES OF TESTS

Project specifications may affect (1) characteristics of the mixture, such as maximum size of aggregate, aggregate proportions, or minimum cement content; (2) characteristics of the cement, water, aggregates, and admixtures; and (3) characteristics of the freshly mixed and hardened concrete, such as temperature, slump, air content, and compressive or flexural strengths.

Cementitious materials are tested for their compliance with CSA (ASTM) standards to avoid any abnormal performance such as early stiffening, delayed setting, or low strengths in concrete. Standards for cementitious materials are contained in CSA A-3000-98, *Cementitious Materials Compendium.* Additional details regarding cementitious materials can be found in Chapters 2 and 3.

Tests of aggregates have two major purposes: (1) to determine the suitability of the material itself for use in concrete, including tests for abrasion, soundness against saturated freeze-thaw cycles, harmful materials by petrographic examination, and potential alkali-aggregate reactivity; and (2) to assure uniformity, such as tests for moisture control, relative density, and gradation of aggregates. Some tests are used for both purposes. Testing aggregates to determine their potential alkali-aggregate reactivity is discussed in Chapter 5, "Aggregates for Concrete." Tests of concrete to evaluate the performance of available materials, to establish mixture proportions, and to control concrete quality during construction include slump, air content, temperature, density, and strength. Slump, air content, and strength tests are usually required in project specifications for concrete quality control, whereas density is more useful in mixture proportioning. CSA A23.1 specifies that slump, air-content, and temperature tests should be made when strength specimens are made.

Following is a discussion of frequency of testing and descriptions of the major control tests to ensure uniformity of materials, desired properties of freshly mixed concrete, and required strength of hardened concrete. Special tests are also described.

ASTM (2000) and Klieger and Lamond (1994) provide extensive discussions of test methods for concrete and concrete ingredients.

FREQUENCY OF TESTING

Frequency of testing is a significant factor in the effectiveness of quality control of concrete. Specified test frequencies are intended for acceptance of the material or one of its components at a random location within the quantity or time period represented by the test. Such frequencies may not occur often enough to control the material within specified limits during production. Process control tests are nonrandom tests performed more often than specified to document trends that allow adjustments to be made before acceptance tests are performed.

The frequency of testing aggregates and concrete for typical batch-plant procedures depends largely upon the uniformity of materials, including the moisture content of aggregates, and the production process. Initially it is advisable to make process control tests several times a day, but as work progresses and the material becomes more predictable, the frequency often can be reduced. ASTM C 1451 provides a standard practice for determining the uniformity of cementitious materials, aggregates, and chemical admixtures used in concrete.

Usually, aggregate moisture tests are made once or twice a day. The first batch of fine aggregate in the morning is often overly wet since moisture will migrate overnight to the bottom of the storage bin. As fine aggregate is drawn from the bottom of the bin and additional aggregate is added, the moisture content should stabilize at a lower level and the first moisture test can be made. Obtaining moisture samples representative of the aggregates being batched is important; a 1% change in moisture content of fine aggregate will change the amount of mix water needed by approximately 8 kg/m³.

Slump tests should be made for the first batch of concrete each day, and whenever the consistency of concrete appears to vary. CSA A23.1 requires that a slump test be made with every strength test and every second or third air test.

Air-content tests should be made often enough at the point of delivery to ensure proper air content, particularly if temperature and aggregate grading change. An air-content test should be performed for each sample of concrete from which cylinders are made; a record of the temperature of each sample of concrete should also be kept. CSA Standard A23.1 requires that, for concrete subjected to frequent cycles of freezing and thawing in the presence of moisture or deicing chemicals, every load or batch of concrete shall be tested until satisfactory control of the air content is established, to a point where fewer air tests are required by the owner. Whenever a test falls outside specified limits, the testing frequency shall revert to one test per load or batch until satisfactory control is reestablished.

The number of strength tests made will depend on the job specifications and the occurrence of variations. CSA Standard A23.1 requires that strength tests of each class of concrete placed each day should be taken not less than once a day, nor less than once for each 100 m³ of concrete. When this frequency of testing results in fewer than three tests for each class of concrete, tests shall be made on at least three randomly selected batches on a project. For standard strength tests, 100 x 200-mm laboratory cured cylinders are used. Accelerated strength tests may also be used for the acceptance of concrete. Specifications may also require that additional specimens be made and field-cured as nearly as practical in the same manner as the concrete in the structure.

A test result is the average of the strength of two cylinders tested at the same age. If any test specimen shows distinct evidence of improper sampling, moulding, handling, curing, or testing, it shall be discarded. The strength of the remaining test cylinder(s) is considered the test result. Unless otherwise specified, cylinders are tested at the age of 28 days. The average strength of two 28-day cylinders is required for each test used for evaluating concrete. A 7-day test cylinder, along with the two 28-day test cylinders, is often made and tested to provide an early indication of strength development. As a rule of thumb, the 7-day strength is about 60% to 75% of the 28-day compressive strength, depending upon the type and amount of cementitious materials, water-cementing materials ratio, curing temperature, and other variables. Additional specimens may be required when high-performance or high-strength concrete is involved or where structural requirements are critical. Specimens should be laboratory cured when tested for acceptance or ultimate performance of the concrete. However, laboratory-cured specimens should not be used as an indication of in-place concrete strengths.

In-place concrete strengths are typically documented by casting specimens that are field-cured (as nearly as practical) in the same manner as concrete in the structure. Field-cured specimens are commonly used to decide when forms and shores might be removed under a structural slab or to determine when traffic will be allowed on new pavement. CSA Test Method A23.2-3C (ASTM C 31) contains additional instructions regarding the handling of field-cured cylinders. Although field-cured specimens may be tested at any age, 7-day tests are often made for comparison with laboratory tests at the same age; these are useful to judge if curing and protection during cold weather concreting is adequate.

TESTING AGGREGATES

Sampling Aggregates

Methods for obtaining representative samples of aggregates are given in CSA Test Method A23.2-1A (ASTM D 75). Accurate sampling is important. The location in the production process where samples will be obtained must be carefully planned. Sampling from a conveyor belt, stockpile, or aggregate bin may require special sampling equipment. Caution must be exercised to obtain a sample away from stockpile segregation and the sample must be large enough to meet minimum sample size requirements. In addition, samples obtained for moisture content testing

Fig. 16-1. Sample splitter commonly used to reduce coarse aggregate samples. (70012)

should be placed in a sealed container or plastic bag as soon as possible to retain moisture until testing.

Reducing large field samples to small quantities for individual tests must be done with care so that the final samples will be truly representative. For coarse aggregate, this is done by the quartering method: The sample, thoroughly mixed, is spread on a piece of canvas in an even layer 75 or 100 mm thick. It is divided into four equal parts. Two opposite parts are then discarded. This process is repeated until the desired size of sample remains. A similar procedure is sometimes used for moist, fine aggregate. Sample splitters are desirable for dry aggregate (Fig. 16-1) but should not be used for samples that are more than saturated surface dry.

Organic Impurities

Organic impurities in fine aggregate should be determined in accordance with CSA Test Method A23.2-7A (ASTM C 40). A sample of fine aggregate is placed in a sodium hydroxide solution and shaken. The following day the colour of the sodium hydroxide solution is compared with a glass colour standard or standard colour solution. If the colour of the solution containing the sample is darker than the standard colour solution or Organic Glass Plate No. 3, the fine aggregate should not be used for important concrete work without further investigation.

Some fine aggregates contain small quantities of coal or lignite that give the liquid a dark colour. The quantity may be insufficient to reduce the strength of the concrete appreciably. If surface appearance of the concrete is not important, CSA A23.1 (ASTM C 33) states that fine aggregate is acceptable if the amount of coal and lignite does not exceed 0.5% of the total fine aggregate mass when tested in accordance with CSA Test Method A23.2-4A. A fine aggre-

gate failing the CSA A23.2-7A test may be used if, when tested in mortar, in accordance with CSA Test Method A23.2-8A (ASTM C 87), the mortar develops a compressive strength at 7 and 28 days of not less than 95% of that developed by a similar mortar made from another portion of the same sample that has been washed in a 3% solution of sodium hydroxide and then thoroughly rinsed in water. It should be realized that appreciable quantities of coal or lignite in aggregates can cause popouts and staining of the concrete and can reduce durability when concrete is exposed to weathering. Local experience is often the best indication of the durability of concrete made with such aggregates.

Objectionable Fine Material

Large amounts of clay and silt in aggregates can adversely affect durability, increase water requirements, and increase shrinkage. Specifications usually limit the amount of material passing the 80 µm sieve to 2% or 3% in fine aggregate and to 1% or less in coarse aggregate. Testing for material finer than the 80 µm sieve should be done in accordance with CSA Test Method A23.2-5A (ASTM C 117). Testing for clay lumps should be in accordance with CSA Test Method A23.2-3A (ASTM C 142).

Grading

Gradation of aggregates significantly affects concrete mixture proportioning and workability. Hence, gradation tests are an important element in the assurance of concrete quality. The grading or particle size distribution of an aggregate is determined by a sieve analysis test in which the particles are divided into their various sizes by standard sieves. The analysis should be made in accordance with CSA Test Method A23.2-2A (ASTM C 136).

Results of sieve analyses are used in three ways: (1) to determine whether or not the materials meet specifications; (2) to select the most suitable material if several aggregates are available; and (3) to detect variations in grading that are sufficient to warrant blending selected sizes or an adjustment of concrete mix proportions.

The grading requirements for concrete aggregate are shown in Chapter 5 and CSA Standard A23.1 (ASTM C 33). Materials containing too much or too little of any one size should be avoided. Some specifications require that mixture proportions be adjusted if the average fineness modulus of fine aggregate changes by more than ±0.20. Other specifications require an adjustment in mixture proportions if the amount retained on any two consecutive sieves changes by more than 10% by mass of the total fine-aggregate sample. A small quantity of clean particles that pass a 160 µm sieve but are retained on a 80 µm sieve is desirable for workability. For this reason, most specifications permit up to 10% of this material in fine aggregate.

Well-graded aggregates contain particles on each sieve size. Well-graded aggregates enhance numerous factors that result in greater workability and durability. The more well-graded an aggregate is, the more it will pack together efficiently, thus reducing the volume between aggregate particles that must be filled by paste. On the other hand, gap-graded aggregates—those having either a large quantity or a deficiency of one or more sieve sizes—can result in reduced workability during mixing, pumping, placing, consolidation and finishing. Durability can suffer too as a result of using more fine aggregate and water to produce a workable mix. See Chapter 5 and Galloway (1994) for additional information on aggregate grading.

Moisture Content of Aggregates

Several methods are used for determining the amount of moisture in aggregate samples. The Hotplate Method of CSA Test Method A23.2-11A (ASTM C 566) is an approximate method for determining the surface (free) moisture of sand and coarse aggregate. In this method a measured sample of damp aggregate is dried either in a ventilated conventional oven, microwave oven, or over an electric or gas hotplate. From the mass before and after drying, the total moisture content can be calculated as follows:

$$P = 100(M - D)/D$$

where

P = moisture content of sample, percent
M = mass of original sample
D = mass of dried sample

The surface moisture can be calculated if the percentage of absorbed moisture in the aggregate is known. The surface moisture content is equal to the total moisture content minus the absorbed moisture. Historic information for an aggregate source can be used to obtain absorbed moisture content data if the mineral composition of the pit or quarry has not changed significantly. However, if recent data is not available, absorbed moisture contents can be determined using CSA Test Method A23.2-12A (ASTM C 127) for coarse aggregate and CSA Test Method A23.2-6A (ASTM C 128) for fine aggregate.

Only the surface moisture, not the absorbed moisture, becomes part of the mixing water in concrete. Surface moisture percentages are used to calculate the amount of water in the aggregates to reduce the amount of mix water used for batching. In addition, the batch quantity of aggregates should be increased by the percentage of surface moisture present in each type of aggregate. If adjustments are not made during batching, surface water will replace a portion of the aggregate mass and the mix will not yield properly.

Another method to determine moisture content, which is not as accurate, is to evaporate the moisture by burning alcohol. In this method: (1) a measured sample of damp fine aggregate is placed in a shallow pan; (2) about 310 ml of alcohol for each kilogram is then poured over the sample; (3) the mixture is stirred with a rod and spread in a thin layer over the bottom of the pan; (4) the alcohol is then ignited and allowed to burn until the sample is dry; (5) after burning, the sample is cooled for a few minutes and the mass determined; and (6) the percentage of moisture is then calculated.

When drying equipment is not available a field or plant determination of surface (free) moisture in fine aggregate can be made in accordance with CSA Test Method A23.2-11A (ASTM C 70). The same procedure can be used for coarse aggregate with appropriate changes in the size of sample and dimensions of the container. This test depends on displacement of water by a known mass of

Table 16-1. Example of Adjustment in Batch Quantities for Moisture in Aggregates

Aggregate data	Absorption, %	Moisture content, %
Fine aggregate	1.2	5.8
Coarse aggregate	0.4	0.8

Concrete ingredients	Mix design quantities (aggregates in dry BOD condition),* kg/m³	Aggregate quantities (SSD condition),** kg/m³ BOD·(Absorbed %)/100	Aggregate quantities (in moist condition), kg/m³ BOD·(Moisture %)/100	Mix water correction for surface moisture in aggregates, kg/m³ BOD·(Moist%-Absorb%)/100	Adjusted batch quantities, kg/m³
Cement	355				355
Fine aggregate	695	703	735	32	735
Coarse aggregate	1060	1064	1068	4	1068
Water	200				164
Total	2310			36	2322 †

* An aggregate in a bulk-oven dry (BOD) condition is one with its permeable voids completely dry so that it is fully absorbent.
** An aggregate in a saturated, surface-dry (SSD) condition is one with its permeable voids filled with water and with no surface moisture on it. Concrete suppliers often request mix design proportions on a SSD basis because of batching software requirements.
† Total adjusted batch quantity is higher than total mix design quantity by the amount of water absorbed in the aggregate.

moist aggregate; therefore, the relative density of the aggregate must be known accurately.

Electric moisture meters are used in many concrete batching plants primarily to monitor the moisture content of fine aggregates, but some plants also use them to check coarse aggregates. They operate on the principle that the electrical resistance of damp aggregate decreases as moisture content increases, within the range of dampness normally encountered. The meters measure the electrical resistance of the aggregate between electrodes protruding into the batch hopper or bin. Moisture meters using the microwave-absorption method are gaining popularity because they are more accurate than the electric meters. However, both methods measure moisture contents accurately and rapidly, but only at the level of the probes. These meters require frequent calibration and must be maintained properly. The variable nature of moisture contents in aggregates cause difficulty in obtaining representative samples for comparison to electric moisture meters. Several oven-dried moisture content tests should be performed to verify the calibration of these meters before trends in accuracy can be established.

Table 16-1 illustrates a method of adjusting batch quantities for moisture in aggregates.

TESTING FRESHLY MIXED CONCRETE

Sampling Freshly Mixed Concrete

The importance of obtaining truly representative samples of freshly mixed concrete for control tests must be emphasized. Unless the sample is representative, test results will be misleading. Samples should be obtained and handled in accordance with CSA Test Method A23.2-1C (ASTM C 172).

Except for routine slump and air-content tests performed for process control, CSA A23.2-1C (ASTM C 172) requires that sample size used for acceptance purposes be at least 30 L in size. The time allowed to obtain the grab samples, transport them to the place where plastic concrete tests are being performed–or where test specimens are being moulded–and remix them with a shovel to ensure uniformity is 10 minutes. When sampling from a mixer, the sample should be taken between the 10 and 90% points of the discharge. The sample should be protected from sunlight, wind, and other sources of rapid evaporation during sampling and testing.

Consistency

The slump test, CSA Test Method A23.2-5C (ASTM C 143), is the most generally accepted method used to measure the consistency of concrete (Fig. 16-2). The test for slump must be completed within 10 minutes after sampling is completed. The test equipment consists of a slump cone (a metal conical mould 300 mm high, with a 200-mm diameter base and 100-mm diameter top) and a steel rod 16 mm in diameter and between 450 mm and 600 mm long with a hemispherically shaped tip. The dampened slump cone, placed upright on a flat, nonabsorbent rigid surface, should be filled in three layers of approximately equal volume. Therefore, the cone should be filled to a depth of about 70 mm for the first layer, a depth of about 160 mm for the second layer, and overfilled for the third layer. Each layer is rodded 25 times. Following rodding, the last layer is struck off and the cone is slowly raised vertically 300 mm in approximately 5 seconds. As the concrete subsides or

Fig. 16-2. Slump test for consistency of concrete. Figure A illustrates a lower slump, Figure B a higher slump. (69786, 69787)

settles to a new height, the empty slump cone is then inverted and gently placed next to the settled concrete. The slump is the vertical distance the concrete settles, measured to the nearest 10 mm; a ruler is used to measure from the top of the slump cone (mould) to the displaced original center of the subsided concrete (see Fig. 16-2).

A higher slump value is indicative of a more fluid concrete. The entire test through removal of the cone should be completed in 2 minutes, as concrete loses slump with time. If a falling away or shearing off occurs from a portion of the concrete, another test should be run on a different portion of the sample.

Another test method for flow of fresh concrete involves the use of the K-Slump Tester (ASTM C 1362). This is a probe-type instrument that is thrust into the concrete in any location where there is a minimum depth of 175 mm of concrete a 75-mm radius of concrete around the tester. The amount of mortar flowing into openings in the tester is reported as a measure of flow.

Additional consistency tests include: the FHWA vibrating slope apparatus (Wong and others 2001 and Saucier 1966); British compacting factor test (BS 1881); Powers remoulding test (Powers 1932); German flow table test (DIN 1048-1); Vebe consistometer for roller-compacted concrete (ASTM C 1170); Kelly ball penetration test (ASTM C 360-92 now discontinued); Thaulow tester; the inverted slump cone for fiber-reinforced concrete (ASTM C 995); Powers and Wiler plastometer (Powers and Wiler 1941); Tattersall (1971) workability device; Colebrand test; BML viscometer (Wallevik 1996); BTRHEOM rheometer for fluid concrete (de Larrard, Szitkar, Hu, and Joly 1993); free-orifice rheometer (Bartos 1978); delivery chute torque meter (US patent 4,332,158 [1982]); delivery-chute vane (US patent 4,578,989 [1986]); Angles flow box (Angles 1974); ring penetration test (Teranishs and others 1994); and the Wigmore (1948) consistometer. The Vebe test and the Thaulow test are especially applicable to stiff and extremely dry mixes while the flow table is especially applicable to flowing concrete (Scanlon 1994).

Temperature Measurement

Because of the important influence concrete temperature has on the properties of freshly mixed and hardened concrete, many specifications place limits on the temperature of fresh concrete. Glass or armored thermometers are available (Figs. 16-3). The thermometer should be accurate to plus or minus 0.5°C and should remain in a representative sample of concrete for a minimum of 2 minutes or until the reading stabilizes. A minimum of 75 mm of concrete should surround the sensing portion of the thermometer. Electronic temperature meters with precise digital readouts are also available. The temperature measurement (ASTM C 1064 or AASHTO T 309) should be completed within 5 minutes after obtaining the sample.

Fig. 16-3. A thermometer is used to take the temperature of fresh concrete. (69885A)

Density and Yield

The density and yield of freshly mixed concrete (Fig. 16-4) are determined in accordance with CSA Test Method A23.2-6C (ASTM C 138). The results should be sufficiently accurate to determine the volumetric quantity (yield) of concrete produced per batch (see Chapter 9). The test also can give indications of air content provided the relative densities of the ingredients are known. A balance or scale sensitive to 0.5 kg is required. The size of the container used to determine density and yield varies with the size of aggregate; if in good condition, the 7 L air metre container is commonly used with nominal maximum size aggregates up to 28 mm; a 14 L container is used with nominal maximum size aggregates up to 56 mm. The container should

Fig. 16-4. Fresh concrete is measured in a container of known volume to determine density. (69785)

be calibrated at least annually (ASTM C 1077). Care is needed to consolidate the concrete adequately by either rodding or internal vibration. Strike off the top surface using a flat plate so that the container is filled to a flat smooth finish. The density is expressed in kilograms per cubic metre and the yield, the volume of concrete produced by batch, in cubic metres.

The density of unhardened as well as hardened concrete can also be determined by nuclear methods, ASTM C 1040.

Air Content

A number of methods for measuring air content of freshly mixed concrete can be used. CSA Test Methods include the pressure method, CSA Test Method A23.2-4C (ASTM C 231), and the volumetric method, CSA Test Method A23.2-7C (ASTM C 173). ASTM also includes a standard for the gravimetric method (ASTM C 138). Variations of the first two methods can also be used.

The pressure method (Fig. 16-5) is based on Boyle's law, which relates pressure to volume. Many commercial air meters of this type are calibrated to read air content directly when a predetermined pressure is applied. The applied pressure compresses the air within the concrete sample, including the air in the pores of aggregates. For this reason, tests by this method are not suitable for determining the air content of concretes made with some low-density aggregates or other very porous materials. Aggregate correction factors that compensate for air trapped in normal-density aggregates are relatively constant and, though small, should

be subtracted from the pressure meter gauge reading to obtain the correct air content. The instrument should be calibrated for various elevations above sea level if it is to be used in localities having considerable differences in elevation. Some meters utilize change in pressure of a known volume of air and are not affected by changes in elevation. Pressure meters are widely used because the mix proportions and relative density of the concrete ingredients need not be known. Also, a test can be conducted in less time than is required for other methods.

The volumetric method (Fig. 16-6) outlined in CSA Test Method A23.2-7C (ASTM C 173) requires removal of air from a known volume of concrete by agitating the concrete in an excess of water. This method can be used for concrete containing any type of aggregate, including low density or porous materials. An aggregate correction factor is not necessary with this test. The volumetric test is not affected by atmospheric pressure, and relative density of the concrete ingredients need not be known. Care must be taken to agitate the sample sufficiently to remove all air. The addition of 500 mL of isopropyl alcohol accelerates the removal of air, thus shortening test times; it also dispels most of the foam and increases the precision of the test, including those performed on high-air-content or high-cement-content concretes.

The gravimetric method utilizes the same test equipment used for determining the density of concrete. The measured density of concrete is subtracted from the theoretical density as determined from the absolute volumes of the ingredients, assuming no air is present (see ASTM C 138). This difference, expressed as a percentage of the theoretical density, is the air content. Mixture proportions and relative densities of the ingredients must be accurately

Fig. 16-5. Pressure-type meter for determining air content. (69766)

Fig. 16-6. Volumetric air meter. (69886)

known; otherwise results may be in error. Consequently, this method is suitable only where laboratory-type control is exercised. Significant changes in density can be a convenient way to detect variability in air content.

A pocket-size, Chace air indicator (AASHTO T 199) can be used as a quick check for the presence of low, medium, or high levels of air in concrete, but it is not a substitute for the other more accurate methods. A representative sample of mortar from the concrete is placed in the container. The container is then filled with alcohol and rolled with the thumb over the open end to remove the air from the mortar. The indicated air content is determined by comparing the drop in the level of the alcohol with a calibration chart. The test can be performed in a few minutes. It is especially useful in checking for the presence of air in concrete near the surface that may have suffered reductions in air because of faulty finishing procedures.

With any of the above methods, air-content tests should be started within 10 minutes after the sampling is completed.

Studies into the effect of fly ash on the air-void stability of concrete have resulted in the development of the foam-index test. The test can be used to measure the relative air-entraining admixture requirements for concrete mixtures containing fly ash. The fly ash to be tested is placed in a wide mouth jar along with the air-entraining admixture and shaken vigorously. Following a waiting period of 45 seconds, a visual determination of the stability of the foam or bubbles is made (Gebler and Klieger 1983).

Fig. 16-7. Preparing test specimens for compressive strength of concrete. (69790)

Strength Specimens

Premoulded specimens for field and laboratory strength tests should be made and cured in accordance with CSA Test Method A23.2-3C (ASTM C 31, ASTM C 192). Molding of strength specimens should be completed within 20 minutes after sampling.

Unlike in the United States, where 150 x 300-mm cylinders are used for compressive strength tests, Canadian standards have adopted the 100 x 200-mm cylinder for standard strength tests. The cylinder size, however, must meet the following aggregate size criteria: the diameter of a cylindrical specimen (compressive strength test) or the minimum cross-sectional dimension of a beam (flexural strength test) must be at least three times the maximum nominal size of the coarse aggregate in the concrete. Also, the height of the cylinders should be twice the diameter (Fig. 16-7). For autogenous curing strength tests, 150 x 300-mm cylinders may be necessary (ASTM C 684). Moulds for compression test specimens must be cylindrical, have non-absorbent surfaces, and meet the requirements of CSA Test Method A23.2-1D (ASTM C 470). Cardboard moulds should not be used unless there is satisfactory documentation available indicating that the cylinders produced from them will have compressive strengths equivalent to those obtained from rigid, nonabsorbent moulds when tested

under the same conditions. Cardboard moulds are not to be used for specified concrete strengths above 35 MPa. Moulds should be placed on a smooth, level, rigid surface and filled carefully to avoid distortion of their shape.

A 100-mm diameter by 200-mm high cylinder mould has been commonly used in the past with high-strength concrete (Burg and Ost 1994, Forstie and Schnormeier 1981, and Date and Schnormeier 1984). The 100 x 200-mm cylinder used in Canada is easier to cast, requires less sample, has considerably less mass than a 150 x 300-mm concrete cylinder, (standard in many areas of the United States), and is therefore easier to handle, and requires less moist-curing storage space. In addition, the smaller cross-sectional area allows higher compressive strengths to be reached by a testing machine that has a smaller load capacity. The difference in indicated strength between the two cylinder sizes is insignificant as illustrated in Fig. 16-8.

The cross-section of beams used for the flexural strength test should not be less than 150 x 150 mm, or three times the maximum size of aggregate, whichever is larger. The length of beams should be at least three times the depth of the beam plus 50 mm. For example, if the maximum size of aggregate in the concrete is 40 mm, the total length should be not less than 500 mm for a 150 x 150-mm beam.

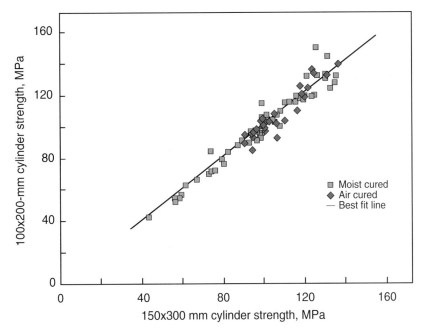

Fig. 16-8. Comparison of 100 x 200-mm and 150 x 300-mm cylinder strengths (Burg and Ost 1994).

For Concrete with a Specified Slump Greater Than 40 mm:

- Test cylinders are to be rodded. The cylinder moulds should be filled in three approximately equal layers with each layer rodded 20 times for 100-mm diameter cylinders. Additional information on rodding for other size cylinders is provided in CSA A23.2-3C. Superplasticized concrete with a slump greater than 180 mm requires only 40% of the number of strokes needed for normal slump concrete.

- Beam specimens 200 mm or less in depth should be filled in two equal layers, with each layer rodded evenly over its area once with a 16-mm rod for each 10 cm^2 of the specimen's top surface area. If the rodding leaves holes, the sides of the mould should be lightly tapped with a mallet or open hand. Beam moulds over 200 mm in depth are to be filled in three approximately equal layers with each layer rodded evenly over its area once for each 10 cm^2 of area.

For Concrete with a Specified Slump Equal to or Less Than 40 mm:

- Test cylinders are to be vibrated. The cylinder moulds should be filled in two equal layers, with three insertions of the vibrator at different points for each layer. The duration of vibration required will depend upon the workability of the concrete and the effectiveness of the vibrator. Usually, sufficient vibration has been applied as soon as the surface of the concrete becomes smooth in appearance and the egress of entrapped air bubbles has ceased. Internal vibrators should have a maximum diameter of no more than ¼ the diameter of the cylinder.

- Beam moulds 200 mm or less in depth are to be filled in approximately two equal layers and consolidated by vibration. Beam moulds more than 200 mm in depth are to be filled in three approximately equal layers and consolidated by vibration. The vibrator should be inserted at intervals not exceeding 150 mm along the centerline of the long dimension of the specimen. For specimens wider than 150 mm, alternating insertions along two lines should be used. Internal vibrators should have a maximum diameter of no more than ⅓ the width of beams.

Immediately after casting, the tops of the specimens should be (1) covered with an oiled glass or steel plate, (2) sealed with a plastic bag, or (3) sealed with a plastic cap. Additional information on the placing and consolidation of concrete in strength test specimens, both compressive and flexure, is given in CSA Test Method A23.2-3C.

The strength of a test specimen can be greatly affected by jostling, changes in temperature, and exposure to drying, particularly within the first 24 hours after casting. Thus, test specimens should be cast in locations where subsequent movement is unnecessary and where protection is possible. Cylinders and test beams should be protected from rough handling at all ages. Remember to identify specimens on the exterior of the mould to prevent confusion and errors in reporting. Do not etch identification numbers into the surface of fresh concrete test specimens. Use tape or identification tags that do not damage the sample.

Standard testing procedures require that specimens be cured under controlled conditions, either in the laboratory (Fig. 16-9) or in the field. Controlled laboratory curing in a moist room or in a limewater storage tank gives an accurate indication of the quality of the concrete as delivered. Limewater must be saturated with hydrated lime, not agricultural lime, in accordance with ASTM C 511 to prevent leaching of lime from concrete specimens.

Specimens cured in the field in the same manner as the structure more closely represent the actual strength of concrete in the structure at the time of testing; however, they give little indication of whether a deficiency is due to the quality of the concrete as delivered or to improper handling and curing. On some projects, field-cured specimens are made in addition to those destined for controlled laboratory curing; these are especially useful when the weather is unfavourable, to determine when forms can be removed, or when the structure can be put into use. For more infor-

Fig. 16-9. Controlled moist curing in the laboratory for standard test specimens with free water maintained on all surfaces at all times and stored at a temperature of 23 ± 2°C as per CSA A23.2-3C (ASTM C 511). (8974)

mation see "Strength Tests of Hardened Concrete" in this chapter, CSA A23.2-14C, and ASTM (2000).

In-place concrete strength development can also be evaluated by maturity testing (ACI Committee 306 and ASTM C 1074), which was discussed in Chapter 14.

Time of Setting

Test method ASTM C 403 is used to determine the time of setting of concrete by means of penetration resistance measurements made at regular time intervals on mortar sieved from the concrete mixture (Fig. 16-10). The initial and final time of setting is determined as the time when the

penetration resistance equals 3.4 MPa and 27.6 MPa, respectively. Typically, initial set occurs between 2 and 6 hours after batching and final set occurs between 4 and 12 hours. The rate of hardening of concrete greatly influences the rate of which construction progresses. Temperature, water-cementing materials ratio, and admixtures all affect setting time.

Accelerated Compression Tests to Project Later-Age Strength

The need to assess the quality of concrete at early ages in comparison to traditional 28-day tests has received much attention due to the quickening pace of today's construction. Accelerated strength tests may be used for acceptance of concrete on the basis of strength, and also to expedite quality control in the production process and for the acceptance of structural concrete where adequate data correlated with standard 28-day compressive strength tests are available. The boiling water test and autogenous accelerated curing methods for such purposes are described in CSA Test Method A23.2-10C (ASTM C 684). Other accelerated curing systems have also been used for predicting the strength of concrete. These include the warm water method, with water at 35°C ± 3°C; high temperature method, 150°C ± 3°C (ASTM C 684); and the polystyrene mould system (Bisaillon 1975).

Accelerated strength tests are performed at ages ranging between 5 and 49 hours, depending on the curing procedure used. Later-age strengths are estimated using previously established relationships between accelerated strength and standard 28-day compressive strength tests.

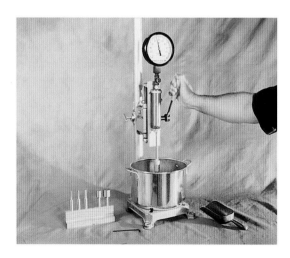

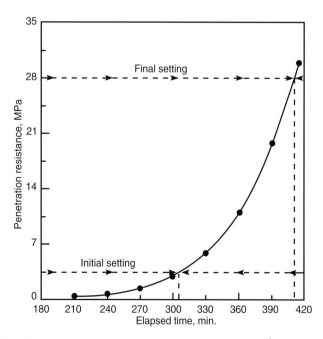

Fig. 16-10. (left) Time of setting equipment. (right) Plot of test results. (69788)

ASTM C 918 uses the maturity method of monitoring temperature of cylinders cured in accordance with standard methods outlined in ASTM C 31. Cylinders are tested at early ages beyond 24 hours, and the concrete temperature history is used to compute the maturity index at the time of test. Using historic data, a prediction equation is developed to project the strength at later ages based on the maturity index and early-age strength tests. See Carino (1994).

Chloride Content

The chloride content of concrete and its ingredients should be checked to make sure it is below the limit necessary to avoid corrosion of reinforcing steel. An approximation of the water-soluble chloride content of freshly mixed concrete, aggregates, and admixtures can be made using a method initiated by the National Ready Mixed Concrete Association (NRMCA 1986). A determination of the total chloride content of freshly mixed concrete may be made by summing up the chloride contents of all of the individual constituents of the mix. The NRMCA method gives a quick approximation and should not be used to determine compliance.

Portland Cement Content, Water Content, and Water-Cement Ratio

Test methods are available for estimating the portland cement and water content of freshly mixed concrete. These test results can assist in determining the strength and durability potential of concrete prior to setting and hardening and can indicate whether or not the desired cement and water contents were obtained. ASTM test methods C 1078-87 and C 1079-87 (discontinued in 1998), based on the Kelly-Vail method, can be used to determine cement content and water content, respectively. Experimental methods using microwave absorption have been developed to estimate water to cement ratio. The disadvantage of these test methods is they require sophisticated equipment and special operator skills, which may not be readily available.

Other tests for determining cement or water contents can be classified into four categories: chemical determination, separation by settling and decanting, nuclear related, and electrical. The Rapid Analysis Machine and nuclear cement gage have been used to measure cement contents (Forester, Black, and Lees 1974 and PCA 1983). The microwave oven and neutron-scattering methods have been used to measure water contents. For an overview of these and other tests from all four categories, see Lawrence (1994). A combination of these tests can be run independently of each other to determine either cement content or water content to calculate the water-cement ratio.

Supplementary Cementing Materials Content

Standard test methods are not available for determining the supplementary cementing materials content of freshly mixed concrete. However, the presence of certain supplementary cementing materials, such as fly ash, can be determined by washing a sample of the concrete's mortar over a 45 µm sieve and viewing the residue retained with a stereo microscope (150 to 250X) (Fig. 16-11). Fly ash particles appear as spheres of various colours. Sieving the mortar through a 160 or 80 µm sieve is helpful in removing sand grains.

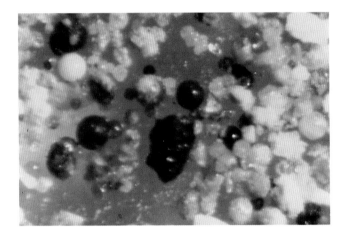

Fig. 16-11. Fly ash particles retained on a 45 µm sieve after washing, as viewed through a microscope at 200X. (58949)

Bleeding of Concrete

The bleeding properties of fresh concrete can be determined by two methods described in ASTM C 232 . One method consolidates the specimen by tamping without further disturbance; the other method consolidates the specimen by vibration after which the specimen is vibrated intermittently throughout the test. The amount of bleed water at the surface is expressed as the volume of bleed water per unit area of exposed concrete, or as a percentage of the net mixing water in the test specimen. Typical values range from 0.01 to 0.08 mL/cm² or 0.1% to 2.5% of mix water. The bleeding test is rarely used in the field (Fig. 16-12). Bleeding was also discussed in Chapter 1.

Fig. 16-12. ASTM C 232 test for bleeding of concrete; Method A without vibration. The container has an inside diameter of about 255 mm and height of about 280 mm. The container is filled to a height of about 255 mm and covered to prevent evaporation of the bleed water. (69780)

Fig. 16-13. Concrete cylinders cast in place in cylindrical moulds provide a means for determining the in-place compressive strength of concrete. (69781)

TESTING HARDENED CONCRETE

Premoulded specimens described in the previous section, "Strength Specimens" (CSA A23.2-3C [ASTM C 31 or ASTM C 192]), or samples of hardened concrete obtained from construction or laboratory work (CSA A23.2-14C [ASTM C 42, ASTM C 823, or ASTM C 873, can be used]), in tests on hardened concrete. Separate specimens should be obtained for each test performed because specimen preconditioning for certain tests can make the specimen unusable for other tests.

Strength Tests of Hardened Concrete

Strength tests of hardened concrete can be performed on the following: (1) cured specimens moulded in accordance with CSA A23.2-3C (ASTM C 31 or ASTM C 192) from samples of freshly mixed concrete; (2) in-situ specimens cored or sawed from hardened concrete in accordance with CSA A23.2-14C (ASTM C 42); or (3) specimens made from cast-in-place cylinder moulds, ASTM C 873 (Fig. 16-13).

Cast-in-place cylinders can be used in concrete that is 125 to 300 mm in depth. The mould is filled in the normal course of concrete placement. The specimen is then cured in place and in the same manner as the rest of the concrete section. The specimen is removed from the concrete and mould immediately prior to testing to determine the in-place concrete strength. This method is particularly applicable in cold-weather concreting, post-tensioning work, slabs, or any concrete work where a minimum in-place strength must be achieved before construction can continue.

For all methods, cylindrical samples should have a diameter at least three times the maximum size of coarse aggregate in the concrete and a length as close to twice the diameter as possible. Correction factors are available in CSA A23.2-14C (ASTM C 42) for samples with lengths of 1 to 2 times the diameter. Cores and cylinders with a height of less than 95% of the diameter before or after capping should not be tested. Use of a minimum core diameter of 100 mm is suggested where a length to diameter (L/D) ratio equal to or greater than one is possible.

Drilled cores should not be taken until the concrete can be sampled without disturbing the bond between the mortar and the coarse aggregate. For horizontal surfaces, cores should be taken vertically and not near formed joints or edges. For vertical or sloped faces, cores should be taken perpendicular to the central portion of the concrete element. Although diamond-studded coring bits can cut through reinforcing steel, it should be avoided if possible when obtaining compression test specimens. A covermeter (electromagnetic device) or a surveyor's magnetic locator can be used to locate reinforcing steel. Cores taken from structures should be tested in a moisture condition as near as that of the in-place concrete as possible. Conditioning options for preparing specimens are described in CSA A23.2-14C (ASTM C 42). Cores taken from structures that are normally wet or moist in service should be submerged in water at room temperature for 40 to 48 hours immediately prior to testing. Those from structures normally dry in service should be conditioned in an atmosphere approximating their service conditions and tested dry after at least 24 hours at room temperature and humidity.

Fig. 16-14 shows the effects of core conditioning on the strength of drilled core samples. Forty-eight hour water immersion of the specimens prior to testing yields significantly lower test results than air-drying specimens for seven days prior to testing. Measured strengths varied by up to 25%, depending upon the time and type of conditioning prior to testing.

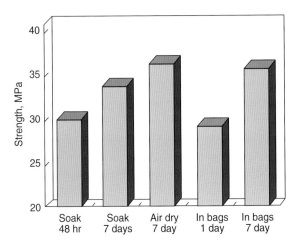

Fig. 16-14. Effect of core conditioning on strength of drilled cores (Fiorato, Burg and Gaynor 2000).

Flexure test specimens that are saw-cut from in-place concrete are always immersed in lime-saturated water at 23.0°C ± 2.0°C for at least 40 hours immediately prior to testing.

Test results are greatly influenced by the condition of the ends of cylinders and cores. For compression testing, specimens should be ground or capped in accordance with the requirements of CSA Test Method A23.2-9C (ASTM C 617 or ASTM C 1231). Various commercially available materials can be used to cap compressive test specimens. Any capping material, when tested at the same age and composition as the caps of the concrete cylinder, should develop a compressive strength greater than 35 MPa. ASTM C 617 outlines methods for using sulphur mortar capping. Caps must be allowed to harden at least two hours before the specimens are tested. Unbonded neoprene caps can be used to test moulded cylinders if quick results are needed. Sulphur mortar caps should be made as thin as is practical to avoid a cap failure that might reduce test results.

ASTM C 1231 describes the use of unbonded neoprene caps that are not adhered or bonded to the ends of the specimen. This method of capping uses a disk-shaped 13 ± 2-mm thick neoprene pad that is approximately the diameter of the specimen. The pad is placed in a cylindrical steel retainer with a cavity approximately 25 mm deep and slightly smaller than the diameter of the pad. A cap is placed on one or both ends of the cylinder; the specimen is then tested in accordance with CSA A23.2-9C (ASTM C 39) with the added step to stop the test at 10% of the anticipated load to check that the axis of the cylinder is vertical within a tolerance of 0.5°. If either the perpendicularity of the cylinder end, or the vertical alignment during loading are not met, the load applied to the cylinder may be concentrated on one side of the specimen. This can cause a short shear fracture in which the failure plane intersects the end of the cylinder. This type of fracture usually indicates cylinder failed prematurely, yielding results lower than the

actual strength of the concrete. If perpendicularity requirements are not met, the cylinder can be saw-cut, ground, or capped with a sulphur mortar compound in accordance with ASTM C 617.

Short shear fractures can also be reduced by: dusting the pad and end of cylinder with corn starch or talcum powder, preventing excess water from cylinders or burlap from draining into the retainer and below the pad, and checking bearing surfaces of retainers for planeness and indentations. In addition, annually clean and lubricate the spherically seated block and adjacent socket on the compression machine.

Testing of specimens (Fig. 16-15) should be done in accordance with (1) CSA Test Method A23.2-9C (ASTM C 39) for compressive strength, (2) CSA Test Method A23.2-8C (ASTM C 78) for flexural strength using third-point loading, (3) ASTM C 293 for flexural strength using center-point loading, and (4) CSA Test Method A23.2-13C (ASTM C 496) for splitting tensile strength. Fig. 16-16 shows the

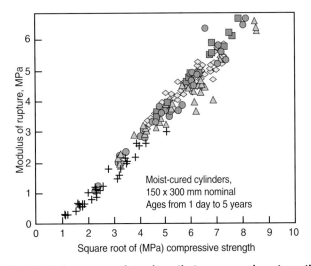

Fig. 16-15. Testing hardened concrete specimens: (left) cylinder, (right) beam (44178, 69684).

Fig. 16-16. Long-term data show that compressive strength is proportional to the square root of flexural strength (measured by third-point loading) over a wide range of strength levels (Wood 1992).

correlation between compressive strength and flexural strength test results.

For both pavement thickness design and pavement mixture proportioning, the modulus of rupture (flexural strength) should be determined by the third-point loading test CSA A23.2-8C (ASTM C 78). However, modulus of rupture by center-point loading (ASTM C 293) or cantilever loading can be used for job control if empirical relationships to third-point loading test results are determined before construction starts.

The moisture content of the specimen has considerable effect on the resulting strength (Fig. 16-15). Beams for flexural tests are especially vulnerable to moisture gradient effects. A saturated specimen will show lower compressive strength and higher flexural strength than those for companion specimens tested dry. This is important to consider when cores taken from hardened concrete in service are compared with moulded specimens tested as taken from the moist-curing room or water storage tank. Cylinders used for acceptance testing for a specified strength must be cured in accordance with CSA A23.2-3C (ASTM C 31 and ASTM C 511) to accurately represent the quality of the concrete. However, cores are subject to workmanship, variable environmental site conditions, and variable conditioning after extraction. Cores are tested in either a dry or moist condition, but rarely in a saturated condition similar to lab-cured cylinders. Because cores and cylinders are handled in very different ways, they cannot be expected to yield the same results.

The amount of variation in compressive-strength testing is far less than for flexural-strength testing. To avoid the extreme care needed in field flexural-strength testing, compressive-strength tests can be used to monitor concrete quality; however, a laboratory-determined empirical relationship (Fig. 16-16) must be developed between the compressive and flexural strength of the concrete used (Kosmatka 1985). Because of this empirical relationship and the economics of testing cylinders instead of beams, most departments of transportation are now utilizing compression tests of cylinders to monitor concrete quality for their pavement and bridge projects.

Evaluation of Compression Test Results. CSA Standard A23.1 states that the compressive strength of concrete can be considered satisfactory if the following conditions are met: the averages of all sets of three consecutive strength tests equal or exceed the specified 28-day strength, f'_c, and no individual strength test (average of two cylinders) is more than 3.5 MPa below the specified strength. If the results of the cylinder tests indicate that the concrete is not of the specified strength, the owner has the right to require one or more of the following:

(a) changes in the mix proportions for the remainder of the work;
(b) additional curing on those portions of the structure represented by the test specimens that failed to meet specified requirements;

(c) nondestructive testing;
(d) that cores be drilled from the portions of the structure in question and tested in accordance with CSA Test Method A23.2-14C;
(e) load testing of the structure or structural elements in accordance with the requirements of CSA Standard A23.3; and
(f) such other tests as the owner may specify.

In addition to the two 28-day cylinders, job specifications often require one or two 7-day cylinders and one or more "hold" cylinders. The 7-day cylinders monitor early strength gain. Hold cylinders are commonly used to provide additional information in case the 28-day cylinders are damaged or do not meet the required compressive strength. For low 28-day test results, the hold cylinders are typically tested at 56 days of age.

Protection and curing procedures should also be evaluated to judge if they are adequate when field-cured cylinders have a strength of less than 85% of that of companion laboratory-cured cylinders. The 85% requirement may be waived if the field-cured strength exceeds f'_c by more than 3.5 MPa.

When necessary, the in-place concrete strength should be determined by testing three cores for each strength test taken in the portion of the structure where the laboratory-cured cylinders did not meet acceptance criteria. Moisture conditioning of cores prior to compression testing should follow the guidelines in CSA Test Method A23.2-14C (ASTM C 42).

Nondestructive test methods are not a substitute for core tests, CSA Test Method A23.2-14C (ASTM C 42). If the average strength of three cores is at least 85% of f'_c, and if no single core is less than 75% of f'_c, the concrete in the area represented by the cores is considered structurally adequate. For high-strength concrete, the compressive strength values are 90% and 80%, respectively, of the specified strength, unless the values are determined by preconstruction trials. Refer to NRMCA (1979), ACI Committee 214 (1997), and CSA Standard A23.3 for load test procedures if required.

Air Content

The air-content and air-void-system parameters of hardened concrete can be determined by ASTM C 457. The hardened air-content test is performed to assure that the air-void system is adequate to resist damage from a freeze-thaw environment. The test is also used to determine the effect different admixtures and methods of placement and consolidation have on the air-void system. The test can be performed on premoulded specimens or samples removed from the structure. Using a polished section of a concrete sample, the air-void system is documented by making measurements using a microscope. The information obtained from this test includes the volume of entrained and entrapped air, its specific surface (surface area of the

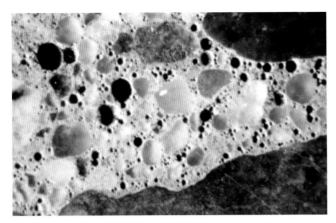

Fig. 16-17. View of concrete air-void system under a microscope. (67840)

air voids), the spacing factor, and the number of voids per lineal distance (Fig. 16-17). See Chapter 8, "Air-Entrained Concrete," for more information.

Density, Relative Density, Absorption, and Voids

The density, relative density, absorption, and voids content of hardened concrete can be determined in accordance with ASTM C 642 procedures (Table 16-2). It should be noted that the boiling procedure in ASTM C 642 can render the specimens useless for certain additional tests, especially strength tests. The density can be obtained by multi-

plying the relative density by the density of water (1000 kg/m³).

Saturated, surface-dry density (SSD) is often required for specimens to be used in other tests. In this case, the density can be determined by soaking the specimen in water for 48 hours and then determining its mass both in air (when SSD) and immersed in water. The SSD density is then calculated as follows:

$$D_{SSD} = \frac{M_1 \, \rho}{M_1 - M_2}$$

where

D_{SSD} is density in the SSD condition

M_1 is the SSD mass in air, kg

M_2 is the apparent mass immersed in water, kg

ρ is the density of water, 1000 kg/m³

The SSD density provides a close indication of the freshly mixed density of concrete. The density of hardened concrete can also be determined by nuclear methods (ASTM C 1040).

Portland Cement Content

The portland cement content of hardened concrete can be determined by ASTM C 1084 standard test method. Although not frequently performed, the cement content test is valuable in determining the cause of lack of strength gain or poor durability of concrete. Aggregate content can also be determined by this test. However, the user of this

Table 16-2. Permeability and Absorption of Concretes Moist Cured 7 Days and Tested After 90 Days.

Mix No.	Cement, kg/m³	w/cm	Compressive strength at 90 days, MPa	Permeability				Porosity, %†	Vol. of permeable voids, %	Absorption after immersion, %	Absorption after immersion and boiling, %
				RCPT, coulombs	90 days ponding, % Cl	Water, m/s**	Air, m/s**				
			ASTM C 39	ASTM C 1202	AASHTO T 259	API RP 27	API RP 27		ASTM C 642	ASTM C 642	ASTM C 642
1	445	0.26*	104.1	65	0.013	—	2.81×10^{-10}	7.5	6.2	2.43	2.56
2	445	0.29*	76.7	852	0.022	—	3.19×10^{-10}	8.8	8.0	3.13	3.27
3	381	0.40*	46.1	3242	0.058	2.61×10^{-13}	1.16×10^{-9}	11.3	12.2	4.96	5.19
4	327	0.50	38.2	4315	0.076	1.94×10^{-12}	1.65×10^{-9}	12.5	12.7	5.45	5.56
5	297	0.60	39.0	4526	0.077	2.23×10^{-12}	1.45×10^{-9}	12.7	12.5	5.37	5.49
6	245	0.75	28.4	5915	0.085	8.32×10^{-12}	1.45×10^{-9}	13.0	13.3	5.81	5.90

* Admixtures: 59.4 kg/m³ silica fume and 25.4 ml/kg of cement HRWR (Mix 1); 13.0 ml/kg HRWR (Mix 2); 2.2 ml/kg WR (Mix 3).
** To convert from m/s to Darcy, multiply by 1.03×10^5, from m/s to m², multiply by 1.02×10^{-7}.
† Measured with helium porosimetrie.
Adapted from Whiting (1988).

test method should be aware of certain admixtures and aggregate types that can alter test results. The presence of supplementary cementing materials would be reflected in the test results.

Supplementary Cementing Material and Organic Admixture Content

The presence and amount of certain supplementary cementing materials, such as fly ash, can be determined by petrographic techniques (ASTM C 856). A sample of the supplementary cementing material used in the concrete is usually necessary as a reference to determine the type and amount of the supplementary cementing material present. The presence and possibly the amount of an organic admixture (such as a water reducer) can be determined by infrared spectrophotometry (Hime, Mivelaz, and Connolly 1966).

Chloride Content

Concern with chloride-induced corrosion of reinforcing steel has led to the need to monitor and limit the chloride content of reinforced concrete. Limits on the water-soluble chloride-ion content of hardened reinforced concrete are given in CSA Standard A23.1. The water-soluble chloride-ion content of hardened concrete can be determined in accordance with procedures outlined in CSA Test Method A23.2-4B (ASTM C 1218). It should be noted that Note 2 to Clause 1 of this test method observes that the test method determines the total water-soluble chloride ion content of the concrete, including the chloride ions in the aggregate that may not be free to move into the cement paste. In cases where chloride in the aggregate is suspected, the soluble chloride-ion content of the aggregate should be determined separately. The water-soluble chloride-ion content of the concrete should then be adjusted accordingly. The philosophy behind this concept is that the test method requires pulverizing of the concrete, which releases any locked-up chlorides in aggregate particles that would otherwise not be released. However, any chlorides present on the surface of the aggregate particles would be free to move into the concrete. It follows then that aggregates to be tested by this method for purposes of determining the reduction permitted should be thoroughly washed before testing. It should also be noted that CSA Standard A23.1 limits on permissible water-soluble chloride content of concrete are expressed in terms of "mass of cementing material" in concrete. Because of this, in cases where difficulty is being experienced in meeting the CSA A23.1 limits, increasing the cementing materials content of the concrete may enable these limits to be met without changing material sources. The water-soluble chloride content is about 50% to 80% of the total (acid-soluble) chloride content. ASTM Test C 1152 is used to determine the acid-soluble chloride

content of concrete, which in most cases is equivalent to total chloride.

There is currently an ASTM provisional standard (ASTM PS 118 to be redesignated as ASTM C 1500) for the analysis of aggregate for water-extractable chloride (Soxhlet method). It is used when chloride contents have been found to be significantly high in aggregates, concretes, or mortars when tested by either ASTM C 1152 or C 1218 (CSA A23.2-4B). Because ASTM PS 118 does not pulverize the aggregates as other tests do, it theoretically measures more closely the chloride-ions available for corrosion. ACI 222.1 is also a Soxhlet procedure that tests chunks of concrete for water-extractable chloride. The true meaning of results from the Soxhlet procedures is still being debated.

Petrographic Analysis

Petrographic analysis uses microscopic techniques described in ASTM C 856 to determine the constituents of concrete, concrete quality, and the causes of inferior performance, distress, or deterioration. Estimating future performance and structural safety of concrete elements can be facilitated. Some of the items that can be reviewed by a petrographic examination include paste, aggregate, supplementary cementing materials, and air content; frost and sulphate attack; alkali-aggregate reactivity; degree of hydration and carbonation; water-cementing materials ratio; bleeding characteristics; fire damage; scaling; popouts; effect of admixtures; and several other aspects. Almost any kind of concrete failure can be analyzed by petrography (St. John, Poole, and Sims 1998). However, a standard petrographic analysis is sometimes accompanied by "wet" chemical analyses, infrared spectroscopy, X-ray diffractometry, scanning electron microscopy with attendant elemental analysis, differential thermal analysis, and other analytical tools.

The Annex to ASTM C 856 describes a technique for field and laboratory detection of alkali-silica gel. Using this method, a uranyl-acetate solution is applied to a broken or roughened concrete surface that has been dampened with distilled or deionized water. After one minute, the solution is rinsed off and the treated surface is viewed under ultraviolet light. Areas of gel fluoresce bright yellow-green. It must be recognized, however, that several materials not related to ASR in concrete can fluoresce and interfere with an accurate indication of ASR gel. Materials that fluoresce like gel include: naturally fluorescent minerals, carbonated paste, opal, and some other rock ingredients, and reactions from fly ash, silica fume, and other pozzolans. ASTM C 856 includes a prescreening procedure that gives a visual impression to compensate for the effects of these materials. However, this test is considered ancillary to more definitive petrographic examinations and other tests. In addition, the toxicity and radioactivity of uranyl acetate warrants special handling and disposal procedures regarding the solution and treated concrete. Caution regarding potential eye damage from ultraviolet light also merits attention.

The Los Alamos method is a staining technique that does not require ultraviolet light or uranyl-acetate solution. Instead, solutions of sodium cobaltinitrite and rhodamine B are used to condition the specimen and produce a dark pink stain that corresponds to calcium-rich ASR gel. It should be noted that these methods can produce evidence of ASR gel without causing damage to concrete. ASR gel can be present when other mechanisms such as freeze-thaw action, sulphate attack, and other deterioration mechanisms have caused the damage. These rapid methods for detecting the presence of ASR gel are useful but their limitations must be understood. Neither of the rapid procedures is a viable substitute for petrographic examination coupled with proper field inspection (Powers 1999).

Volume and Length Change

Volume or length change limits are sometimes specified for certain concrete applications. Volume change is also of concern when a new ingredient is added to concrete to make sure there are no significant adverse effects. ASTM C 157 (water and air storage methods) determines length change in concrete due to drying shrinkage, chemical reactivity, and forces other than those intentionally applied. Determination of early volume change of concrete before hardening can be performed using ASTM C 827. Creep can be determined in accordance with ASTM C 512. The static modulus of elasticity and Poisson's ratio of concrete in compression can be determined by methods outlined in ASTM C 469 and dynamic values of these parameters can be determined by using ASTM C 215.

Durability

Durability refers to the ability of concrete to resist deterioration from the environment or from the service in which it is placed. Properly designed concrete should endure without significant distress throughout its service life. In addition to tests for air content and chloride content described previously, the following tests are used to measure the durability of concrete:

Frost Resistance. The freeze-thaw resistance of concrete is usually determined in accordance with ASTM C 666. Samples are monitored for changes in dynamic modulus, mass, and volume over a period of 300 or more cycles of freezing and thawing. ASTM C 617 and ASTM C 682 are also available to evaluate frost resistance. Concrete that will be exposed to deicers as well as saturated freezing should be tested according to ASTM C 672 for deicer-scaling resistance. Although ASTM C 672 requires that only surface scaling be monitored, many practitioners also measure mass loss, as is done in Canada (Fig. 16-18). Concrete mixtures that perform well in ASTM C 666 do not always perform well in ASTM C 672. ASTM C 666 and

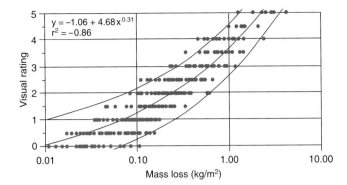

Fig. 16-18. Correlation between mass loss and visual rating for each specimen tested according to ASTM C 672 (Pinto and Hover 2001).

ASTM C 672 are often used to evaluate innovative mix designs, or new materials such as chemical admixtures, supplementary cementitious materials, and aggregates to determine their effect on frost and deicer resistance.

Sulphate Resistance. The sulphate resistance of concrete materials can be evaluated by using a saturated mortar bar test, ASTM C 1012. This test is valuable in assessing the sulphate resistance of concrete that will be continuously wet, but it does not evaluate the more aggressive wet-dry cycling environment. The test can be modified to include wet-dry cycling or the U.S. Bureau of Reclamation's wet-dry concrete prism test for sulphate attack can be used. CSA Test Methods A23.2-2B and 3B (ASTM D 516) or the Bureau's method (U.S. Bureau of Reclamation 1975) can be used to test soil and water for sulphate ion content to determine the severity of the sulphate exposure (ASTM is currently developing a new test).

Alkali-Silica Reactivity. Alkali-silica reaction is best controlled at the design stage when selecting materials for use in a specific concrete mixture. Suitable CSA test methods for evaluating the potential alkali-silica reactivity (ASR) of aggregates in concrete are: CSA Test Method A23.2-25A (ASTM C 1260), *Rapid Mortar Bar Method*; and CSA Test Method A23.2-14A (ASTM C 1293), *Concrete Prism Test*. Alkali-silica reactivity can also be analyzed by ASTM C 227 (*Mortar-Bar Method*), and ASTM C 289 (*Chemical Method*). CSA A23.2-14A and CSA A23.2-25A (ASTM C 441) can be used to determine effectiveness of supplementary cementing material inhibitors of alkali-silica reaction. A rapid 13-week version of CSA A23.2-14A (ASTM C 1293) is being developed by the University of Texas at Austin through the International Center for Aggregate Research (Touma, Fowler, Folliard, and Nelson 2001). Existing concrete structures can be evaluated for alkali-silica reaction using ASTM C 856.

Alkali-Carbonate Reactivity. Alkali-carbonate reactivity is more rare than alkali-silica reactivity. Potential reactivity of aggregates can be evaluated by using CSA Test Method A23.2-26A (*Chemical Composition*), CSA Test Method A23.2-14A (*Concrete Prisms*), ASTM C 295, ASTM C 586 (*Rock Cylinder Method*), and ASTM C 1105. Existing concrete structures can be evaluated for alkali-carbonate reaction using ASTM C 856.

Corrosion Resistance. The corrosion resistance of reinforced concrete is rarely tested unless unusual materials are used, concrete will be used in a very severe environment, or there is a need to evaluate the potential for in-place corrosion. Corrosion activity can be evaluated using ASTM C 876.

Abrasion Resistance. Abrasion resistance can be determined by using ASTM C 418 (sandblasting), ASTM C 779 (revolving disk, dressing wheel, and ball bearing methods), ASTM C 944 (rotating cutter), and ASTM C 1138 (underwater test).

Moisture Testing

The in-place moisture content, water vapour emission rate, and relative humidity of hardened concrete are useful indicators in determining if concrete is dry enough for application of floor-covering materials and coatings. The moisture content of concrete should be low enough to avoid spalling when exposed to temperatures above the boiling point of water. Moisture related test methods fall into two general categories: qualitative or quantitative. Qualitative tests provide an indication of the presence or absence of moisture while quantitative tests measure the amount of moisture. Qualitative tests may give a strong indication that excessive moisture is present and the floor is not ready for floor-covering materials. Quantitative tests are performed to assure that the floor is dry enough for these materials.

Qualitative moisture tests include: plastic sheet, mat bond, electrical resistance, electrical impedance, and nuclear moisture gauge tests. The plastic sheet test (ASTM D 4263) uses a square sheet of clear plastic film that is taped to the slab surface and left for 24 hours to see if moisture develops under it. The plastic sheet test is unreliable. In the mat bond test, a 1-m² sheet of floor covering is glued to the floor with the edges taped to the concrete for 72 hours. The force needed to remove the flooring is an indication of the slab moisture condition. Electrical resistance is measured using a moisture meter through two probes placed in contact with the concrete. Electrical impedance uses an electronic signal that is influenced by the moisture in the concrete. Nuclear moisture gauges contain high-speed neutrons that are slowed by the hydrogen atoms in water. The affect of these encounters is a measure of the moisture content of the concrete. Although the last three tests each yield a numeric test result, their value is quite limited. Experience and skill are needed to judge the trustworthiness of the devices and the test results produced by them.

Quantitative test methods include: gravimetric moisture content, moisture vapour emission rate, and relative humidity probe tests. The most direct method for determining moisture content is to dry cut a specimen from the concrete element in question, place it in a moisture proof container, and transport it to a laboratory for testing. After obtaining the specimen's initial mass, dry the specimen in an oven at about 105°C for 24 hours or until constant mass is achieved. The difference between the two masses divided by the dry mass, times 100, is the moisture content in percent. The moisture vapour emission rate (ASTM F 1869) is the most commonly used test for measuring the readiness of concrete for application of floor coverings. The emission rate is expressed as kilograms of moisture emitted from 93 m² in 24 hours. See Kosmatka (1985), and PCA (2000) for more information.

Relative humidity tests are used in several countries outside North America for measuring moisture in concrete slabs. Two British standards, BS 5325: 1996 and BS 8203: 1996 use a hygrometer or relative humidity probe sealed under an insulated, impermeable box to trap moisture in an air pocket above the floor. The probe is allowed to equilibrate for at least 72 hours or until two consecutive readings at 24 hours intervals are within the precision of the instrument (typically ± 3% RH). Acceptable relative humidity limits for the installation of floor coverings range from a maximum of 60% to 90%. It can require several months of air-drying to achieve the desired relative humidity. A method for estimating drying time to reach a specified relative humidity based on water-cementing materials ratio, thickness of structure, number of exposed sides, relative humidity, temperature and curing conditions can be found in Hedenblad (1997), Hedenblad (1998), and Farny (2001).

Carbonation

The depth or degree of carbonation can be determined by petrographic techniques (ASTM C 856) through the observation of calcium carbonate—the primary chemical product of carbonation. In addition, a phenolphthalein colour test can be used to estimate the depth of carbonation by testing the pH of concrete (carbonation reduces pH). Upon application of the phenolphthalein solution to a freshly fractured or freshly cut surface of concrete, noncarbonated areas turn red or purple while carbonated areas remain colourless. (Fig. 16-19). The phenolphthalein indicator when observed against hardened paste changes colour at a pH of 9.0 to 9.5. The pH of good quality noncarbonated concrete without admixtures is usually greater than 12.5. For more information, see "pH Testing Methods" below, and see Verbeck (1958), Steinour (1964), and Campbell, Sturm, and Kosmatka (1991).

Fig. 16-19. The depth of carbonation is determined by spraying phenolphthalein solution on a freshly broken concrete surface. Noncarbonated areas turn red or purple, carbonated areas stay colourless. (69804)

pH Testing Methods

There are three practical methods for measuring the pH of hardened concrete in the field. The first uses litmus paper designed for the alkaline range of pH readings. Place a few drops of distilled water on the concrete, wait 60 ± 5 seconds and immerse an indicator strip (litmus paper) in the water for 2 to 3 seconds. After removing the strip, compare it to the standard pH colour scale supplied with the indicator strips. A second method uses a pH "pencil." The pencil is used to make a 25 mm long mark after which 2 to 3 drops of distilled water are placed on the mark. After waiting 20 seconds, the colour is compared to a standard colour chart to judge the pH of the concrete. Finally, the third method uses a wide-range liquid pH indicator on a freshly fractured surface of the concrete or a core obtained from the concrete. After several minutes, the resulting colour is compared to a colour chart to determine the pH of the concrete. This method is also effective for measuring the depth of carbonation present on the concrete surface. See PCA (2000) for more information.

Permeability

Both direct and indirect methods of determining permeability are used. Table 16-2 shows typical concrete permeabilities. Resistance to chloride-ion penetration, for example, can be determined by ponding chloride solution on a concrete surface and, at a later age, determining the chloride content of the concrete at particular depths (AASHTO T 259). The rapid chloride permeability test (RCPT) (ASTM C 1202), also called the Coulomb or electrical resistance test, is often specified for concrete bridge

and parking garage decks. Various absorption methods, including ASTM C 642, are used. Direct water permeability data can be obtained by using the U.S. Army Corp of Engineers CRC C 163-92 test method for water permeability of concrete using a triaxial cell. A test method recommended by the American Petroleum Institute for determining the permeability of rock is also available. ASTM is in the process of developing a standard method for hydraulic permeability of concrete. All the above test methods have limitations. For more information, see American Petroleum Institute (1956), Tyler and Erlin (1961), Whiting (1981), Pfeifer and Scali (1981), and Whiting (1988).

Nondestructive Test Methods

Nondestructive tests (NDT) can be used to evaluate the relative strength and other properties of hardened concrete. The most widely used are the rebound hammer, penetration, pullout, and dynamic or vibration tests. Other techniques for testing the strength and other properties of hardened concrete include X-rays, gamma radiography, neutron moisture gages, magnetic cover meters, electricity, microwave absorption, and acoustic emissions. Each method has limitations and caution should be exercised against acceptance of nondestructive test results as having a constant correlation to the traditional compression test; for example, empirical correlations must be developed prior to use (Malhotra 1976, NRMCA 1979, Malhotra 1984, Clifton 1985, Malhotra and Carino 1991).

An NDT program may be undertaken for a variety of purposes regarding the strength or condition of hardened concrete, including:

- Determination of in-place concrete strength
- Monitoring rate of concrete strength gain
- Location of nonhomogeneity, such as voids or honeycombing in concrete
- Determination of relative strength of comparable members
- Evaluation of concrete cracking and delaminations
- Evaluation of damage from mechanical or chemical forces
- Steel reinforcement location, size, and corrosion activity
- Member dimensions

Irrespective of the type of NDT test used, adequate and reliable correlation data with standard 28-day compressive strength data is usually necessary to evaluate the accuracy of the NDT method. In addition, correlation to in-place compressive strengths using drilled cores from one or two locations can provide guidance in interpreting NDT test results; these can then be used to survey larger portions of the structure. Care should be taken to consider the influence that varying sizes and locations of structural elements can have on the NDT test being used.

Rebound Hammer Tests. The Schmidt rebound hammer (Fig. 16-20) is essentially a surface-hardness tester that pro-

Fig. 16-20. The rebound hammer gives an indication of the compressive strength of concrete. (69782)

vides a quick, simple means of checking concrete uniformity. It measures the rebound of a spring-loaded plunger after it has struck a smooth concrete surface. The rebound number reading gives an indication of the compressive strength and stiffness of the concrete. Two different concrete mixtures having the same strength but different stiffnesses will yield different readings. In view of this, an understanding of the factors influencing the accuracy of the test is required.

The results of a Schmidt rebound hammer test (ASTM C 805) are affected by surface smoothness, size, shape, and rigidity of the specimen; age and moisture condition of the concrete; type of coarse aggregate; and degree of carbonation of the concrete surface. When these limitations are recognized, and the hammer is calibrated for the particular materials used in the concrete (Fig. 16-21) by comparison

with cores or cast specimens, then this instrument can be useful for determining the relative compressive strength and uniformity of concrete in the structure.

Penetration Tests. The Windsor probe (ASTM C 803), like the rebound hammer, is basically a hardness tester that provides a quick means of determining the relative strength of the concrete. The equipment consists of a powder-actuated gun that drives a hardened alloy probe into the concrete (Fig. 16-22). The exposed length of the probe is measured and related by a calibration table to the compressive strength of the concrete.

The results of the Windsor-probe test will be influenced by surface smoothness of the concrete and the type and hardness of aggregate used. Therefore, to improve accuracy, a calibration table or curve for the particular concrete to be tested should be made, usually from cores or cast specimens.

Both the rebound hammer and the probe damage the concrete surface to some extent. The rebound hammer leaves a small indentation on the surface; the probe leaves a small hole and may cause minor cracking and small craters similar to popouts.

Maturity Tests. The maturity principle is that strength gain is a function of time and temperature. ASTM C 1074 generates a maturity index that is based on temperature and time factors. The estimated strength depends on properly determining the strength-maturity function for a particular concrete mixture. The device uses thermocouples or thermistors placed in the concrete and connected to strip-

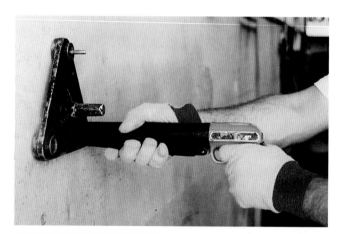

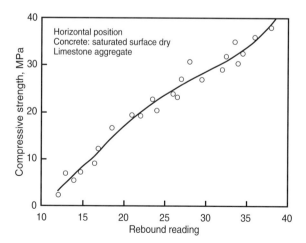

Fig. 16-21. Example of a calibration chart for an impact (rebound) test hammer.

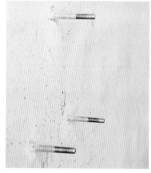

Fig. 16-22. The Windsor-probe technique for determining the relative compressive strength of concrete.
(top) Powder-actuated gun drives hardened alloy probe into concrete. (69783)
(left) Exposed length of probe is measured and relative compressive strength of the concrete then determined from a calibration table. (69784)

chart recorders or digital data-loggers that record concrete temperature as a function of time. The temperature in relation to time data is correlated to compression tests performed on cylindrical specimens to generate a temperature-time versus strength curve that is used to estimate in-place concrete strength.

Pullout Tests. A pullout test (ASTM C 900) involves casting the enlarged end of a steel rod in the concrete to be tested and then measuring the force required to pull it out (Fig. 16-23). The test measures the direct shear strength of the concrete. This in turn is correlated with the compressive strength; thus a measurement of the in-place compressive strength is made.

Fig. 16-23. Pullout test equipment being used to measure the in-place strength of concrete. (44217)

Break-Off Tests. The break-off test (ASTM C 1150) determines the in-place strength of the concrete by breaking off an in situ cylindrical concrete specimen at a failure plane parallel to the finished surface of the concrete element. A break-off number is generated and assessed in relation to the strength of the concrete. Similar to pullout tests, the relationship between break-off test numbers and compression tests must also be developed prior to obtaining final test results.

For more information on test methods used to estimate in-place concrete strengths, see *Appendix A of CSA A23.2-00–Nondestructive Methods for Testing Concrete*, and ACI Committee Report 228–*In-Place Methods to Estimate Concrete Strength* (1995).

Dynamic or Vibration Tests. A dynamic or vibration (ultrasonic pulse velocity) test (ASTM C 597) is based on the principle that velocity of sound in a solid can be measured by (1) determining the resonant frequency of a specimen, or (2) recording the travel time of short pulses of vibrations through a sample. High velocities are indicative of good concrete and low velocities are indicative of poor concrete.

Resonant frequency methods employ low-frequency vibration to impart mechanical energy used to detect, locate, and record discontinuities within solids. Resonant frequency is a function of the dynamic modulus of elasticity, poisson ratio, density, and geometry of the structural element. The presence and orientation of surface and internal cracking can be determined. In addition, fundamental transverse, longitudinal, and torsional frequencies of concrete specimens can also be determined by ASTM C 215, a method frequently used in laboratory durability tests such as freezing and thawing (ASTM C 666).

Stress wave propagation methods using impact-echo tests are employed in ASTM C 1383 to measure P-wave speeds and the thickness of concrete elements such as slabs, pavements, bridge decks and walls. The advantage of the test is that its not only nondestructive, but also access is only required to one side of the structure. Other stress-wave methods not yet mentioned include: ultrasonic-echo and spectral analysis of surface waves.

Other Tests. The use of X-rays for testing concrete properties is limited due to the costly and dangerous high-voltage equipment required, as well as radiation hazards.

Gamma-radiography equipment can be used in the field to determine the location of reinforcement, density, and perhaps honeycombing in structural concrete units. ASTM C 1040 procedures use gamma radiation to determine the density of unhardened and hardened concrete in place.

Battery-operated magnetic detection devices, like the pachometer or covermeter, are available to measure the depth of reinforcement in concrete and to detect the position of rebars. Electrical-resistivity equipment is being developed to estimate the thickness of concrete pavement slabs.

A microwave-absorption method has been developed to determine the moisture content of porous building materials such as concrete. Acoustic-emission techniques show promise for studying load levels in structures and for locating the origin of cracking.

Ground-penetrating (short-pulse) radar is a rapid technique for nondestructive detection of delaminations and other types of defects in overlaid reinforced concrete decks. It also shows potential for monitoring strength development in concrete, measuring the thickness of concrete members, and locating reinforcement.

Infrared thermographic techniques are used to detect and show, both large and small, internal voids, delaminations and cracks in bridges, highway pavements, garage decks, buildings and other structural elements exposed to direct sunlight. For more information, see Malhotra and Carino (1991). ACI Committee 228 (1998) presents additional information on these and other nondestructive test methods.

Finally, acoustic impact methods also employ simple hammer and chain drag soundings that are low-cost accurate tests used to identify delaminated areas of concrete. Hammer soundings can be used on either vertical or horizontal surfaces, but are usually limited to small areas of delaminations. These areas are identified by striking the surface of the concrete with a hammer while listening for either a ringing or hollow sound. Dragging either a single chain, in small areas, or for larger areas, a T-bar with or without wheels having four or more chains attached are also used to identify delaminated concrete (ASTM D 4580). Approximately one metre of chain is in contact with the concrete during chain drag soundings. The sound emitted indicates whether the concrete is delaminated or not.

Chain drag soundings are usually limited to horizontal surfaces that have a relatively rough texture. Smooth concrete may not bounce the chain links enough to generate adequate sound to detect delaminated areas. Note that corrosion of reinforcing bars in the area of delaminated concrete will probably extend beyond the boundary identified as delaminated.

Table 16-3 lists several nondestructive test methods along with main applications.

Table 16-3. Nondestructive Evaluation (NDE) Methods for Concrete Materials

Concrete properties	Recommended NDE methods	Possible NDE properties
Strength	Penetration probe Rebound hammer Pullout methods	
General quality and uniformity	Penetration probe Rebound hammer Ultrasonic pulse velocity Gamma radiography	Ultrasonic pulse echo Visual examination
Thickness		Radar Gamma radiography Ultrasonic pulse echo
Stiffness	Ultrasonic pulse velocity	Proof loading (load-deflection)
Density	Gamma radiography Ultrasonic pulse velocity	Neutron density gage
Rebar size and location	Covermeter (pachometer) Gamma radiography	X-ray radiography Ultrasonic pulse echo Radar
Corrosion state of reinforcing steel	Electrical potential measurement	
Presence of subsurface voids	Acoustic impact Gamma radiography Ultrasonic pulse velocity	Infrared thermography X-ray radiography Ultrasonic pulse echo Radar Resonant frequency testing
Structural integrity of concrete structures	Proof loading (load-deflection)	Proof testing using acoustic emission

Adapted from ACI Subcommittee 364 (1994) and Clifton (1985).

REFERENCES

AASHTO, *Resistance of Concrete to Chloride Ion Penetration,* AASHTO T 259-80, American Association of State Highway and Transportation Officials, Washington, D.C., 2000, pages 896 to 897.

AASHTO, *Standard Method of Test for Air Content of Freshly Mixed Concrete by the Chace Indicator,* AASHTO T 199-00, American Association of State Highway and Transportation Officials, 2000, pages 600 to 602.

AASHTO, *Rapid Determination of the Chloride Permeability of Concrete,* AASHTO T 277-96, American Association of State Highway and Transportation Officials, 2000, pages 956 to 961.

ACI Committee 214, *Recommended Practice for Evaluation of Strength Test Results of Concrete,* ACI 214-77, reapproved 1997, American Concrete Institute, Farmington Hills, Michigan, 1997, 14 pages.

ACI Committee 222, *Provisional Standard Test Method for Water Soluble Chloride Available for Corrosion of Embedded Steel in Mortar and Concrete Using the Soxhlet Extractor,* ACI 222.1-96, American Concrete Institute, Farmington Hills, Michigan, 1996, 3 pages.

ACI Committee 228, *In-Place Methods to Estimate Concrete Strength,* ACI 228.1R-95, American Concrete Institute, Farmington Hills, Michigan, 1995, 41 pages.

ACI Committee 228, *Nondestructive Test Methods for Evaluation of Concrete in Structures,* ACI 228.2R-98, American Concrete Institute, Farmington Hills, Michigan, 1998, 62 pages.

ACI Committee 318, *Building Code Requirements for Structural Concrete and Commentary,* ACI 318-99, American Concrete Institute, Farmington Hills, Michigan, 1999, 369 pages.

ACI Committee 364, *Guide for Evaluation of Concrete Structures Prior to Rehabilitation,* ACI 364.1 R-94, American Concrete Institute, Farmington Hills, Michigan, 1994, 22 pages.

American Petroleum Institute, *Recommended Practice for Determining Permeability of Porous Media,* API RP 27, American Petroleum Institute, Washington, D.C., 1956.

Angles, J., "Measuring Workability," 8(12), *Concrete,* 1974, page 26.

ASTM, *Manual of Aggregate and Concrete Testing,* American Society for Testing and Materials, West Conshohocken, Pennsylvania, 2000.

Bartos, P., "Workability of Flowing Concrete—Assessment by a Free Orifice Rheometer," 12(10), *Concrete,* 1978, pages 28 to 30.

Bisaillon, André; Frichette, Guy; and Keyser, J. Hode, "Field Evaluation of Expanded Polystyrene Molds for Self-Cured Accelerated Strength Testing of Concrete," *Transportation Research Record 558,* Transportation Research Board, Washington, D.C., 1975, pages 50 to 60.

Bois, Karl J.; Mubarak, Khalid; and Zoughi, Reza, *A Simple, Robust and On-Site Microwave Inspection Technique for Determining Water-to-Cement-Based Materials,* PCA R&D Serial No. 2405, Portland Cement Association, 1999, 46 pages.

Burg, R. G.; Caldarone, M. A.; Detwiler, G.; Jansen, D. C.; and Willems, T. J., "Compression Testing of HSC: Latest Technology." *Concrete International,* American Concrete Institute, Farmington Hills, Michigan, August 1999, pages 67 to 76.

Burg, Ron G., and Ost, Borje W., *Engineering Properties of Commercially Available High-Strength Concrete (Including Three-Year Data),* Research and Development Bulletin RD104, Portland Cement Association, 1994, 58 pages

Campbell, D. H.; Sturm, R. D.; and Kosmatka, S. H., "Detecting Carbonation," *Concrete Technology Today,* PL911, http://www.portcement.org/pdf_files/PL911.pdf, Portland Cement Association, March 1991, pages 1 to 5.

Carino, Nicholas J., "Prediction of Potential Concrete Strength at Later Ages," *Significance of Tests and Properties of Concrete and Concrete-Making Materials,* STP 169C, American Society for Testing and Materials, West Conshohocken, Pennsylvania, 1994, pages 140 to 152.

Clear, K. C., and Harrigan, E. T., *Sampling and Testing for Chloride Ion in Concrete,* FHWA-RD-77-85, Federal Highway Administration, Washington, D.C., August 1977.

Clemena, Gerardo G., *Determination of the Cement Content of Hardened Concrete by Selective Solution,* PB-213 855, Virginia Highway Research Council, Federal Highway Administration, National Technical Information Service, U.S. Department of Commerce, Springfield, Virginia, 1972.

Clifton, James R., "Nondestructive Evaluation in Rehabilitation and Preservation of Concrete and Masonry Materials," SP-85-2, *Rehabilitation, Renovation, and Preservation of Concrete and Masonry Structures,* SP-85, American Concrete Institute, Farmington Hills, Michigan, 1985, pages 19 to 29.

CSA Standard A23.1-00/ A23.2-00, *Concrete Materials and Methods of Concrete Construction / Methods of Test for Concrete,* Canadian Standards Association, Toronto, 2000.

The Concrete Society, *Permeability of Concrete and Its Control,* The Concrete Society, London, 1985.

Date, Chetan G., and Schnormeier, Russell H., "Day-to-Day Comparison of 4 and 6 Inch Diameter Concrete Cylinder Strengths," *Concrete International,* American Concrete Institute, Farmington Hills, Michigan, August 1984, pages 24 to 26.

de Larrard, F.; Szitkar, J.; Hu, C.; and Joly, M., "Design and Rheometer for Fluid Concretes," *Proceedings, International RILEM Workshop,* Paisley, Scotland, March 2 to 3, 1993, pages 201 to 208.

Farny, James A., *Concrete Floors on Ground,* EB075, Portland Cement Association, 2001, 140 pages.

Fiorato A. E.; Burg, R. G.; and Gaynor, R. D., "Effects of Conditioning on Measured Compressive Strength of Concrete Cores" *Concrete Technology Today,* CT003, Portland Cement Association, http://www.portcement.org/pdf_files/CT003.pdf, 2000, pages 1 to 3.

Forester, J. A.; Black, B. F.; and Lees, T. P., *An Apparatus for Rapid Analysis Machine,* Technical Report, Cement and Concrete Association, Wexham Springs, Slough, England, April 1974

Forstie, Douglas A., and Schnormeier, Russell, "Development and Use of 4 by 8 Inch Concrete Cylinders in Arizona," *Concrete International,* American Concrete Institute, Farmington Hills, Michigan, July 1981, pages 42 to 45.

Galloway, Joseph E., "Grading, Shape, and Surface Properties," *Significance of Tests and Properties of Concrete and Concrete-Making Materials,* STP 169C, American Society for Testing and Materials, West Conshohocken, Pennsylvania, 1994, pages 401 to 410

Gebler, S. H., and Klieger, P., *Effects of Fly Ash on the Air-Void Stability of Concrete,* Research and Development Bulletin RD085T, Portland Cement Association, http://www.portcement.org/pdf_files/RD085.pdf, 1983.

Hedenblad, Göran, *Drying of Construction Water in Concrete—Drying Times and Moisture Measurement,* T9, Swedish Council for Building Research, Stockholm, 1997, 54 pages. [Available from Portland Cement Association as LT229]

Hedenblad, Göran, "Concrete Drying Time," *Concrete Technology Today,* PL982, Portland Cement Association, http://www.portcement.org/pdf_files/PL982.pdf, 1998, pages 4 and 5.

Hime, W. G.; Mivelaz, W. F.; and Connolly, J. D., *Use of Infrared Spectrophotometry for the Detection and Identification of Organic Additions in Cement and Admixtures in Hardened Concrete,* Research Department Bulletin RX194, Portland Cement Association, http://www.portcement.org/pdf_files/RX194.pdf, 1966, 22 pages.

Kelly, R. T., and Vail, J. W., "Rapid Analysis of Fresh Concrete," *Concrete,* The Concrete Society, Palladian Publications, Ltd., London, Volume 2, No. 4, April 1968, pages 140-145; Volume 2, No. 5, May 1968, pages 206 to 210.

Klieger, Paul, and Lamond, Joseph F., *Significance of Tests and Properties of Concrete and Concrete-Making Materials,* STP 169C, American Society for Testing and Materials, West Conshohocken, Pennsylvania, 1994.

Kosmatka, Steven H., "Compressive versus Flexural Strength for Quality Control of Pavements," *Concrete Technology Today,* PL854, Portland Cement Association, http://www.portcement.org/pdf_files/PL854.pdf, 1985, pages 4 and 5.

Kosmatka, Steven H., "Floor-Covering Materials and Moisture in Concrete," *Concrete Technology Today,* PL853, Portland Cement Association, http://www.portcement.org/pdf_files/PL853.pdf, September 1985, pages 4 and 5.

Kosmatka, Steven H., "Petrographic Analysis of Concrete," *Concrete Technology Today,* PL862, Portland Cement Association, http://www.portcement.org/pdf_files/PL862.pdf, 1986, pages 2 and 3.

Lawrence, Deborah J., "Cement and Water Content of Fresh Concrete," *Significance of Tests and Properties of Concrete and Concrete-Making Materials,* STP 169C, American Society for Testing and Materials, West Conshohocken, Pennsylvania, 1994, pages 112 to 120.

Malhotra, V. M., *Testing Hardened Concrete, Nondestructive Methods,* ACI Monograph No. 9, American Concrete Institute-Iowa State University Press, Farmington Hills, Michigan, 1976.

Malhotra, V. M., *In Situ/Nondestructive Testing of Concrete,* SP-82, American Concrete Institute, Farmington Hills, Michigan, 1984.

Malhotra, V. M., and Carino, N. J., *Handbook on Nondestructive Testing of Concrete,* ISBN 0-8493-2984-1, CRC Press, Boca Raton, Florida, 1991, 343 pages.

Mor, Avi, and Ravina, Dan, "The DIN Flow Table," *Concrete International,* American Concrete Institute, Farmington Hills, Michigan, December 1986.

NRMCA, *In-Place Concrete Strength Evaluation—A Recommended Practice,* NRMCA Publication 133, revised 1979, National Ready Mixed Concrete Association, Silver Spring, Maryland, 1979.

NRMCA, "Standard Practice for Rapid Determination of Water Soluble Chloride in Freshly Mixed Concrete, Aggregate and Liquid Admixtures," *NRMCA Technical Information Letter No. 437,* National Ready Mixed Concrete Association, March 1986.

Parry, James M., *Wisconsin Department of Transportation QC/QA Concept,* Portland Cement Concrete Technician I/IA Course Manual, Wisconsin Highway Technician Certification Program, Platteville, WI, 2000, pages B-2 to B-4.

PCA, "Rapid Analysis of Fresh Concrete," *Concrete Technology Today,* PL832, Portland Cement Association, http://www.portcement.org/pdf_files/PL832.pdf, June 1983, pages 3 and 4.

PCA, *Understanding Concrete Floors and Moisture Issues,* CD014, Portland Cement Association, 2000.

Pistilli, Michael F., and Willems, Terry, "Evaluation of Cylinder Size and Capping Method in Compression Testing of Concrete," *Cement, Concrete and Aggregates,* American Society for Testing and Materials, West Conshohocken, Pennsylvania, Summer 1993.

Powers, Laura J., "Developments in Alkali-Silica Gel Detection," *Concrete Technology Today,* PL991, Portland Cement Association, http://www.portcement.org/pdf_files/PL991.pdf, April 1999, pages 5 to 7.

Powers, T. C., "Studies of Workability of Concrete," *Journal of the American Concrete Institute,* Volume 28, American Concrete Institute, Farmington Hills, Michigan, 1932, page 419.

Powers, T. C., *The Properties of Fresh Concrete,* Wiley, New York, 1968.

Powers, T. C., and Wiler, E. M., "A Device for Studying the Workability of Concrete," *Proceedings of ASTM,* Volume 41, American Society for Testing and Materials, West Conshohocken, Pennsylvania, 1941.

Saucier, K. L., *Investigation of a Vibrating Slope Method for Measuring Concrete Workability,* Miscellaneous Paper 6-849, U.S. Army Engineer Waterways Experiment Station, Vicksburg, Mississippi, 1966.

Scanlon, John M., "Factors Influencing Concrete Workability," *Significance of Tests and Properties of Concrete and Concrete-Making Materials,* STP 169C, American Society for Testing and Materials, West Conshohocken, Pennsylvania, 1994, pages 49 to 64.

St. John, Donald A.; Poole, Alan W.; and Sims, Ian, *Concrete Petrography—A handbook of investigative techniques,* Arnold, London, http://www.arnoldpublishers.com, PCA LT226, 485 pages.

Steinour, Harold H., *Influence of the Cement on Corrosion Behavior of Steel in Concrete,* Research Department Bulletin RX168, Portland Cement Association, http://www.portcement.org/pdf_files/RX168.pdf, 1964, 22 pages.

Tabikh, A. A.; Balchunas, M. J.; and Schaefer, D. M., "A Method Used to Determine Cement Content in Concrete," *Concrete,* Highway Research Record Number 370, Transportation Research Board, National Research Council, Washington, D.C., 1971.

Tattersall, G. H., *Measurement of Workability of Concrete,* East Midlands Region of the Concrete Society of Nottingham, England, 1971.

Teranishs, K.; Watanabe, K.; Kurodawa, Y.; Mori, H.; and Tanigawa, Y., "Evaluation of Possibility of High-Fluidity Concrete," *Transactions of the Japan Concrete Institute (16),* Japan Concrete Institute, Tokyo, 1994, pages 17 to 24.

Touma, Wissam E.; Fowler, David W.; Folliard, Kevin; and Nelson, Norm, "Expedited Laboratory Testing and Mitigation Procedures for Alkali-Silica Reaction," *ICAR– 9th Annual Symposium Proceedings,* 2001, 21 pages.

Tyler, I. L., and Erlin, Bernard, *A Proposed Simple Test Method for Determining the Permeability of Concrete,* Research Department Bulletin RX133, Portland Cement Association, http://www.portcement.org/pdf_files/RX133.pdf, 1961.

U.S. Bureau of Reclamation, *Concrete Manual,* 8th Edition, Denver, 1975, page 11.

Verbeck, G. J., *Carbonation of Hydrated Portland Cement,* Research Department Bulletin RX087, Portland Cement Association, http://www.portcement.org/pdf_files/RX087.pdf, 1958.

Wallevik, O., "The Use of BML Viscometer for Quality Control of Concrete," *Concrete Research,* Nordic Concrete Research, Espoo, Finland, 1996, pages 235 to 236.

Wong, G. Sam; Alexander, A. Michel; Haskins, Richard; Poole, Toy S.; Malone, Phillip G.; and Wakeley, Lillian, *Portland-Cement Concrete Rheology and Workability: Final Report,* FHWA—RD-00-025, Federal Highway Administration, Washington, D.C., 2001, 117 pages.

Wood, Sharon L., *Evaluation of the Long-Term Properties of Concrete,* Research and Development Bulletin RD102, Portland Cement Association, 1992, 99 pages.

Whiting, David, *Rapid Determination of the Chloride Permeability of Concrete,* FHWA-RD-81-119, Federal Highway Administration, Washington, D.C., 1981.

Whiting, David, "Permeability of Selected Concretes," *Permeability of Concrete,* SP-108, American Concrete Institute, Farmington Hills, Michigan, 1988.

CHAPTER 17
High-Performance Concrete

High-performance concrete (HPC) exceeds the properties and constructability of normal concrete. Normal and special materials are used to make these specially designed concretes that must meet a combination of performance requirements. Special mixing, placing, and curing practices may be needed to produce and handle high-performance concrete. Extensive performance tests are usually required to demonstrate compliance with specific project needs (ASCE 1993, Russell 1999, and Bickley and Mitchell 2001). High-performance concrete has been primarily used in tunnels, bridges, and tall buildings for its strength, durability, and high modulus of elasticity (Fig. 17-1). It has also been used in shotcrete repair, poles, parking garages, and agricultural applications.

High-performance concrete characteristics are developed for particular applications and environments; some of the properties that may be required include:

- High strength
- High early strength
- High modulus of elasticity
- High abrasion resistance
- High durability and long life in severe environments
- Low permeability and diffusion
- Resistance to chemical attack
- High resistance to frost and deicer scaling damage
- Toughness and impact resistance
- Volume stability
- Ease of placement
- Compaction without segregation
- Inhibition of bacterial and mould growth

High-performance concretes are made with carefully selected high-quality ingredients and optimized mixture designs; these are batched, mixed, placed, compacted and cured to the highest industry standards. Typically, such concretes will have a low water-cementing materials ratio of 0.20 to 0.45. Plasticizers are usually used to make these concretes fluid and workable.

High-performance concrete almost always has a higher strength than normal concrete. However, strength is not always the primary required property. For example, a normal strength concrete with very high durability and very low permeability is considered to have high-performance properties. Bickley and Fung (2001) demonstrated that 40 MPa high-performance concrete for bridges could be economically made while meeting

Fig. 17-1. High-performance concrete is often used in bridges (left) and tall buildings (right). (70017, 70023)

Table 17-1. Materials Used in High-Performance Concrete

Material	Primary contribution/Desired property
Portland cement	Cementing material/durability
Blended cement	Cementing material/durability/high strength
Fly ash	Cementing material/durability/high strength
Slag	Cementing material/durability/high strength
Silica fume	Cementing material/durability/high strength
Calcined clay*	Cementing material/durability/high strength
Metakaolin*	Cementing material/durability/high strength
Calcined shale*	Cementing material/durability/high strength
Superplasticizers	Flowability
High-range water reducers	Reduce water to cement ratio
Hydration control admixtures	Control setting
Retarders	Control setting
Accelerators	Accelerate setting
Corrosion inhibitors	Control steel corrosion
Water reducers	Reduce cement and water content
Shrinkage reducers	Reduce shrinkage
ASR inhibitors	Control alkali-silica reactivity
Polymer/latex modifiers	Durability
Optimally graded aggregate	Improve workability and reduce paste demand

* No significant use in Canada at this time.

durability factors for air-void system and resistance to chloride penetration.

Table 17-1 lists materials often used in high-performance concrete and why they are selected. Table 17-2 lists properties that can be selected for high-performance concrete. Not all properties can be achieved at the same time. High-performance concrete specifications ideally should be performance oriented. Unfortunately, many specifications are a combination of performance requirements (such as permeability or strength limits) and prescriptive requirements (such as air content limits or dosage of supplementary cementing material (Ferraris and Lobo 1998). Table 17-3 provides examples of high-performance concrete mixtures used in a variety of structures. Selected high-performance concretes are presented in this chapter.

HIGH-EARLY-STRENGTH CONCRETE

High-early-strength concrete, also called fast-track concrete, achieves its specified strength at an earlier age than normal concrete. The time period in which a specified strength should be achieved may range from a

Table 17-2. Selected Properties of High-Performance Concrete

Property	Test method	Criteria that may be specified
High strength	CSA A23.2-9C (ASTM C 39)	40 to 140 MPa at 28 to 91 days
High-early compressive strength	CSA A23.2-9C (ASTM C 39)	20 to 30 MPa at 3 to 12 hours or 1 to 3 days
High-early flexural strength	CSA A23.2-8C (ASTM C 78)	2 to 4 MPa at 3 to 12 hours or 1 to 3 days
Abrasion resistance	ASTM C 944	0 to 1 mm depth of wear
Low permeability	ASTM C 1202	500 to 2000 coulombs
Chloride penetration	AASHTO T 259 & T 260	Less than 0.07% Cl at 6 months
High resistivity	ASTM G 59	
Low absorption	ASTM C 642	2% to 5%
Low diffusion coefficient	Wood, Wilson and Leek (1989) Test under development by ASTM	1000 x 10^{-13} m/s
Resistance to chemical attack	Expose concrete to saturated solution in wet/dry environment	No deterioration after 1 year
Sulphate attack	ASTM C 1012	0.10% max. expansion at 6 months for moderate sulphate exposures or 0.5% max. expansion at 6 months for severe sulphate exposure
High modulus of elasticity	ASTM C 469	More than 40 GPa (Aitcin, 1998)
High resistance to freezing and thawing damage	ASTM C 666, Procedure A	Durability factor of 95 to 100 at 300 to 1000 cycles (max. mass loss or expansion can also be specified)
High resistance to deicer scaling	ASTM C 672	Scale rating of 0 to 1 or mass loss of 0 to 0.5 kg/m³ after 50 to 300 cycles
Low shrinkage	ASTM C 157	Less than 400 millionths. (Aitcin, 1998)
Low creep	ASTM C 512	Less than normal concrete

Table 17-3. Typical High-Performance Concretes Used in Structures

Mixture number	1	2	3	4	5	6
Water, kg/m³	151	145	135	145	130	130
Cement, kg/m³	311	398*	500	335*	513	315
Fly ash, kg/m³	31	45	—	—	—	40
Slag, kg/m³	47	—	—	125	—	—
Silica fume, kg/m³	16	32*	30	40*	43	23
Coarse aggregate, kg/m³	1068	1030	1100	1130	1080	1140
Fine aggregate, kg/m³	676	705	700	695	685	710
Water reducer, L/m³	1.6	1.7	—	1.0	—	1.5
Retarder, L/m³		—	1.8	—	—	—
Air content, (%)	7 ± 1.5	5 – 8	—	—	—	5.5
HRWR or plasticizer, L/m³	2.1	3	14	6.5	15.7	5.0
Water to cementing materials ratio	0.37	0.30	0.27	0.29	0.25	0.34
Comp. strength at 28 days, MPa	59	—	93	99	119	—
Comp. strength at 91 days, MPa	—	60	107	104	145	N/A

1. Wacker Drive bi-level roadway, Chicago, 2001.
2. Confederation Bridge, Northumberland Strait, Prince Edward Island/New Brunswick, 1997.
3. La Laurentienne Building, Montreal, 1984.
4. BCE Place Phase 2, Toronto, 1993.
5. Two Union Square, Seattle, 1988.
6. Great Belt Link, East Bridge, Denmark, 1996.
 * Originally used a blended cement containing silica fume. Portland cement and silica fume quantities have been separated for comparison purposes.

few hours (or even minutes) to several days. High-early-strength can be attained by using traditional concrete ingredients and concreting practices, although sometimes special materials or techniques are needed.

High-early-strength can be obtained by using one or a combination of the following, depending on the age at which the specified strength must be achieved and on job conditions:

1. Type 30 (III or HE high-early-strength cement)
2. High cement content (400 to 600 kg/m³)
3. Low water-cementing materials ratio (0.20 to 0.45 by mass)
4. Higher freshly mixed concrete temperature
5. Higher curing temperature
6. Chemical admixtures
7. Silica fume (or other supplementary cementing materials)
8. Steam or autoclave curing
9. Insulation to retain heat of hydration
10. Special rapid hardening cements.

High-early-strength concrete is used for prestressed concrete to allow for early stressing; precast concrete for rapid production of elements; high-speed cast-in-place construction; rapid form reuse; cold-weather construction; rapid repair of pavements to reduce traffic downtime; fast-track paving; and several other uses.

In fast-track paving, use of high-early-strength mixtures allows traffic to open within a few hours after concrete is placed. An example of a fast-track concrete mixture used for a bonded concrete highway overlay consisted of 380 kg of Type 30 (III) cement, 42 kg of Type C fly ash, 6½% air, a water reducer, and a water-to-cementing materials ratio of 0.4. Strength data for this 40-mm slump concrete are given in Table 17-4. Figs. 17-2 and 17-3 illustrate early strength development of concretes designed to open to traffic within 4 hours after placement. Fig. 17-4 illustrates the benefits of blanket curing to develop early strength for patching or fast-track applications.

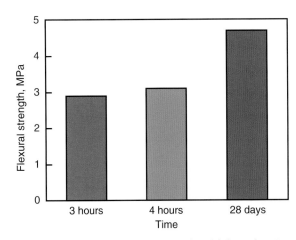

Fig. 17-2. Strength development of a high-early strength concrete mixture using 390 kg/m³ of rapid hardening cement, 676 kg/m³ of sand, 1115 kg/m³ of 25 mm nominal max. size coarse aggregate, a water to cement ratio of 0.46, a slump of 100 to 200 mm, and a plasticizer and retarder. Initial set was at one hour (Pyle 2001).

Table 17-4. Strength Data for Fast-Track Bonded Overlay

Age	Compressive strength, MPa	Flexural strength, MPa	Bond strength, MPa
4 hours	1.7	0.9	0.9
6 hours	7.0	2.0	1.1
8 hours	13.0	2.7	1.4
12 hours	17.6	3.4	1.6
18 hours	20.1	4.0	1.7
24 hours	23.9	4.2	2.1
7 days	34.2	5.0	2.1
14 days	36.5	5.7	2.3
28 days	40.7	5.7	2.5

Adapted from Knutson and Riley 1987

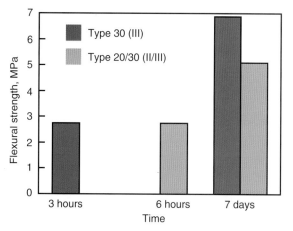

Fig. 17-3. Strength development of high-early strength concrete mixtures made with 504 to 528 kg/m³ of Type 30 (III) or Type 20/30 (II/III) cement, a nominal maximum size coarse aggregate of 25 mm, a water to cement ratio of 0.30, a plasticizer, a hydration control admixture, and an accelerator. Initial set was at one hour (Pyle 2001).

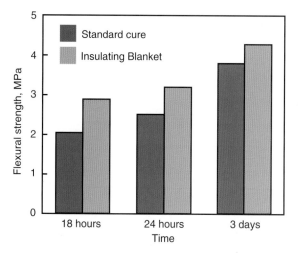

Fig. 17-4. Effect of blanket insulation on fast-track concrete. The concrete had a Type 10 (I) cement content of 421kg/m³ and a water to cement ratio of 0.30 (Grove 1989).

When designing early-strength mixtures, strength development is not the only criteria that should be evaluated; durability, early stiffening, autogenous shrinkage, drying shrinkage, temperature rise, and other properties also should be evaluated for compatibility with the project. Special curing procedures, such as fogging, may be needed to control plastic shrinkage cracking.

HIGH-STRENGTH CONCRETE

The definition of high strength changes over the years as concrete strength used in the field increases. This publication considers high-strength concrete (HSC) to have a strength significantly beyond what is used in normal practice. For example, today about 90% of ready mixed concrete has a 28-day specified compressive strength ranging from 20 MPa to 40 MPa, with most of it between 30 MPa and 35 MPa. Therefore, HSC is considered here, and as defined in CSA Standard A23.1, as concrete having a specified 28-day compressive strength of at least 70 MPa.

Most high-strength concrete applications are designed for compressive strengths of 70 MPa or greater as shown in Tables 17-3 and 17-5. For strengths of 70 MPa and higher, stringent application of the best practices is required. Compliance with the guidelines and recommendations for preconstruction laboratory and field-testing procedures described in ACI 363.2 are essential. Concrete with a design strength of 131 MPa has been used in buildings (Fig. 17-5).

Traditionally, the specified strength of concrete has been based on 28-day test results. However, in high-rise concrete structures, the process of construction is such that the structural elements in lower floors are not fully loaded for periods of a year or more. For this reason, compressive strengths based on 56- or 91-day test results are commonly specified in order to achieve significant economy in material costs. When later ages are specified, supplementary cementing materials are usually incorporated into the

Table 17-5. Mixture Proportions and Properties of Commercially Available High-Strength Concrete (Burg and Ost 1994)

Units per m³	Mix number					
	1	2	3	4	5	6
Cement, Type 10 (I), kg	564	475	487	564	475	327
Silica fume, kg	—	24	47	89	74	27
Fly ash, kg	—	59	—	—	104	87
Coarse aggregate SSD (14 mm crushed limestone), kg	1068	1068	1068	1068	1068	1121
Fine aggregate SSD, kg	647	659	676	593	593	742
HRWR Type F, litres	11.6	11.6	11.22	20.11	16.44	6.3
HRWR Type G, litres	—	—	—	—	—	3.24
Retarder, Type D, litres	1.12	1.05	0.97	1.46	1.5	—
Water to cementing materials ratio	0.28	0.29	0.29	0.22	0.23	0.32
Fresh concrete properties						
Slump, mm	197	248	216	254	235	203
Density, kg/ m³	2451	2453	2433	2486	2459	2454
Air content, %	1.6	0.7	1.3	1.1	1.4	1.2
Concrete temp., °C	24	24	18	17	17	23
Compressive strength, 100 x 200-mm moist-cured cylinders						
3 days, MPa	57	54	55	72	53	43
7 days, MPa	67	71	71	92	77	63
28 days, MPa	79	92	90	117	100	85
56 days, MPa	84	94	95	122	116	—
91 days, MPa	88	105	96	124	120	92
182 days, MPa	97	105	97	128	120	—
426 days, MPa	103	118	100	133	119	—
1085 days, MPa	115	122	115	150	132	—
Modulus of elasticity in compression, 100 x 200-mm moist-cured cylinders						
91 days, GPa	50.6	49.9	50.1	56.5	53.4	47.9
Drying shrinkage, 75 by 75 x 285-mm prisms						
7 days, millionths	193	123	100	87	137	—
28 days, millionths	400	287	240	203	233	—
90 days, millionths	573	447	383	320	340	—
369 days, millionths	690	577	520	453	467	—
1075 days, millionths	753	677	603	527	523	—

concrete mixture. This produces additional benefits in the form of reduced heat generation during hydration.

With use of low-slump or no-slump mixes, high-compressive-strength concrete is produced routinely under careful control in precast and prestressed concrete plants. These stiff mixes are placed in ruggedly-built forms and consolidated by prolonged vibration or shock methods. However, cast-in-place concrete uses more fragile forms that do not permit the same compaction procedures, hence more workable concretes are necessary to achieve the required compaction and to avoid segregation and honeycomb. Superplasticizing admixtures are invariably added to HPC mixtures to produce workable and often flowable mixtures.

Production of high-strength concrete may or may not require the purchase of special materials. The producer must know the factors affecting compressive strength and know how to vary those factors for best results. Each variable should be analyzed separately in developing a mix design. When an optimum or near optimum is established for each variable, it should be incorporated as the remaining variables are studied. An optimum mix design is then developed keeping in mind the economic advantages of using locally available materials. Many of the items discussed below also apply to most high-performance concretes.

Fig. 17-5. The Two Union Square building in Seattle used concrete with a designed compressive strength of 131 MPa in its steel tube and concrete composite columns. High-strength concrete was used to meet a design criteria of 41 GPa modulus of elasticity. (59577)

Cement

Today in Canada there is almost universal use of blended cements for cast-in-place HPC. The most commonly used is Type 10E-SF. Ternary blended cements meeting CSA A362 have also been used. For precast concrete Type 30 (III) and Type 10E-SF are most common. Blended cements containing fly ash, silica fume or slag can be used to make high-strength concrete with or without the addition of supplementary cementing materials.

Selection of cement for high-strength concrete should not be based only on mortar-cube tests but should also include tests of comparative strengths of concrete at 28, 56, and 91 days. A cement that yields the highest concrete compressive strength at extended ages (91 days) is preferable. For high-strength concrete, a cement should produce a minimum 7-day mortar-cube strength of approximately 30 MPa.

Trial mixtures with cement contents between 400 and 550 kg/m³ should be made for each cement being considered for the project. Amounts will vary depending on target strengths. Other than decreases in sand content as the cement content increases, the trial mixtures should be as nearly identical as possible.

Supplementary Cementing Materials

Fly ash, silica fume, or slag are often mandatory in the production of high-strength concrete; the strength gain obtained with these supplementary cementing materials cannot be attained by using additional cement alone. These supplementary cementing materials are usually added at dosage rates of 5% to 20% or higher by mass of cementing material. Some specifications only permit use of up to 10% silica fume, unless evidence is available indicating that concrete produced with a larger dosage rate will have satisfactory strength, durability, and volume stability. The water-to-cementing materials ratio should be adjusted so that equal workability becomes the basis of comparison between trial mixtures. For each set of materials, there will be an optimum cement-plus-supplementary cementing materials content at which strength does not continue to increase with greater amounts and the mixture becomes too sticky to handle properly. Blended cements containing fly ash, silica fume, slag, or calcined clay can be used to make high-strength concrete with or without the addition of supplementary cementing materials.

Aggregates

In high-strength concrete, careful attention must be given to aggregate size, shape, surface texture, mineralogy, and cleanness. For each source of aggregate and concrete strength level there is an optimum-size aggregate that will yield the most compressive strength per unit of cement. To find the optimum size, trial batches should be made with 20 mm and smaller coarse aggregates and varying cement contents. Many studies have found that 10 mm to 14 mm nominal maximum-size aggregates give optimum strength.

In high-strength concretes, the strength of the aggregate itself and the bond or adhesion between the paste and aggregate become important factors. Tests have shown that crushed-stone aggregates produce higher compressive strength in concrete than gravel aggregate using the same size aggregate and the same cementing materials content; this is probably due to a superior aggregate-to-paste bond when using rough, angular, crushed material. For specified concrete strengths of 70 MPa or higher, the potential of the aggregates to meet design requirements must be established prior to use.

Coarse aggregates used in high-strength concrete should be clean, that is, free from detrimental coatings of dust and clay. Removing dust is important since it may affect the quantity of fines and consequently the water demand of a concrete mix. Clay may affect the aggregate-paste bond. Washing of coarse aggregates may be necessary. Combining single sizes of aggregate to produce the required grading is recommended for close control and reduced variability in the concrete.

The quantity of coarse aggregate in high-strength concrete should be the maximum consistent with required workability. Because of the high percentage of cementitious material in high-strength concrete, an increase in coarse-aggregate content beyond values recommended in standards for normal-strength mixtures is necessary and allowable.

In high-rise buildings and in bridges, the stiffness of the structure is of interest to structural designers. On certain projects a minimum static modulus of elasticity has been specified as a means of increasing the stiffness of a structure (Fig. 17-5). The modulus of elasticity is not necessarily proportional to the compressive strength of a concrete. There are code formulas for normal-strength concrete and suggested formulas for high-strength concrete. The modulus achievable is affected significantly by the properties of the aggregate and also by the mixture proportions (Baalbaki and others 1991). If an aggregate has the ability to produce a high modulus, then the optimum modulus in concrete can be obtained by using as much of this aggregate as practical, while still meeting workability and cohesiveness requirements. If the coarse aggregate being used is a crushed rock, and manufactured fine aggregate of good quality is available from the same source, then a combination of the two can be used to obtain the highest possible modulus.

Due to the high amount of cementitious material in high-strength concrete, the role of the fine aggregate (sand) in providing workability and good finishing characteristics is not as crucial as in conventional strength mixes. Sand with a fineness modulus (FM) of about 3.0—considered a coarse sand—has been found to be satisfactory for producing good workability and high compressive strength. For specified strengths of 70 MPa or greater, FM should be between 2.8 and 3.2 and not vary by more than 0.10 from the FM selected for the duration of the project. Finer sand, say with a FM of between 2.5 and 2.7, may produce lower-strength, sticky mixtures.

Admixtures

The use of chemical admixtures such as water reducers, retarders, high-range water reducers or superplasticizers is necessary. They make more efficient use of the large amount of cementitious material in high-strength concrete and help to obtain the lowest practical water to cementing materials ratio. Chemical admixture efficiency must be evaluated by comparing strengths of trial batches. Also, compatibility between cement and supplementary cementing materials, as well as water-reducing and other admixtures, must be investigated by trial batches. From these trial batches, it will be possible to determine the workability, setting time, and amount of water reduction for given admixture dosage rates and times of addition.

The use of air-entraining admixtures is not necessary or desirable in high-strength concrete that is protected from the weather, such as interior columns and shearwalls of high-rise buildings. However, for bridges, concrete piles, piers, or parking structures, where durability in a freeze-thaw environment is required, entrained air is mandatory. Because air entrainment decreases concrete strength of rich mixtures, testing to establish optimum air contents and spacing factors may be required. Certain high-strength concretes may not need as much air as normal-strength concrete to be frost resistant. Pinto and Hover (2001) found that non-air-entrained, high-strength concretes had good frost and deicer-scaling resistance at a water to portland cement ratio of 0.25. Burg and Ost (1996) found good frost resistance with non-air-entrained concrete containing silica fume at a water to cementing materials ratio of 0.22 (Mix No. 4 in Table 17-5); however, this was not the case with other mixtures, including a portland-only mixture with a water to cement ratio of 0.28.

Current belief is that mixes with water to cementing materials ratios of 0.30 and greater require air entrainment and those less than 0.25 do not. It is uncertain at this time if concrete exposed to a freeze-thaw environment with a water-to-cementing materials ratio between these two values requires air entrainment for durability. However, on a recent Toronto project requiring 75 MPa concrete exposed to a freeze-thaw environment, Relative Durability Factors of over 90 % were obtained without air entrainment after 400 cycles of rapid freeze-thaw tests using ASTM C 666 procedure A. The water-cementing materials ratio of the 75 MPa mix was 0.27. It has also been shown that the spacing factor for HPC need not be as small as traditionally specified (Aïtcin, 1998). Consideration of the adoption of this approach must be preceded by extensive testing to confirm the durability of the concrete on a project by project basis. This characteristic of HPC has recently been recognized by the adoption in CSA A23.1 of slightly relaxed limits for spacing factors for any type of HPC exposed to freezing and thawing.

Proportioning

The trial mixture approach is best for selecting proportions for high-strength concrete. To obtain high strength, it is necessary to use a low water to cementing materials ratio and a high portland cement content. The unit strength obtained for each unit of cement used in a cubic metre of concrete can be plotted as strength efficiency to assist with mix designs.

The water requirement of concrete increases as the fine aggregate content is increased for any given size of coarse aggregate. Because of the high cementing materials content of these concretes, the fine aggregate content can be kept low. However, even with well-graded aggregates, a low water-cementing materials ratio may result in concrete that is not sufficiently workable for the job. If a superplasticizer is not already being used, this may be the time to consider one. A slump of around 200 mm will provide adequate workability for most applications. ACI Committee 211

(1993), Farny and Panarese (1994), and Nawy (2001) provide additional guidance on proportioning.

Mixing

High-strength concrete has been successfully mixed in transit mixers and central mixers; however, many of these concretes tend to be sticky and cause build-up in these mixers. Where dry, uncompacted silica fume has been batched into a mix, "balling" of the mix has occurred and mixing has been less than complete. In such instances it has been found necessary to experiment with the sequence in which solids and liquids are added, and the percentage of each material added at each step in the batching procedure. Batching and mixing sequences should be optimized during the trial mix phase. Where truck mixing is unavoidable, the best practice is to reduce loads to 90% of the rated capacity of the trucks.

Where there is no recent history of HSC or HPC mixtures that meet specified requirements, it is essential to first make laboratory trial mixes to establish optimum proportions. At this stage, the properties of the mix, such as workability, air content, density, strength and modulus of elasticity can be determined. Once laboratory mixture proportions have been determined, field trials using full loads of concrete are essential; they should be delivered to the site or to a mock-up to establish and confirm the suitability of the batching, mixing, transporting and placing systems to be used.

Prequalification of concrete suppliers for high-strength concrete projects is recommended (Bickley 1993). In a prequalification procedure, one or more loads of the proposed mixture is cast into a trial mock-up. The fresh concrete is tested for slump, air content, temperature, and density. Casting the mock-up provides the opportunity to assess the suitability of the mix for placing and compaction. The mock-up can be instrumented to record temperatures and temperature gradients; it can also be cored and tested to provide correlation with standard cylinder test results. The cores can be tested to provide the designer with in-place strength and modulus values for reference during construction. The heat characteristics of the mixture can also be determined using a computer interactive program, and the data used to determine how curing technology should be applied to the project.

Placing, Consolidation, and Curing

Close liaison between the contractor and the concrete producer allows concrete to be discharged rapidly after arrival at the jobsite. Final adjustment of the concrete should be supervised by the concrete producer's technicians at the site, by a concrete laboratory, or by a consultant familiar with the performance and use of high-strength concrete.

Delays in delivery and placing must be eliminated; sometimes it may be necessary to reduce batch sizes if placing procedures are slower than anticipated. Rigid surveillance must be exercised at the jobsite to prevent any addition of retempering water. Increases in workability should only be achieved by the addition of a superplasticizer. Any addition should be by the supplier's technician. The contractor must be prepared to receive the concrete and understand the consequences of exceeding the specified slump and water-cementing materials ratio.

Consolidation is very important in achieving the potential strengths of high-strength concrete. Concrete must be vibrated as quickly as possible after placement in the forms. High-frequency vibrators should be small enough to allow sufficient clearance between the vibrating head and reinforcing steel. Over-vibration of workable normal-strength concrete often results in segregation, loss of entrained air, or both. On the other hand, high-strength concrete without a superplasticizer, will be relatively stiff and contain little air. Consequently, inspectors should be more concerned with under-vibration rather than over-vibration. Most high-strength concrete, particularly very high-strength-concrete, is placed at slumps of 180 mm to 220 mm. Even at these slumps, some vibration is required to ensure compaction. The amount of compaction should be determined by onsite trials.

High-strength concrete is often difficult to finish because of its sticky nature. High cementing materials contents, large dosages of admixtures, low water contents, and air entrainment all contribute to the difficulty of finishing these concretes. Because the concrete sticks to the trowels and other finishing equipment, finishing activities should be minimized. The finishing sequence should be modified from that used for normal concrete.

Curing of high-strength concrete is even more important than curing normal-strength concrete. Providing adequate moisture and favourable temperature conditions is recommended for a prolonged period, particularly when 56- or 91-day concrete strengths are specified.

Additional curing considerations apply with HSC and HPC. Where very low water-cementing materials ratios are used in flatwork (slabs and overlays), and particularly where silica fume is used in the mixture, there will be little if any bleeding before or after finishing. In these situations it is imperative that fog curing or evaporation retarders be applied to the concrete immediately after the surface has been struck off. This is necessary to avoid plastic shrinkage cracking of horizontal surfaces and to minimize crusting. Fog curing, followed by 7 days of wet curing, has proven to be very effective.

It is inevitable that some vertical surfaces, such as columns, may be difficult to cure effectively. Where projects are fast-tracked, columns are often stripped at an early age to allow raising of self-climbing form systems. Concrete is thus exposed to early drying, sometimes within eleven hours after casting. Because of limited access, providing further curing is difficult and impractical.

Tests were conducted on column concrete to determine if such early exposure and lack of curing have any harmful effects. The tests showed that for a portland cement-slag-silica fume mixture with a specified strength of 70 MPa, the matrix was sound and a very high degree of impermeability to water and chloride ions had been achieved (Bickley and others 1994). Nevertheless, the best curing possible is recommended for all HPC.

The temperature history of HPC is an integral part of its curing process. Advantage should also be taken of recent developments in curing technology. Temperature increases and gradients that will occur in a concrete placement can be predicted by procedures that provide data for this purpose. With this technique, measures to heat, cool, or insulate a concrete placement can be determined and applied to significantly reduce both micro- and macro-cracking of the structure and assure durability. The increasing use of these techniques will be required in most structures using HPC to assure that the cover concrete provides long term protection to the steel, and results in the intended service life of the structure.

Quality Control

A comprehensive quality-control program is required at both the concrete plant and onsite to guarantee consistent production and placement of high-strength concrete. Inspection of concreting operations from stockpiling of aggregates through completion of curing is important. Closer production control than is normally obtained on most projects is necessary. Also, routine sampling and testing of all materials is particularly necessary to control uniformity of the concrete.

While tests on concrete should always be made in strict accordance with standard procedures, some additional requirements are recommended for specified strengths of 55 MPa and higher, and are required by CSA Standard A23.1 where specified strengths are 70 MPa or higher. In testing high-strength concrete, some changes and more attention to detail are required. For example, cardboard cylinder molds, which can cause lower strength-test results, should be replaced with reusable steel or plastic molds. Capping of cylinders must be done with great care using appropriate capping compounds. Lapping (grinding) the cylinder ends is an alternative to capping. For specified strengths of 70 MPa or greater, CSA Standard A23.1 requires that the ends of cylinders be ground flat prior to testing and that the ends not depart from a plane by more than 0.025 mm. While not a requirement, it is also recommended for HPC test cylinders. The initial curing requirements for test specimens is that they be stored in water or a fog room at 23°C ± 2°C from the time of casting to the time they are transported to the laboratory.

The physical characteristics of a testing machine can have a major impact on the result of a compression test. It has been shown that the dimensions of the upper platen assembly are critical (ACI Committee 363 1998). Also, it is recommended that testing machines be extremely stiff, both longitudinally and laterally.

The quality control necessary for the production of high compressive strength concrete will, in most cases, lead to low variance in test results. Strict vigilance in all aspects of quality control on the part of the producer and quality testing on the part of the laboratory are necessary on high-strength concrete projects. For concretes with specified strengths of 70 MPa, or greater, the coefficient of variation is the preferred measure of quality control.

HIGH-DURABILITY CONCRETE

Most of the attention in the 1970s and 1980s was directed toward high strength HPC; today the focus is more on concretes with high durability in severe environments resulting in structures with long life. For example, the Confederation Bridge across the Northumberland Strait between Prince Edward Island and New Brunswick has a 100-year design life (see Mix No. 2 in Table 17-3). This bridge contains HPC designed to efficiently protect the embedded reinforcement. The concrete had a diffusion coefficient of 4.8×10^{-13} at six months (a value 10 to 30 times lower than that of conventional concrete). The electrical resistivity was measured at 470 to 530 ohm-m, compared to 50 for conventional concrete. The design required that the concrete be rated at less than 1000 coulombs. The high concrete resistivity in itself will result in a rate of corrosion that is potentially less than 10 percent of the corrosion rate for conventional concrete (Dunaszegi 1999). The following sections review durability issues that high-performance concrete can address:

Abrasion Resistance

Abrasion resistance is directly related to the strength of concrete. This makes high strength HPC ideal for abrasive environments. The abrasion resistance of HPC incorporating silica fume is especially high. This makes silica-fume concrete particularly useful for spillways and stilling basins, and concrete pavements or concrete pavement overlays subjected to heavy or abrasive traffic.

Holland and others (1986) describe how severe abrasion-erosion had occurred in the stilling basin of a dam; repairs using fibre-reinforced concrete had not proven to be durable. The new HPC mix used to repair the structure the second time contained 386 kg/m³ of cement, 70 kg/m³ of silica fume, admixtures, and had a water to cementing materials ratio of 0.28, and a 90-day compressive strength exceeding 103 MPa.

Berra, Ferrara, and Tavano (1989) studied the addition of fibres to silica fume mortars to optimize abrasion resistance. The best results were obtained with a mix using slag cement, steel fibres, and silica fume. Mortar strengths ranged from 75 MPa to 100 MPa. In addition to better erosion resistance, less drying shrinkage, high freeze-thaw resistance, and good bond to the substrate were achieved.

In Norway steel studs are allowed in tires; this causes severe abrasion wear on pavement surfaces, with resurfacing required within one to two years. Tests using an accelerated road-wear simulator showed that in the range of 100 MPa to 120 MPa, concrete had the same abrasion resistance as granite (Helland 1990). Abrasion-resistant highway mixes usually contain between 320 and 450 kg/m³ of cement, plus silica fume or fly ash. They have water to cementing materials ratios of 0.22 to 0.36 and compressive strengths in the range of 85 to 130 MPa. Applications have included new pavements and overlays to existing pavements.

Blast Resistance

High-performance concrete can be designed to have excellent blast resistance properties. These concretes often have a compressive strength exceeding 120 MPa and contain steel fibres. Blast-resistant concretes are often used in bank vaults and military applications.

Permeability

The durability and service life of concrete exposed to weather is related to the permeability of the cover concrete protecting the reinforcement. HPC typically has very low permeability to air, water, and chloride ions. Low permeability is often specified through the use of a coulomb value; a maximum of 1000 coulombs is common across Canada. It appears from recent experience that for precast application, an average or even a maximum of 500 coulombs is attainable.

Test results obtained on specimens from a concrete column specified to be 70 MPa at 91 days and which had not been subjected to any wet curing were as follows (Bickley and others):

Water permeability of vacuum-saturated specimens:

Age at test:	7 years
Applied water pressure:	0.69 MPa
Permeability:	7.6×10^{-13} cm/s

Rapid chloride permeability (ASTM C 1202):

Age at test, years	Coulombs
1	303
2	258
7	417

The dense pore structure of high-performance concrete, which makes it so impermeable, gives it characteristics that make it eminently suitable for uses where a high quality concrete would not normally be considered. Latex-modified HPC is able to achieve these same low levels of permeability at normal strength levels without the use of supplementary cementing materials.

A large amount of concrete is used in farm structures. It typically is of low quality and often porous and with a rough surface, either when placed or after attack by farmyard wastes.

Gagne, Chagnon, and Parizeau (1994) provided a case history of the successful application of high performance concrete for agricultural purposes. In one case a farmer raising pigs on a large scale was losing about 1 kg per pig through diarrhea. This problem was resolved by reconstructing the pig pens with high performance concrete. Cited as beneficial properties in this application were:

- Surface smoothness that is compatible with the sensitive skin of a piglet
- Non-slip surface
- Good thermal conductivity resulting in uniform distribution of heat
- Impermeable surface to resist the growth of bacteria and viruses
- Easy to place
- Resistant to attack by acid (farmyard) wastes
- Price effective

The mix used had a water to portland cement ratio of 0.33 and had a 7-day compressive strength of 50 MPa.

Diffusion

Aggressive ions, such as chloride, in contact with the surface of concrete will diffuse through the concrete until a state of equilibrium in ion concentration is achieved. If the concentration of ions at the surface is high, diffusion may result in corrosion-inducing concentrations at the level of the reinforcement.

The lower the water-cementing materials ratio the lower the diffusion coefficient will be for any given set of materials. Supplementary cementing materials, particularly silica fume, further reduce the diffusion coefficient. Typical values for diffusion for HPC are as follows:

Type of Concrete	Diffusion Coefficient
Portland cement-fly-ash silica fume mix:	1000×10^{-15} m²/s
Portland cement-fly ash mix:	1600×10^{-15} m²/s

Carbonation

HPC has a very good resistance to carbonation due to its low permeability. It was determined that after 17 years the concrete in the CN Tower in Toronto had carbonated to an average depth of 6 mm (Bickley, Sarkar, and Langlois 1992). The concrete mixture in the CN Tower had a water-cementing materials ratio of 0.42. For a cover to the reinforcement of 35 mm, this concrete would provide corrosion protection for 500 years. For the lower water-cementing materials ratios common to HPC, significantly longer times to corrosion would result, assuming a crack free structure. In practical terms, uncracked HPC cover concrete is immune to carbonation to a depth that would cause corrosion.

Temperature Control

The quality, strength, and durability of HPC is highly dependent on its temperature history from the time of delivery to the completion of curing. In principle, favourable construction and placing methods will enable: (1) a low temperature at the time of delivery; (2) the smallest possible maximum temperature after placing; (3) minimum temperature gradients after placing; and (4) a gradual reduction to ambient temperature after maximum temperature is reached. Excessively high temperatures and gradients can cause excessively fast hydration and micro- and macro-cracking of the concrete.

It has been a practice on major high-rise structures incorporating concretes with specified strengths of 70 MPa to 85 MPa to specify a maximum delivery temperature of 18°C (Ryell and Bickley 1987). In summertime it is possible that this limit could only be met by using liquid nitrogen to cool the concrete. Experience with very-high-strength concrete suggests that a delivery temperature of no more than 25°C, preferably 20°C, should be allowed. In addition to liquid nitrogen, measures to cool HPC in the summer may involve using ice or chilled water as part of the mix water. The specifier should state the required delivery temperature.

In HPC applications such as high-rise buildings, column sizes are large enough to be classed as mass concrete. Normally, excessive heat generation in mass concrete is controlled by using a low cement content. When high-cement-content HPC mixes are used under these conditions, other methods of controlling maximum concrete temperature must be employed. Burg and Ost (1994) recorded temperature rise for 1220-mm concrete cubes using the mixtures in Table 17-5. A maximum temperature rise of 9.4°C to 11.7°C for every 100 kg of cement per cubic metre of concrete was measured. Burg and Fiorato (1999) monitored temperature rise in high-strength concrete caissons; they determined that in-place strength was not affected by temperature rise due to heat of hydration.

Freeze-Thaw Resistance

Because of its very low water-cementing materials ratio (less than 0.25), it is widely believed that HPC should be highly resistant to both scaling and physical breakup due to freezing and thawing. There is ample evidence that properly air-entrained high performance concretes are highly resistant to freezing and thawing and to scaling.

Gagne, Pigeon, and Aitcin (1990) tested 27 mixes using cement and silica fume with water-cementing materials ratios of 0.30, 0.26, and 0.23 and a wide range of quality in air-voids systems. All specimens performed exceptionally well in salt-scaling tests, confirming the durability of high-per-

formance concrete, and suggesting that air-entrainment is not needed. Tachitana and others (1990) conducted ASTM C 666 (Procedure A) tests on non-air-entrained high performance concretes with water-cementing materials ratios between 0.22 and 0.31. All were found to be extremely resistant to freeze-thaw damage and again it was suggested that air-entrainment is not needed.

Pinto and Hover (2001) found that non-air-entrained concrete with a water to portland cement ratio of 0.25 was deicer-scaling resistant with no supplementary cementing materials present. They found that higher strength portland cement concretes needed less air than normal concrete to be frost and scale resistant.

Burg and Ost (1994) found that of the six non-air-entrained mixes tested in Table 17-5 using ASTM C 666, only the silica fume concrete (Mix 4) with a water to cementing materials ratio of 0.22 was frost resistant.

Sidewalks constructed in Chicago in the 1920s used 25-mm thick toppings made of no-slump dry-pack mortar that had to be rammed into place. The concrete contained no air entrainment. Many of these sidewalks are still in use today; they are in good condition (minus some surface paste exposing fine aggregate) after 60 years of exposure to frost and deicers. No documentation exists on the water to cement ratio; however, it can be assumed that the water to cement ratio was comparable to that of modern HPCs.

While the above experiences prove the excellent durability of certain high-performance concretes to freeze-thaw damage and salt scaling, it is considered prudent to use air-entrainment. No well-documented field experiments have been made to prove that air-entrainment is not needed. Until such data are available, current practice for air-entrainment should be followed. It has been shown that the prime requirement of an air-void system for HPC is a preponderance of air bubbles of 200 μm size and smaller. If the correct air bubble size and spacing can be assured, then a moderate air content will ensure durability and minimize strength loss. CSA Standard A23.1 states that for high-performance concrete with a water-to-cementing materials ratio of 0.36 or less, the average spacing factor should not exceed 250 μm with no single value greater than 300 μm. The best measure of air-entrainment is the spacing factor.

Chemical Attack

For resistance to chemical attack on most structures, HPC offers a much improved performance. Resistance to various sulphates is achieved primarily by the use of a dense, strong concrete of very low permeability and low water-to-cementing materials ratio; these are all characteristics of HPC. Similarly, as discussed by Gagne and others (1994), resistance to acid from wastes is also much improved.

Alkali-Silica Reactivity (ASR)

Reactivity between certain siliceous aggregates and alkali hydroxides can affect the long-term performance of concrete. Two characteristics of HPC that help combat alkali-silica reactivity are:

(1) HPC concretes at very low water to cement ratios can self desiccate (dry out) to a level that does not allow ASR to occur (relative humidity less than 80%). Burg and Ost (1994) observed relative humidity values ranging from 62% to 72% for their six mixes in Table 17-5. The low permeability of HPC also minimizes external moisture from entering the concrete.

(2) HPC concretes can use significant amounts of supplementary cementing materials that may have the ability to control alkali-silica reactivity. However, this must be demonstrated by test. HPC concretes can also use ASR inhibiting admixtures to control ASR.

HPC concretes are not immune to alkali-silica reactivity and appropriate precautions must be taken.

Resistivity

HPC, particularly that formulated with silica fume, has very high resistivity, up to 20 to 25 times that of normal concrete. This increases resistance to the flow of electrical current and reduces corrosion rates. Particularly if dry, HPC acts as an effective dielectric. Where cracking occurs in HPC, the corrosion is localized and minor; this is due to the high resistivity of the concrete which suppresses the development of a macro corrosion cell.

SELF-COMPACTING CONCRETE

Self-compacting concrete (SCC), also referred to as self-consolidating concrete, is able to flow and consolidate on its own. At the same time it is cohesive enough to fill spaces of almost any size and shape without segregation or bleeding. This makes SCC particularly useful wherever placing is difficult, such as in heavily-reinforced concrete members or in complicated formwork.

This technology, developed in Japan in the 1980s, is based on increasing the amount of fine material, for example fly ash or limestone filler, without changing the water content compared to common concrete. This changes the rheological behavior of the concrete. SCC has to have a low yield value to ensure high flowability; a low water content ensures high viscosity, so the coarse aggregate can float in the mortar without segregating. To achieve a balance between deformability and stability, the total content of particles finer than the 160 μm sieve has to be high, usually about 520 to 560 kg/m³. High-range water reducers based on polycarboxylate ethers are typically used to plasticize the mixture. SCC is very sensitive to fluctuation in water content; therefore, stabilizers such as polysaccharides are used. Fig. 17-6 shows an example of mix

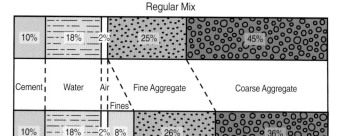

Fig. 17-6. Examples of materials used in regular concrete and self-compacting concrete by absolute volume.

proportions used in self-compacting concrete as compared to a regular concrete mix.

In Japan, self-compacting concretes are divided into three different types according to the composition of the mortar:

- Powder type
- Viscosity agent (stabilizer) type
- Combination type

For the powder type, a high proportion of fines produces the necessary mortar volume. In the stabilizer type, the fines content can be in the range admissible for vibrated concrete. The viscosity required to inhibit segregation will then be adjusted by using a stabilizer. The combination type is created by adding a small amount of stabilizer to the powder type to balance the moisture fluctuations in the manufacturing process.

Since SCC is characterized by special fresh concrete properties, many new tests have been developed to measure flowability, viscosity, blocking tendency, self-leveling, and stability of the mixture (Skarendahl and Peterson 1999 and Ludwig and others 2001). A simple test to measure the stable and unblocked flow is the J-Ring test, which is a modified slump test. The J-Ring—300 mm diameter with circular rods—is added to the slump test (Fig. 17-7). The number of rods has to be adjusted depending on the maximum size aggregate in the SCC mix. The SCC has to pass through the obstacles in the J-Ring without separation of paste and coarse aggregates. The slump diameter of a well-proportioned SCC is approximately the same with and without the J-Ring. It is usually about 750 mm, therefore, the test surface has to be at least 1000 mm in diameter.

The J-Ring is generally not used in current Canadian practice to measure slump flow (refer to CSA A23.1, *Appendix K, Commentary on the Slump Flow Test*).

Strength and durability of well-designed SCC are almost similar to conventional concrete. Without proper curing, SCC tends to have higher plastic shrinkage cracking than conventional concrete (Grube and Rickert 2001). Research indicates greater tensile creep for SCC, resulting in a reduced tendency to crack (Bickley and

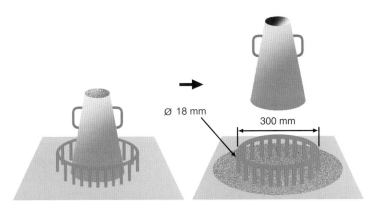

Ø 18 mm

300 mm

Fig. 17-7. J-ring test. Photo courtesy of VDZ.

Mitchell 2001). The use of fly ash as a filler seems to be advantageous compared to limestone filler; it results in higher strength and higher chloride resistance (Bouzoubaa and Lachemi 2001 and Ludwig and others 2001).

The production of SCC is more expensive than regular concrete and it is difficult to keep SCC in the desired consistency over a long period of time. However, construction time is shorter and production of SCC is environmentally friendly (no noise, no vibration). Furthermore, SCC produces a good surface finish. These advantages make SCC particularly interesting for use in precasting plants. SCC has been successfully used in a number of rehabilitation projects in Canada (Bickley and Mitchell 2001).

REACTIVE-POWDER CONCRETE

Reactive-powder concrete (RPC) was first patented by a French construction company in 1994. It is characterized by high strength and very low porosity, which is obtained by optimized particle packing and low water content.

The properties of RPC are achieved by: (1) eliminating the coarse aggregates; just very fine powders are used such as sand, crushed quartz, and silica fume, all with particle sizes between 0.02 and 300 µm; (2) optimizing the grain size distribution to densify the mixture; (3) post-set heat-treatment to improve the microstructure; (4) addition of steel and synthetic fibres (about 2% by volume); and (5) use of superplasticizers to decrease the water to cement ratio—usually to less than 0.2—while improving the rheology of the paste. See Fig. 17-8 for a typical fresh RPC.

The compressive strength of reactive-powder concrete is typically around 200 MPa, but can be produced with compressive strengths up to 810 MPa (Semioli 2001). However, the low comparative tensile strength requires prestressing reinforcement in severe structural service. Table 17-6 compares hardened concrete properties of RPC with those of an 80-MPa concrete.

Fig. 17-8. Freshly-mixed reactive-powder concrete.

RPC has found some applications in pedestrian bridges (Fig. 17-9) (Bickley and Mitchell 2001 and Semioli 2001). Also, the low porosity of RPC gives excellent durability and transport properties, which makes it a suitable material for the storage of nuclear waste (Matte and Moranville 1999). A low-heat type of reactive-powder concrete has been developed to meet needs for mass concrete pours for nuclear reactor foundation mats and underground containment of nuclear wastes (Gray and Shelton 1998).

Table 17-6. Typical Mechanical Properties of Reactive Powder Concrete (RPC) Compared to an 80-MPa Concrete (Perry 1998)

Property	Unit	80 MPa	RPC
Compressive strength	MPa	80	200
Flexural strength	MPa	7	40
Tensile strength	MPa		8
Modulus of Elasticity	GPa	40	60
Fracture Toughness	10^3 J/m^2	<1	30
Freeze-thaw, ASTM C 666	RDF	90	100
Carbonation depth: 36 days in CO_2	mm	2	0
Abrasion	10^{-12}m^2/s	275	1.2

Fig. 17-9. The Sherbrooke footbridge in Quebec, built in 1997, is North America's first reactive-powder concrete structure. (68300)

REFERENCES

AASHTO, *Resistance of Concrete to Chloride Ion Penetration*, AASHTO T 259-80, American Association of State Highway and Transportation Officials, Washington, D.C., 2000, pages 896 to 897.

AASHTO, *Sampling and Testing for Chloride Ion in Concrete and Concrete Raw Materials*, AASHTO T 260-97, American Association of State Highway and Transportation Officials, Washington, D.C., 2000, pages 898 to 904.

ACI Committee 211, *Guide for Selecting Proportions for High-Strength Concrete with Portland Cement and Fly Ash*, ACI 211.4-93, reapproved 1998, American Concrete Institute, Farmington Hills, Michigan, 1998, 13 pages.

ACI Committee 363, *Guide to Quality Control and Testing of High-Strength Concrete*, 363.2R-98, American Concrete Institute, Farmington Hills, Michigan, 1998, 18 pages.

ACI Committee 363, *State-of-the-Art Report on High-Strength Concrete*, 363R-92, reapproved 1997, American Concrete Institute, Farmington Hills, Michigan, 1997, 55 pages.

ACPA, *Fast-Track Concrete Pavements*, TB004, American Concrete Pavement Association, Skokie, Illinois, 1994, 32 pages.

Aitcin, P.-C., *High-Performance Concrete*, Modern Concrete Technology 5, E & FN Spon, London, 1998, 591 pages.

Aitcin, P.-C. and Richard, P., "The Pedestrian/Bikeway Bridge of Sherbrooke", *Proceedings of the 4th International Symposium on Utilization of High-Strength/High-Performance Concrete*, Paris, 1996, pages 1399 to 1406.

ASCE, *High-Performance Construction Materials and Systems*, Technical Report 93-5011, American Society of Civil Engineers, New York, April 1993.

Baalbaki, W.; Benmokrane, B.; Chaallal, O.; and Aitcin, P. C., "Influence of Coarse Aggregate on Elastic Properties of High-Performance Concrete," *ACI Materials Journal*, American Concrete Institute, Farmington Hills, Michigan, September/October 1991, page 499.

Bentz, Dale, *CIKS for High Performance Concrete*, http://ciks.cbt.nist.gov/bentz/welcome.html, National Institute of Standards and Technology, Gaithersburg, Maryland, 2001.

Bentz, Dale, Concrete *Optimization Software Tool*, http://ciks.cbt.nist.gov/bentz/welcome.html, National Institute of Standards and Technology, Gaithersburg, Maryland, 2001.

Bickley, John A., and Fung, Rico, *Optimizing the Economics of High-Performance Concrete*, A Concrete Canada and Canadian Cement Industry Joint Research Project, Cement Association of Canada, Ottawa, Ontario, 2001, 21 pages.

Bickley, John A., and Mitchell, Denis, *A State-of-the-Art Review of High Performance Concrete Structures Built in Canada: 1990—2000*, Cement Association of Canada, Ottawa, Ontario, May 2001, 122 pages.

Bickley, J. A.; Ryell, J.; Rogers, C.; and Hooton, R. D., "Some Characteristics of High-Strength Structural Concrete: Part 2," *Canadian Journal for Civil Engineering,* National Research Council of Canada, December 1994.

Bickley, J. A., Prequalification Requirements for the Supply and Testing of Very High Strength Concrete, *Concrete International,* American Concrete Institute, Farmington Hills, Michigan, February 1993, pages 62 to 64.

Bickley, J. A.; Sarkar, S.; and Langlois, M., The CN Tower, *Concrete International,* American Concrete Institute, Farmington Hills, Michigan, August 1992, pages 51 to 55.

Bonneau, O.; Poulin, C.; Dugat, J.; Richard, P.; and Aitcin, P.-C., "Reactive Powder Concrete: From Theory to Practice", *Concrete International,* American Concrete Institute, Farmington Hills, Michigan, April 1996, pages 47 to 49.

Bouzoubaa, N., and Lachemi, M., "Self-compacting concrete incorporating high volumes of class F fly ash. Preliminary results," Vol. 31, *Cement and Concrete Research,* Pergamon-Elsevier Science, Oxford, 2001, pages 413 to 420.

Burg, R. G., and Fiorato, A. E., *High-Strength Concrete in Massive Foundation Elements,* Research and Development Bulletin RD117, Portland Cement Association, 1999, 28 pages.

Burg, R. G., and Ost, B. W., *Engineering Properties of Commercially Available High-Strength Concretes (Including Three-Year Data),* Research and Development Bulletin RD104, Portland Cement Association, 1994, 62 pages.

CSA Standard A23.1-00, *Concrete Materials and Methods of Concrete Construction including Appendix J—High-Performance Concrete,* Canadian Standards Association, Toronto, 2000.

Dunaszegi, Laszlo, "HPC for Durability of the Confederation Bridge," *HPC Bridge Views,* Federal Highway Administration and National Concrete Bridge Council, Portland Cement Association, September/October 1999, page 2.

Farny, James A., and Panarese, William C., *High-Strength Concrete,* EB114, Portland Cement Association, 1994, 60 pages.

Ferraris, Chiara F., and Lobo, Colin L., "Processing of HPC," *Concrete International,* American Concrete Institute, Farmington Hills, Michigan, April 1998, pages 61 to 64.

Gagne, R.; Chagnon, D.; and Parizeau, R., "L'Utilization du Beton a Haute Performance dans l'Industrie Agricole," *Proceedings of Seminar at Concrete Canada Annual Meeting,* Sherbrooke, Quebec, October 1994, pages 23 to 33.

Gagne, R.; Pigeon, M.; and Aitcin, P. C., "Durabilite au gel des betons de hautes performance mecaniques," *Materials and Structures,* Vol. 23, 1990, pages 103 to 109.

Goodspeed, Charles H.; Vanikar, Suneel; and Cook, Raymond A., "High-Performance Concrete Defined for Highway Structures," *Concrete International,* American Concrete Institute, Farmington Hills, Michigan, February 1996, pages 62 to 67.

Gray, M. N., and Shelton, B. S., "Design and Development of Low-Heat, High-Performance Reactive Powder Concrete," *Proceedings, International Symposium on High-Performance and Reactive Powder Concretes,* Sherbrooke, Quebec, August 1998, pages 203 to 230.

Grove, James D., "Blanket Curing to Promote Early Strength Concrete," *Concrete and Construction—New Developments and Management,* Transportation Research Record, 1234, Transportation Research Board, Washington, D.C., 1989.

Grube, Horst, and Rickert, Joerg, "Self compacting concrete—another stage in the development of the 5-component system of concrete," *Betontechnische Berichte (Concrete Technology Reports),* Verein Deutscher Zementwerke, Düsseldorf, 2001, pages 39 to 48.

Helland, S., "High Strength Concrete Used in Highway Pavements," *Proceedings of the Second International Symposium on High Strength Concrete,* SP-121, American Concrete Institute, Farmington Hills, Michigan, 1990, pages 757 to 766.

Holland, T. C.; Krysa, Anton; Luther, Mark D.; and Liu, Tony C., "Use of Silica-Fume Concrete to Repair Abrasion-Erosion Damage in the Kinzua Dam Stilling Basin," *Proceedings of the Second International Conference on Fly Ash, Silica Fume, Slag and Natural Pozzolans in Concrete,* SP-91, American Concrete Institute, Farmington Hills, Michigan, 1986, pages 841 to 864.

Khayat, K.H.; Bickley, J.A.; and Lessard, M., "Performance of Self-Consolidating Concrete for Casting Basement and Foundation Walls", *ACI Materials Journal,* American Concrete Institute, Farmington Hills, Michigan, May-June 2000, pages 374 to 380.

Knutson, Martin, and Riley, Randall, "Fast-Track Concrete Paving Opens Door to Industry Future," *Concrete Construction,* Addison, Illinois, January 1987, pages 4 to 13.

Ludwig, Horst-Michael; Weise, Frank; Hemrich, Wolfgang; and Ehrlich, Norbert, "Self compacting concrete—principles and practice," *Betonwerk- und Fertigteil-Technik (Concrete Plant+Precast Technology),* Wiesbaden, Germany, June 2001, pages 58 to 67.

Matte, V., and Moranville, M., "Durability of Reactive Powder Composites: influence of silica fume on leaching properties of very low water/binder pastes," *Cement and Concrete Composites 21,* 1999, pages 1 to 9.

McCullough, B. Frank, and Rasmussen, Robert Otto, *Fast-Track Paving: Concrete Temperature Control and Traffic Opening Criteria for Bonded Concrete Overlays,* FHWA RD 98-167, Federal Highway Administration, Washington, D.C., October 1999, 205 pages.

Meschino, M., and Ryell, J., "*Bottom-Up Construction of 31 Meter High Reinforced Concrete-Filled Steel Columns for the Greater Toronto Airports Authority New Terminal Building,*" presented at the ACI Convention, Toronto, October 2000.

Nawy, Edward G., *Fundamentals of High-Performance Concrete,* John Wiley and Sons, New York, 2001.

NIST, *Bridge LCC,* http://www.bfrl.nist.gov/bridgelcc, National Institute of Standards and Technology, 2001.

NIST, *Partnership for High Performance Concrete Technology,* http://ciks.cbt.nist.gov/phpct, National Institute of Standards and Technology, Gaithersburg, Maryland, 2001.

Okamura, H., and Ozawa, K., "Mix Design for Self-Compacting Concrete," *Concrete Library of JSCE No. 25,* (Translation of Proceedings of JSCE, No. 496/V-24, 1994.8), Tokyo, June 1995, pages 107 to 120.

Okamura, H.; Ozawa, K.; and Ouchi, M., "Self-Compacting Concrete," *Structural Concrete 1,* 2000, pages 3 to 17.

Perenchio, W. F., *An Evaluation of Some of the Factors Involved in Producing Very-High-Strength Concrete,* Research and Development Bulletin RD014, Portland Cement Association, http://www.portcement.org/pdf_files/RD014.pdf, 1973.

Perenchio, William F., and Klieger, Paul, *Some Physical Properties of High-Strength Concrete,* Research and Development Bulletin RD056, Portland Cement Association, http://www.portcement.org/pdf_files/RD056.pdf, 1978.

Perry, V., "Industrialization of Ultra-High Performance Ductile Concrete," *Symposium on High-Strength/High-Performance Concrete,* University of Calgary, Alberta, November 1998.

Pinto, Roberto C. A., and Hover, Kenneth C., *Frost and Scaling Resistance of High-Strength Concrete,* Research and Development Bulletin RD122, Portland Cement Association, 2001, 75 pages.

Pyle, Tom, Caltrans, Personal communication on early strength concretes used in California, 2001.

Ryell, J., and Bickley, J.A., "Scotia Plaza: High Strength Concrete for Tall Buildings, "*Proceedings International Conference on Utilization of High Strength Concrete,* Stavanger, Norway, June 1987, pages 641 to 654.

Russell, Henry G., "ACI Defines High-Performance Concrete," *Concrete International,* American Concrete Institute, Farmington Hills, Michigan, February 1999, pages 56 to 57.

Saak, Aaron W.; Jennings, Hamlin M.; and Shah, Surendra P., "New Methodology for Designing Self-Compacting Concrete," *ACI Materials Journal,* American Concrete Institute, Farmington Hills, Michigan, November-December 2001, pages 429 to 439.

Semioli, William J., "The New Concrete Technology," *Concrete International,* American Concrete Institute, Farmington Hills, Michigan, November 2001, pages 75 to 79.

Skarendahl, Å., and Peterson, Ö., *Self-compacting concrete — Proceedings of the first international RILEM-Symposium,* Stockholm, 1999, 786 pages.

Tachitana, D.; Imai, M.; Yamazaki, N.; Kawai, T.; and Inada, Y., "High Strength Concrete Incorporating Several Admixtures," *Proceedings of the Second International Symposium on High Strength Concrete,* SP-121, American Concrete Institute, Farmington Hills, Michigan, 1990, pages 309 to 330.

Wood, J. G. M.; Wilson, J. R.; and Leek, D. S., "Improved Testing of Chloride Ingress Resistance of Concretes and Relation of Results to Calculated Behaviour", *Proceedings of the Third International Conference of Deterioration and Repair of Reinforced Concrete in the Arabian Gulf,* Bahrain Society of Engineers and CIRIA, 1989.

CHAPTER 18
Special Types of Concrete

Special types of concrete are those with out-of-the-ordinary properties or those produced by unusual techniques. Concrete is by definition a composite material consisting essentially of a binding medium and aggregate particles, and it can take many forms. Table 18-1 lists many special types of concrete made with portland cement and some made with binders other than portland cement. In many cases the terminology of the listing describes the use, property, or condition of the concrete. Brand names are not given. Some of the more common concretes are discussed in this chapter.

STRUCTURAL LOW-DENSITY CONCRETE

Low-density concretes are concretes that have substantially less mass per unit volume than those made with gravel or with crushed stone aggregates. They range in ovendry

Table 18-1. Some Special Types of Concrete

Special types of concrete made with portland cement		
Architectural concrete	High-density concrete	Recycled concrete
Autoclaved cellular concrete	High-early-strength concrete	Roller-compacted concrete
Centrifugally cast concrete	High-performance concrete	Sawdust concrete
Colloidal concrete	High-strength concrete	Self-compacting concrete
Colored concrete	Insulating concrete	Shielding concrete
Controlled-density fill	Latex-modified concrete	Shotcrete
Cyclopean (rubble) concrete	Low-density concrete	Shrinkage-compensating concrete
Dry-packed concrete	Mass concrete	Silica-fume concrete
Epoxy-modified concrete	Moderate-strength low-density concrete	Soil-cement
Exposed-aggregate concrete	Nailable concrete	Stamped concrete
Ferrocement	No-slump concrete	Structural low-density concrete
Fibre concrete	Polymer-modified concrete	Superplasticized concrete
Fill concrete	Pervious (porous) concrete	Terrazzo
Flowable fill	Pozzolan concrete	Tremie concrete
Flowing concrete	Precast concrete	Vacuum-treated concrete
Fly-ash concrete	Prepacked concrete	Vermiculite concrete
Gap-graded concrete	Preplaced aggregate concrete	White concrete
Geopolymer concrete	Reactive-powder concrete	Zero-slump concrete
Special types of concrete not using portland cement		
Acrylic concrete	Furan concrete	Polyester concrete
Aluminum phosphate concrete	Gypsum concrete	Polymer concrete
Asphalt concrete	Latex concrete	Potassium silicate concrete
Calcium aluminate concrete	Magnesium phosphate concrete	Sodium silicate concrete
Epoxy concrete	Methyl methacrylate (MMA) concrete	Sulphur concrete

Most of the definitions of these types of concrete appear in *Cement and Concrete Terminology,* ACI 116.

density from 240 to 1850 kg/m³ and range in compressive strength from 0.7 MPa to greater than 40 MPa.

Structural low-density concrete is defined in CSA Standard A23.1 as a concrete with a 28-day compressive strength in excess of 20 MPa and an air-dry density not exceeding 1850 kg/m³. Structural semi-low-density concrete has the same minimum compressive strength, but an air-dry density of between 1850 and 2150 kg/m³. These concretes are made with normal-density fine aggregate and low-density coarse aggregate. For comparision, normal-density concrete containing regular sand, gravel, or crushed stone has an air-dry density between 2150 and 2500 kg/m³. Low-density concrete is usually chosen for structural applications where its use will lead to a lower overall cost of the structure than the cost expected using normal-density concrete. Low-density aggregates are more expensive to produce than gravel or crushed stone aggregates. The generally higher cost of structural low-density concrete is usually offset by reduced dead loads on the structure resulting in less reinforcing steel, lower foundation costs, and therefore lower overall cost. There may also be advantages in using structural low-density concrete to expand or renovate existing buildings. Furthermore, Chapter 2 of the *Supplement to the National Building Code of Canada—Fire Performance Ratings* assigns enhanced fire-resistance ratings to concretes made with low-density aggregates.

Structural Low-Density Aggregates

Structural low-density aggregates are usually classified according to their production process because various processes produce aggregates with somewhat different properties. Processed structural low-density aggregates should meet the requirements of ASTM C 330, which includes:

- Rotary kiln expanded clays (Fig. 18-1), shales, and slates
- Sintering grate expanded shales and slates

Fig. 18-1. Expanded clay. (69793)

- Pelletized or extruded fly ash
- Expanded slags

Structural low-density aggregates can also be produced by processing other types of material, such as naturally occurring pumice and scoria.

Structural low-density aggregates have densities significantly lower than normal-density aggregates, ranging from 560 to 1120 kg/m³, compared to 1200 to 1760 kg/m³ for normal-density aggregates. These aggregates may absorb 5% to 20% water by mass of dry material. To control the uniformity of structural low-density concrete mixtures, the aggregates are prewetted (but not saturated) prior to batching.

Compressive Strength

The compressive strength of structural low-density concrete is usually related to the cement content at a given slump and air content, rather than to a water-to-cementing materials ratio. This is due to the difficulty in determining how much of the total mix water is absorbed into the aggregate and thus not available for reaction with the cement. ACI 211.2 provides guidance on the relationship between compressive strength and cement content. Typical compressive strengths range from 20 to 35 MPa. High-strength concrete can also be made with structural low-density aggregates.

In well-proportioned mixtures, the cementing materials content and strength relationship is fairly constant for a particular source of low-density aggregate. However, the relationship will vary from one aggregate source or type to another. When information on this relationship is not available from the aggregate manufacturer, trial mixtures with varying cementing materials contents are required to develop a range of compressive strengths, including the strength specified. Fig. 18-2 shows the relationship between cementing materials content and compressive strength. An example of a 30-MPa structural low-density concrete mixture with an air-dry density of about 1800 kg/m³, a combination of natural sand and gravel, and a low-density rotary kiln expanded clay coarse aggregate follows:

- 356 kg Type 10 (I) portland cement
- 534 kg sand, ovendry
- 320 kg gravel (14 to 2.5 mm), ovendry
- 356 kg low-density aggregate (10 mm to 630 μm), ovendry
- 172 kg mix water added
- 0.7 L water-reducing admixture
- 0.09 L air-entraining admixture
- 1 m³ yield
- Slump—75 mm
- Air content—6%

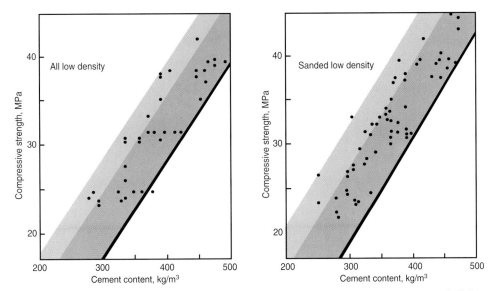

Fig. 18-2. Relationship between compressive strength and cement content of field structural low-density concrete using (left) low-density fine aggregate and coarse aggregate or (right) low-density coarse aggregate and normal-density fine aggregate (data points represent actual project strength results using a number of cement and aggregate sources) (ACI 211.2).

Material proportions vary significantly for different materials and strength requirements.

Entrained Air

As with normal-density concrete, entrained air in structural low-density concrete ensures resistance to freezing and thawing and to deicer applications. It also improves workability, reduces bleeding and segregation, and may compensate for minor grading deficiencies in the aggregate.

The amount of entrained air should be sufficient to provide good workability to the plastic concrete and adequate freeze-thaw resistance to the hardened concrete. Air contents are generally between 4% and 9%, depending on the maximum size of coarse aggregate (paste content) used and the exposure conditions. Testing for air content should be performed by the volumetric method, CSA Test Method A23.2-7C (ASTM C 173). For concrete that will be exposed to a freeze-thaw environment, air entrainment should be provided in accordance with CSA Standard A23.1. The freeze-thaw durability is also significantly improved if structural low-density concrete is allowed to dry before exposure to a freeze-thaw environment.

Specifications

Many suppliers of low-density aggregates for use in structural low-density concrete have information on suggested specifications and mixture proportions pertaining to their product. The usual specifications for structural concrete state a minimum compressive strength, a maximum density, a maximum slump, and an acceptable range in air content.

The contractor should also be concerned with the bleeding, workability, and finishing properties of structural low-density concrete.

Mixing

In general, mixing procedures for structural low-density concrete are similar to those for normal-density concrete; however, some of the more absorptive aggregates may require prewetting before use. Water added at the batching plant should be sufficient to produce the specified slump at the jobsite. Measured slump at the batch plant will generally be appreciably higher than the slump at the site. Pumping can especially aggravate slump loss.

Workability and Finishability

Structural low-density concrete mixtures can be proportioned to have the same workability, finishability, and general appearance as a properly proportioned normal-density concrete mixture. Sufficient cement paste must be present to coat each particle, and coarse-aggregate particles should not separate from the mortar. Enough fine aggregate is needed to keep the freshly mixed concrete cohesive. If aggregate is deficient in minus 630 µm sieve material, finishability may be improved by using a portion of natural sand, by increasing cementing materials content, or by using satisfactory mineral fines. Since entrained air improves workability, it should be used regardless of exposure.

Slump

Due to lower aggregate density, structural low-density concrete does not slump as much as normal-density concrete with the same workability. A low-density air-entrained mixture with a slump of 50 to 75 mm can be placed under conditions that would require a slump of 75 to 125 mm for normal-density concrete. It is seldom necessary to exceed slumps of 125 mm for normal placement of structural low-density concrete. With higher slumps, the large aggregate particles tend to float to the surface, making finishing difficult.

Vibration

As with normal-density concrete, vibration can be used effectively to consolidate low-density concrete; the same frequencies commonly used for normal-density concrete are recommended. The length of time for proper consolidation varies, depending on mix characteristics. Excessive vibration causes segregation by forcing large aggregate particles to the surface.

Placing, Finishing, and Curing

Structural low-density concrete is generally easier to handle and place than normal-density concrete. A slump of 50 to 100 mm produces the best results for finishing. Greater slumps may cause segregation, delay finishing operations, and result in rough, uneven surfaces.

If pumped concrete is being considered, the specifier, suppliers, and contractor should all be consulted about performing a field trial using the pump and mixture planned for the project. Adjustments to the mixture may be necessary; pumping pressure causes the aggregate to absorb more water, thus reducing the slump and increasing the density of the concrete.

Finishing operations should be started earlier than for comparable normal-density concrete, but finishing too early may be harmful. A minimum amount of floating and troweling should be done; magnesium finishing tools are preferred.

The same curing practices should be used for low-density concrete as for normal-density concrete. The two methods commonly used in the field are water curing (ponding, sprinkling, or using wet coverings) and preventing loss of moisture from the exposed surfaces (covering with waterproof paper, plastic sheets, or sealing with liquid membrane-forming compounds). Generally, 7 days of curing are adequate for ambient air temperatures above 10°C. Refer to Chapter 12, "Curing Concrete," for specific requirements.

INSULATING AND MODERATE-STRENGTH LOW-DENSITY CONCRETES

Insulating concrete is a low-density concrete with an ovendry density of 800 kg/m³ or less. It is made with cementing materials, water, air, and with or without aggregate and chemical admixtures. The ovendry density ranges from 240 to 800 kg/m³ and the 28-day compressive strength is generally between 0.7 and 7 MPa. Cast-in-place insulating concrete is used primarily for thermal and sound insulation, roof decks, fill for slab-on-grade subbases, leveling courses for floors or roofs, firewalls, and underground thermal conduit linings.

Moderate-strength low-density concrete has a density of 800 to 1900 kg/m³ ovendry and has a compressive strength of approximately 7 to 15 MPa. It is made with cementing materials, water, air, and with or without aggregate and chemical admixtures. At lower densities, it is used as fill for thermal and sound insulation of floors, walls, and roofs and is referred to as fill concrete. At higher densities it is used in cast-in-place walls, floors and roofs, and precast wall and floor panels. See ACI documents for more information.

For discussion purposes, insulating and moderate-strength low-density concretes can be grouped as follows:

Group I is made with expanded aggregates such as perlite, vermiculite, or expanded polystyrene beads. Ovendry concrete densities using these aggregates generally range between 240 to 800 kg/m³. This group is used primarily in insulating concrete. Some moderate-strength concretes can also be made from aggregates in this group.

Group II is made with aggregates manufactured by expanding, calcining, or sintering materials such as blast-furnace slag, clay, diatomite, fly ash, shale, or slate, or by processing natural materials such as pumice, scoria, or tuff. Ovendry concrete densities using these aggregates can range between 720 to 1440 kg/m³. Aggregates in this group are used in moderate-strength low-density concrete and some of these materials (expanded slag, clay, fly ash, shale, and slate) are also used in both moderate-strength and structural low-density concrete (up to about 1900 kg/m³ air-dry).

Group III concretes are made by incorporating into a cement paste or cement-sand mortar a uniform cellular structure of air voids that is obtained with preformed foam (ASTM C 869), formed-in-place foam, or special foaming agents. Ovendry densities ranging between 240 to 1900 kg/m³ are obtained by substitution of air voids for some or all of the aggregate particles; air voids can consist of up to 80% of the volume. Cellular concrete can be made to meet the requirements of both insulating and moderate strength low-density concrete.

Aggregates used in Groups I and II should meet the requirements of ASTM C 332, *Standard Specification for Lightweight Aggregates for Insulating Concrete*. These aggregates have dry densities in the range of from 96 to 1120 kg/m³ down to 16 kg/m³ for expanded polystyrene beads.

Mixture Proportions

Examples of mixture proportions for Group I and III concretes appear in Table 18-2. In Group I, air contents may be as high as 25% to 35%. The air-entraining agent can be prepackaged with the aggregate or added at the mixer. Because of the absorptive nature of the aggregate, the volumetric method CSA Test Method A23.2-7C (ASTM C 173) should be used to measure air content.

Water requirements for insulating and fill concretes vary considerably, depending on aggregate characteristics, entrained air, and mixture proportions. An effort should be made to avoid excessive amounts of water in insulating

Table 18-2. Examples of Low-Density Insulating Concrete Mixtures

Type of concrete	Ratio: portland cement to aggregate by volume	Ovendry density, kg/m³	Type 10 (I) portland cement, kg/m³	Water-cement ratio, by mass	28-day compressive strength, MPa, 150 x 300-mm cylinders
Perlite*	1:4	480 to 608	362	0.94	2.75
	1:5	416 to 576	306	1.12	2.24
	1:6	352 to 545	245	1.24	1.52
	1:8	320 to 512	234	1.72	1.38
Vermiculite*	1:4	496 to 593	380	0.98	2.07
	1:5	448 to 496	295	1.30	1.17
	1:6	368 to 464	245	1.60	0.90
	1:8	320 to 336	178	2.08	0.55
Polystyrene:**					
0 kg sand	1:3.4	545‡	445	0.40	2.24
73 kg sand/m³	1:3.1	625‡	445	0.40	2.76
154 kg sand/m³	1:2.9	725‡	445	0.40	3.28
200 kg sand/m³	1:2.5	769‡	474	0.40	3.79
Cellular*	—	625	524	0.57	2.41
(neat cement)	—	545	468	0.56	1.45
	—	448	396	0.57	0.90
	—	368	317	0.65	0.34
Cellular†	1:1	929	429	0.40	3.17
(sanded)††	1:2	1250	73	0.41	5.66
	1:3	1602	360	0.51	15.10

  * Reichard (1971).
 ** Source: Hanna (1978). The mix also included air entrainment and a water-reducing agent.
  † Source: Gustaferro (1970).
 †† Dry-rodded sand with a bulk density of 1600 kg/m³.
  ‡ Air-dry density at 28 days, 50% relative humidity.

concrete used in roof fills. Excessive water causes high drying shrinkage and cracks that may damage the water-proofing membrane. Accelerators containing calcium chloride should not be used where galvanized steel will remain in permanent contact with the concrete because of possible corrosion problems.

Mixture proportions for Group II concretes usually are based on volumes of dry, loose materials, even when aggregates are moist as batched. Satisfactory proportions can vary considerably for different aggregates or combinations of aggregates. Mixture proportions ranging from 0.24 to 0.90 cubic metres of aggregate per 100 kg of cementing material can be used in low-density concretes that are made with pumice, expanded shale, and expanded slag. Some mixtures, such as those for no-fines concretes, are made without fine aggregate but with total void contents of 20% to 35%. Cementing materials contents for Group II concretes range between 120 to 360 kg per cubic metre depending on air content, aggregate gradation, and mixture proportions.

No-fines concretes containing pumice, expanded slag, or expanded shale can be made with 150 to 170 kg of water

per cubic metre, total air voids of 20% to 35%, and a cementing materials content of about 280 kg per cubic metre.

Workability

Because of their high air content, low-density concretes having a mass less than 800 kg/m³ generally have excellent workability. Slumps of up to 250 mm usually are satisfactory for Group I and Group III concretes; appearance of the mix, however, may be a more reliable indication of consistency. Cellular concretes are handled as liquids; they are poured or pumped into place without further consolidation.

Mixing and Placing

All concrete should be mechanically mixed to produce a uniform distribution of materials of proper consistency and required density. In batch-mixing operations, various sequences can be used for introducing the ingredients; the preferred sequence is to first introduce the required amount of water into the mixer, then add the cementing materials, air-entraining or foaming agent, aggregate, pre-formed foam, and any other ingredients.

Excessive mixing and handling should be avoided because they tend to break up aggregate particles, thereby changing density and consistency. Segregation is not usually a problem (though it could be for Group II) because of the relatively large amounts of entrained air in these mixtures.

Pumping is the most common method of placement, but other methods can be used. Finishing operations should be kept to a minimum; smoothing with a darby or bullfloat is usually sufficient. Placement of insulating concretes should be done by workers experienced with these special concretes.

Periodic plastic (wet) density tests, CSA Test Method A23.2-6C (ASTM C 138) at the jobsite can be performed to check the uniformity of the concrete. Variations in density generally should not exceed plus or minus 32 kg/m³. A close approximation of the ovendry density can be determined from the freshly mixed density.

Thermal Resistance

ASTM C 177, *Test Method for Steady-State Heat Flux Measurements and Thermal Transmission Properties by Means of the Guarded-Hot-Plate Apparatus,* is used to determine values of thermal conductivity. Fig. 18-3 shows an approximate relationship between thermal resistance and density. The thermal conductivity of concrete increases with an increase in moisture content and density. See Brewer (1967) for additional density and conductivity relationships.

Strength

Strength requirements depend on the intended use of the concrete. For example, a compressive strength of 0.7 MPa, or even less, may be satisfactory for insulation of underground steam lines. Roof-fill insulation requires sufficient early strength to withstand foot traffic. Compressive strengths of 0.7 to 1.5 MPa are usually adequate for roof fills, but strengths up to 3.5 MPa are sometimes specified. In general, the strength of insulating concrete is of minor importance. Compressive strength of low-density insulating concrete should be determined by the methods specified in ASTM C 495 or C 513.

Table 18-2 and Fig. 18-4 give examples of the relationship between density and strength for low-density insulating concretes. Fig. 18-5 shows examples for cellular concrete containing sand. Mixtures with strengths outside the ranges shown can be made by varying the mixture proportions. Strengths comparable to those at 28 days would be obtained at 7 days with high-early-strength cement. The relationships shown do not apply to autoclaved products.

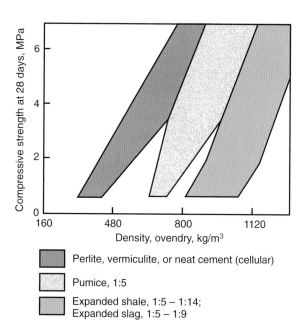

Fig. 18-4. Approximate relationship between ovendry bulk density and compressive strength of 150 x 300-mm cylinders tested in an air-dry condition for some insulating and fill concretes. For the perlite and vermiculite concretes, mix proportions range from 1:3 to 1:10 by volume.

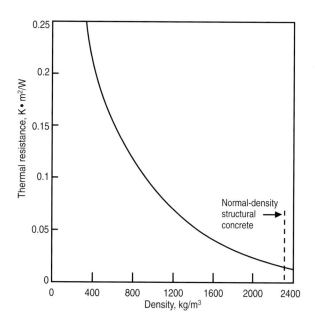

Fig. 18-3. Thermal resistance of concrete versus density (PCA 1980).

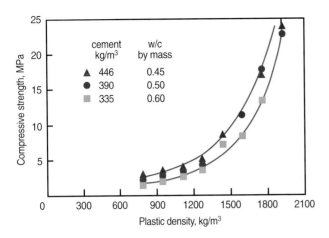

Fig. 18-5. Plastic (wet) density versus compressive strength for sanded cellular concretes. Compressive strength was determined with 150 x 300-mm cylinders that were cured for 21 days in a 100% relative humidity moist room followed by 7 days in air at 50% RH (McCormick 1967 and ACI 523.3R).

Resistance to Freezing and Thawing

Insulating and moderate-strength low-density concretes normally are not required to withstand freeze-thaw exposure in a saturated condition. In service they are normally protected from the weather; thus little research has been done on their resistance to freezing and thawing.

Drying Shrinkage

The drying shrinkage of insulating or moderate-strength low-density concrete is not usually critical when it is used for insulation or fill; however, excessive shrinkage can cause curling. In structural use, shrinkage should be considered. Moist-cured cellular concretes made without aggregates have high drying shrinkage. Moist-cured cellular concretes made with sand may shrink from 0.1% to 0.6%, depending on the amount of sand used. Autoclaved cellular concretes shrink very little on drying. Insulating concretes made with perlite or pumice aggregates may shrink 0.1% to 0.3% in six months of drying in air at 50% relative humidity; vermiculite concretes may shrink 0.2% to 0.45% during the same period. Drying shrinkage of insulating concretes made with expanded slag or expanded shale ranges from about 0.06% to 0.1% in six months.

Expansion Joints

Where insulating concrete is used on roof decks, a 25-mm expansion joint at the parapets and all roof projections is often specified. Its purpose is to accommodate expansion caused by the heat of the sun so that the insulating concrete can expand independently of the roof deck. Transverse expansion joints should be placed at a maximum of 30 m in any direction for a thermal expansion of 1 mm per metre. A fibreglass material that will compress to one-half its thickness under a stress of 0.17 MPa is generally used to form these joints.

AUTOCLAVED CELLULAR CONCRETE

Autoclaved cellular concrete (also called autoclaved aerated concrete) is a special type of low-density building material. It is manufactured from a mortar consisting of pulverized siliceous material (sand, slag, or fly ash), cement and/or lime, and water; to this a gas forming admixture, for example aluminum powder, is added. The chemical reaction of aluminum with the alkaline water forms hydrogen, which expands the mortar as macropores with a diameter of 0.5 mm to 1.5 mm form. The material is then pressure steam cured (autoclaved) over a period of 6 to 12 hours using a temperature of 190°C and a pressure of 1.2 MPa. This forms a hardened mortar matrix, which essentially consists of calcium silicate hydrates.

This porous mineral building material has densities between 300 and 1000 kg/m³ and compressive strengths between 2.5 and 10 MPa. Due to the high macropore content—up to 80 percent by volume—autoclaved cellular concrete has a thermal conductivity of only 0.15 to 0.20 W/(m•K).

Autoclaved cellular concrete is produced in block or panel form for construction of residential or commercial buildings (Fig. 18-6).

Additional information can be found in ACI 523.2R, *Guide for Precast Cellular Concrete.*

Fig. 18-6. (top) Residential building constructed with autoclaved cellular concrete blocks. (bottom) Autoclaved cellular concrete block floating in water. (70025, 44109)

HIGH-DENSITY CONCRETE

High-density concrete and has a density of up to about 6400 kg/m³. High-density concrete is used principally for radiation shielding but is also used for counterweights and other applications where high-density is important. As a shielding material, high-density concrete protects against the harmful effects of X-rays, gamma rays, and neutron radiation. Selection of concrete for radiation shielding is based on space requirements and on the type and intensity of radiation. Where space requirements are not important, normal-density concrete will generally produce the most economical shield; where space is limited, high-density concrete will allow for reductions in shield thickness without sacrificing shielding effectiveness.

Type and intensity of radiation usually determine the requirements for density and water content of shielding concrete. Effectiveness of a concrete shield against gamma rays is approximately proportional to the density of the concrete—the higher the density, the more effective the shield. On the other hand, an effective shield against neutron radiation requires both high and low-density elements. The hydrogen in water provides an effective low-density element in concrete shields. Some aggregates contain crystallized water, called fixed water, as part of their structure. For this reason, high-density aggregates with high fixed-water contents often are used if both gamma rays and neutron radiation are to be attenuated. Boron glass (boron frit) is also added to attenuate neutrons.

High-Density Aggregates

High-density aggregates such as barite, ferrophosphorus, goethite, hematite, ilmenite, limonite, magnetite, and degreased steel punchings and shot are used to produce high-density concrete. Where high fixed-water content is desirable, serpentine (which has a slightly higher density than normal-density aggregate) or bauxite can be used (see ASTM C 637 and C 638).

Table 18-3 gives typical bulk density, relative density, and percentage of fixed water for some of these materials. The values are a compilation of data from a wide variety of tests or projects reported in the literature. Steel punchings and shot are used where concrete with a density of more than 4800 kg/m³ is required.

In general, selection of an aggregate is determined by physical properties, availability, and cost. High-density aggregates should be reasonably free of fine material, oil, and foreign substances that may affect either the bond of paste to aggregate particle or the hydration of cement. For good workability, maximum density, and economy, aggregates should be roughly cubical in shape and free of excessive flat or elongated particles.

Additions

Boron additions such as colemanite, boron frits, and borocalcite are sometimes used to improve the neutron shielding properties of concrete. However, they may adversely affect setting and early strength of concrete; therefore, trial mixes should be made with the addition under field conditions to determine suitability. Admixtures such as pressure-hydrated lime can be used with coarse-sand sizes to minimize any retarding effect.

Properties of High-Density Concrete

The properties of high-density concrete in both the freshly mixed and hardened states can be tailored to meet job conditions and shielding requirements by proper selection of materials and mixture proportions.

Except for density, the physical properties of high-density concrete are similar to those of normal-density concrete. Strength is a function of water-cementing materials ratio; thus, for any particular set of materials, strengths

Table 18-3. Physical Properties of Typical High-Density Aggregates and Concrete

Type of aggregate	Fixed-water,* percent by mass	Aggregate relative density	Aggregate bulk density, kg/m³	Concrete density, kg/m³
Goethite	10–11	3.4–3.7	2080–2240	2880–3200
Limonite**	8–9	3.4–4.0	2080–2400	2880–3360
Barite	0	4.0–4.6	2320–2560	3360–3680
Ilmenite	†	4.3–4.8	2560–2700	3520–3850
Hematite	†	4.9–5.3	2880–3200	3850–4170
Magnetite	†	4.2–5.2	2400–3040	3360–4170
Ferrophosphorus	0	5.8–6.8	3200–4160	4080–5290
Steel punchings or shot	0	6.2–7.8	3860–4650	4650–6090

* Water retained or chemically bound in aggregates.
** Test data not available.
† Aggregates may be combined with limonite to produce fixed-water contents varying from about ½% to 5%.

comparable to those of normal-density concretes can be achieved. Typical densities of concretes made with some commonly used high-density aggregates are shown in Table 18-3. Because each radiation shield has special requirements, trial mixtures should be made with job materials and under job conditions to determine suitable mixture proportions.

Proportioning, Mixing, and Placing

The procedures for selecting mix proportions for high-density concrete are the same as those for normal-density concrete. However, additional mixture information and sample calculations are given in ACI 211.1. Following are the most common methods of mixing and placing high-density concrete:

Conventional methods of mixing and placing often are used, but care must be taken to avoid overloading the mixer, especially with very high-density aggregates such as steel punchings. Batch sizes should be reduced to about 50% of the rated mixer capacity. Because some high-density aggregates are quite friable, excessive mixing should be avoided to prevent aggregate breakup with resultant detrimental effects on workability and bleeding.

Preplaced aggregate methods can be used for placing normal and high-density concrete in confined areas and around embedded items; this will minimize segregation of coarse aggregate, especially steel punchings or shot. The method also reduces drying shrinkage and produces concrete of uniform density and composition. With this method, the coarse aggregates are preplaced in the forms and grout made of cementing material, sand, and water is then pumped through pipes to fill the voids in the aggregate.

Pumping of high-density concrete through pipelines may be advantageous in locations where space is limited. High-density concretes cannot be pumped as far as normal-density concretes because of their higher densities.

Puddling is a method whereby a 50-mm layer or more of mortar is placed in the forms and then covered with a layer of coarse aggregate that is rodded or internally vibrated into the mortar. Care must be taken to ensure uniform distribution of aggregate throughout the concrete.

MASS CONCRETE

Mass concrete is defined by ACI Committee 116 as "Any large volume of cast-in-place concrete with dimensions large enough to require that measures be taken to cope with the generation of heat and attendant volume change to minimize cracking." Mass concrete includes not only low-cement-content concrete used in dams and other massive structures but also moderate- to high-cement-content concrete in structural members that require special considerations to reduce heat of hydration and temperature rise (Fig. 18-7).

Fig. 18-7. Large foundation placements as shown require mass-concrete precautions. (50918)

In mass concrete, temperature rise is caused by heat of hydration (Fig. 18-8). As the interior concrete increases in temperature, the surface concrete may be cooling and contracting. This causes tensile stress and cracks at the surface if the temperature differential is too great. The width and depth of cracks depends upon the temperature gradient between the hot internal concrete and the cooler outer surface.

A definite member size beyond which a concrete structure should be classified as mass concrete is not readily available. Many large structural elements may be massive enough that heat generation should be considered; this is particularly critical when the minimum cross-sectional dimensions of a solid concrete member approach or exceed 1 metre or when cement contents exceed 355 kg/m³. Temperature rise in mass concrete is related to the initial concrete temperature (Fig. 18-9), ambient temperature, size of the concrete element (volume to surface ratio), and the amount of reinforcement. Small concrete members are of little concern as the generated heat is rapidly dissipated.

To avoid cracking, the internal concrete temperature for dams and other nonreinforced mass concrete structures of relatively low compressive strength should not be allowed to rise more than 11°C to 14°C above the mean annual ambient temperature (ACI 308). Internal concrete temperature gain can be controlled a number of ways: (1) a low cement content—120 to 270 kg/m³; (2) large aggregate size—75 to 150 mm; (3) high coarse aggregate content—up

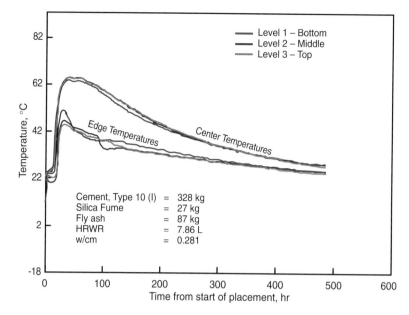

Fig. 18-8. (left) A drilled shaft (caisson), 3 m in diameter and 12.2 m in depth in which "low-heat" high-strength concrete is placed and (right) temperatures of this concrete measured at the center and edge and at three different levels in the caisson (Burg and Fiorato 1999). (57361)

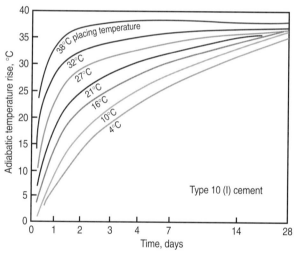

Fig. 18-9. The effect of concrete-placing temperature on temperature rise in mass concrete with 223 kg/m³ of cement. Higher placing temperatures accelerate temperature rise (ACI 207.2R).

to 80% of total aggregate; (4) low-heat-of-hydration cement; (5) pozzolans—where heat of hydration of a pozzolan is approximately 25% to 50% that of cement; (6) reductions in the initial concrete temperature to approximately 10°C by cooling the concrete ingredients; (7) cooling the concrete through the use of embedded cooling pipes; (8) steel forms for rapid heat dissipation; (9) water curing; and (10) low lifts—1.5 m or less during placement. In massive structures of high volume-to-surface ratio, an estimate of the adiabatic temperature rise can be made using equations in PCA (1987).

Massive structural reinforced concrete members with high cement contents (300 to 600 kg/m³) cannot use many of the placing techniques and controlling factors mentioned above to maintain low temperatures to control cracking. For these concretes (often used in mat foundations and power plants), a good technique is to (1) place the entire concrete section in one continuous pour, (2) avoid external restraint from adjacent concrete elements, and (3) control internal differential thermal strains by preventing the concrete from experiencing excessive temperature differential between the internal concrete and the surface. The latter is done by keeping the concrete warm through use of insulation (tenting, quilts, or sand on polyethylene sheeting). Studies and experience have shown that the maximum temperature differential between the interior and exterior concrete should not exceed about 20°C to avoid surface cracking (FitzGibbon 1977 and Fintel and Ghosh 1978). Internal cracking is also reduced. Some sources indicate that the maximum temperature differential (MTD) for concrete containing granite or limestone (low-thermal-coefficient aggregates) should be 25°C and 31°C, respectively (Bamforth 1981). However, an MTD of 20°C should be assumed unless tests on the actual concrete mix to be used show that higher MTD values are allowable.

By reducing the temperature differential to 20°C or less, the concrete will cool slowly to ambient temperature with little or no surface cracking; however, this is true only if the member is not restrained by continuous reinforcement crossing the interface of adjacent or opposite sections of hardened concrete. Restrained concrete will tend to crack due to eventual thermal contraction after the cool down. Unrestrained concrete will not crack if proper procedures are followed and the temperature differential is monitored and controlled. If there is any concern over excess temperature differentials in a concrete member, the element should be considered as mass concrete and appropriate precautions taken.

Fig. 18-10 illustrates the relationship between temperature rise, cooling, and temperature differentials for a section of mass concrete. As can be observed, if the forms (which are providing adequate insulation in this case) are removed too early, cracking will occur once the difference between interior and surface concrete temperatures exceeds the critical temperature differential of 20°C.

The maximum temperature rise can be estimated by methods previously mentioned, or by an approximation, if the concrete contains 300 to 600 kg of cement per cubic metre and the least dimension of the member is 1.8 m or more. This approximation (under normal, not adiabatic conditions) would be 12°C for every 100 kg of cement per cubic metre. For example, the maximum temperature of such an element made with concrete having 535 kg of cement per cubic metre and cast at 16°C would be about

$$16°C + (12°C \times 535/100) \text{ or } 80°C$$

The slow rate of heat exchange between concrete and its surroundings is caused by concrete's low thermal conductivity. Heat escapes from concrete at a rate that is inversely proportional to the square of its least dimension. A 150-mm thick wall cooling from both sides will take approximately 1½ hours to dissipate 95% of its developed heat. A 1.5-m thick wall would take an entire week to dissipate the same amount of heat (ACI 207). Inexpensive thermocouples can be used to monitor concrete temperature.

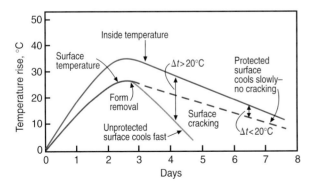

Fig. 18-10. Potential for surface cracking after form removal, assuming a critical temperature differential, Δ*t*, of 20°C. No cracking should occur if concrete is cooled slowly and Δ*t* is less than 20°C (Fintel and Ghosh 1978 and PCA 1987).

PREPLACED AGGREGATE CONCRETE

Preplaced aggregate concrete is produced by first placing coarse aggregate in a form and later injecting a cement-sand grout, usually with admixtures, to fill the voids. Properties of the resulting concrete are similar to those of comparable concrete placed by conventional methods; however, considerably less thermal and drying shrinkage can be expected because of the point-to-point contact of aggregate particles.

Coarse aggregates should meet requirements of CSA Standard A23.1 (ASTM C 33). In addition, most specifications limit both the maximum and minimum sizes; for example, 75-mm maximum and 13-mm minimum. Aggregates are generally graded to produce a void content of 35% to 40%. Fine aggregate used in the grout is generally graded to a fineness modulus of between 1.2 and 2.0, with nearly all of the material passing a 1.25 mm sieve.

Although the preplaced aggregate method has been used principally for restoration work and in the construction of reactor shields, bridge piers, and underwater structures, it has also been used in buildings to produce unusual architectural effects. Since the forms are completely filled with coarse aggregate prior to grouting, a dense and uniform exposed-aggregate facing is obtained when the surface is sandblasted, tooled, or retarded and wire-brushed at an early age.

Tests for preplaced aggregate concrete are given in ASTM C 937 through C 943. Preplaced aggregate concrete is discussed in more detail in ACI 304-00, *Guide for Measuring, Transporting, and Placing Concrete.*

NO-SLUMP CONCRETE

No-slump concrete is defined as concrete with a consistency corresponding to a slump of 6 mm or less. Such concrete, while very dry, must be sufficiently workable to be placed and consolidated with the equipment to be used on the job. The methods referred to here do not necessarily apply to mixtures for concrete masonry units or for compaction by spinning techniques.

Many of the basic laws governing the properties of higher-slump concretes are applicable to no-slump concrete. For example, the properties of hardened concrete depend principally on the ratio of water to cementing materials, provided the mix is properly consolidated.

Measurement of the consistency of no-slump concrete differs from that for higher-slump concrete because the slump cone is impractical for use with the drier consistencies. ACI 211.3, *Standard Practice for Selecting Proportions for No-Slump Concrete*, describes three methods for measuring the consistency of no-slump concrete: (1) the Vebe apparatus; (2) the compacting-factor test; and (3) the Thaulow drop table. In the absence of the above test equipment, workability can be adequately judged by a trial mixture that is placed and compacted with the equipment and methods to be used on the job.

Intentionally entrained air is recommended for no-slump concrete where durability is required. The amount of air-entraining admixture usually recommended for higher-slump concretes will not produce air contents in no-slump concretes that are as high as those in the higher-slump concretes. The lower volume of entrained air, however, generally provides adequate durability for no-slump concretes; while the volume of entrained air is not there, sufficient small air voids are present. This departure from the usual methods of designing and controlling entrained air is necessary for no-slump concretes.

For a discussion of water requirements and computation of trial mixtures, see ACI 211.3.

ROLLER-COMPACTED CONCRETE

Roller-compacted concrete (RCC) is a zero slump mixture of aggregate, cement, and water, (supplementary cementing materials such as fly ash also have been used) that is compacted in place by vibratory rollers or plate compaction equipment (Fig. 18-11). Cement contents range from 60 to 360 kg per cubic metre. Mixing is done in continuous flow pugmills, twin-shaft mixers, conventional batch mixers, tilting-drum truck mixers, and in some instances for small jobs, ready-mix trucks.

Fig. 18-11. Vibratory rollers are used to compact roller-compacted concrete. (64401)

Applications for RCC fall into two distinct categories—water control structures (dams) and pavements. While the same term is used to describe both types of concrete use, the design and construction processes are different.

Water Control Structures

RCC can be used for the entire dam structure, or as an overtopping protection on the upper section and on the downstream face. The nominal maximum aggregate size can range up to 150 mm. The zero slump mix is produced in a high-capacity central-mixing plant near the site and delivered by truck or by conveyor belt. Cement content is usually lower than that used in a conventional concrete mix, but similar to that of mass concrete. Compressive strengths ranging from 7 to 30 MPa have been obtained from roller-compacted concrete in dam projects. The RCC mix is transported by trucks and conveyor belts and spread by grader or bulldozer, followed by rolling with vibratory compactors. No forms are used. On some projects the upstream face is surfaced with higher strength conventional air-entrained concrete or precast panels for improved durability.

RCC dams have the advantage of allowing much steeper slopes on both faces than an earth fill dam. In addition to the advantage of using less material, the dam is completed and placed in service earlier, usually at a significant savings in overall cost compared to an earth fill structure.

Other water control RCC applications include use as an emergency spillway or overtopping protection for embankment dams, low permeable liner for settling ponds, bank protection, and grade control structure for channels and riverbeds.

Pavements

The uses for RCC paving range from pavements as thick as one metre for the mining industry to city streets, paved surfaces for composting operations, logging, truck staging areas, and warehouse floors. The procedures for construction of an RCC pavement require tighter control than for dam construction (Arnold and Zamensky 2000). Cement content is in the same range as conventional concrete, 300 to 350 kg/m³, and compressive strength is of the same order, 30 to 40 MPa. New mix designs with supplementary cementing materials have provided strengths in the 50 to 60 MPa range and above. The nominal maximum aggregate size is limited to 20 mm to provide a smooth, dense surface.

The zero slump mix is usually produced in a continuous flow pugmill mixer at production rates as high as 350 tonnes per hour. It is possible to mix RCC in a central batch plant, but the plant must be dedicated to RCC production exclusively, because the material tends to stick to the inside drums. Specifications usually require that the mix be transported, placed, and compacted within 60 minutes of the start of mixing; although ambient weather conditions may increase or decrease that time window.

RCC is typically placed in layers 125 to 250 mm in thickness using an asphalt-type paving machine. High-density paving equipment is preferred for layers thicker than 150 mm since the need for subsequent compaction by rollers is reduced. Where a design calls for pavement thickness greater than 250 mm, or the equipment being utilized is not able to properly compact 250 mm lifts, the RCC should be placed in multiple layers. In this type of construction, it is important that there be a minimum time delay in placing subsequent layers so that good bond is assured. Following placement by a paver, RCC can be compacted with a combination of vibratory steel-wheeled rollers and rubber-tired equipment.

Curing is vitally important in RCC pavement construction. The very low water content at the initial mixing stage means that an RCC mix will dry out very quickly once it is in place. Continuous water curing is the recommended method, although sprayed on asphaltic emulsion, plastic sheeting, and concrete curing compounds have been used in some cases. Pavement projects have had design compressive strengths of about 35 MPa with field strengths in the range of 35 to 70 MPa (Hansen 1987)

High-performance roller compacted concrete for areas subjected to high impact and abrasive loading were developed in the mid-1990's. These mixes are based on obtaining the optimum packing of the various sizes of aggregate particles, and the addition of silica fume to the mix by the use of a 10E-SF blended cement. Field test specimens having compressive strengths on the order of 55 MPa in 7 days are being achieved. (Marchard and others 1997 and Reid and others 1998).

ACI addresses roller-compacted concrete in two guides—ACI 207.5, *Roller Compacted Mass Concrete,* deals with RCC for water control structures and ACI 325.10, *Roller Compacted Concrete Pavements,* covers new developments in RCC pavements. Holderbaum and Schweiger (2000) provide a guide for developing RCC specifications and commentary.

SOIL-CEMENT

Soil-cement is a mixture of pulverized soil or granular material, cement, and water. Some other terms applied to soil-cement are "cement-treated base or subbase," "cement stabilization," "cement-modified soil," and "cement-treated aggregate." The mixture is compacted to a high density, and as the cement hydrates the material becomes hard and durable.

Soil-cement is primarily used as pavement base course for roads, streets, airports, and parking areas. A bituminous or portland cement concrete wearing course is usually placed over the base. Soil-cement is also used as slope protection for earth dams and embankments, reservoir and ditch linings, deep-soil mixing, and foundation stabilization (Fig. 18-12).

The soil material in soil-cement can be almost any combination of sand, silt, clay, and gravel or crushed stone. Local granular materials (such as slag, caliche, limerock, and scoria) plus a wide variety of waste materials (such as cinders, ash, and screenings from quarries and gravel pits) can be used to make soil-cement. Also, old granular-base

Fig. 18-12. A dozer places a plant-mixed soil-cement on a slope for construction a of wastewater lagoon. (59819)

roads, with or without their bituminous surfaces, can be recycled to make soil-cement.

Soil-cement should contain sufficient portland cement to resist deterioration from freeze-thaw and wet-dry cycling and sufficient moisture for maximum compaction. Cement contents range from 80 to 255 kg per cubic metre. The soil-cement and water are mixed in a central mix plant or in-place using transverse shaft mixers or travelling pugmills. There are four basic steps in constructing soil-cement by in-place methods: spreading cement, mixing, compaction, and curing. The proper quantity of cement must be spread on the in-place soil; the cement and soil material, and the necessary amount of water are mixed thoroughly using any of several types of travelling mixing machines; and finally, the mixture is compacted with conventional road-building equipment to 96% to 100% of maximum density (ASTM D 558) and PCA 1992. A light coat of bituminous material is commonly used to prevent moisture loss; it also helps bond and forms part of the bituminous surface. A common type of wearing surface for light traffic is one or two surface treatments of bituminous material and chips 13 to 20 mm thick. For heavy-duty use a 40-mm asphalt mat is used.

Depending on the soil used, the 7-day compressive strengths of saturated specimens at the minimum cement content meeting soil-cement criteria are generally between 2 to 5 MPa. Like concrete, soil-cement continues to gain strength with age; compressive strengths in excess of 17 MPa have been obtained after many years of service.

See ACI Committee 230 (1997) and PCA (1995) for detailed information on soil-cement construction.

SHOTCRETE

Shotcrete is mortar or small-aggregate concrete that is pneumatically projected onto a surface at high velocity

Fig. 18-13. Shotcrete. (70018)

(Fig. 18-13). Also known as "gunite" and "sprayed concrete," shotcrete was developed in 1911 and its concept is essentially unchanged even in today's use. The relatively dry mixture is consolidated by the impact force and can be placed on vertical or horizontal surfaces without sagging.

Shotcrete is applied by either a dry or wet process. In the dry process, a premixed blend of cement and damp aggregate is propelled through a hose by compressed air to a nozzle. Water is added to the cement and aggregate mixture at the nozzle and the intimately mixed ingredients are projected onto the surface. In the wet process, all the ingredients are premixed. The wet mixture is pumped through a hose to the nozzle, where compressed air is added to increase the velocity and propel the mixture onto the surface.

As the shotcrete mixture hits the surface, some coarser aggregates ricochet off the surface until sufficient paste builds up, providing a bed into which the aggregate can stick. To reduce overspray (mortar that attaches to nearby surfaces) and rebound (aggregate that ricochets off the receiving surface) the nozzle should be held at a 90° angle to the surface. The appropriate distance between nozzle and surface is usually between 0.5 m and 1.5 m, depending on the velocity.

Shotcrete is used for both new construction and repair work. It is especially suited for curved or thin concrete structures and shallow repairs, but can be used for thick members. The hardened properties of shotcrete are very operator dependent. Shotcrete has a density and compressive strength similar to normal- and high-strength concrete. Aggregate sizes up to 20 mm can be used, however most mixtures contain aggregates only up to 10 mm; 25% to 30% pea gravel are commonly used for wet mixes (Austin and Robbins 1995).

Supplementary cementing materials, such as fly ash and silica fume, can also be used in shotcrete. They improve workability, chemical resistance, and durability. The use of accelerating admixtures allows build-up of thicker layers of shotcrete in a single pass. They also reduce the time of initial set. However, using rapid-set accelerators often increases drying shrinkage and reduces later-age strength (Gebler and others 1992).

Steel fibres are used in shotcrete to improve flexural strength, ductility, and toughness; they can be used as a replacement for wire mesh reinforcement in applications like rock slope stabilization and tunnel linings (ACI 506.1R). Steel fibres can be added up to 2 percent by volume of the total mix. Polypropylene fibres if used are normally added to shotcrete at a rate of 0.9 to 2.7 kg/m^3, but dosages up to 9 kg/m^3 have also been used (The Aberdeen Group 1996).

Guidelines for the use of shotcrete are described in ACI 506R, ASCE (1995), and Balck and others (2000).

SHRINKAGE-COMPENSATING CONCRETE

Shrinkage-compensating concrete is concrete containing expansive cement or an expansive admixture, which produces expansion during hardening and thereby offsets the contraction occurring during drying (drying shrinkage). Shrinkage-compensating concrete is used in concrete slabs, pavements, structures, and repair work to minimize drying shrinkage cracks. Expansion of concrete made with shrinkage-compensating cement should be determined by the method specified in ASTM C 878.

Reinforcing steel in the structure restrains the concrete and goes into tension as the shrinkage compensating concrete expands. Upon shrinking due to drying contraction caused by moisture loss in hardened concrete, the tension in the steel is relieved; as long as the resulting tension in the concrete does not exceed the tensile strength of the concrete, no cracking should result. Shrinkage-compensating concrete can be proportioned, batched, placed, and cured similarly to normal concrete with some precautions; for example, it is necessary to assure the expected expansion by using additional curing. More information can be found in Chapter 2 and in ACI 223-98, *Standard Practice for the Use of Shrinkage-Compensating Concrete.*

PERVIOUS CONCRETE

Pervious (porous or no-fines) concrete contains a narrowly graded coarse aggregate, little to no fine aggregate, and insufficient cement paste to fill voids in the coarse aggregate. This low water-cementing materials ratio, low-slump concrete resembling popcorn is primarily held together by the paste at the contact points of the coarse aggregate particles; this produces a concrete with a high volume of voids

(20% to 35%) and a high permeability that allows water to flow through it easily.

Pervious concrete is used in hydraulic structures as drainage media, and in parking lots, pavements, and airport runways to reduce storm water run off. It also recharges the local groundwater supply by allowing water to penetrate the concrete to the ground below. Pervious concretes have also been used in tennis courts and greenhouses.

As a paving material, porous concrete is raked or slip-formed into place with conventional paving equipment and then roller compacted. Vibratory screeds or hand rollers can be used for smaller jobs. In order to maintain its porous properties, the surfaces of pervious concrete should not be closed up or sealed; therefore, troweling and finishing are not desired. The compressive strength of different mixes can range from 3.5 to 30 MPa. Drainage rates commonly range from 100 to 900 litres per minute per square metre.

No-fines concrete is used in building construction (particularly walls) for its thermal insulating properties. For example, a 250-mm thick porous-concrete wall can have an *RSI* value of 0.9 compared to 0.125 for normal concrete. No-fines concrete is also low-density, 1600 to 1900 kg/m^3, and has low drying shrinkage properties (Malhotra 1976 and Concrete Construction 1983).

WHITE AND COLOURED CONCRETE

White Concrete

White portland cement is used to produce white concrete, a widely used architectural material (Fig. 18-14). It is also used in mortar, plaster, stucco, terrazzo, and portland cement paint. White portland cement is manufactured from raw materials of low iron content; it conforms to CSA Standard A5 (ASTM C 150) even though these specifications do not specifically mention white portland cement.

White concrete is made with aggregates and water that contain no materials that will discolour the concrete. White or light-coloured aggregates can be used. Oil that could stain concrete should not be used on the forms. Care must be taken to avoid rust stains from tools and equipment. Curing

Fig. 18-14. Office building constructed with white cement concrete. (55828)

materials that could cause stains must be avoided. Refer to Farny (2001) and http://www.portcement.org/white for more information.

Coloured Concrete

Coloured concrete can be produced by using coloured aggregates or by adding colour pigments (ASTM C 979) or both. When coloured aggregates are used, they should be exposed at the surface of the concrete. This can be done several ways; for example, casting against a form that has been treated with a retarder. Unhydrated paste at the surface is later brushed or washed away. Other methods involve removing the surface mortar by sandblasting, waterblasting, bushhammering, grinding, or acid washing. If surfaces are to be washed with acid, a delay of approximately two weeks after casting is necessary. Coloured aggregates may be natural rock such as quartz, marble, and granite, or they may be ceramic materials.

Pigments for colouring concrete should be pure mineral oxides ground finer than cement; they should be insoluble in water, free of soluble salts and acids, colourfast in sunlight, resistant to alkalies and weak acids, and virtually free of calcium sulphate. Mineral oxides occur in nature and are also produced synthetically; synthetic pigments generally give more uniform results.

The amount of colour pigments added to a concrete mixture should not be more than 10% of the mass of the cement. The amount required depends on the type of pigment and the colour desired. For example, a dose of pigment equal to 1.5% by mass of cement may produce a pleasing pastel colour, but 7% may be needed to produce a deep colour. Use of white portland cement with a pigment will produce cleaner, brighter colours and is recommended in preference to gray cement, except for black or dark gray colours (Fig. 18-15).

To maintain uniform colour, do not use calcium chloride, and batch all materials carefully by mass. To prevent streaking, the dry cement and colour pigment must be thoroughly blended before they are added to the mixer. Mixing time should be longer than normal to ensure uniformity.

In air-entrained concrete, the addition of pigment may require an adjustment in the amount of air-entraining admixture to maintain the desired air content.

Dry-Shake Method. Slabs or precast panels that are cast horizontally can be coloured by the dry-shake method. Prepackaged, dry colouring materials consisting of mineral oxide pigment, white portland cement, and specially graded silica sand or other fine aggregate are marketed ready for use by various manufacturers.

After the slab has been bullfloated once, two-thirds of the dry colouring material should be broadcast evenly by hand over the surface. The required amount of colouring material can usually be determined from previously cast sections. After the material has absorbed water from the

Fig. 18-15. Coloured concrete, used for a rapid transit platform. (67328)

fresh concrete, it should be floated into the surface. Then the rest of the material should be applied immediately at right angles to the initial application, so that a uniform colour is obtained. The slab should again be floated to work the remaining material into the surface.

Other finishing operations may follow depending on the type of finish desired. *Exterior concrete should not be troweled because troweled surfaces can be slippery when wet and troweling can lead to a loss of entrained air by overworking the surface. Floating and brooming should be sufficient for exterior flatwork.* Curing should begin immediately after finishing; take precautions to prevent discolouring the surface. See Kosmatka (1991) for more information.

POLYMER-PORTLAND CEMENT CONCRETE

Polymer-portland cement concrete (PPCC), also called polymer-modified concrete, is basically normal portland cement concrete to which a polymer or monomer has been added during mixing to improve durability and adhesion. Thermoplastic and elastomeric latexes are the most commonly used polymers in PPCC, but epoxies and other polymers are also used. In general, latex improves

ductility, durability, adhesive properties, resistance to chloride-ion ingress, shear bond, and tensile and flexural strength of concrete and mortar. Latex-modified concretes (LMC) also have excellent freeze-thaw, abrasion, and impact resistance. Some LMC materials can also resist certain acids, alkalies, and organic solvents. Polymer-portland cement concrete is primarily used in concrete patching and overlays, especially bridge decks. See ACI 548.3R for more information on polymer-modified concrete and ACI 548.4 for LMC overlays.

FERROCEMENT

Ferrocement is a special type of reinforced concrete composed of closely spaced layers of continuous relatively thin metallic or nonmetallic mesh or wire embedded in mortar. It is constructed by hand plastering, shotcreting, laminating (forcing the mesh into fresh mortar), or a combination of these methods.

The mortar mixture generally has a sand-cement ratio of 1.5 to 2.5 and a water-cementing materials ratio of 0.35 to 0.50. Reinforcement makes up about 5% to 6% of the ferrocement volume. Fibres and admixtures may also be used to modify the mortar properties. Polymers or cement-based coatings are often applied to the finished surface to reduce porosity.

Ferrocement is considered easy to produce in a variety of shapes and sizes; however, it is labour intensive. Ferrocement is used to construct thin shell roofs, swimming pools, tunnel linings, silos, tanks, prefabricated houses, barges, boats, sculptures, and thin panels or sections usually less than 25 mm thick (ACI 549R and ACI 549.1R).

REFERENCES

ACI Committee 116, *Cement and Concrete Technology*, ACI 116R-00, ACI Committee 116 Report, American Concrete Institute, Farmington Hills, Michigan, 2000, 73 pages.

ACI Committee 207, *Cooling and Insulating Systems for Mass Concrete*, ACI 207.4R-93, reapproved 1998, ACI Committee 207 Report, American Concrete Institute, Farmington Hills, Michigan, 1998, 22 pages.

ACI Committee 207, *Effect of Restraint, Volume Change, and Reinforcement on Cracking of Mass Concrete*, ACI 207.2R-95, ACI Committee 207 Report, American Concrete Institute, Farmington Hills, Michigan, 1995, 26 pages.

ACI Committee 207, *Mass Concrete*, ACI 207.1R-96, ACI Committee 207 Report, American Concrete Institute, Farmington Hills, Michigan, 1996, 42 pages.

ACI Committee 207, *Roller-Compacted Mass Concrete*, ACI 207.5R-99, ACI Committee 207 Report, American Concrete Institute, Farmington Hills, Michigan, 1999, 46 pages.

ACI Committee 211, *Guide for Selecting Proportions for No-Slump Concrete,* ACI 211.3R-97, ACI Committee 211 Report, American Concrete Institute, Farmington Hills, Michigan, 1997, 26 pages.

ACI Committee 211, *Standard Practice for Selecting Proportions for Structural Lightweight Concrete,* ACI 211.2-98, ACI Committee 211 Report, American Concrete Institute, Farmington Hills, Michigan, 1998, 18 pages.

ACI Committee 211, *Standard Practice for Selecting Proportions for Normal, Heavyweight, and Mass Concrete,* ACI 211.1-91, reapproved 1997, ACI Committee 211 Report, American Concrete Institute, Farmington Hills, Michigan, 1997, 38 pages.

ACI Committee 213, *Guide for Structural Lightweight Aggregate Concrete,* ACI 213R-87, reapproved 1999, ACI Committee 213 Report, American Concrete Institute, Farmington Hills, Michigan, 1999, 27 pages.

ACI Committee 223, *Standard Practice for the Use of Shrinkage-Compensating Concrete,* ACI 223-98, ACI Committee 223 Report, American Concrete Institute, Farmington Hills, Michigan, 1998 28 pages.

ACI Committee 224, *Control of Cracking in Concrete Structures,* ACI 224R-90, ACI Committee 224 Report, American Concrete Institute, Farmington Hills, Michigan, 1990, 43 pages.

ACI Committee 230, *State-of-of-the-Art Report on Soil Cement,* ACI 230.1R-90, reapproved 1997, ACI Committee 230 Report, American Concrete Institute, Farmington Hills, Michigan, 1997, 23 pages. Also available through PCA as LT120.

ACI Committee 304, *Guide for Measuring, Mixing, Transporting, and Placing Concrete,* ACI 304R-00, ACI Committee 304 Report, American Concrete Institute, Farmington Hills, Michigan, 2000, 41 pages.

ACI Committee 308, *Standard Practice for Curing Concrete, ACI 308-92 reapproved 1997, ACI Committee 308* Report, American Concrete Institute, Farmington Hills, Michigan, 1997, 11 pages.

ACI Committee 318, *Building Code Requirements for Structural Concrete and Commentary,* ACI 318-99, ACI Committee 318 Report, American Concrete Institute, Farmington Hills, Michigan, 1999, 369 pages.

ACI Committee 325, *State-of-the-Art Report on Roller-Compacted Concrete Pavements,* ACI 325.10R-95, ACI Committee 325 Report, American Concrete Institute, Farmington Hills, Michigan, 1995, 32 pages.

ACI Committee 363, *State-of-the-Art Report on High-Strength Concrete,* ACI 363R-92, reapproved 1997, ACI Committee 363 Report, American Concrete Institute, Farmington Hills, Michigan, 1997, 55 pages.

ACI Committee 506, *Committee Report on Fiber Reinforced Shotcrete,* ACI 506.1R-98, ACI Committee 506 Report, American Concrete Institute, Farmington Hills, Michigan, 1998, 11 pages.

ACI Committee 506, *Guide to Shotcrete,* ACI 506R-90, re-approved 1995, ACI Committee 506 Report, American Concrete Institute, Farmington Hills, Michigan, 1995, 41 pages.

ACI Committee 523, *Guide for Cast-in-Place Low-Density Concrete,* ACI 523.1R-92 ACI Committee 523 Report, American Concrete Institute, Farmington Hills, Michigan, 1992, 8 pages.

ACI Committee 523, *Guide for Precast Cellular Concrete Floor, Roof, and Wall Units,* ACI 523.2R-96, ACI Committee 523 Report, American Concrete Institute, Farmington Hills, Michigan, 1996, 5 pages.

ACI Committee 523, *Guide for Cellular Concretes Above 50 pcf and for Aggregate Concretes Above 50 pcf with Compressive Strengths Less Than 2500 psi,* ACI 523.3R-93, ACI Committee 523 Report, American Concrete Institute, Farmington Hills, Michigan, 1993, 16 pages.

ACI Committee 544, *Guide for Specifying, Proportioning, Mixing, Placing, and Finishing Steel Fiber Reinforced Concrete,* ACI 544.3R-93, reapproved 1998, ACI Committee 544 Report, American Concrete Institute, Farmington Hills, Michigan, 1998, 10 pages.

ACI Committee 544, *State-of-the-Art Report on Fiber Reinforced Concrete,* ACI 544.1R-96, ACI Committee 544 Report, American Concrete Institute, Farmington Hills, Michigan, 1996, 66 pages.

ACI Committee 548, *Guide for the Use of Polymers in Concrete,* ACI 548.1R-97, ACI Committee 548 Report, American Concrete Institute, Farmington Hills, Michigan, 1997, 29 pages.

ACI Committee 548, *Guide for Mixing and Placing Sulfur Concrete in Construction,* ACI 548.2R-93, reapproved 1998, ACI Committee 548 Report, American Concrete Institute, Farmington Hills, Michigan, 1998, 12 pages.

ACI Committee 548, *State-of-the-Art Report on Polymer-Modified Concrete,* ACI 548.3R-95, ACI Committee 548 Report, American Concrete Institute, Farmington Hills, Michigan, 1995, 47 pages.

ACI Committee 548, *Standard Specification for Latex-Modified Concrete (LMC) Overlays,* ACI 548.4-93, reapproved 1998, ACI Committee 548 Report, American Concrete Institute, Farmington Hills, Michigan, 1998, 6 pages.

ACI Committee 549, *State-of-the-Art Report on Ferrocement,* ACI 549R-97, ACI Committee 549 Report, American Concrete Institute, Farmington Hills, Michigan, 1997.

ACI Committee 549, *Guide for the Design, Construction, and Repair of Ferrocement*, ACI 549.1R-93, reapproved 1999, ACI Committee 549 Report, American Concrete Institute, Farmington Hills, Michigan, 1999, 30 pages.

ACI, *Concrete for Nuclear Reactors*, SP-34, three volumes, 73 papers, American Concrete Institute, Farmington Hills, Michigan, 1972, 1766 pages.

ACI, *Concrete for Radiation Shielding*, Compilation No. 1, American Concrete Institute, Farmington Hills, Michigan, 1962.

ACI, *Fiber Reinforced Concrete*, SP-81, American Concrete Institute, Farmington Hills, Michigan, 1984, 460 pages.

Arnold, Terry, and Zamensky, Greg, *Roller-Compacted Concrete: Quality Control Manual*, EB215, Portland Cement Association, 2000, 58 pages.

ASCE, *Standard Practice for Shotcrete*, American Society of Civil Engineers, New York, 1995, 65 pages.

Austin, Simon, and Robbins, Peter, *Sprayed Concrete: Properties, Design and Application*, Whittles Publishing, U.K., 1995, 382 pages.

Balck, Lars; Gebler, Steven; Isaak, Merlyn; and Seabrook Philip, *Shotcrete for the Craftsman (CCS-4)*, American Concrete Institute, Farmington Hills, Michigan, 2000, 59 pages. Also available from PCA as LT242.

Bamforth, P. B., "Large Pours," letter to the editor, *Concrete*, Cement and Concrete Association, Wexham Springs, Slough, England, February 1981.

Brewer, Harold W., *General Relation of Heat Flow Factors to the Unit Weight of Concrete*, Development Department Bulletin DX114, Portland Cement Association, http://www.portcement.org/pdf_files/DX114.pdf, 1967.

Bureau of Reclamation, *Concrete Manual*, 8th ed., U.S. Department of the Interior, Bureau of Reclamation, Denver, 1981.

Burg, R. G., and Fiorato, A. E., *High-Strength Concrete in Massive Foundation Elements*, Research and Development Bulletin RD117, Portland Cement Association, 1999, 22 pages.

Carlson, Roy W., and Thayer, Donald P., "Surface Cooling of Mass Concrete to Prevent Cracking," *Journal of the American Concrete Institute*, American Concrete Institute, Farmington Hills, Michigan, August 1959, page 107.

Cordon, William A., and Gillespie, Aldridge H., "Variables in Concrete Aggregates and Portland Cement Paste which Influence the Strength of Concrete," *Proceedings of the American Concrete Institute*, vol. 60, no. 8, American Concrete Institute, Farmington Hills, Michigan, August 1963, pages 1029 to 1050; and Discussion, March 1964, pages 1981 to 1998.

Farny, J. A., *White Cement Concrete*, EB217, Portland Cement Association, 2001.

Fintel, Mark, and Ghosh, S. K., "Mass Reinforced Concrete Without Construction Joints," presented at the Adrian Pauw Symposium on Designing for Creep and Shrinkage, Fall Convention of the American Concrete Institute, Houston, Texas, November 1978.

FitzGibbon, Michael E., "Large Pours for Reinforced Concrete Structures," Current Practice Sheets No. 28, 35, and 36, *Concrete*, Cement and Concrete Association, Wexham Springs, Slough, England, March and December 1976 and February 1977.

Gebler, Steven H.; Litvin, Albert; McLean, William J.; and Schutz, Ray, "Durability of Dry-Mix Shotcrete Containing Rapid-Set Accelerators," *ACI Materials Journal*, May-June 1992, pages 259 to 262.

Gustaferro, A. H.; Abrams, M. S.; and Litvin, Albert, *Fire Resistance of Lightweight Insulating Concretes*, Research and Development Bulletin RD004, Portland Cement Association, http://www.portcement.org/pdf_files/RD004.pdf, 1970.

Hanna, Amir N., *Properties of Expanded Polystyrene Concrete and Applications for Pavement Subbases*, Research and Development Bulletin RD055, Portland Cement Association, http://www.portcement.org/pdf_files/RD055.pdf, 1978.

Hansen, Kenneth D., "A Pavement for Today and Tomorrow," *Concrete International*, American Concrete Institute, Farmington Hills, Michigan, February 1987, pages 15 to 17.

Holderbaum, Rodney E., and Schweiger, Paul G., *Guide for Developing RCC Specifications and Commentary*, EB214, Portland Cement Association, 2000, 80 pages.

Klieger, Paul, "Proportioning No-Slump Concrete," *Proportioning Concrete Mixes*, SP-46, American Concrete Institute, Farmington Hills, Michigan, 1974, pages 195 to 207.

Klieger, Paul, *Early-High-Strength Concrete for Prestressing*, Research Department Bulletin RX091, Portland Cement Association, http://www.portcement.org/pdf_files/RX091.pdf, 1958.

Kosmatka, Steven H., *Finishing Concrete Slabs with Color and Texture*, PA124, Portland Cement Association, 1991, 40 pages.

Malhotra, V. M., "No-Fines Concrete—Its Properties and Applications," *Journal of the American Concrete Institute*, American Concrete Institute, Farmington Hills, Michigan, November 1976.

Marchand, J.; Gagne, R.; Ouellet, E.; and Lepage, S., "Mixture Proportioning of Roller Compacted Concrete—A Review," *Proceedings of the Third CANMET/ACI International Conference,* Auckland, New Zealand, Available as: *Advances in Concrete Technology,* SP 171, American Concrete Institute, Farmington Hills, Michigan, 1997, pages 457 to 486.

Mather, K., "High-Strength, High-Density Concrete," *Proceedings of the American Concrete Institute,* vol. 62, American Concrete Institute, Farmington Hills, Michigan, August 1965, pages 951 to 960.

McCormick, Fred C., "Rational Proportioning of Preformed Foam Cellular Concrete," *Journal of the American Concrete Institute,* American Concrete Institute, Farmington Hills, Michigan, February 1967, pages 104 to 110.

PCA, *Concrete Energy Conservation Guidelines,* EB083, Portland Cement Association, 1980.

PCA, *Concrete for Massive Structures,* IS128, Portland Cement Association, http://www.portcement.org/pdf_files/IS128.pdf, 1987, 24 pages.

PCA, *Fiber Reinforced Concrete,* SP039, Portland Cement Association, 1991, 54 pages.

PCA, *Soil-Cement Construction Handbook,* EB003, Portland Cement Association, 1995, 40 pages.

PCA, *Soil-Cement Laboratory Handbook,* EB052, Portland Cement Association, 1992, 60 pages.

PCA, *Structural Lightweight Concrete,* IS032, Portland Cement Association, http://www.portcement.org/pdf_files/IS032.pdf, 1986, 16 pages.

"Porous Concrete Slabs and Pavement Drain Water," *Concrete Construction,* Concrete Construction Publications, Inc., Addison, Illinois, September 1983, pages 685 and 687 to 688.

Reichard, T. W., "Mechanical Properties of Insulating Concretes," *Lightweight Concrete,* American Concrete Institute, Farmington Hills, Michigan, 1971, pages 253 to 316.

Reid, E., and Marchand, J., "High-Performance Roller Compacted Concrete Pavements: Applications and Recent Developments", *Proceedings of the Canadian Society for Civil Engineering 1998 Annual Conference,* Halifax, Nova Scotia, 1998.

The Aberdeen Group, *Shotcreting—Equipment, Materials, and Applications,* Addison, Illinois, 1996, 46 pages.

Townsend, C. L., "Control of Cracking in Mass Concrete Structures," *Engineering Monograph No. 34,* U.S. Department of Interior, Bureau of Reclamation, Denver, 1965.

Tuthill, Lewis H., and Adams, Robert F., "Cracking Controlled in Massive, Reinforced Structural Concrete by Application of Mass Concrete Practices," *Journal of the American Concrete Institute,* American Concrete Institute, Farmington Hills, Michigan, August 1972, page 481.

Valore, R. C., Jr., "Cellular Concretes, Parts 1 and 2," *Proceedings of the American Concrete Institute,* vol. 50, American Concrete Institute, Farmington Hills, Michigan, May 1954, pages 773 to 796 and June 1954, pages 817 to 836.

Valore, R. C., Jr., "Insulating Concrete," *Proceedings of the American Concrete Institute,* vol. 53, American Concrete Institute, Farmington Hills, Michigan, November 1956, pages 509 to 532.

VanGeem, Martha G.; Litvin, Albert; and Musser, Donald W., *Insulative Lightweight Concrete for Building Walls,* presented at the Annual Convention and Exposition of the American Society of Civil Engineers in Detroit, October 1985.

Walker, Stanton, and Bloem, Delmar L., "Effects of Aggregate Size on Properties of Concrete," *Proceedings of the American Concrete Institute,* vol. 57, no. 3, American Concrete Institute, Farmington Hills, Michigan, September 1960, pages 283 to 298, and Discussion by N. G. Zoldner, March 1961, pages 1245 to 1248.

Appendix

GLOSSARY OF TERMS USED IN DESIGN AND CONTROL OF CONCRETE MIXTURES

The intent of this *Glossary* is to clarify terminology used in concrete construction, with special emphasis on those terms used in *Design and Control of Concrete Mixtures.* Additional terminology that may not be in this book is included in the *Glossary* for the convenience of our readers. Other sources for terms include CSA Standard A23.1, ACI Committee 116, and ASTM standards.

A

Absorption—see *Water absorption.*

Accelerating admixture—admixture that speeds the rate of hydration of hydraulic cement, shortens the normal time of setting, or increases the rate of hardening, of strength development, or both, of portland cement, concrete, mortar, grout, or plaster.

Addition—substance that is interground or blended in limited amounts into a hydraulic cement during manufacture—not at the jobsite—either as a "processing addition" to aid in manufacture and handling of the cement or as a "functional addition" to modify the useful properties of the cement.

Admixture—material, other than water, aggregate, and hydraulic cement, used as an ingredient of concrete, mortar, grout, or plaster and added to the batch immediately before or during mixing.

Aggregate—granular mineral material such as natural sand, manufactured sand, gravel, crushed stone, air-cooled blast-furnace slag, vermiculite, or perlite.

Air content—total volume of air voids, both entrained and entrapped, in cement paste, mortar, or concrete. Entrained air adds to the durability of hardened mortar or concrete and the workability of fresh mixtures.

Air entrainment—intentional introduction of air in the form of minute, disconnected bubbles (generally smaller than 1 mm) during mixing of portland cement concrete, mortar, grout, or plaster to improve desirable characteristics such as cohesion, workability, and durability.

Air-entraining admixture—admixture for concrete, mortar, or grout that will cause air to be incorporated into the mixture in the form of minute bubbles during mixing, usually to increase the material's workability and frost resistance.

Air-entraining portland cement—portland cement containing an air-entraining addition added during its manufacture.

Air void—entrapped air pocket or an entrained air bubble in concrete, mortar, or grout. Entrapped air voids usually are larger than 1 mm in diameter; entrained air voids are smaller. Most of the entrapped air voids should be removed with internal vibration, power screeding, or rodding.

Alkali-aggregate reactivity—production of expansive gel caused by a reaction between aggregates containing certain forms of silica or carbonates and alkali hydroxides in concrete.

Architectural concrete—concrete that will be permanently exposed to view and which therefore requires special care in selection of concrete ingredients, forming, placing, consolidating, and finishing to obtain the desired architectural appearance.

Autoclaved cellular concrete—concrete containing very high air content resulting in low density, and cured at high temperature and pressure in an autoclave.

B

Batching—process of determining quantities either by mass or volumetric measurement and introducing into the mixer the ingredients for a batch of concrete, mortar, grout, or plaster.

Blast-furnace slag—nonmetallic byproduct of steel manufacturing, consisting essentially of silicates and aluminum silicates of calcium that are developed in a molten condition simultaneously with iron in a blast furnace.

Bleeding—flow of mixing water from a newly placed concrete mixture caused by the settlement of the solid materials in the mixture.

nded hydraulic cement—cement containing ...binations of portland cement, pozzolans, slag, and/or other hydraulic cement.

Bulking—increase in volume of a quantity of sand when in a moist condition compared to its volume when in a dry state.

━━━━━━━━━ **C** ━━━━━━━━━

Calcined clay—clay heated to high temperature to alter its physical properties for use as a pozzolan or cementing material in concrete.

Calcined shale—shale heated to high temperature to alter its physical properties for use as a pozzolan or cementing material in concrete.

Carbonation—reaction between carbon dioxide and a hydroxide or oxide to form a carbonate.

Cellular concrete—high air content or high void ratio concrete resulting in low density.

Cement—see *Portland cement* and *Hydraulic cement.*

Cement paste—constituent of concrete, mortar, grout, and plaster consisting of cement and water.

Cementing material—any material having cementing properties or contributing to the formation of hydrated calcium silicate compounds. When proportioning concrete, the following are considered cementing materials: portland cement, blended hydraulic cement, fly ash, ground granulated blast-furnace slag, silica fume, calcined clay, metakaolin, calcined shale, and rice husk ash.

Chemical admixture—see *Admixture.*

Chemical bond—bond between materials resulting from cohesion and adhesion developed by chemical reaction.

Clinker—end product of a portland cement kiln; raw cementing material prior to grinding

Chloride (attack)—chemical compounds containing chloride ions, which promote the corrosion of steel reinforcement. Chloride deicing chemicals are primary sources.

Coarse aggregate—natural gravel, crushed stone, or iron blast-furnace slag, usually larger than 5 mm and commonly ranging in size between 10 mm and 40 mm.

Cohesion—mutual attraction by which elements of a substance are held together.

Coloured concrete—concrete containing white cement and/or mineral oxide pigments to produce colours other than the normal gray hue of traditional gray cement concrete.

Compaction—process of inducing a closer arrangement of the solid particles in freshly mixed and placed concrete, mortar, or grout by reduction of voids, usually by vibration, tamping, rodding, puddling, or a combination of these techniques. Also called consolidation.

Compressive strength—maximum resistance that a concrete, mortar, or grout specimen will sustain when loaded axially in compression in a testing machine at a specified rate; expressed as force per unit of cross sectional area, such as megapascals (MPa).

Concrete—mixture of binding materials and coarse and fine aggregates. Portland cement and water are commonly used as the binding medium for normal concrete mixtures, but may also contain pozzolans, slag, and/or chemical admixtures.

Consistency—degree of fluidity of freshly mixed concrete, mortars, or grout. (Also see *Slump and Workability.*)

Construction joint—a stopping place in the process of construction. A true construction joint allows for bond between new concrete and existing concrete and permits no movement. In structural applications their location must be determined by the structural engineer. In slab on grade applications, construction joints are often located at contraction (control) joint locations and are constructed to allow movement and perform as contraction joints.

Contraction joint (control joint)—provides for movement in the plane of the slab or wall and to induce cracking caused by drying and thermal shrinkage at preselected locations. Joint may be grooved, sawed, or formed.

Corrosion—deterioration of metal by chemical, electrochemical, or electrolytic reaction.

Creep—time-dependent deformation of concrete, or of any material, due to a sustained load.

Curing—maintenance of a satisfactory moisture content and temperature in concrete, mortar, grout, or plaster for a suitable period of time during its early stages (immediately following placing and finishing) so that the desired properties of the material can develop. Curing assures satisfactory hydration and hardening of the cementing materials.

━━━━━━━━━ **D** ━━━━━━━━━

Dampproofing—treatment of concrete, mortar, grout, or plaster to retard the passage or absorption of water, or water vapor.

Density—mass per unit volume; expressed in kg/m³.

Durability—ability of portland cement concrete, mortar, grout, or plaster to resist weathering action and other conditions of service, such as chemical attack, freezing and thawing, and abrasion.

━━━━━━━━━ **E** ━━━━━━━━━

Early stiffening—rapidly developing rigidity in freshly mixed hydraulic cement paste, mortar, grout, plaster, or concrete.

Entrapped air—irregularly shaped, unintentional air voids in fresh or hardened concrete 1 mm or larger in size.

Entrained air—spherical microscopic air bubbles—from 10 µm to 1000 µm in diameter—majority from 10 µm to 100 µm—intentionally incorporated into concrete to provide resistance to freezing and thawing when exposed to water and deicing chemicals and/or improve workability.

Epoxy resin—class of organic chemical bonding systems used in the preparation of special coatings or adhesives for concrete or masonry or as binders in epoxy-resin mortars and concretes.

Ettringite—needle like crystalline compound produced by the reaction of C_3A, gypsum, and water within a portland cement concrete.

Expansion joint—a separation provided between adjoining parts of a structure to allow movement.

F

Ferrocement—one or more layers of steel or wire reinforcement encased in portland cement mortar creating a thin-section composite material.

Fibres—thread or thread like material ranging from 0.05 to 4 mm in diameter and from 10 to 150 mm in length and made of steel, glass, synthetic (plastic), carbon, or natural materials.

Fibre concrete—concrete containing randomly oriented fibres in 2 or 3 dimensions through out the concrete matrix.

Fine aggregate—aggregate that passes the 10-mm sieve, almost entirely passes the 5-mm sieve, and is predominantly retained on the 80-µm sieve.

Fineness modulus (FM)—factor obtained by adding the cumulative percentages of material in a sample of aggregate retained on each of a specified series of sieves and dividing the sum by 100.

Finishing—mechanical operations like screeding, consolidating, floating, troweling, or texturing that establish the final appearance and texture of any concrete surface.

Fire resistance—that property of a building material, element, or assembly to withstand fire or give protection from fire; it is characterized by the ability to confine a fire or to continue to perform a given structural function during a fire, or both.

Flexural strength—ability of solids to resist bending.

Fly ash—residue from coal combustion, which is carried in flue gases, and is used as a pozzolan or cementing material in concrete.

Freeze-thaw resistance—ability of concrete to withstand cycles of freezing and thawing. (See also *Air entrainment* and *Air-entraining admixture*.)

Fresh concrete—concrete that has been recently mixed and is still workable and plastic.

G

Grading—size distribution of aggregate particles, determined by separation with standard screen sieves.

Grout—mixture of cementing material with or without aggregate or admixtures to which sufficient water is added to produce a pouring or pumping consistency without segregation of the constituent materials.

H

Hardened concrete—concrete that is in a solid state and has developed a certain strength.

High-density concrete—normally designed by the use of high-density aggregates and having an air-dry density exceeding 2500 kg/m³.

High-strength concrete—concrete with a design strength of at least 70 MPa.

Honeycomb—term that describes the resulting condition when mortar fails to completely surround coarse aggregates in concrete, leaving empty spaces (voids) between them.

Hydrated lime—dry powder obtained by treating quicklime with sufficient water to satisfy its chemical affinity for water; consists essentially of calcium hydroxide or a mixture of calcium hydroxide and magnesium oxide or magnesium hydroxide, or both.

Hydration—in concrete, mortar, grout, and plaster, the chemical reaction between hydraulic cement and water in which new compounds with strength-producing properties are formed.

Hydraulic cement—cement that sets and hardens by chemical reaction with water, and is capable of doing so under water. (See also *Portland cement*.)

I

Isolation Joint—a joint that allows relative movement to take place between adjoining parts of a structure to prevent cracking and spalling of the concrete.

J

Joint—see *Construction joint, Contraction joint, Isolation joint,* and *Expansion joint*.

K

Kiln—rotary furnace used in cement manufacture to heat and chemically combine raw inorganic materials, such as limestone, sand and clay, into calcium silicate clinker.

L

Lime—general term that includes the various chemical and physical forms of quicklime, hydrated lime, and hydraulic lime. It may be high-calcium, magnesian, or dolomitic.

Low-density aggregate—used to produce low-density concrete. Could be expanded or sintered clay, slate, diatomaceous shale, perlite, vermiculite, or slag; natural pumice, scoria, volcanic cinders, tuff, or diatomite; sintered fly ash or industrial cinders.

-density concrete—having an oven dry density 240 to 1850 kg/m³ and a compressive strength ranging from 0.7 MPa to 40 MPa. Structural low-density concrete has a 28-day compressive strength in excess of 20 MPa and an air-dry density not exceeding 1850 kg/m³. Structural semi-low-density concrete has a specified compressive strength of 20 MPa or greater and an air-dry density between 1850 and 2150 kg/m³.

M

Masonry—concrete masonry units, clay brick, structural clay tile, stone, terra cotta, and the like, or combinations thereof, bonded with mortar, dry-stacked, or anchored with metal connectors to form walls, building elements, pavements, and other structures.

Masonry cement—hydraulic cement, primarily used in masonry and plastering construction, consisting of a mixture of portland or blended hydraulic cement and plasticizing materials (such as limestone, hydrated or hydraulic lime) together with other materials introduced to enhance one or more properties such as setting time, workability, water retention, and durability.

Mass concrete—cast-in-place concrete in volume large enough to require measures to compensate for volume change caused by temperature rise from heat of hydration in order to keep cracking to a minimum.

Metakaolin—highly reactive pozzolan made from kaolin clays.

Mineral admixtures—see *Supplementary cementing materials.*

Modulus of elasticity—ratio of normal stress to corresponding strain for tensile or compressive stress below the proportional limit of the material; also referred to as elastic modulus, Young's modulus, and Young's modulus of elasticity; denoted by the symbol E.

Moist-air curing—curing with moist air (no less than 95% relative humidity) at atmospheric pressure and a temperature of about 23°C.

Mortar—mixture of cementing materials, fine aggregate, and water, which may contain admixtures, and is usually used to bond masonry units.

Mortar cement—hydraulic cement, primarily used in masonry construction, consisting of a mixture of portland or blended hydraulic cement and plasticizing materials (such as limestone, hydrated or hydraulic lime) together with other materials introduced to enhance one or more properties such as setting time, workability, water retention, and durability. Mortar cement and masonry cement are similar in use and function. However, specifications for mortar cement usually require lower air contents and they include a flexural bond strength requirement.

N

Normal-density concrete—class of concrete made with normal density aggregates, usually crushed stone or gravel, having a density between 2150 and 2500 kg/m³ (approximate value of 2400 kg/m³ often quoted). (See also *Low-density concrete* and *High-density concrete.*)

No-slump concrete—concrete having a slump of less than 6 mm.

O

Overlay—layer of concrete or mortar placed on or bonded to the surface of an existing pavement or slab. Normally done to repair a worn or cracked surface. Overlays are seldom less than 25 mm thick.

P

Pavement (concrete)—highway, road, street, path, or parking lot surfaced with concrete. Although typically applied to surfaces that are used for travel, the term also applies to storage areas and playgrounds.

Permeability—property of allowing passage of fluids or gases.

Pervious concrete (no-fines or porous concrete)—concrete containing insufficient fines or no fines to fill the voids between aggregate particles in a concrete mixture. The coarse aggregate particles are coated with a cement and water paste to bond the particles at their contact points. The resulting concrete contains an interconnected pore system allowing storm water to drain through the concrete to the subbase below.

pH—chemical symbol for the logarithm of the reciprocal of hydrogen ion concentration in gram atoms per litre, used to express the acidity or alkalinity (base) of a solution on a scale of 0 to 14, where less than 7 represents acidity, and more than 7 alkalinity

Plastic cement—special hydraulic cement product manufactured for plaster and stucco application. One or more inorganic plasticizing agents are interground or blended with the cement to increase the workability and molding characteristics of the resultant mortar, plaster, or stucco.

Plasticity—that property of freshly mixed cement paste, concrete, mortar, grout, or plaster that determines its workability, resistance to deformation, or ease of molding.

Plasticizer—admixture that increases the plasticity of portland cement concrete, mortar, grout, or plaster.

Polymer-portland cement concrete—fresh portland cement concrete to which a polymer is added for improved durability and adhesion characteristics, often used in overlays for bridge decks; also referred to as polymer-modified concrete and latex-modified concrete.

Popout—shallow depression in a concrete surface resulting from the breaking away of pieces of concrete due to internal pressure.

Portland blast-furnace slag cement—hydraulic cement consisting of: (1) an intimately interground mixture of portland-cement clinker and granulated blast-furnace slag; (2) an intimate and uniform blend of portland cement and fine granulated blast-furnace slag; or (3) finely ground blast-furnace slag with or without additions.

Portland cement—Calcium silicate hydraulic cement produced by pulverizing portland-cement clinker, and usually containing calcium sulphate and other compounds. (See also *Hydraulic cement.*)

Portland cement plaster—a combination of portland cement-based cementing material(s) and aggregate mixed with a suitable amount of water to form a plastic mass that will adhere to a surface and harden, preserving any form and texture imposed on it while plastic. See also *Stucco.*

Pozzolan—siliceous or siliceous and aluminous materials, like fly ash or silica fume, which in itself possess little or no cementing value but which will, in finely divided form and in the presence of moisture, chemically react with calcium hydroxide at ordinary temperatures to form compounds possessing cementing properties.

Precast concrete—concrete cast in forms in a controlled environment and allowed to achieve a specified strength prior to placement on location.

Prestressed concrete—concrete in which compressive stresses are induced by high-strength steel tendons or bars in a concrete element before loads are applied to the element which will balance the tensile stresses imposed in the element during service. This may be accomplished by the following: **Post-tensioning**—a method of prestressing in which the tendons/bars are tensioned after the concrete has hardened; or **Pretensioning**—a method of prestressing in which the tendons are tensioned before the concrete is placed.

Q

Quality control—actions taken by a producer or contractor to provide control over what is being done and what is being provided so that applicable standards of good practice for the work are followed.

R

Reactive-powder concrete—high-strength, low-water and low-porosity concrete with high silica content and aggregate particle sizes of less than 0.3 mm.

Ready-mixed concrete—concrete manufactured for delivery to a location in a fresh state

Recycled concrete—hardened concrete that has been processed for reuse, usually as an aggregate.

Reinforced concrete—concrete to which tensile bearing materials such as steel rods or metal wires are added for tensile strength.

Relative density—a ratio relating the density of a material to that of water.

Relative humidity—the ratio of the quantity of water vapor actually present in the atmosphere to the amount of water vapor present in a saturated atmosphere at a given temperature, expressed as a percentage.

Retarder—an admixture that delays the setting and hardening of concrete.

Roller-compacted concrete (RCC)—a zero slump mix of aggregates, cementitious materials and water that is consolidated by rolling with vibratory compactors; typically used in the construction of dams, industrial pavements, storage and composting areas, and as a component of composite pavements for highways and streets.

S

Scaling—disintegration and flaking of a hardened concrete surface, frequently due to repeated freeze-thaw cycles and application of deicing chemicals.

Segregation—separation of the components (aggregates and mortar) of fresh concrete, resulting in a nonuniform mixture.

Self-compacting concrete—concrete of high workability that require little or no vibration or other mechanical means of consolidation.

Set—the degree to which fresh concrete has lost its plasticity and hardened.

Silica fume—very fine noncrystalline silica which is a byproduct from the production of silicon and ferrosilicon alloys in an electric arc furnace; used as a pozzolan in concrete.

Shotcrete—a process in which compressed air forces mortar or concrete through a hose and nozzle onto a surface at a high velocity and forms structural or non-structural components. The relatively dry mixture is consolidated by the force of impact and develops a compressive strength similar to normal and high strength concrete. Also known as "gunite" and "sprayed concrete."

Shrinkage—decrease in either length or volume of a material resulting from changes in moisture content, temperature, or chemical changes.

Shrinkage-compensating concrete—concrete containing expansive cement, or an admixture, which produces expansion during hardening and thereby offsets the contraction occurring later during drying (drying shrinkage).

Slag cement—hydraulic cement consisting mostly of an intimate and uniform blend of ground, granulated blast-furnace slag with or without portland cement or hydrated lime.

Slump—measure of the consistency of freshly mixed concrete, equal to the immediate subsidence of a specimen molded with a standard slump cone.

Slurry—thin mixture of an insoluble substance, such as portland cement, slag, or clay, with a liquid, such as water.

Soil cement—mixture of pulverized soil or granular material, cement and water compacted to a high density and as the cement hydrates the material becomes hard and durable; primarily used as a base course for roads, streets, airports and parking areas; also also referred to as "cement-treated base," "cement stabilization," "cement-modified soil," and "cement-treated aggregate."

Stucco—portland cement plaster and stucco are the same material. The term "stucco" is widely used to describe the cement plaster used for coating exterior surfaces of buildings. However, in some geographical areas, "stucco" refers only to the factory-prepared finish coat mixtures. (See also *Portland cement plaster.*)

Sulphate attack—common form of chemical attack on concrete caused by sulphates in the groundwater or soil manifested by expansion and disintegration of the concrete.

Superplasticizer (plasticizer)—admixture that increases the flowability of a fresh concrete mixture.

Supplementary cementing materials—Cementing material other than portland cement or blended cement. See *also Cementing material.*

T

Tensile strength—maximum resistance to cracking that concrete can sustain when loaded axially in tension; expressed as force per unit of cross sectional area (MPa).

U

V

Vibration—high-frequency agitation of freshly mixed concrete through mechanical devices, for the purpose of consolidation.

Volume change—Either an increase or a decrease in volume due to any cause, such as moisture changes, temperature changes, or chemical changes. (See also *Creep.*)

W

Water absorption—(1) The process by which a liquid (water) is drawn into and its tendencies to fill permeable pores in a porous solid. (2) The increase in the mass of a material due to water in the pores of the material (but not including water adhering to the outside of the material) expressed as a percentage of the dry mass.

Water to cementing materials ratio—ratio by mass of the amount of water to the mass of the total amount of cementing material in a freshly mixed batch of concrete or mortar, stated as a decimal. The water is exclusive to that absorbed by the aggregate.

Water to cement ratio (water-cement ratio and w/c) —ratio of mass of water to mass of cement in concrete.

Water reducer—admixture whose properties permit a reduction of water required to produce a concrete mix of a certain slump, reduce water-cement ratio, reduce cement content, or increase slump.

White portland cement—cement manufactured from raw materials of low iron content.

Workability—That property of freshly mixed concrete, mortar, grout, or plaster that determines its working characteristics, that is, the ease with which it can be mixed, placed, molded, and finished. (See also *Slump* and *Consistency.*)

X

Y

Yield—volume per batch of concrete expressed in cubic metres.

Z

Zero-slump concrete—concrete without measurable slump (see also *No-slump concrete*).

CSA STANDARDS

CSA Standards and Test Methods related to aggregates, cement, and concrete that are relevant to or referenced in the text are listed as follows and can be obtained at http://www.csa.ca:

CAN/CSA A3000-98   Cementitious Materials Compendium

A5	Portland Cement
A8	Masonry Cement
A23.5	Supplementary Cementing Materials
A362	Blended Hydraulic Cement
A363	Cementitious Hydraulic Slag
A456.1	Chemical Test Methods for Hydraulic Cement, Supplementary Cementing Materials, and Cementitious Hydraulic Slag
A456.2	Physical Test Methods for Hydraulic Cements, Supplementary Cementing Materials, and Cementitious Hydraulic Slag
A456.3	Test Equipment and Materials for Hydraulic Cement, Supplementary Cementing Materials, and Cementitious Hydraulic Slag
A23.1-00	Concrete Materials and Methods of Concrete Construction
A23.2-00	Methods of Test for Concrete
A23.3-94 (R2000)	Design of Concrete Structures
A23.4-00	Precast Concrete —Materials and Construction
A251-00	Qualification Code for Architectural and Structural Precast Concrete Products
A438-00	Concrete Construction for Housing and Small Buildings

CSA A23.2-00   Methods of Test for Concrete

A23.2-1A	Sampling Aggregates for Use in Concrete
A23.2-2A	Sieve Analysis of Fine and Coarse Aggregate
A23.2-3A	Clay Lumps in Natural Aggregates
A23.2-4A	Low-Density Granular Material in Aggregate
A23.2-5A	Amount of Material Finer than 80 µm in Aggregate
A23.2-6A	Relative Density and Absorption of Fine Aggregate
A23.2-7A	Test for Organic Impurities in Fine Aggregates for Concrete
A23.2-8A	Measuring Mortar-Strength Properties of Fine Aggregate
A23.2-9A	Soundness of Aggregate by Use of Magnesium Sulphate
A23.2-10A	Density of Aggregate
A23.2-11A	Surface Moisture in Fine Aggregate
A23.2-12A	Relative Density and Absorption of Coarse Aggregate
A23.2-13A	Flat and Elongated Particles in Coarse Aggregate
A23.2-14A	Potential Expansivity of Aggregates (Procedure for Length Change Due to Alkali-Aggregate Reaction in Concrete Prisms)
A23.2-16A	Resistance to Degradation of Small-Size Coarse Aggregate by Abrasion and Impact in the Los Angeles Machine
A23.2-17A	Resistance to Degradation of Large-Size Coarse Aggregate by Abrasion and Impact in the Los Angeles Machine
A23.2-23A	Test Method for the Resistance of Fine Aggregate to Degradation by Abrasion in the Micro-Deval Apparatus
A23.2-24A	Test Method for Resistance of Unconfined Coarse Aggregate to Freezing and Thawing

A23.2-25A	Test Method for Detection of Alkali-Silica Reactive Aggregate by Accelerated Expansion of Mortar Bars
A23.2-26A	Determination of Potential Alkali-Carbonate Reactivity of Quarried Carbonate Rocks by Chemical Composition
A23.2-27A	Standard Practice to Identify Degree of Alkali-Reactivity of Aggregates and to Identify Measures to Avoid Deleterious Expansion in Concrete
A23.2-28A	Standard Practice for Laboratory Testing to Demonstrate the Effectiveness of Supplementary Cementing Materials and Chemical Admixtures to Prevent Alkali-Silica Reaction in Concrete
A23.2-29A	Test Method for the Resistance of Coarse Aggregate to Degradation by Abrasion in the Micro-Deval Apparatus
A23.2-1B	Viscosity, Bleeding, Expansion, and Compressive Strength of Flowable Grout
A23.2-2B	Determination of Sulphate Ion Content in Groundwater
A23.2-3B	Determination of Total or Water-Soluble Sulphate Ion Content of Soil
A23.2-4B	Sampling and Determination of Water-Soluble Chloride Ion Content in Hardened Grout or Concrete
A23.2-6B	Method of Test to Determine Adhesion by Tensile Load
A23.2-7B	Random Sampling of Construction Materials
A23.2-1C	Sampling Plastic Concrete
A23.2-2C	Making Concrete Mixes in the Laboratory
A23.2-3C	Making and Curing Concrete Compression and Flexural Test Specimens
A23.2-4C	Air Content of Plastic Concrete by the Pressure Method
A23.2-5C	Slump of Concrete
A23.2-6C	Density, Yield, and Cementing Materials Factor of Plastic Concrete
A23.2-7C	Air Content of Plastic Concrete by the Volumetric Method
A23.2-8C	Flexural Strength of Concrete (Using Simple Beam with Third-Point Loading)
A23.2-9C	Compressive Strength of Cylindrical Concrete Specimens
A23.2-10C	Accelerating the Curing of Concrete Cylinders and Determining Their Compressive Strength
A23.2-11C	Water Absorption of Concrete
A23.2-12C	Making, Curing, and Testing Compression Test Specimens of No-Slump Concrete
A23.2-13C	Splitting Tensile Strength of Cylindrical Concrete Specimens
A23.2-14C	Obtaining and Testing Drilled Cores for Compressive Strength Testing
A23.2-1D	Moulds for Forming Concrete Test Cylinders Vertically

ASTM STANDARDS

American Society for Testing and Materials (ASTM) documents related to aggregates, cement, and concrete that are relevant to or referred to in the text are listed as follows and can be obtained at www.astm.org.

A 820-96	Standard Specification for Steel Fibers for Fiber Reinforced Concrete
C 29	Standard Test Method for Bulk Density ("Unit Weight") and Voids in Aggregate
C 31	Standard Practice for Making and Curing Concrete Test Specimens in the Field
C 33-01	Standard Specification for Concrete Aggregates
C 39-01	Standard Test Method for Compressive Strength of Cylindrical Concrete Specimens
C 40-99	Standard Test Method for Organic Impurities in Fine Aggregates for Concrete
C 42-99	Standard Test Method for Obtaining and Testing Drilled Cores and Sawed Beams of Concrete
C 70-01	Standard Test Method for Surface Moisture in Fine Aggregate
C 78-00	Standard Test Method for Flexural Strength of Concrete (Using Simple Beam with Third Point Loading)
C 85	Test Method for Cement Content of Hardened Portland Cement Concrete (Discontinued 1989–Replaced by C 1084)
C 87-83 (1995)	Standard Test Method for Effect of Organic Impurities in Fine Aggregate on Strength of Mortar
C 88-99	Standard Test Method for Soundness of Aggregates by Use of Sodium Sulfate or Magnesium Sulfate
C 91-99	Standard Specification for Masonry Cement
C 94-00	Standard Specification for Ready Mixed Concrete
C 109-99	Standard Test Method for Compressive Strength of Hydraulic Cement Mortars (using 50 mm Cube Specimens)
C 114-00	Standard Test Methods for Chemical Analysis of Hydraulic Cement
C 115-96	Standard Test Method for Fineness of Portland Cement by the Turbidimeter
C 117-95	Standard Test Method for Materials Finer than 75-µm (No. 200) Sieve in Mineral Aggregates by Washing
C 123-98	Standard Test Method for Lightweight Particles in Aggregate
C 125-00	Standard Terminology Relating to Concrete and Concrete Aggregates
C 127-01	Standard Test Method for Density, Relative Density (Specific Gravity), and Absorption of Coarse Aggregate
C 128-01	Standard Test Method for Density, Relative Density (Specific Gravity), and Absorption of Fine Aggregate
C 131-96	Standard Test Method for Resistance to Degradation of Small Size Coarse Aggregate by Abrasion and Impact in the Los Angeles Machine
C 136-01	Standard Test Method for Sieve Analysis of Fine and Coarse Aggregates
C 138-01	Standard Test Method for Density (Unit Weight), Yield, and Air Content (Gravimetric) of Concrete
C 141-97	Standard Specification for Hydraulic Hydrated Lime for Structural Purposes
C 142-97	Standard Test Method for Clay Lumps and Friable Particles in Aggregates
C 143-00	Standard Test Method for Slump of Hydraulic Cement Concrete
C 150-00	Standard Specification for Portland Cement
C 151-00	Standard Test Method for Autoclave Expansion of Portland Cement
C 156-98	Standard Test Method for Water Retention by Concrete Curing Materials
C 157-99	Standard Test Method for Length Change of Hardened Hydraulic Cement, Mortar, and Concrete
C 171-97	Standard Specification for Sheet Materials for Curing Concrete
C 172-99	Standard Practice for Sampling Freshly Mixed Concrete
C 173-01	Standard Test Method for Air Content of Freshly Mixed Concrete by the Volumetric Method
C 174-97	Standard Test Method for Measuring Thickness of Concrete Elements Using Drilled Concrete Cores
C 177-97	Standard Test Method for Steady State Heat Flux Measurements and Thermal Transmission Properties by Means of the Guarded-Hot-Plate Apparatus
C 183-97	Standard Practice for Sampling and the Amount of Testing of Hydraulic Cement
C 184-94	Standard Test Method for Fineness of Hydraulic Cement by the 150-µm (No. 100) and 75-µm (No. 200) Sieves
C 185-01	Standard Test Method for Air Content of Hydraulic Cement Mortar

C 186-98	Standard Test Method for Heat of Hydration of Hydraulic Cement
C 187-98	Standard Test Method for Normal Consistency of Hydraulic Cement
C 188-95	Standard Test Method for Density of Hydraulic Cement
C 191-99	Standard Test Method for Time of Setting of Hydraulic Cement by Vicat Needle
C 192-00	Standard Practice for Making and Curing Concrete Test Specimens in the Laboratory
C 204-00	Standard Test Method for Fineness of Hydraulic Cement by Air Permeability Apparatus
C 215-97	Standard Test Method for Fundamental Transverse, Longitudinal, and Torsional Frequencies of Concrete Specimens
C 219-01	Standard Terminology Relating to Hydraulic Cement
C 226-96	Standard Specification for Air-Entraining Additions for Use in the Manufacture of Air-Entraining Hydraulic Cement
C 227-97	Standard Test Method for Potential Alkali Reactivity of Cement
C 230-98	Standard Specification for Flow Table for Use in Tests of Hydraulic Cement
C 231-97	Standard Test Method for Air Content of Freshly Mixed Concrete by the Pressure Method
C 232-99	Standard Test Methods for Bleeding of Concrete
C 233-01	Standard Test Method for Air-Entraining Admixtures for Concrete
C 235-68	Standard Method of Test for Scratch Hardness of Coarse Aggregate Particles (Discontinued 1978)
C 243-99	Standard Test Method for Bleeding of Cement Pastes and Mortars
C 260-00	Standard Specification for Air-Entraining Admixtures for Concrete
C 265-99	Standard Test Method for Calcium Sulfate in Hydrated Portland Cement Mortar
C 266-99	Standard Test Method for Time of Setting of Hydraulic-Cement Paste by Gillmore Needles
C 270-00	Standard Specification for Mortar for Unit Masonry
C 289-01	Standard Test Method for Potential Alkali-Silica Reactivity of Aggregates (Chemical Method)
C 293-00	Standard Test Method for Flexural Strength of Concrete (Using Simple Beam With Center-Point Loading
C 294-98	Standard Descriptive Nomenclature for Constituents of Concrete Aggregates

C 295-98	Standard Guide for Petrographic Examination of Aggregates for Concrete
C 305-99	Standard Practice for Mechanical Mixing of Hydraulic Cement Pastes and Mortars of Plastic Consistency
C 309-98	Standard Specification for Liquid Membrane
C 311-00	Standard Test Methods for Sampling and Testing Fly Ash or Natural Pozzolans for Use as a Mineral Admixture in Portland-Cement Concrete
C 330-00	Standard Specification for Lightweight Aggregates for Structural Concrete
C 332-99	Standard Specification for Lightweight Aggregates for Insulating Concrete
C 341-96	Standard Test Method for Length Change of Drilled or Sawed Specimens of Hydraulic-Cement Mortar and Concrete
C 342-97	Standard Test Method for Potential Volume Change of Cement-Aggregate Combinations (Discontinued 2001)
C 348-97	Standard Test Method for Flexural Strength of Hydraulic-Cement Mortars
C 359-99	Standard Test Method for Early Stiffening of Portland Cement (Mortar Method)
C 360-92	Test Method for Ball Penetration in Fresh Portland Cement Concrete (Discontinued 1999)
C 384-98	Standard Test Method for Impedance and Absorption of Acoustical Materials by the Impedance Tube Method
C 387-00	Standard Specification for Packaged, Dry, Combined Materials for Mortar and Concrete
C 403M-99	Standard Test Method for Time of Setting of Concrete Mixtures by Penetration Resistance
C 418-98	Standard Test Method for Abrasion Resistance of Concrete by Sandblasting
C 430-96	Standard Test Method for Fineness of Hydraulic Cement by the 45-μm (No. 325) Sieve
C 441-97	Standard Test Method for Effectiveness of Mineral Admixtures or Ground Blast-Furnace Slag in Preventing Excessive Expansion of Concrete Due to the Alkali-Silica Reaction
C 451-99	Standard Test Method for Early Stiffening of Hydraulic Cement (Paste Method)
C 452-95	Standard Test Method for Potential Expansion of Portland-Cement Mortars Exposed to Sulfate
C 457-98	Standard Test Method for Microscopical Determination of Parameters of the Air-Void Content and Parameters of the Air-Void System in Hardened Concrete

C 465-99	Standard Specification for Processing Additions for Use in the Manufacture of Hydraulic Cements
C 469-94	Standard Test Method for Static Modulus of Elasticity and Poisson's Ratio of Concrete in Compression
C 470-98	Standard Specification for Molds for Forming Concrete Test Cylinders Vertically
C 490-00	Standard Practice for Use of Apparatus for the Determination of Length Change of Hardened Cement Paste, Mortar, and Concrete
C 494-99	Standard Specification for Chemical Admixtures for Concrete
C 495-99	Standard Test Method for Compressive Strength of Lightweight Insulating Concrete
C 496-96	Standard Test Method for Splitting Tensile Strength of Cylindrical Concrete Specimens
C 511-98	Standard Specification for Moist Cabinets, Moist Rooms, and Water Storage Tanks Used in the Testing of Hydraulic Cements and Concretes
C 512-87 (1994)	Standard Test Method for Creep of Concrete in Compression
C 513-89 (1995)	Standard Test Method for Obtaining and Testing Specimens of Hardened Lightweight Insulating Concrete for Compressive Strength
C 535-96	Standard Test Method for Resistance to Degradation of Large-Size Coarse Aggregate by Abrasion and Impact in the Los Angeles Machine
C 566-97	Standard Test Method for Total Evaporable Moisture Content of Aggregate by Drying
C 567-00	Standard Test Method for Density of Structural Lightweight Concrete
C 586-99	Standard Test Method for Potential Alkali Reactivity of Carbonate Rocks for Concrete Aggregates (Rock Cylinder Method)
C 595-00	Standard Specification for Blended Hydraulic Cements
C 597-97	Standard Test Method for Pulse Velocity Through Concrete
C 617-98	Standard Practice for Capping Cylindrical Concrete Specimens
C 618-00	Standard Specification for Coal Fly Ash and Raw or Calcined Natural Pozzolan for Use as a Mineral Admixture in Concrete
C 637-98	Standard Specification for Aggregates for Radiation-Shielding Concrete
C 638-92 (1997)	Standard Descriptive Nomenclature of Constituents of Aggregates for Radiation-Shielding Concrete

C 641-98	Standard Test Method for Staining Materials in Lightweight Concrete Aggregates
C 642-97	Standard Test Method for Density, Absorption, and Voids in Hardened Concrete
C 666-97	Standard Test Method for Resistance of Concrete to Rapid Freezing and Thawing
C 671-94	Standard Test Method for Critical Dilation of Concrete Specimens Subjected to Freezing
C 672-98	Standard Test Method for Scaling Resistance of Concrete Surfaces Exposed to Deicing Chemicals
C 682-94	Standard Practice for Evaluation of Frost Resistance of Coarse Aggregates in Air-Entrained Concrete by Critical Dilation Procedures
C 684-99	Standard Test Method for Making, Accelerated Curing, and Testing Concrete Compression Test Specimens
C 685-00a	Standard Specification for Concrete Made By Volumetric Batching and Continuous Mixing
C 688-00	Standard Specification for Functional Additions for Use in Hydraulic Cements
C 702-98	Standard Practice for Reducing Samples of Aggregate to Testing Size
C 778-00	Standard Specification for Standard Sand
C 779-00	Standard Test Method for Abrasion Resistance of Horizontal Concrete Surfaces
C 786-96	Standard Test Method for Fineness of Hydraulic Cement and Raw Materials by the 300-μm (No. 50), 150-μm (No. 100), and 75-μm (No. 200) Sieves by Wet Methods
C 796-97	Standard Test Method for Foaming Agents for Use in Producing Cellular Concrete Using Preformed Foam
C 801-98	Standard Test Method for Determining the Mechanical Properties of Hardened Concrete Under Triaxial Loads
C 803-97	Standard Test Method for Penetration Resistance of Hardened Concrete
C 805-97	Standard Test Method for Rebound Number of Hardened Concrete
C 806-95	Standard Test Method for Restrained Expansion of Expansive Cement Mortar
C 807-99	Standard Test Method for Time of Setting of Hydraulic Cement Mortar by Modified Vicat Needle
C 823-00	Standard Practice for Examination and Sampling of Hardened Concrete in Constructions

C 869-91 (1999)	Standard Specification for Foaming Agents Used in Making Preformed Foam for Cellular Concrete
C 873-99	Standard Test Method for Compressive Strength of Concrete Cylinders Cast in Place in Cylindrical Molds
C 876-91 (1999)	Standard Test Method for Half-Cell Potentials of Uncoated Reinforcing Steel in Concrete
C 878-95a	Standard Test Method for Restrained Expansion of Shrinkage-Compensating Concrete
C 881-99	Standard Specification for Epoxy-Resin-Base Bonding Systems for Concrete
C 900-99	Standard Test Method for Pullout Strength of Hardened Concrete
C 917-98	Standard Test Method for Evaluation of Cement Strength Uniformity From a Single Source
C 918-97	Standard Test Method for Measuring Early-Age Compressive Strength and Projecting Later-Age Strength
C 926-98	Standard Specification for Application of Portland Cement-Based Plaster
C 928-00	Standard Specification for Packaged, Dry, Rapid-Hardening Cementitious Materials for Concrete Repairs
C 937-97	Standard Specification for Grout Fluidifier for Preplaced-Aggregate Concrete
C 938-97	Standard Practice for Proportioning Grout Mixtures for Preplaced-Aggregate Concrete
C 939-97	Standard Test Method for Flow of Grout for Preplaced-Aggregate Concrete (Flow Cone Method)
C 940-98	Standard Test Method for Expansion and Bleeding of Freshly Mixed Grouts for Preplaced-Aggregate Concrete in the Laboratory
C 941-96	Standard Test Method for Water Retentivity of Grout Mixtures for Preplaced-Aggregate Concrete in the Laboratory
C 942-99	Standard Test Method for Compressive Strength of Grouts for Preplaced-Aggregate Concrete in the Laboratory
C 943-96	Standard Practice for Making Test Cylinders and Prisms for Determining Strength and Density of Preplaced-Aggregate Concrete in the Laboratory
C 944-99	Standard Test Method for Abrasion Resistance of Concrete or Mortar Surfaces by the Rotating-Cutter Method
C 953-87 (1997)	Standard Test Method for Time of Setting of Grouts for Preplaced-Aggregate Concrete in the Laboratory
C 979-99	Standard Specification for Pigments for Integrally Coloured Concrete

C 989-99	Standard Specification for Ground Granulated Blast-Furnace Slag for Use in Concrete and Mortars
C 995-94	Standard Test Method for Time of Flow of Fiber-Reinforced Concrete Through Inverted Slump Cone
C 1012-95	Standard Test Method for Length Change of Hydraulic-Cement Mortars Exposed to a Sulfate Solution
C 1017-98	Standard Specification for Chemical Admixtures for Use in Producing Flowing Concrete
C 1018-97	Standard Test Method for Flexural Toughness and First-Crack Strength of Fiber-Reinforced Concrete (Using Beam With Third-Point Loading)
C 1038-95	Standard Test Method for Expansion of Portland Cement Mortar Bars Stored in Water
C 1040-93 (2000)	Standard Test Methods for Density of Unhardened and Hardened Concrete In Place By Nuclear Methods
C 1059-99	Standard Specification for Latex Agents for Bonding Fresh To Hardened Concrete
C 1064-99	Standard Test Method for Temperature of Freshly Mixed Portland Cement Concrete
C 1073-97	Standard Test Method for Hydraulic Activity of Ground Slag by Reaction with Alkali
C 1074-98	Standard Practice for Estimating Concrete Strength by the Maturity Method
C 1077-00	Standard Practice for Laboratories Testing Concrete and Concrete Aggregates for Use in Construction and Criteria for Laboratory Evaluation
C 1078-87 (1992)	Test Method for Determining Cement Content of Freshly Mixed Concrete
C 1084-97	Standard Test Method for Portland-Cement Content of Hardened Hydraulic-Cement Concrete
C 1105-95	Standard Test Method for Length Change of Concrete Due to Alkali-Carbonate Rock Reaction
C 1116-00	Standard Specification for Fiber-Reinforced Concrete and Shotcrete
C 1137-97	Standard Test Method for Degradation of Fine Aggregate Due to Attrition
C 1138-97	Standard Test Method for Abrasion Resistance of Concrete (Underwater Method)
C 1150-96	Standard Test Method for The Break-Off Number of Concrete
C 1152-97	Standard Test Method for Acid-Soluble Chloride in Mortar and Concrete

C 1157-00	Standard Performance Specification for Hydraulic Cement
C 1170-91 (1998)	Standard Test Methods for Determining Consistency and Density of Roller-Compacted Concrete Using a Vibrating Table
C 1202-97	Standard Test Method for Electrical Indication of Concrete's Ability to Resist Chloride Ion Penetration
C 1218-99	Standard Test Method for Water-Soluble Chloride in Mortar and Concrete
C 1231-00	Standard Practice for Use of Unbonded Caps in Determination of Compressive Strength of Hardened Concrete Cylinders
C 1240-01	Standard Specification for Use of Silica Fume for Use as a Mineral Admixture in Hydraulic-Cement Concrete, Mortar, and Grout
C 1252-98	Standard Test Methods for Uncompacted Void Content of Fine Aggregate (as Influenced by Particle Shape, Surface Texture, and Grading)
C 1260-94	Standard Test Method for Potential Alkali Reactivity of Aggregates (Mortar-Bar Method)
C 1263-95 (2000)	Standard Test Method for Thermal Integrity of Flexible Water Vapor Retarders
C 1293-01	Standard Test Method for Determination of Length Change of Concrete Due to Alkali-Silica Reaction
C 1315-00	Standard Specification for Liquid Membrane-Forming Compounds Having Special Properties for Curing and Sealing Concrete
C 1328-00	Standard Specification for Plastic (Stucco) Cement
C 1329-00	Standard Specification for Mortar Cement
C 1356-96	Standard Test Method for Quantitative Determination of Phases in Portland Cement Clinker by Microscopical Point-Count Procedure
C 1362-97	Standard Test Method for Flow of Freshly Mixed Hydraulic Cement Concrete
C 1365-98	Standard Test Method for Determination of the Proportion of Phases in Portland Cement and Portland-Cement Clinker Using X-Ray Powder Diffraction Analysis
C 1383-98	Standard Test Method for Measuring the P-Wave Speed and the Thickness of Concrete Plates Using the Impact-Echo Method
C 1437-99	Standard Test Method for Flow of Hydraulic Cement Mortar

C 1451-99	Standard Practice for Determining Uniformity of Ingredients of Concrete From a Single Source
C 1500	Standard Test Method for Water-Extractable Chloride in Aggregate (Soxhlet Method)
D 75-97	Standard Practice for Sampling Aggregates
D 98-98	Standard Specification for Calcium Chloride
D 345-97	Standard Test Method for Sampling and Testing Calcium Chloride for Roads and Structural Applications
D 448-98	Standard Classification for Sizes of Aggregate for Road and Bridge Construction
D 512-89 (1999)	Standard Test Methods for Chloride Ion in Water
D 516-90 (1995)	Standard Test Method for Sulfate Ion in Water
D 558-96	Standard Test Methods for Moisture-Density Relations of Soil-Cement Mixtures
D 632-01	Standard Specification for Sodium Chloride
D 2240-00	Standard Test Method for Rubber Property-Durometer Hardness
D 3042-97	Standard Test Method for Insoluble Residue in Carbonate Aggregates
D 3398-00	Standard Test Method for Index of Aggregate Particle Shape and Texture
D 3963-01	Standard Specification for Fabrication and Jobsite Handling of Epoxy-Coated Steel Reinforcing Bars
D 4263-83 (1999)	Standard Test Method for Indicating Moisture in Concrete by the Plastic Sheet Method
D 4580-86 (1997)	Standard Practice for Measuring Delaminations in Concrete Bridge Decks by Sounding
E 11-01	Standard Specification for Wire Cloth and Sieves for Testing Purposes
F 1869-98	Standard Test Method for Measuring Moisture Vapor Emission Rate of Concrete Subfloor Using Anhydrous Calcium Chloride
G 59-97	Standard Test Method for Conducting Potentiodynamic Polarization Resistance Measurements
PS 118-99	Standard Test Method for Water-Extractable Chloride in Aggregate (Soxhlet Method)

CEMENT AND CONCRETE RESOURCES AND RELATED ORGANIZATIONS

AASHTO	American Association of State Highway and Transportation Officials http://www.aashto.org
ACAA	American Coal Ash Association http://www.acaa-usa.org
ACBM	Center for Advanced Cement Based Materials (Northwestern University) http://www.civil.northwestern.edu/ACBM
ACI	American Concrete Institute http://www.aci-int.org
ACPA	American Concrete Pavement Association http://www.pavement.com
ACPA	American Concrete Pipe Association http://www.concrete-pipe.org
ACPA	American Concrete Pumping Association http://www.concretepumpers.com/
ACPPA	American Concrete Pressure Pipe Association http://www.acppa.org
ACS	American Ceramic Society http://www.acers.org
AFNOR	Association Française de Normalisation (French Standards Institution) http://www.afnor.fr
AIT	International Ferrocement Society http://www.ferrocement.org
APA	Architectural Precast Association http://www.archprecast.org
APCA	American Portland Cement Alliance http://www.apca.org
API	American Petroleum Institute http://www.api.org
ANSI	American National Standards Institute http://www.ansi.org
ASCC	American Society of Concrete Contractors http://www.assconc.org
ASCE	American Society of Civil Engineers http://www.asce.org
ASHRAE	American Society of Heating, Refrigerating, and Air-Conditioning Engineers, Inc. http://www.ashrae.org
ASI	American Shotcrete Association http://www.shotcrete.org
ASTM	American Society for Testing and Materials http://www.astm.org
BCA	British Cement Association http://www.bca.org.uk
BDZ	Bundesverband der Deutschen Zementindustrie e. V. http://www.bdzement.de
BFRL	Building and Fire Research Laboratory (NIST) http://www.bfrl.nist.gov

BOCA	Building Officials and Code Administrators International http://www.bocai.org
BRE	Building Research Establishment (UK) http://www.bre.co.uk
BSI	British Standards Institution http://www.bsi-global.com
CABO	Council of American Building Officials http://www.cabo.org
CAC	Cement Association of Canada http://www.cement.ca
CC	Cement Careers http://www.cementcareers.org
CCA	Canadian Construction Association http://www.cca-acc.com
CCAA	Cement and Concrete Association of Australia http://www.concrete.net.au
CCRI	Concrete Corrosion Inhibitors Foundation http://www.corrosioninhibitors.org/
CERF	Civil Engineering Research Foundation http://www.cerf.org
CFA	Concrete Foundations Association http://www.cfana.org
CH	Concrete Homes http://www.concretehomes.com
CHBA	Canadian Home Builders Association http://www.chba.ca
CMHC	Canada Mortgage and Housing Corporation http://www.cmhc.ca
CPCI	Canadian Precast/Prestressed Concrete Institute http://cpci.ca
CRMCA	Canadian Ready-Mixed Concrete Association http://www.crmca.ca
CRSI	Concrete Reinforcing Steel Institute http://www.crsi.org
CSA	Canadian Standards Association http://www.csa.ca
CSCE	Canadian Society for Civil Engineering http://www.csce.ca
CSDA	Concrete Sawing and Drilling Association http://www.csda.org
CSI	Construction Specifications Institute http://www.csinet.org
CSI	Cast Stone Institute http://www.caststone.org
DIN	Deutsches Institut für Normung e.V. (German Standards Institution) http://www.din.de

ECCO	Environmental Council of Concrete Organizations http://www.ecco.org
EPRI	Electric Power Research Institute http://www.epri.com
ESCSI	Expanded Shale, Clay and Slate Institute http://www.escsi.org
EUC	European Concrete http://www.europeanconcrete.com
FHWA	Federal Highway Administration http://www.fhwa.dot.gov/
FICEM	Federacion Interamericana del Cemento http://www.ficem.org
HPC-CAN	High Performance Concrete Network of Centres of Excellence (Canada) http://www.usherb.ca/concrete
IBC	International Building Code http://www.intlcode.org
ICAR	International Center for Aggregates Research http://www.ce.utexas.edu/org/icar/
ICBO	International Conference of Building Officials http://www.icbo.org
ICC	International Code Council http://www.intlcode.org
ICFA	Insulating Concrete Forms Association http://www.forms.org/
ICMA	International Cement Microscopy Association http://www.cemmicro.org
ICPI	Interlocking Concrete Pavement Institute http://www.icpi.org
ICRI	International Concrete Repair Institute http://www.icri.org
IEEE	Institute of Electrical & Electronics Engineers http://www.ieee.org
IGGA	International Grinding & Grooving Association http://www.igga.net
IMCYC	Instituto Mexicano del Cemento y del Concreto A.C. http://www.imcyc.com
IPRF	Innovative Pavement Research Foundation http://www.iprf.org
ISO	International Standards Organization http://www.iso.ch
JCI	Japan Concrete Institute http://www.jci-net.or.jp/
LTPP	Long Term Pavement Performance http://www.tfhrc.gov/pavement/ltpp/ltpp.htm
MERL	U.S. Bureau of Reclamation Materials and Engineering Research Laboratory http://www.usbr.gov/merl
MRS	Materials Research Society http://www.mrs.org
NACE	National Association of Corrosion Engineers http://www.nace.org
NAHB	National Association of Home Builders http://www.nahb.com
NCMA	National Concrete Masonry Association http://www.ncma.org

NIBS	National Institute of Building Sciences http://www.nibs.org
NPCA	National Precast Concrete Association http://www.precast.org
NIST	National Institute of Standards and Technology http://www.nist.gov
NRC	National Research Council http://www.nas.edu/nrc/
NRCC	National Research Council of Canada http://www.nrc.ca/
NRMCA	National Ready Mixed Concrete Association http://www.nrmca.org
NTMA	National Terrazzo & Mosaic Association Inc. http://www.ntma.com
NSSGA	National Stone, Sand & Gravel Association http://www.nssga.org
NSA	National Slag Association http://www.nationalslagassoc.org
OP&CMIA	Operative Plasterers' and Cement Masons' International Association http://www.opcmia.org
PCA	Portland Cement Association http://www.portcement.org
PCI	Precast/Prestressed Concrete Institute http://www.pci.org
PI	Perlite Institute http://www.perlite.org
PTI	Post-Tensioning Institute http://www.post-tensioning.org/
REMR	Repair, Evaluation, Maintenance, and Rehabilitation Research Program http://www.wes.army.mil/REMR
RILEM	International Union of Testing and Research Laboratories for Materials and Structures http://www.rilem.org
SCA	Slag Cement Association
SFA	Silica Fume Association http://www.silicafume.org
TCA	Tilt-Up Concrete Association http://www.tilt-up.org
TRB	Transportation Research Board http://www.nas.edu/trb/
USACE	U.S. Army Corps of Engineers http://www.usace.army.mil
VDZ	Verein Deutscher Zementwerke (German Cement Works Association) http://www.vdz-online.de
TVA	The Vermiculite Association http://www.vermiculite.org/
VCCTL	Virtual Cement and Concrete Testing Laboratory http://ciks.cbt.nist.gov/vcctl
WES	U.S. Army Engineer Waterways Experiment Station http://www.wes.army.mil
WRI	Wire Reinforcement Institute http://www.bright.net/~wwri

347

Index